INTRODUCTION TO HORTICULTURAL SCIENCE

INTRODUCTION TO HORTICULTURAL SCIENCE

Richard N. Arteca

THOMSON
DELMAR LEARNING

Australia Canada Mexico Singapore Spain United Kingdom United States

Introduction to Horticultural Science
Richard N. Arteca

Vice President, Career Education Strategic Business Unit:
Dawn Gerrain

Director of Learning Solutions:
Sherry Dickinson

Managing Editor:
Robert Serenka

Acquisitions Editor:
David Rosenbaum

Product Manager:
Gerald O'Malley

Editorial Assistant:
Christina Gifford

Director of Production:
Wendy A. Troeger

Production Manager:
JP Henkel

Senior Production Editor:
Kathryn B. Kucharek

Director of Marketing:
Wendy Mapstone

Channel Manager:
Gerard McAvey

Cover Image:
Getty Images Inc.

Cover Design:
Suzanne Nelson

Printed in Canada
1 2 3 4 5 XXX 10 09 08 07 06

For more information contact Delmar Learning,
5 Maxwell Drive, PO Box 8007, Clifton Park, NY 12065-2919.

Or you can visit our Internet site at
http://www.delmarlearning.com.

Library of Congress Cataloging-in-Publication Data

Richard N. Arteca, 1950–

Introduction to horticultural science / Richard N. Arteca.

p. cm.

Includes bibliographical references (p.).

ISBN 0-7668-3592-8 (hardcover)

1. Horticulture—Textbooks. I. Title.

SB318.A74 2006

635—dc22

2006001565

NOTICE TO THE READER

Publisher does not warrant or guarantee any of the products described herein or perform any independent analysis in connection with any of the product information contained herein. Publisher does not assume, and expressly disclaims, any obligation to obtain and include information other than that provided to it by the manufacturer.

The reader is expressly warned to consider and adopt all safety precautions that might be indicated by the activities herein and to avoid all potential hazards. By following the instructions contained herein, the reader willingly assumes all risks in connection with such instructions.

The Publisher makes no representation or warranties of any kind, including but not limited to, the warranties of fitness for particular purpose or merchantability, nor are any such representations implied with respect to the material set forth herein, and the publisher takes no responsibility with respect to such material. The Publisher shall not be liable for any special, consequential, or exemplary damages resulting, in whole or part, from the readers' use of, or reliance upon, this material.

Dedication

To my loving wife, Jeannette, for her dedication to the completion of this book and the countless hours spent preparing figures, editing, organizing, and doing whatever it took to get the job done.

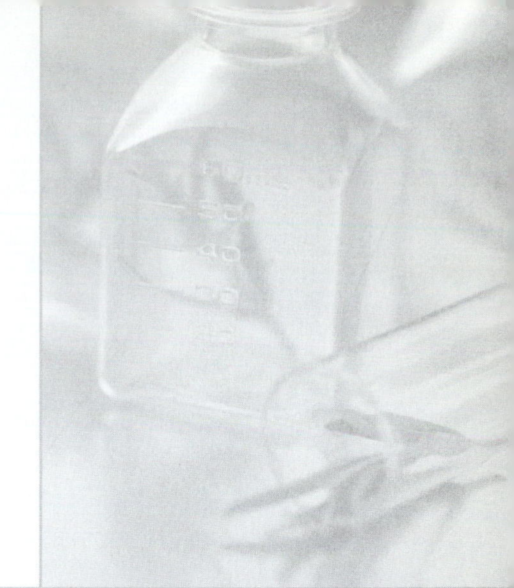

Table of Contents

Chapter 19

Interiorscaping 375

Designing Landscapes 389

Chapter 23 Warm- and Cool Season Turfgrass Selection, Establishment, Care, and Maintenance 434

Chapter 24 Olericulture 454

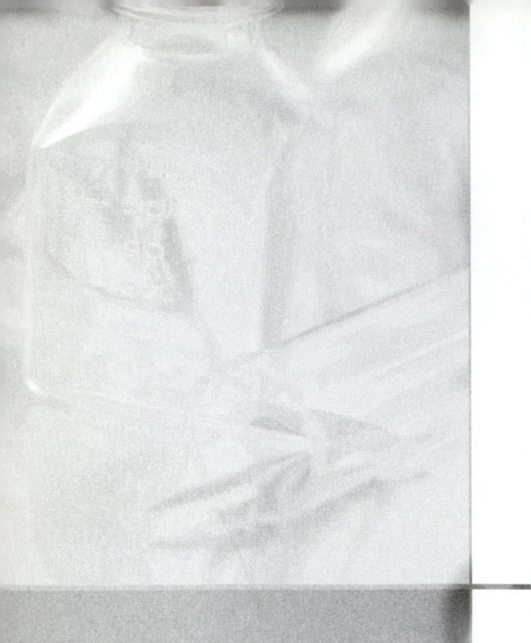

Preface

Presently, no horticultural science textbooks adequately cover the subject in enough detail to be used as a college textbook for a horticultural science class, which is a general science option for students at universities. The main feature that sets this text aside from existing textbooks is its completeness. This *Introduction to Horticultural Science* textbook can be used at the college level as a required textbook for an introductory course in horticulture that serves as a general science option. The numbers of students taking Introductory Horticultural Science courses is growing continually; for example, 400 to 500 students take my Hort 101 course each year, and this number is increasing continually. This textbook provides an excellent survey of all aspects of horticultural science, including gene jockeying, flower arranging, vegetable production, landscape construction, and much more. Although this text is designed for college students, it is also useful for high school students preparing for college.

Chapter 1 discusses the green plant and what an amazing organism it is. This is followed by the origin of agriculture and the domestication of plants in Chapter 2; the horticulture industry in Chapter 3; the fundamental steps leading to success in a horticulture career in Chapter 4; the relationship between horticulture and the environment in Chapter 5; and the classification of plants and plant anatomy in Chapter 6. Chapters 7, 8, 9, 10, 11, and 12 cover plant propagation; media, nutrients, and fertilizers; plants and the environment; plant growth regulators; postharvest physiology; and pest management, respectively. Plant biotechnology and genetically modified organisms are covered in Chapter 13, which is followed by Chapters 14 through 22 covering greenhouse structures; growing crops in the greenhouse; nursery site selection, development, and facilities; producing nursery crops; floral design; interiorscaping; designing landscapes; installing landscapes; and landscape maintenance, respectively. Chapter 23 discusses warm season and cool season turfgrass selection, establishment, care, and maintenance; Chapter 24 covers olericulture; and Chapter 25 discusses pomology. Each chapter contains a variety of different types of review questions, which will be helpful in digesting the information contained in the chapter. A supplement is also available to instructors with answers to questions in each of the chapters. In addition to questions, activities are given at the end of each chapter so students can explore a given topic in more detail. At the end of most chapters, a list of references is included for further study. In addition, this text is an excellent resource for anyone interested in horticultural science.

Richard N. Arteca, PhD

About the Author

The author of this text is Professor of Horticultural Physiology in the Department of Horticulture at The Pennsylvania State University and has been there for more than 25 years. He received his Ph.D. at Washington State University in 1979 and his M.S. and B.S. at Utah State University in 1976 and 1972, respectively. His appointment breakdown at Penn State is 60 percent research and 40 percent teaching. Dr. Arteca has published more than 80 publications in refereed journals and is internationally known for his work in biotechnology and genetic engineering. Some examples of his work include leading one of the initial research groups to develop cell culture systems for the production of the antitumor compound taxol, which is commonly used today in chemotherapy against breast cancer and other forms of cancer. Another example of his work is the biological, molecular, and genetic regulatory mechanisms involved in the plant's response to externally applied stimuli. Dr. Arteca's teaching appointment includes three classes: Introductory Horticulture (Hort 101), Plant Growth Regulators (Hort 420), and Advanced Plant Growth Regulators (Hort 520). Introductory Horticulture is a very popular course taught as a resident education course and as a world campus course (which is taught on-line).

About the Author

CHAPTER

1

The Green Plant, What an Organism!

ABSTRACT

The green plant is an amazing organism. This chapter presents a number of reasons why plants are important and summarizes the main processes in plants, including seed germination, photosynthesis, phototropism, thigmotropism, dormancy, senescence, flowering, abscission, fruit growth and development, growth retardation, weed control, production of important chemicals by plants, and biotechnology.

Objectives

After reading this chapter, you should be able to

- list and discuss the many reasons why plants are important.
- discuss the important plant processes and why they are so special.

Key Terms

abscission
aesthetic beauty
aspirin
biotechnology
dormancy
erosion
flowering
food source
fruit growth and development
fuel
genetically modified organisms
gravitropism
growth retardation
habitat
herbicide
negative gravitropism
pesticide
pharmaceuticals
photosynthesis
phototropism
plant growth regulator
positive gravitropism
purifying air
quiescence
rest
seed
seed germination
senescence
statolith
stress reduction
thigmotropism
viable
weed
weed control
wetland

INTRODUCTION

This chapter discusses a wide range of reasons plants are so important to our existence. Plants provide us with food, shelter, and pharmaceuticals, while purifying the air we breathe. They are aesthetically beautiful and have been scientifically proven to reduce stress in our daily lives. Plants prevent erosion of the limited amount of topsoil on the Earth's surface, provide habitat and cover for animals, and play a key role in wetland water purification.

Exciting areas of plant research are also covered starting with **seed germination,** which is a very complex process even though it does not appear to be very complicated to the naked eye. Just considering how a tiny seed can rapidly turn into a large tree or how seeds know which way is up and which way is down reveals the complexity. The process of **photosynthesis** is another fascinating area of research, and a better knowledge of how this process works will enable us to provide more food for a hungry world.

Plants are able to move in response to environmental cues—such as light (**phototropism**), touch (**thigmotropism**), and gravity (**gravitropism**)—to protect themselves and increase their efficiency. Plants also respond to environmental cues by dropping their leaves through a process called **abscission,** which protects them against winter injury. After the plant sheds its leaves, it can temporarily suspend visible signs of plant growth by a process called **dormancy,** which is another exciting area of research. Currently, researchers are trying to unlock the mystery of dormancy so that humans suffering from serious disease might be put into a dormant state pending a future cure. The regulation of **senescence,** or aging, in plants is currently being studied in hopes that if we can understand how to delay aging in plants, this information might be applied to methods for delaying aging in humans.

The process of **flowering** is another rapidly advancing area of research for a variety of reasons. One of the major economic reasons is the consumer demand for flowering plants at specific times of the year, for example, poinsettias for Christmas. By better understanding the flowering process, it will be easier for growers to have crops come in on specific dates. **Fruit growth and development** is studied in order to produce an adequate supply of high-quality fruits for human consumption. **Plant growth regulator** research receives a considerable amount of attention from scientists throughout the world for a variety of purposes, including reducing plant height and controlling weeds. The production of important plant chemicals for **pharmaceuticals,** fragrances, and a variety of other purposes is a highly competitive area of research because of the potential to make large sums of money. The production of **genetically modified organisms** is also a highly competitive and highly controversial area of plant research because of ethical concerns by the general public and legal battles with patenting. Many of the topics discussed in this chapter will be discussed in more detail later in the book.

WHY ARE PLANTS SO IMPORTANT?

For many years, plants have been thought of as immobile creatures because they do not make any obvious movements; however, they do contribute to the majority of any given area. Plants are able to adapt to a variety of surroundings because they continually monitor their environmental surroundings and respond rapidly to the wide variety of conditions they are subjected to (Figure 1-1). One of the ways to determine intelligence is by measuring the ability to move in response to a given situation. A growing shoot has been shown to use near-infrared light to determine the closeness of other plants and consequently alter its direction of growth. Also documented is the fact that the stilt palm can, by differential growth of its prop roots, move away from plants that will compete with it. From the other angle, the parasitic weed dodder touches another plant to determine whether it can be exploited. After the dodder plant determines that the plant is susceptible, it engulfs the host and uses all of the host plant's resources. These few examples demonstrate that plants are intelligent. Trewavas (2002) states that traditional definitions of intelligence use movement as a criterion, but the adaptive behaviors shown by individual plants indicate that they are intelligent.

Figure 1-1 This tree has grown regardless of adverse conditions, illustrating that plants are amazingly adaptable.

This section summarizes and briefly explains the numerous reasons plants are so important. Plants are important as a **food source;** for true vegetarians, plants are their only source of food. As a food source, plants provide all the essentials to sustain animal life. In addition to being important sources of vitamins, minerals, and fiber, the American Cancer Society and the American Heart Association report that eating fruits and vegetables reduces certain cancer rates and heart disease, respectively (Figure 1-2). Fruit pectins trap dietary cholesterol, keeping the cholesterol, from depositing in the linings of blood vessels and thereby preventing heart attacks. Fruits and vegetables also contain antioxidants that neutralize free radicals involved in aging and some forms of cancer. Plants also provide people a way to establish shelter; lumber from trees is used to build a variety of structures, which protect humans, animals, machinery, and so on from the sometimes harsh external environment. Many special articles of clothing, such as dress shirts, pants, and other clothing, are made from cotton (Figure 1-3), which is derived from plants.

Plants are also responsible for **purifying the air** because they remove the carbon dioxide we create and add the oxygen we need, thereby maintaining a balance in the atmosphere. In fact, a large group of people in a small room creates high levels of carbon dioxide that may cause tiredness. Global warming, which is caused by higher levels of carbon dioxide in Earth's atmosphere, has

Figure 1-2 Produce is important in our daily diet.

Figure 1-3 Cotton bolls harvested from the parent plant

caused concern among today's scientists. This is a controversial topic because the elevated levels of carbon dioxide are thought to be caused by indiscriminately replacing green spaces with buildings, clear cutting forests without proper restoration, and a variety of other factors. Scientists and politicians are now getting together to determine the best way to overcome the problem of global warming before it worsens. Plants are also grown in an orbiting spacecraft as excellent food sources for astronauts and as air purifiers.

Some plants, such as the common household spider plant, remove pollutants and carbon dioxide from the air in homes and other enclosed spaces (Figure 1-4).

Plants also provide **aesthetic beauty,** such as roses for Valentine's Day and colorful fall leaves (Figure 1-5). You have probably seen futuristic movies where everything is asphalt, and the important green spaces and aesthetic beauty of plants are nonexistent.

Figure 1-4 The spider plant is a commonly grown houseplant known to remove pollutants from the air.

Figure 1-5 Leaves changing colors in the fall (top) and a beautiful yellow rose (bottom) show the aesthetic beauty of plants.

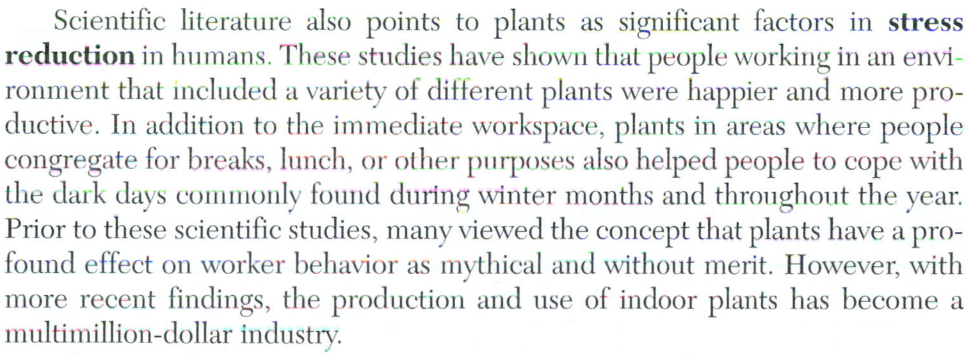

Scientific literature also points to plants as significant factors in **stress reduction** in humans. These studies have shown that people working in an environment that included a variety of different plants were happier and more productive. In addition to the immediate workspace, plants in areas where people congregate for breaks, lunch, or other purposes also helped people to cope with the dark days commonly found during winter months and throughout the year. Prior to these scientific studies, many viewed the concept that plants have a profound effect on worker behavior as mythical and without merit. However, with more recent findings, the production and use of indoor plants has become a multimillion-dollar industry.

Plants are also an important source of pharmaceuticals. For example, acetylsalicylic acid (**aspirin**) comes from willow trees, taxol from the yew plant (Figure 1-6), and aloe from the aloe plant. There is a variety of other examples that can be given (DiCosmo & Misawa, 1996). Products from plants can serve as **fuel** or as a fuel additive, such as ethanol in gasoline.

Plants play an important part in preventing **erosion,** thereby preserving precious topsoil. Today, the availability of high-quality topsoil is limited, and considerable research is underway on ways to prevent erosion. Without topsoil, farmland used to grow crops would be reduced, thereby reducing the food supply. In Pennsylvania, crown vetch along roadways is commonly used to prevent soil erosion (Figure 1-7). Plants provide **habitat** and cover for animals in the wild, protecting them from predators. Plants are an integral part of **wetland** purification of water. Although once destroyed due to agricultural and urban development, today the federal government protects wetlands. Plants are also used in artificial wetlands to purify wastewater from greenhouses (Figure 1-8). These are only some of the important uses of plants, and you can probably think of many more.

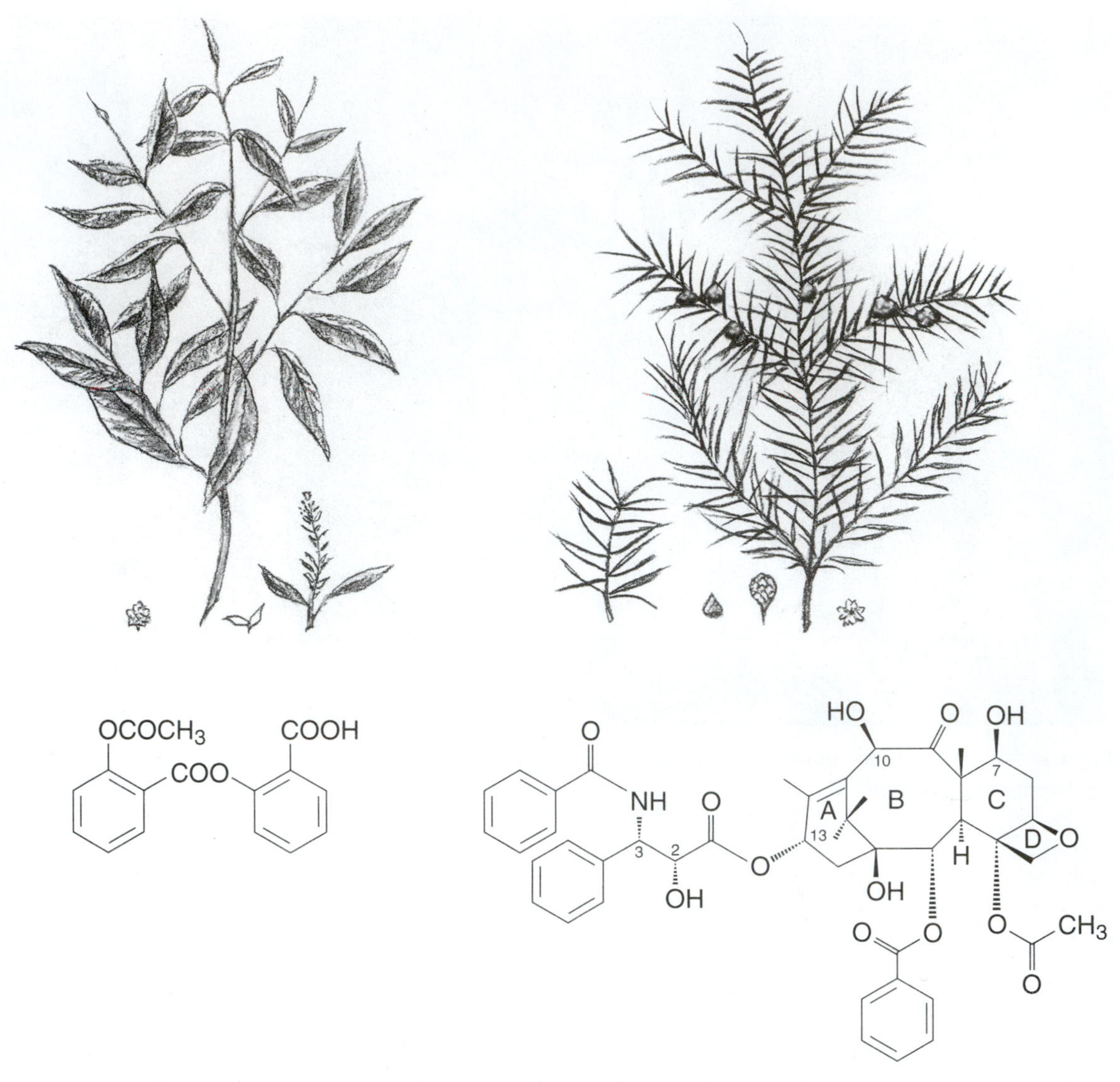

Figure 1-6 The structure of acetylsalicylic acid (aspirin) derived from willow trees is shown on the left, and taxol, an important antitumor compound derived from the yew plant, is shown on the right.

Figure 1-7 Crown vetch is used along a roadside to prevent erosion.

Figure 1-8 This man-made wetland purification system operates in a greenhouse. *Courtesy of Dr. Robert Berghage, Department of Horticulture, The Pennsylvania State University.*

IMPORTANT PROCESSES IN PLANTS

In this section a variety of selected examples of important processes in plants will be briefly covered; these examples will be discussed in more detail later in the book.

Seed Germination

Seed germination or propagation by seeds is the major method of reproduction in nature and the most widely used method in agriculture due to its high efficiency and ease. A seed has all the genetic information to make a whole plant. A **seed** is defined as a ripened ovule, which consists of an embryo, stored food reserves, and a seed coat or covering (Figure 1-9). When placed in the soil, the seed's special sensors enable it to determine which way is up and which way is down (Figure 1-10). Much research is underway to understand this process so that plants can be grown in outer space without disoriented growth.

Certain criteria must be met for germination to occur:

- The seed must be **viable,** which means the embryo is alive and capable of germination.
- The proper environmental conditions must be available, such as water, proper temperature, oxygen, and, in some cases, light.
- Primary dormancy must be overcome. Dormancy acts as a safety mechanism that protects the seed from adverse environmental conditions.

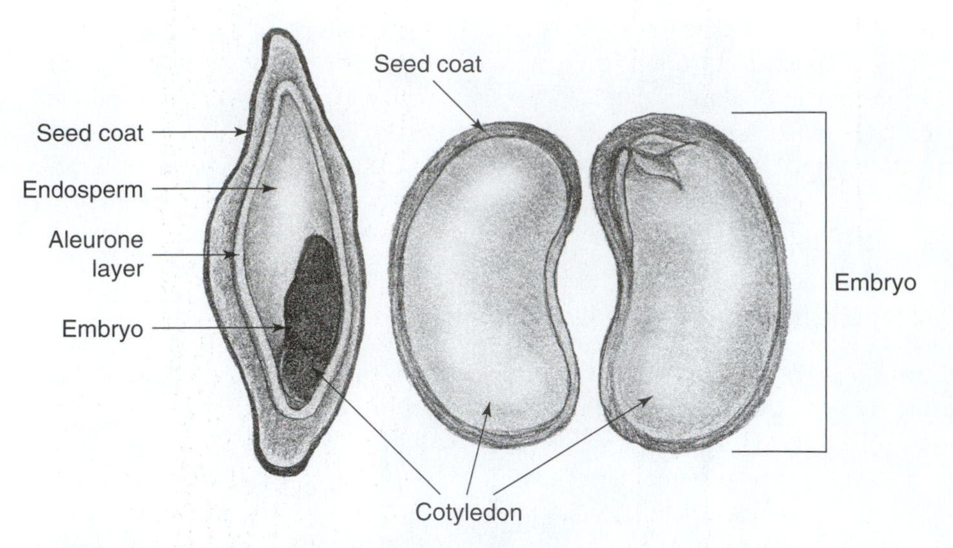

Figure 1-9 The key parts of a seed

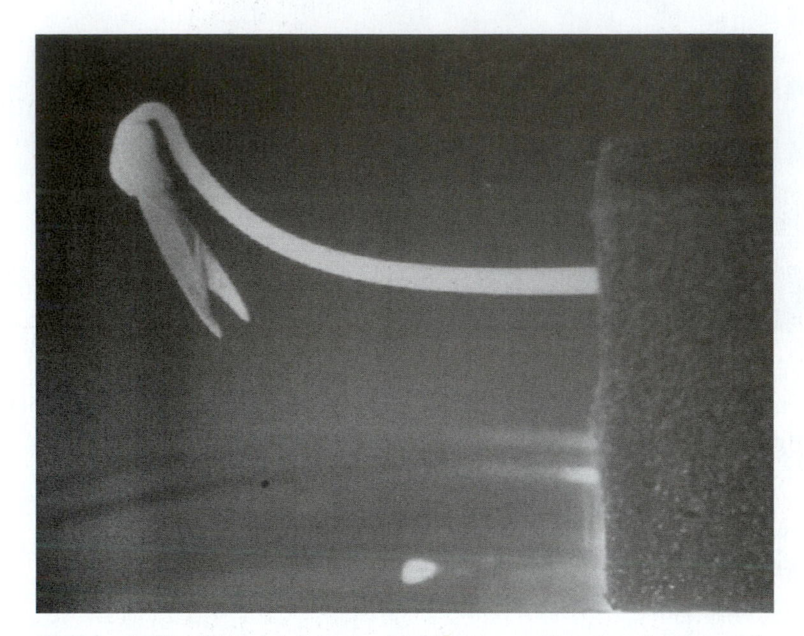

Figure 1-10 This etiolated mungbean seedling bends upward against the force of gravity.

After the criteria for seed germination have been satisfied, the seed goes through three stages of seed germination prior to any visible changes in the seed which include:

1. Imbibition or uptake of water.
2. Formation of enzyme systems.
3. Breakdown or metabolism of storage products for energy and as building blocks.

Plant hormones are responsible for inhibiting or promoting seed germination. Gibberellins (GA) typically stimulate germination, whereas abscisic acid (ABA) inhibits seed germination (Arteca, 1996).

While the first three stages are in motion, there are no visible changes in the seed. The first visible sign of seed germination is the emergence of the radicle followed by seedling growth. After the radicle has emerged and the seedling begins to grow and acts as a subterranium organism with no pigmentation exhibiting exaggerated growth until it reaches the soil surface. How does the seed know that it is close enough to the soil surface to germinate and survive? For instance, pesty weed seeds do not germinate until they know they are close to the soil surface and that their chances of survival are very good. In addition to gravity sensors, seeds also have light sensors to perceive light so that when a certain wavelength of light is perceived, germination occurs.

Photosynthesis

Photosynthesis refers to a series of chemical reactions in which carbon dioxide and water in the presence of light are converted into carbohydrate (sugar) and oxygen (Figure 1-11). Essential to the photosynthetic process is light and chlorophyll, which is a green pigment contained in the chloroplast of plant cells. The purpose of most plant-related research is to explore ways to manipulate growth and increase productivity of plants (Arteca, 1996; Devlin & Witham, 1983). The regulation of photosynthesis and the movement of photosynthetic products from their site of synthesis in the leaf (source) to their sites of accumulation (sink) have a profound effect on the size of the plant and are currently an exciting area of research.

For life as we know it to exist, photosynthesis is essential; however, surprisingly very little research was done in this area until the eighteenth century. One of the main reasons for this lack of research is that the early Greeks believed the plant received its food directly from the Earth, which contained plant and animal debris. They specified that the roots of plants took up everything necessary for plant growth. Early researchers found that adding more plant and animal materials to the soil increased the size of the plant, thereby supporting the Greek theory, which remained uncontested until much later.

Van Helmont in the early 1600s performed a simple yet elegant experiment with willow seedlings. This experiment involved carefully weighing a willow seedling, the tub, and the soil it was planted in, and then growing the plant for five years. At the beginning of the experiment, the seedling weighed 2 kg, and by the end of the five-year period, it had increased to 75 kg. Van Helmont also measured the weight of the soil, which had only lost a few grams in dry weight. Based on these facts, he concluded that water, not soil, was responsible for the growth of the plant. The few grams of soil that were lost were nutrients, which

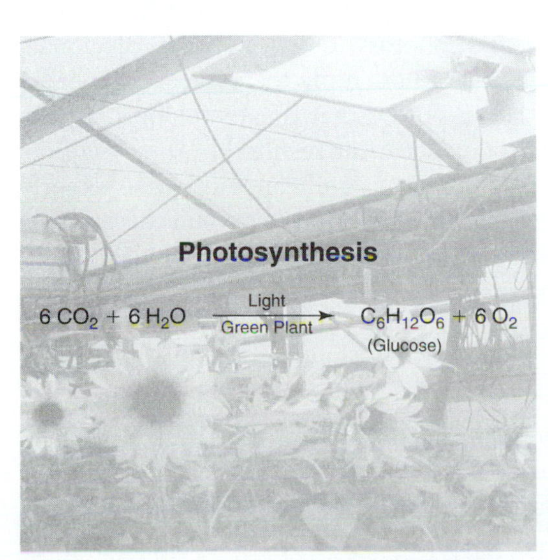

Figure 1-11 A simple chemical equation for photosynthesis

were essential to growth, and water did not contribute appreciably to the overall mass of the willow plant; rather the process of photosynthesis was responsible for the size increase.

In 1699, Woodward reevaluated the work of Van Helmont and found that plants required more than water for growth. He worked with mint plants grown on water from different sources, which included rainwater, river water, and Hyde Park drainage water. From this study, he came to the following conclusion:

> Vegetables are not formed of water but of a certain peculiar terrestrial matter. It has been shown that there is considerable quantity of this matter contained in rain, spring, and river water; that the greatest part of the fluid mass that ascends up into plants does not settle there but passes through their pores and exhales up into the atmosphere; that a great part of the terrestrial matter, mixed with water, passes up into the plant along with it and that the plant is more or less augmented in proportion as the water contains a greater or less quantity of that matter; from all of which we may reasonably infer, that earth, and not water, is the matter that constitutes vegetables. (Quotation taken from W. Loomis, 1960)

In 1772, Priestly studied the gas exchange that accompanies the process of photosynthesis. His experiment involved placing a mouse in a bell jar with a burning candle. He concluded that the mouse could not live in the air contaminated by the burning candle. However, he did note that if a sprig of mint were placed in the chamber with the burning candle, the air would be purified and the mouse would survive under these conditions. Priestly also observed that the mint plants could survive in the contaminated air caused by the burning candle. Based on his scientific findings, Priestly concluded that

> . . . plants, instead of affecting the air in the same manner with animal respiration, reverse the effects of breathing and tend to keep the atmosphere sweet and wholesome when it has become noxious in consequence of animals either living and breathing or dying and putrefying in it. (Quotation taken from W. Loomis, 1960)

In 1779, Ingenhousz reported that plants could only purify the air in the light. In addition, he stated that only the green parts of the plant produced the purifying agent (oxygen); however, nongreen tissues contaminated the air. Ingenhousz was the first to recognize that chlorophyll and light participated in the photosynthetic process.

In 1842, Mayer established the law of the conservation of energy. He stated that the energy used by plants came from the sun and that this energy was converted to chemical energy by the process of photosynthesis. In 1905, Blackman demonstrated that photosynthesis consisted of both a photochemical (light) and biochemical (dark) reaction. In 1937, Hill reported that isolated chloroplasts in the presence of light, water, and a hydrogen acceptor resulted in the evolution of oxygen in the absence of carbon dioxide. The significance of these experiments was that they provided evidence that the evolution of oxygen was a result of photochemical reactions. Today we know that oxygen from photosynthesis comes from water and not from carbon dioxide.

Phototropism

Phototropism is the movement of the plant in response to directional fluxes or gradients in light. Although Darwin is better known for his theory of evolution, he is considered responsible for initiating modern plant hormone research. In his book, *The Power of Movement in Plants* (Darwin, 1880), he described phototropism and gravitropism for the first time. Darwin used coleoptiles to study phototropism. Coleoptiles are specialized leaves in the form of a hollow cylinder that enclose the epicotyl and are attached to the first node in grasses. The coleoptile protects the growing tip of the grass seedling until the more rapidly growing leaf emerges above the ground. In simple yet elegant experiments, Darwin showed that when coleoptiles were exposed to unidirectional light, there was bending toward the source. If the coleoptile tip was removed, phototropic curvature did not occur, which suggested that the phototropic stimulus is in the tip (Figure 1-12). Today we know that the tip contains auxin, which is responsible for phototropic bending in plants, and that plants bend toward higher light levels to maximize growth and, in some cases, to survive. The Cholodny-Went theory describes how the process of phototropism occurs. This theory states that when higher concentrations of the plant hormone auxin are on the shaded side than on the light side, the shaded side experiences accelerated growth, which causes bending toward the light.

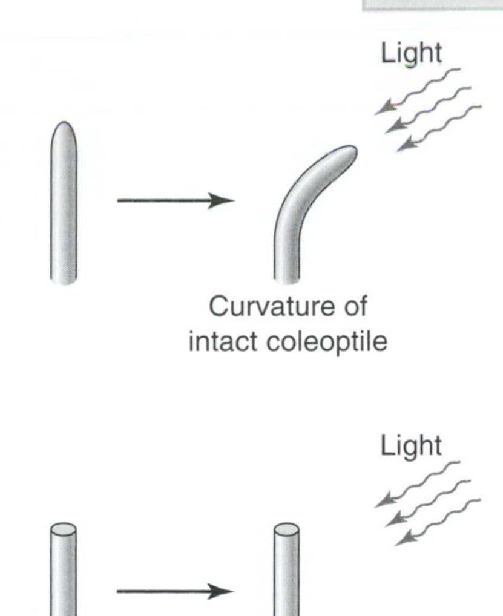

Figure 1-12 Experiments conducted by Darwin demonstrated the phototrophic response in plants. These experiments showed that the phototropic stimulus was found in the tip of the coleoptile.

Thigmotropism

Thigmotropism is the movement of a plant in response to touch; examples include the Venus flytrap, sensitive plant, tendrils from the pea, and many others. The inability of plants to move from an adverse external environment has led to the evolution of adaptive mechanisms, which permit them to respond to environmental changes to survive. Plants are exposed to mechanical forces (touch) through wind, vibrations, rain, plant parts rubbing against one another, and others.

Touch stimulation has positive effects, such as

- shorter and sturdier plants that can withstand certain types of stress (such as wind),
- plants that are more resistant to drought stress, pathogen attack, and more.

Touch stimulation also has negative effects in that it

- causes an inhibition of leaf expansion.
- decreases photosynthesis.
- promotes leaf yellowing.
- delays flowering which reduces overall growth and subsequent crop yields (see Figure 1-13).

A number of researchers are studying touch-induced gene expression. They have shown that when a plant is touched for as little as 30 seconds, genes are turned on as rapidly as 5 minutes (Figure 1-14). Research in this area is an attempt to better understand the effects of touch on plant growth. The eventual goal of touch research is genetically engineering plants to reduce plant height, thereby overcoming the need to use chemicals for height reduction, which is a common practice today.

Figure 1-13 The Arabidopsis plant on the left received no touch treatment, and the one on the right received a 30-second touch treatment daily for four weeks. Picture was taken when the plants were six weeks old.

Gravitropism

Gravitropism is the movement of a plant in response to gravity. Research in this area began with Darwin in 1880 and is still being pursued by many research groups today. Studying plant growth and development in response to gravity is

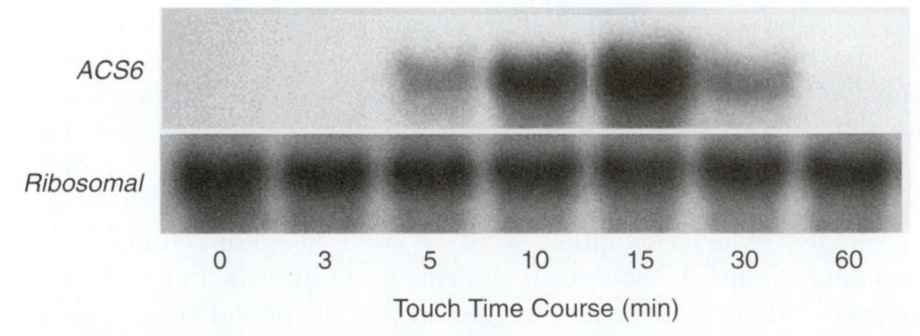

Figure 1-14 Time course of touch-induced ACC synthase (*ACS6*) gene expression. ACC synthase is a gene encoding for an important regulatory enzyme in the ethylene biosynthetic pathway. Three-week-old, light-grown Arabidopsis plants receiving no touch served as a control, and others were touch-stimulated for 30 seconds. At designated time intervals, samples were taken and placed immediately in liquid N_2. Ten micrograms of total RNA was probed with *ACS6,* and a pea ribosomal probe was used as a control to show that equal amounts of RNA were loaded on the gel.

Figure 1-15 Arabidopsis inflorescences prior to gravistimulation (left), and inflorescences placed on their side for one hour (right) showing upward bending

important for many reasons, including how a seed knows which way is up, growth under weightlessness conditions in an orbiting spacecraft, and unsightly bending of flower stalks during shipment and storage. Therefore, how a plant responds to gravitational force is important. When a seedling is placed horizontally, it responds to Earth's gravitational force with an altered growth pattern. The stem bends upward against gravity and this process is thus called **negative gravitropism** (Figure 1-15), whereas the roots bend downward with gravity and this process is called **positive gravitropism.** The Cholodny-Went theory provides a reasonably sound explanation for gravitropism. The theory states that an accumulation of the plant hormone auxin on the lower side of the stem and root results in differential growth. The accumulation of auxin on the lower side of the stem accelerates growth on this side, which causes upward bending. The roots are much more sensitive to auxin; therefore, the growth rate of the bottom side of the root is reduced although the top grows normally. This differential growth results in downward bending.

The Statolith Theory has been used to describe how plants perceive gravity. A **statolith** is an object in the plant cell now thought to be an amyloplast (it belongs to a class of plastids called leucoplasts, which are colorless). When a plant is placed in the horizontal position, statoliths fall to the bottom of the cell, which causes changes in auxin gradients. These changes result in differential growth that enables the plant to perceive that it is in the wrong position and reorient its growth pattern.

Dormancy

Dormancy is the temporary suspension of visible plant growth. Plants have the amazing ability to stop growth functions in order to protect themselves. For example, during the winter months, trees become dormant. In many cases, if the trees did not do this, they would suffer severe damage or die. Plants perceive environmental cues, such as day length, and temperature, which tell them when to become dormant and when to come out of dormancy. In the fall, trees know to drop their leaves to prepare for the winter months; if not, snow loads would cause many trees to fall down. Sometimes plants are fooled; for example, citrus crop buds break dormancy because the temperature warms up in the spring, but then a frost comes that kills the bud and causes orange juice prices to go up. Researchers are currently trying to better understand the process of dormancy in plants and how the process might apply to humans as well. For example, a person dying from a disease might be put into a dormant state to buy time for a future cure.

Plant scientists define dormancy as the temporary suspension of growth and development of seeds, buds, and other plant parts under conditions that appear to be suited for plant growth. Plant growth can be suspended due to adverse environmental conditions; for example, seeds will not germinate if conditions are too dry. Growth can be suspended due to inhibitors contained within the seed coat or mechanically due to the presence of a seed-enclosing structure that does not allow the embryo to expand. In some seeds, germination cannot occur due to the presence of membranes or seed coats that are impermeable to water or oxygen, thereby keeping the seed in the dormant state. More commonly, seeds and buds require special light and temperature conditions. An excellent example of dormancy that is regulated by temperature and light is the deciduous habit of northern temperate zone plants.

Dormancy can be broken down into two categories. The first category is suspension of growth due to the lack of necessary external environmental factors, which is called **quiescence.** The second category of dormancy, called **rest,** is caused by internal limitations in which dormancy cannot be broken even if environmental conditions are favorable.

Seeds that can remain dormant until a sufficient amount of water is available have a better chance of survival than those that germinate at the first sign of water. Dormancy can also be a more complex survival mechanism; for example, the desert shrub guayuale has a material covering the seed that contains an inhibitor causing the seed to stay in the dormant state. Under conditions of very heavy rainfall, the inhibitor is diluted, allowing germination to occur. Species of *Convolvulus*—which grow in arid regions—have seed coats that are impermeable to water. For these seeds to germinate, the seed coats must be mechanically broken. Under natural conditions, this occurs gradually over a long period of time, which has the advantage that not all seeds will germinate at the same time, thereby protecting the species against being wiped out by a single adverse season.

Dormancy is both beneficial and nonbeneficial to humans. For example, due to dormancy, cereal grains can be stored dry and later be used for food. Without dormancy, these grains would germinate and lose their usefulness. The nonbeneficial aspect of dormancy is with respect to weeds. Certain weeds can lie dormant in the soil for many years. Then when a field is plowed, the dormancy is broken in a small percentage of these seeds, which enables them to compete with any economic crop sown in the field. This becomes an annual problem because although some weed seeds are triggered to germinate by plowing, some always remain dormant in the soil. Therefore, growers are faced with the same problem each year: they can only destroy those seeds that germinate and have almost no control of over those lying dormant in the soil.

Senescence

Senescence refers to the general failure of many synthetic reactions that precedes cell death and is characterized by chlorophyll degradation as well as many other factors (Figure 1-16). These internally controlled deteriorative changes are natural causes of plant death. Senescence may be thought of as a natural developmental process, namely, terminal differentiation. The deterioration process involved in the natural termination of the functional life of a particular organism may occur very slowly or very rapidly. Much research is required to better understand the senescence process in plants, and learning how to regulate senescence may have economic benefits. For example, by delaying senescence, crops can grow for longer periods of time and increase yields. In addition to increasing crop yields, delaying senescence increases the keeping quality of flowers, vegetables, fruits, grains, and other crops, which reduces postharvest losses and saves millions of dollars annually. A better understanding of the senescence process in plants may lead to extending human life spans and enhancing the quality of life.

Figure 1-16 The older leaves of this Arabidopsis plant are senescing.

Flowering

The initiation of flowering is very important for a number of reasons. Flowers attract bees and other insects which act as pollinators, resulting in seed formation and perpetuation of the species. Research in horticulture has led to many advances in this area. The pioneering work of Garner and Allard (1920) laid the foundation for the use of photoperiod to manipulate flowering in horticultural crops today. They used the large-leaf tobacco plant mutant called Maryland mammoth. This mutant did not flower during the summer months under long photoperiods; however, when they subjected this mutant to shorter photoperiods, it flowered profusely. Today, flowering is promoted in a number of crops by modifying day length, temperature, and other environmental conditions. For example, chrysanthemums exposed to short days will flower in time for specific holidays, such as Thanksgiving. Other examples are the promotion of flowering of poinsettias for Christmas and lilies for Easter (Figure 1-17). Early work on photoperiod showed that the plant perceived light in the leaves via receptors, which sensed the photoperiodic stimulus and produced a compound that was transported to the apical bud, causing the initiation of floral primordia. Since this pioneering work and despite many years of research, there has been little success in identifying this compound, which has been called Florigen (flowering hormone).

The study of sex in plants started many years ago with Empedocles, Aristotle, and Aristotle's student, Theophrastus. Only about 35 years ago, however, the effects of plant hormones on the modification of sex expression in plants began to be studied. Today, scientists generally accept that plant hormones stimulate flowering in plants; for example, in *Cucurbits* foliar applications of auxin and ethylene promote female flowers, while gibberellins promote male flowers (Figure 1-18). In *Cannabis*, a crop that has been extensively studied by Russian scientists, foliar applications of auxins, ethylene, and cytokinins promote female flowers, while gibberellins promote male flowers. Plant hormones have also been reported to be involved in sex expression in begonia, hops, grape,

Figure 1-17 Lily plants flowering in time for the Easter holiday

Figure 1-18 Female cucumber flower (left) and a male cucumber flower (right)

muskmelon, squash, pumpkin, tomato, and cotton (Weaver, 1972). Using plant hormones to modify sex expression in plants has great potential in seed production and breeding programs.

Abscission

Abscission is the separation of a plant part from the parent plant. Theophrastus, a student of Aristotle, was the first to study abscission. Much research has been focused on this area; in fact, manipulating abscission is a common agricultural practice today. For example, hastening leaf abscission facilitates the mechanical harvesting of cotton, and abscission in cherries reduces the fruit removal force for mechanical harvesters. Delaying abscission also allows fruits to remain on the tree longer to achieve the desired size and to prevent floral abscission. Transgenic plants are being produced that are insensitive to the plant hormone ethylene, thereby decreasing floral abscission and increasing the plant's shelf life (Figure 1-19).

Figure 1-19 A regal Pelargonium plant's floral abscission following treatment with ethylene

Fruit Growth and Development

The use of plant growth regulators to control fruit size, shape, and maturation has become increasingly important in modern agriculture because altering any of these characteristics may increase marketability. In addition to modifying fruit size and shape, the hastening or delaying of maturation offers numerous benefits to the grower, such as utilizing peak demands, avoiding unfavorable environmental conditions, and extending the market period. Auxins have been shown to stimulate strawberry fruit growth. For example, when all seeds are removed and lanolin paste is applied to the fruit, no growth occurs; when seeds are removed and lanolin paste containing auxin is applied, growth occurs; if one seed is left on the receptacle, then growth occurs below it (Figure 1-20). Thompson seedless grapes are all sprayed with the plant hormone gibberellin to elongate the bunch, promoting larger berries and elongated fruit (Figure 1-21). When apple trees are sprayed with growth substances, such as cytokinin and gibberellin, fruits become elongated with defined calyx lobes, making them more salable (Figure 1-22). The manipulation of fruit ripening is also of major economic importance. The plant hormone ethylene is involved in the ripening of many fruits. Today, ethylene is commonly used in agriculture to promote ripening, or if the grower wants to delay ripening, ethylene is blocked using chemical and cultural methods (Figure 1-23).

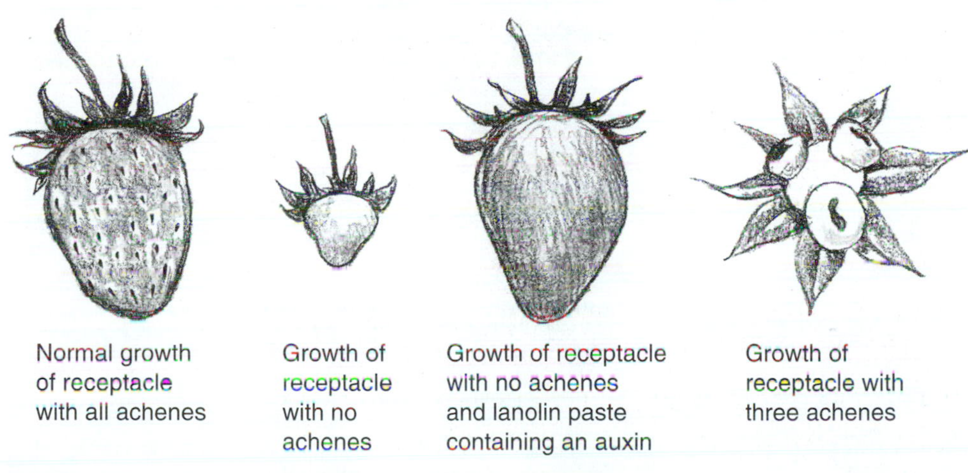

Figure 1-20 The classic work of Nitsch shows how auxin is involved in the growth of strawberry fruits.

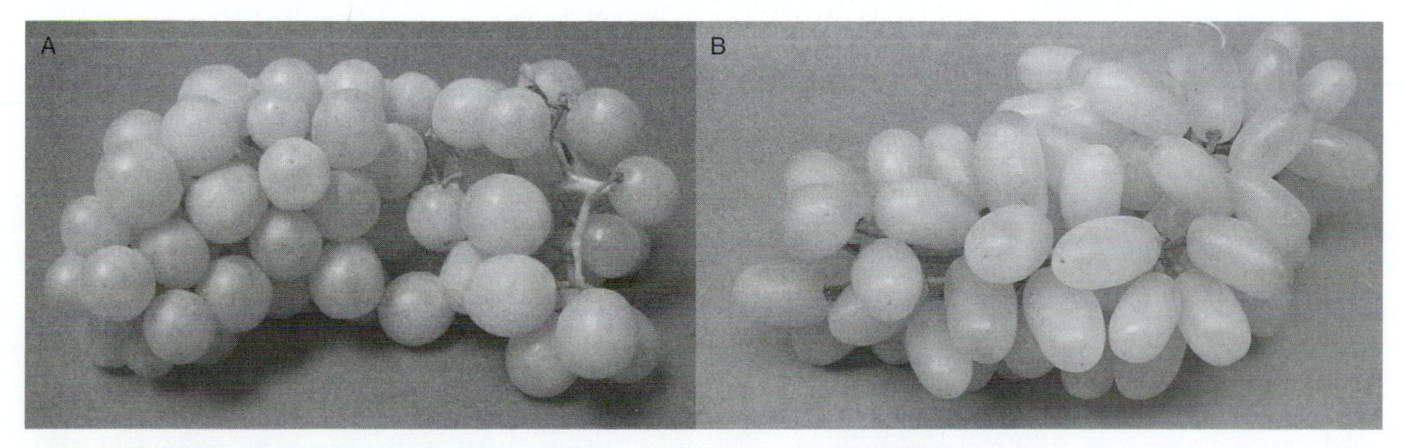

Figure 1-21 Thompson seedless grapes on the left were not treated with gibberellin, whereas the bunch on the right was treated with gibberellin.

Figure 1-22 This Red Delicious apple was harvested from a tree sprayed with a mixture of cytokinin and gibberellin.

Figure 1-23 The tomato on the left was untreated, whereas the tomato on the right was treated with Ethephon, an ethylene releasing compound (top). A tomato plant is shown with fruits at different stages of ripening (bottom).

Figure 1-24 The plant on the left was the control plant, and the next two plants were treated with increasing concentrations of a plant growth retardant. Courtesy of Dr. E. Jay Holcomb, Department of Horticulture, The Pennsylvania State University.

Growth Retardation

Reducing plant size by using plant growth retardants is a common horticultural practice to make the plant compact, which in many cases is more commercially acceptable. Many plant growth retardants are currently available for restricting plant growth. The most commonly used and best understood plant growth retardants inhibit the biosynthesis of the plant hormone gibberellin. Plant growth retardants are commonly used on floricultural crops, such as poinsettias, chrysanthemums and slavia (Figure 1-24), to maintain an optimum height, which is accepted by consumers. In many cases, bedding plants such as zinnia and geranium are treated with plant growth retardants to promote compactness, to maintain quality prior to sale, and to enable longer shelf life because the plants do not rapidly outgrow their containers (Arteca, 1996).

Weed Control

All who work with plants are well acquainted with weeds. A **weed** is basically a plant growing where it is not desired. Weeds are costly because they compete with crop plants for water, nutrients, and light, and they harbor diseases and insects that attack crop plants. Weeds are widespread in their occurrence and cost billions of dollars each year in control efforts and crop losses. Weeds are classified as pests in the same manner as insects, diseases, nematodes, rodents, and other animals. A chemical used to control a pest is called a **pesticide;** a pesticide used specifically for weed control is known as a **herbicide.** The word *herbicide* comes from the Latin *herba,* or plant, and *caedere,* to kill. Since Theophrastus reported that pouring olive oil over their roots could kill trees; many nonselective materials such as salt and petroleum oils have been shown to kill weeds. The first selective herbicide was not found until the early 1900s, and today more then 130 different selective herbicides are used throughout the world. In the United States alone, one billion pounds of pesticides are sold annually, and over 65 percent are herbicides. Note that selectivity is the key to the widespread use of herbicides. Herbicides are effective when used as a part of a responsible, integrated pest-management program. An example of a selective herbicide is 2,4-dichlorophenoxyacetic acid (2,4-D), which is commonly used commercially and residentially. Since 2,4-D was introduced in the mid-1940s, it has had a profound effect on weed control. Today, 2,4-D is included in fertilizers to kill broadleaf weeds found in lawns (Figure 1-25). Current research programs are continually striving to produce better selective herbicides that have minimal effects on the environment.

Figure 1-25 Broadleaf weeds found in lawns are commonly killed by fertilizer containing 2,4-D.

Production of Important Chemicals by Plants

Plants are sources of pharmaceuticals, colorants, flavors, fragrances, and pesticides. Because plants produce so many important chemicals, a huge industry has emerged to screen for new industrially interesting compounds. Thus far, only a small percentage (estimated between 5 and 10 percent) of plants have been screened for useful compounds with the aid of modern scientific tools. Clearly the plant kingdom is an untapped source of novel chemicals to be used for drug development (DiCosmo & Misawa, 1996). Scientists are interested in screening for important chemicals produced by plants for biological activity because products produced by plants are highly complex, which makes them nearly impossible to synthesize economically. With this in mind, how can the untapped potential of plants best be exploited? Two possible paths can be taken: (1) use folklore for potentially useful compounds and (2) randomly screen for specific biological activities of a particular interest. Sources of plant material to extract for novel compounds include the whole plant and/or its individual parts. When plants are rare or slow growing, an alternative is to screen plant cell cultures for interesting compounds (Figure 1-26). Two examples of plant-derived drugs are acetylsalicylic acid (aspirin), which initially came from the willow tree, and taxol from the yew tree.

Herbs, such as basal, pepper, oregano, parsley, and garlic, are commonly used as foods. However, archaeological evidence shows that prehistoric humans used plants to heal, and many cultures from all over the world have used herbs not only as a food source but to treat a wide variety of illnesses. The World Health Organization estimates approximately 80 percent of the world's population use some form of herbal medicine. The percentage of people using plants for medicinal purposes had been considerably less in North America. National surveys have shown that only 3 percent of the American public in

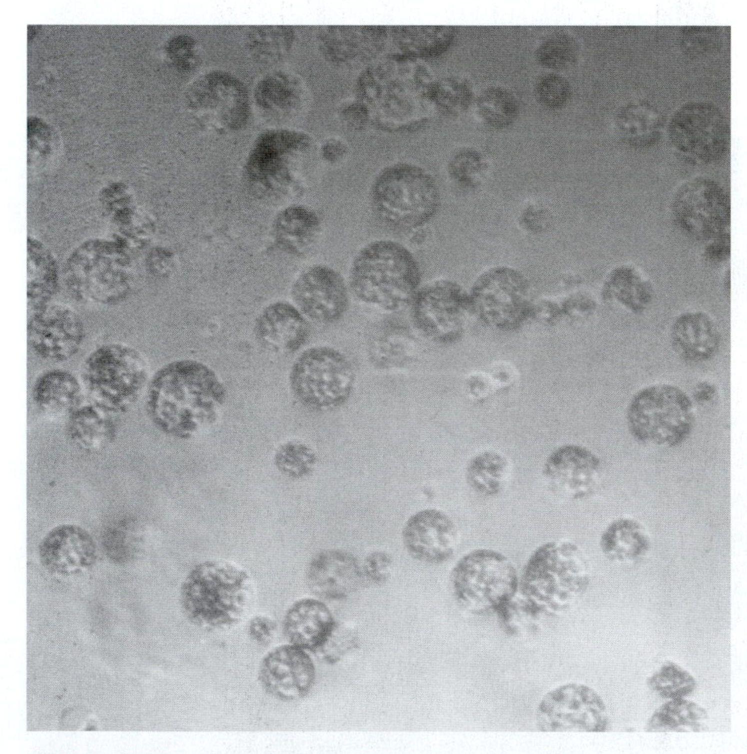

Figure 1-26 Plant protoplasts, plant cells without their cell walls, are screened for interesting compounds.

1991 used plants for medicinal purposes. However, since this time, there has been a huge increase in the interest and use of medicinal plants in North America. Today, more than 37 percent of the U.S. population uses plant medicinal products, and this number is continually increasing (Brevoort, 1998). Sales of plant medicinal products are now approximately 3 billion dollars per year in the North American market (Glaser, 1999). One of the main contributing factors to this increase in plant medicinals in the United States has been the Dietary Supplement Health and Education Act (DSHEA) passed in 1994 by the U.S. federal government. The DSHEA has facilitated the production and marketing of plant medicinal products (Brevoort, 1998). Today, health food and specialty stores offering plant medicinal products are very popular. In addition, several major pharmaceutical companies have thrown their hats into the ring and are now making plant medicinal products available to the North American public and the world.

Biotechnology

Biotechnology is the manipulation of living organisms or substances obtained from living organisms for the benefit of humanity. Earth's population is increasing at rapid rate and is estimated to reach 12 billion people by 2100. Producing food to meet this ever-increasing population while minimizing adverse effects on the environment is extremely important. Recent developments in biotechnology are changing the way food crops are being produced. For example, techniques for producing genetically modified organisms (GMOs) have changed the way new and improved varieties are made. *Biotechnology* is a relatively new term; however, its origin goes back to the beginning of time when farmers set aside a portion of their harvest to select seeds with specific traits. They selected for plants that would be used in future breeding programs, leading to the varieties used today (Chrispeels & Sadava, 1994).

There are many forms of plant biotechnology; one of the simplest forms is the use of hydroponics (Figure 1-27). Other examples of plant biotechnology are

- using plant cell, tissue, and organ culture for micropropagation (Figure 1-28) and production of important chemicals (Figure 1-29).

Figure 1-27 Hydroponically grown yew plants

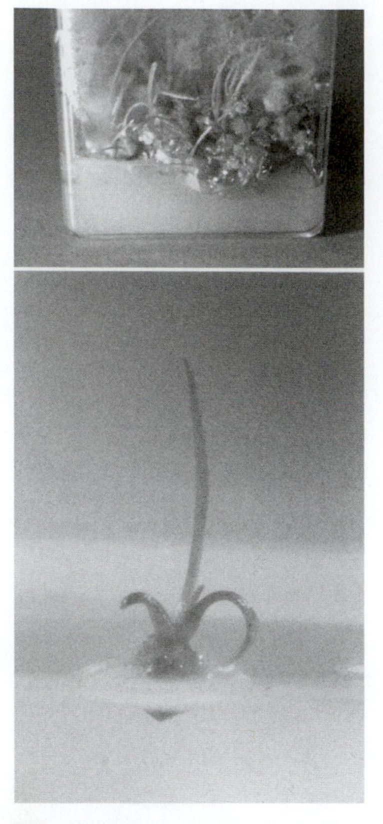

Figure 1-28 Mass of callus producing lily plants (top) and a single lily plant from the above mass grown on agar aseptically (bottom)

Figure 1-29 Plant tissue culture can be grown on a small scale using flasks (top), or on a large scale using specialized bioreactors (bottom).

- producing whole plants from single cells as a new source of variability for plant improvement.
- transfering foreign genes into plants to improve crop plants, for example, genetically modifying plants to block ethylene production or action to enhance the postharvest storage life of tomatoes, potatoes, and other crops.
- producing cold-resistant plants containing an antifreeze gene from flounder.
- incorporating genes enabling fruits and vegetables to produce compounds, thereby serving as a tasty oral vaccine.
- genetically engineering plants that are resistant to herbicides.

SUMMARY

This chapter showed how important plants are from a variety of perspectives. In fact, you should be able to list at least 11 reasons why plants are so important. Plants are beneficial for a food source, shelter, pharmaceuticals, air purification, aesthetic beauty, stress reduction, prevent soil erosion, provide habitat and cover for animals and purification of water in wetlands.

In addition to the important uses of plants, plant research is currently going on in fascinating areas. These important plant processes include seed germination, photosynthesis, phototropism, thigmotropism, gravitropism, abscission, dormancy (including quiescence and rest), senescence, flowering, plant growth regulators, weed control, production of important chemicals, and biotechnology. By understanding how these processes work, we can manipulate them to our benefit. Many of the topics covered in this chapter are discussed in more detail later in the book, and the information presented here is an important foundation for chapters that follow.

Review Questions for Chapter 1

Short Answer

1. List six reasons plants are important and explain why.
2. What are some examples of plant-derived pharmaceuticals?
3. What criteria must be met for germination to occur?
4. What are the four stages of seed germination?
5. Provide two positive and two negative effects of touch stimulation.
6. What are four major reasons weeds cause problems among crop plants?

Define

Define the following terms:

seed	thigmotropism	gravitropism	abscission
viable	dormancy	negative gravitropism	weed
photosynthesis	rest	positive gravitropism	biotechnology
phototropism	quiescence	senescence	

True or False

1. The number one requirement for a seed to germinate is proper environmental conditions.
2. One of the positive effects of touching plants is that they become shorter and more resistant to other forms of stress.
3. Rest is a form of dormancy regulated by internal factors.
4. Plants have been scientifically shown to reduce stress in humans.
5. Asexual propagation is the most common form of reproduction in nature.

Fill in the Blanks

1. Photosynthesis is a series of chemical reactions in which ________ and ________ are converted in the presence of ________ to ________ and ________.
2. Plants have the ability to perceive environmental cues, such as ________ and ________, which tell them when to become dormant and when to come out of dormancy.
3. The first visible sign of seed germination is emergence of the ________.
4. Plants remove ________ from the air and add ________, maintaining a balance of gasses in our atmosphere.
5. Senescence is a general failure of many synthetic reactions that precedes cell death and is visually characterized by ________ degradation as well as many other factors.

Activities

For this activity, you will need to search university Web sites throughout the world to obtain information about horticulture research currently being conducted at those institutions, specifically related to the processes discussed in this chapter. (If you are having difficulty finding information about horticultural research, try checking the faculty links on the home pages.) Choose four universities that are doing research that interests you or would be of value to your classmates. The research may or may not be related to the plant processes presented in this chapter. Write a summary about the research, including the purpose of the research, what is being evaluated, and why you selected it. Be sure to include the name of the university and a title for the research topic.

The Internet's variety of information will be a great resource for this and other activities contained in this book. Bear in mind that URLs (Web addresses) can change, so you may need to be persistent to find the particular information you want. To explore the Internet, use Yahoo!™, Google™, or your favorite search engine.

References

Arteca, R. (1996). *Plant Growth Substances: Principles and Applications*. New York: Chapman & Hall.

Brevoort, P. (1998). The blooming U.S. botanical market: a new overview. *Herbalgram 44*, 33–46.

Chrispeels, M. J., & Sadava, D. E. (1994). *Plants, Genes, and Agriculture*. Boston: Jones and Bartlett.

Darwin, C. (1880). *The Power of Movement in Plants*. New York: Appleton.

Devlin, R. M., & Witham, F. H. (1983). *Plant Physiology* (4th ed.). Boston: Willard Grant Press.

DiCosmo, F., & Misawa M. (1996). *Plant Cell Culture and Secondary Metabolism*. New York: CRC Press.

Glaser, V. (1999). Billion-dollar market blossoms as botanicals take root. *Nature Biotechnology 17*, 17–18.

Loomis, W. (1960). Historical introduction. In W. Ruhland (Ed.), *Encyclopedia of Plant Physiology 5, Part 1* (pp. 85–114). Berlin: Springer.

Trewavas, A. (2002). Mindless mastery. *Nature* 410, 841.

Weaver, R. J. (1972). *Plant Growth Substances in Agriculture.* San Francisco, CA: W. H. Freeman.

CHAPTER

2

Origin of Agriculture and the Domestication of Plants

Objectives

After reading this chapter, you should be able to

- discuss several theories on the origin of agriculture and why it represents one of the most significant achievements by the human civilization.
- provide background information on the selection of edible plants used today.
- discuss the major contributions that the Egyptian, Greek, and Roman civilizations have made to agriculture.
- provide background information on medieval horticulture and agriculture.
- discuss the impact of the Age of Discovery and the New World on agriculture today.
- provide background information on the beginnings of experimental research throughout the world and agricultural research in the United States.
- discuss some theories of the origins of cultivated plants and the 13 major regions of the world where major food crops were domesticated.

Key Terms

agriculture
agricultural adjustment act
agricultural experiment station
Agricultural Marketing Act
agronomy
Charles Darwin
Dark Ages
Dioscorides
Egyptian
forestry
Greek
Gregor Mendel
Hatch Act
horticulture
Joseph Priestley
Linnaeus
Manorial System
Marcello Malpighi
Marco Polo
Morrill Land Grant College Act
Nehemiah Grew
Pliny the Elder
Renaissance
Robert Hooke
Roman Empire
Rudolph Camerarius
Stephen Hales
Theophrastus
U.S. Department of Agriculture

ABSTRACT

A variety of theories are involved in the origin of agriculture and the domestication of plants. This chapter begins with a brief description of five key theories on the origin of agriculture, including agriculture as a divine gift, agriculture as a discovery, agriculture as a result of stress, agriculture as an extension of gathering, and agriculture as a result of no specific model. The basis for selecting edible plants and the resulting major food crops used today is also explained.

A chronological history of agriculture picks up with a discussion of the Egyptian, Greek, and Roman civilizations, which is followed, after the fall of the Roman Empire, by the medieval era. History continues with the Renaissance (a time of rebirth in the arts and sciences), which is followed by the Age of Discovery and the New World. Into modern times, the beginnings of experimental science are covered, along with the beginnings of agricultural research in the United States.

A brief description of the origin of cultivated species shows that earlier scientists, namely de Candolle and Vavilov, believed that there were eight centers of plant origin. More recently, Harlan, an agronomist, states that probably three centers contributed to the propagation of agriculture around the world. The chapter concludes with a discussion of the 13 primary regions of the world where major food crops were domesticated, along with examples of the crops that characterized each of these regions.

INTRODUCTION

Although many theories exist on the origin of **agriculture** (the production of plants and animals to meet basic human needs), one fact remains clear: agriculture provided the human civilization with an available and dependable food supply (Figure 2-1). The advent of agriculture led to a new class of specialists—scientists, artists, engineers, and others—which in turn led to discoveries that increased the quality of human life. Interestingly, in the past, humans used thousands of plant species as food sources, thereby enjoying a highly varied diet. Only a small number of these species were ever domesticated, and with fast food prevalent in many diets, even fewer plant species are now being grown on a large scale.

The major accomplishments of **Egyptian** agriculture were the development of irrigation systems, drainage methodology, and land preparation methods using the hoe and plow. The **Greeks** contributed to agriculture indirectly through their great contributions to the study of botany. The **Roman Empire** soon followed the Greek civilization and lasted for about 1,000 years. Even though the Romans added little new information to agriculture, they made dramatic improvements to existing technology, such as grafting and budding, legume rotation, fertility analysis, and postharvest storage of fruits and vegetables. The Romans also described a prototype greenhouse, which was used to force vegetables. Medieval times were a stagnation period for the arts and sciences in the West. However, during this period, horticulture and other cultural activities were flourishing in the East (India, China, Japan, and others). Following the medieval times, the **Renaissance** blossomed and is described as the rebirth of the arts and sciences. Works by Leonardo da Vinci, Michelangelo, and others were key to this period. Experimental science began with da Vinci, Galileo, and Newton, and because of their innovative findings, there was a dramatic rise in botanical research.

Agricultural research in the United States began in 1862 when Congress passed bills constituting the U.S. **Morrill Land Grant College Act;** also that same year, the **United States Department of Agriculture** was created. Since that time, numerous studies in agricultural research have been conducted in the United States.

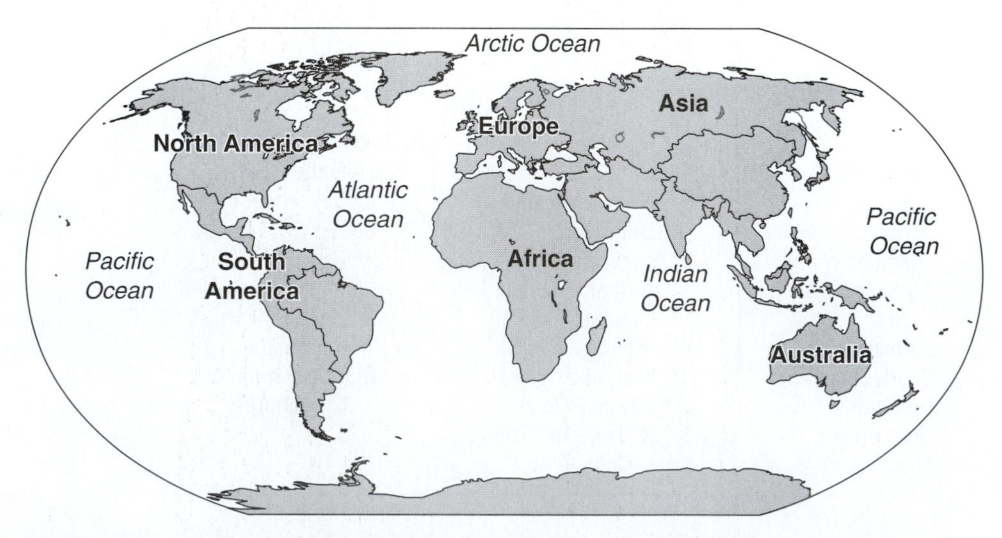

Figure 2-1 As civilizations developed throughout the world humans developed agriculture to meet their food supply demands.

Clearly, centers of plant origin from which agriculture was propagated occurred around the world; however, the actual number of centers is still debated. Many believe there were 13 major regions of the world where plants were domesticated: North America, Meso-America, Highland South America, Lowland South America, Europe, Africa, Near East, Central Asia, India, Southeast Asia, China, South Pacific, and Australia (Figure 2-2). Plants are thought to have been domesticated toward the end of the ice age or 11,000 to 15,000 years ago, and this occurred over a wide geographical range. Considering that some believe humans have been on Earth for approximately 4 million years, the domestication of plants is relatively new. However, the question of where plants were domesticated is much more complex than it appears.

THEORIES ON THE ORIGIN OF AGRICULTURE

In *Crops and Man* (Harlan, 1992), the many theories on the origin of agriculture theories are presented in detail; however, for the purpose of this introductory text, following is a brief summary of the five major theories.

- **Agriculture as a divine gift.** Although there are many descriptions on the origin of agriculture based on classical histories from a variety of civilizations, the same general theme is that agriculture was given as a divine gift.
- **Agriculture as a discovery.** The origin of agriculture based on invention or discovery has received a great deal of attention. Although there are many theories on how agriculture was discovered, the theories proposed by Sauer (1952) and Anderson (1954) are the most widely accepted.
- **Agriculture as a result of stress.** A considerable amount of support exists for the theory proposed by Cohen (1977), which suggests the increased stress caused by an increase in population and depleted resources led to the adoption of agriculture as a means of procuring food.
- **Agriculture as an extension of gathering.** In primitive societies based on hunting/gathering, humans had the knowledge to develop agriculture but elected not to. Basically, hunter/gatherers did not farm because food in the wild was plentiful. A quote taken from Berndt and Berndt (1970) by the Aborigines sums this point up nicely:

  > You people go to all that trouble, working and planting seeds, but we don't have to do that. All these things are there for us; the Ancestral Beings left them for us. In the end, you depend on the sun and the rain just the same as we do, but the difference is that we just have to go and collect the food when it is ripe. We don't have all this other trouble.

- **Agriculture as a result of no specific model.** All existing models proposed thus far have been refuted in one way or another. The no-specific-model concept allows for a wide array of possibilities in which agriculture was developed. Basically, this theory recognizes that human civilization is very different in all parts of the world, and no single model can explain the origin of agriculture (Harlan, 1992).

Agriculture (together with fire) is one of the greatest discoveries made by human civilization. In primitive societies, each individual had to be involved with procuring food for a particular group to survive. Therefore, the major

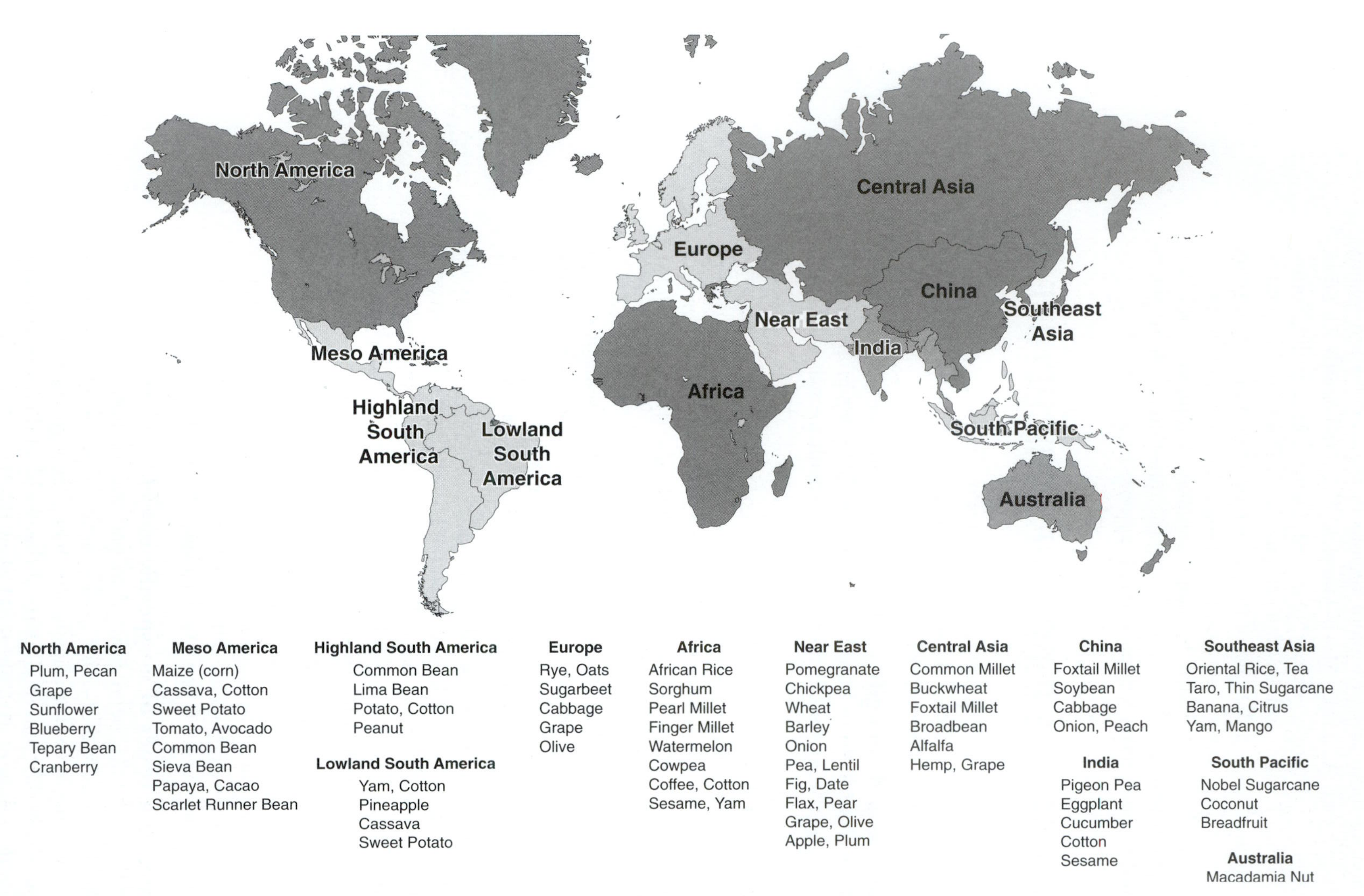

Figure 2-2 The 13 major regions of the world where major food crops were domesticated and examples of each

limiting factor in primitive societies was the availability and dependability of a food supply. With the discovery of agriculture, this limiting factor was minimized. A surplus supply of food released people who would normally be occupied with food production to contribute to society in other ways, leading to new classes of specialists—such as scientists, artists, engineers, priests, and others—thereby contributing to an increase in the standard of living for a society. Table 2-1 outlines human progress in the fields of agriculture and horticulture over a broad period of time.

TABLE 2-1 BROAD DATES OUTLINING HUMAN PROGRESS IN THE AREAS OF AGRICULTURE AND HORTICULTURE

Date	Event
8000 B.C.	Domestication of edible plants.
3500 B.C.	Egyptians employed irrigation, pruning, and field cultivation using an oxen-pulled plow.
300 B.C.	Theophrastus (ca. 372–287 BC), the Father of Greek Botany, taught about plants from his own working knowledge of them. Text covers 550 kinds of plants. The rediscovery of his works in 1483 rekindle an interest in botany.
77 AD	Dioscorides, Father of Medicinal Botany, wrote a medical reference.
50–500	Grafting, budding, legume rotation, fertility appraisals, and cold storage of fruit practiced by the Romans.
500–1200	Dark Ages in Europe—Horticulture technologies preserved and practiced only in monasteries while flourishing in the Far East. Interest in horticulture began in Renaissance Italy as feudalism died out.
1492	Discovery of the New World. Broadening trade routes aid the transport and transplantation of new plant species.
1554	First written record of the tomato. Italians grew the plant by about 1550. Thomas Jefferson was the first American to grow tomatoes in 1781. George W. Carver dedicated to promoting the tomato and peanut.
1665	Robert Hooke detailed the structure of cork cells using a newly invented microscope.
1727	Stephen Hales's work in *Vegetable Staticks* represented the first significant publication in plant physiology. He explained some aspects of water uptake by roots, movement of liquid through plants, and evaporation of water from leaves. Hales was one of the first to use the equipment and methods of the physical sciences to study plants.
1739	About 500,000 people died in Ireland due to widespread crop failure of potatoes.
1747	Dr. James Lind experimented with sailors who had scurvy and discovered that consuming lemons and oranges for six days improved their health. Scurvy is a nutritional disease caused by inadequate vitamin C.
1774	Joseph Priestley reported that burning a candle in a closed container changes the quality of the atmosphere so that the flame is extinguished. Animals placed in that environment quickly die. A living mint sprig renews the air so a candle will burn.
1830	The first machine for cutting lawns was introduced by Edwin Budding. The machine was imported to the U.S. 25 years later.

continued

1839	Salicylic acid (chemically related to salicin, the pain-relieving compound named for its source, Salix-willow) was isolated from the flowerbuds of a member of the rose family. In 1853, a number of synthetically prepared derivatives of this compound were prepared, one of which was acetylsalicylic acid. The Bayer Company selected that chemical as a substitute for salicylic acid and named it aspirin.
1847	Chocolate candy was first created.
1850	The mechanization of agriculture began. Mechanical reapers, the internal combustion engine, and the tractor altered the face of the world and the growth and increasing urbanization of the world population. Between 1860 and 1920, about 1 billion acres of new land were brought under cultivation, with another 1 billion acres coming into production during the following six decades. Improvements in shipping, refrigeration, and processing further industrialized this process.
1859	Charles Darwin published *On the Origin of Species* by means of natural selection. The impact of Darwin's work has been significant in all areas of biology, including the search for natural relationships of plants and interpretations of plant adaptations and ecology.
1862	Congress passed bills constituting the U.S. Morrill Land Grant College Act and at this same time created the U.S. Department of Agriculture. Over 13 million acres of federal land were given to states to support the establishment of colleges for the agricultural and mechanical arts.
1866	Gregor Mendel discovered and published the basic patterns of inheritance and his understanding of the hereditary nature of variation between individuals in a population.
1869	A biologist imported European gypsy moth to the United States for study. A few of those insects escaped and established populations that have caused great devastation to Eastern forests.
1870	The Red Delicious apple is discovered in Iowa. The Golden Delicious apple originated on a farm in West Virginia in 1910.
1875	The first agricultural experiment station in the United States was established in Connecticut.
1880	Darwin published *Power of Movement in Plants.*
1882	Professor Millardet invented the Bordeaux mixture by adding lime to a copper sulfate spray, which caused the copper to precipitate and stick to the leaves. When applied to grapes, it deterred downy mildew.
1886	John S. Pemberton created Coca-Cola, a beverage using water, caramel, kola nut, sugar, vanilla, cinnamon, lime, and coca leaf extractions. By 1903, the makers began purging the coca leaf extract of its cocaine component.
1887	The Hatch Act established a yearly grant to support an agricultural experiment station in each state. The experiment station system became the basis for the U.S. Agricultural Extension Service.
1900	Liberty H. Bailey completes the compilation of a four-volume *Cyclopedia of American Horticulture.*
1917	Ford's Fordson tractor was introduced at $397.
1919	The publication of *Inbreeding and Outbreeding* by E. M. East and D. F. Jones gave scientific underpinnings to corn breeding and introduced Jones's system of double crossing through the use of four inbred lines. This work was one of the most significant early accomplishments of modern agricultural science.

1922	W. J. Robbins initiated plant tissue culture studies.
1922	Knudson published his asymbiotic method of seed germination "Nonsymbiotic Germination of Orchid Seeds" in *Botanical Gazette*. This technique revolutionized the propagation of orchids, both sexually and vegetatively. It led to techniques of mericloning and meristemming that are used widely for production of many crops today.
1939	Chemist Paul Müller discovered the insecticidal qualities of DDT, a compound first synthesized by German chemist Othmar Zeidler in 1874.
1947	Developed during World War II, the herbicide 2,4-D was introduced for weed control.
1961	Melvin Calvin was awarded the Nobel Prize for his work describing the light-independent reactions (often called the dark reactions, or the Calvin cycle) of the photosynthetic system.
1962	Rachael Carson published *Silent Spring,* spurring a new era of environmental concern and awareness.
1970	Norman Borlaug, Father of the Green Revolution, was awarded the Nobel Peace Prize for developing high-yielding dwarf strains of wheat that allowed tropical countries to double their wheat productivity.
1972	DDT (dichloro-diphenyl-trichloroethane) usage was banned in the United States.
1982	The first genetically engineered crop was developed, and by 1994, the Flavr-Savr tomato became the first such plant approved for commercial marketing. The Flavr-Savr tomato was designed for slow fruit ripening and increased shelf life.
1983	Kary B. Mullis devised the polymerase chain reaction, a system to replicate large quantities of DNA from a small initial sample. The ability to create a large sample of DNA for testing and study had extraordinary impact on various fields of study, from areas of paleobiology to forensic analysis.
1983	Barbara McClintock received the Nobel Prize for her work with the complex color patterns of Indian corn, studies that revealed moveable genetic elements termed jumping genes.

BACKGROUND ON THE SELECTION OF EDIBLE PLANTS USED TODAY

In the past, humans used thousands of plant species as a food source; however, only a small number of these species were ever domesticated. In the early stages of agriculture, farmers unknowingly selected species that grew easily, were good producers, and required very little input. As time passed, farmers sold only the most profitable crops. In recent years, with the development of large supermarkets and fast food, the number of crops produced has decreased even further. Of the 30 major food crops cultivated today, the top 4 crops produce more annual tonnage than the remaining 26 on the list (Harlan, 1976) (Table 2-2).

This is a dangerous trend because as our society moves faster and faster, the limited number of crops currently used puts us in the position that if any one of the crops that we depend on fails, millions of people could go hungry or starve to death. For example, the potato famine, which occurred in Ireland in 1845 and

TABLE 2-2 THIRTY MAJOR FOOD CROPS CULTIVATED TODAY AND THE ANNUAL TONNAGE PRODUCED BY EACH

*Crop	Annual Production (Millions of Metric Tons)
1. Wheat	360
2. Rice	320
3. Maize	300
4. Potato	300
5. Barley	170
6. Sweet Potato	130
7. Cassava	100
8. Grapes	60
9. Soybean	60
10. Oats	50
11. Sorghum	50
12. Sugarcane	50
13. Millets	45
14. Banana	35
15. Tomato	35
16. Sugar Beet	30
17. Rye	30
18. Orange	30
19. Coconut	30
20. Cottonseed Oil	25
21. Apples	20
22. Yam	20
23. Peanut	20
24. Watermelon	20
25. Cabbage	15
26. Onion	15
27. Bean	10
28. Pea	10
29. Sunflower Seed	10
30. Mango	10

*Thirty major crops used by humans (Harlan, 1976).

1846, claimed the lives of more than 1 million people. The majority of the population of Ireland was almost entirely dependent on the potato for its existence less than 50 years after its introduction, which allowed the failure of the potato crop to cause a national disaster. In fact, prior to the 1845 famine, the potato crop had failed in several areas, causing considerable hardship. In 1845 and 1846, crop failure was nationwide. The late blight disease, *Phytophora infestans*, which came from America, caused the destruction of vines and decay of the tubers before or shortly after harvest. The disease infected potato crops in other parts of Europe at the time, but those countries (unlike parts of Ireland) were not solely dependent upon the potato.

HISTORY OF AGRICULTURE

Before agriculture, people survived by hunting and gathering. Life flourished when herds and plants were plentiful but people suffered when the food sources diminished. This led to the eventual domestication of animals and cultivation of plants. The gradual transition from a hunter/gatherer society to an agricultural society predates the invention of writing and the details are unknown.

Egyptian Civilization

The beginnings of agriculture can be traced to where the Indus, Tigris, Euphrates, or Nile rivers were located (Figure 2-3). Some of the major accomplishments of Egyptian agriculture were the development of drainage methodology (removal of excess water), irrigation systems (adding an artificial supply of water) through the use of hydraulic engineering, and land preparation by refining the hoe and perfecting the plow. The Egyptians also developed a variety of technologies associated with the culinary arts, such as ceramics, the process of baking, wine making, and food storage (for example, pickling and drying).

Because of their ability to provide water artificially through irrigation and remove it via drainage systems, the Egyptians were the first to establish exotic gardens, which were carefully planned and planted for both ornamental and utilitarian purposes. The first records date back to about 2200 B.C. to a walled garden of an official in Pharaoh Amenhotep's government. The plan for this

Figure 2-3 The Egyptian Empire included the area surrounding the Nile River and the major rivers of the Middle East.

Figure 2-4 This Ancient Egyptian garden scene shows the harvesting of pomegranates in a formal planting together with ornamental columns.

garden was symmetrical, and shade trees surrounded the pool. All the Egyptian plantings were formal, orderly, small, and required irrigation. Typical elements found in Egyptian gardens were pools for fish; trees bearing figs, pomegranates, and dates; grapevine-covered trellises; and beds of flowers, including roses, jasmine, and myrtle (Figure 2-4). The Egyptian gardens were generally for the pharaohs and government officials, or they were sites for religious or sacred services. Egyptians developed gardens to provide an artificial oasis for privacy and a cool, leafy, shady place, for refreshment from the hot desert sun. Some suggest that the Egyptian gardens and their design were associated with the Formalism School of Landscape Design. The ancient cultures of Mesopotamia, Babylonia, and Asyria, which were located east of Egypt, added to Egyptian technology by establishing innovations such as irrigated terraces, gardens, and parks. These cultures established irrigation canals lined with brick and asphalt-sealed joints to maintain more than 10,000 square miles under cultivation, which were responsible for feeding more than 15 million people.

Greek Civilization

The Greeks were involved in agriculture in only a minor way; however, they did dedicate much attention to the study of botany and were concerned about the nature of things. When we think of Greek cultures, the names of the great philosophers Plato and Aristotle come to mind because they greatly influenced modern political and education structures. **Theophrastus,** Aristotle's student, is considered the most significant horticulturist. The botanical writings of Aristotle have been lost; however, the writings of Theophrastus remain intact and cover a wide range of topics, including taxonomy, plant physiology, and natural history. One of his books, *History of Plants and Causes of Plants,* includes topics such as plant classification, propagation, forestry, horticulture, pharmacology, viticulture, plant pests, and flavors from plants. Theophrastus has been referred to as the father of botany. The excellence of his work exemplifies the contribution the Greeks made to the science and contrasts with the period of scientific stagnation during the Middle Ages. Some speculate that the fall of the Greek Empire was not only due to war but also to a decrease in resources. Although the culture was rich in scientific thinking, the Greek mind did not show much interest in the day-to-day problems of agriculture, which can be boring but extremely important. A variety of factors, such as shallow soil and poor conservation practices, led to the decline of the Greek agricultural base. These reasons, coupled with the 27-year Peloponnesian wars, led to the fall of the Greek Empire. Although the Greek culture added only incidentally to practical agriculture, their curiosity and analytical nature had a profound effect on the future of technological advances.

Roman Civilization

The Roman Empire (Figure 2-5) soon followed the Greek civilization and then lasted for about 1,000 years. The Roman Empire was a conglomeration of a wide range of people from different lands. The Romans were fascinated with practical agriculture, and it was an integral part of their economy. Even though the Romans produced little information in the way of new discoveries, they did make great improvements. They copied and borrowed anything they thought was good from the peoples they conquered. Roman agricultural writings included information detailing grafting and budding, using many kinds of cultivated varieties of fruits and vegetables, legume rotation, fertility analysis, and methods for placing fruits in cold storage. They described a prototype greenhouse, which was used to force vegetables. A major book describing Roman agriculture, *Historia Naturalis,* was written by **Pliny the Elder.**

The Romans are thought to have first developed ornamental horticulture to a high level. Their sophisticated ornamental gardens were extensions of their houses and contained pools, statues, walkways, and vine-covered trellises. The

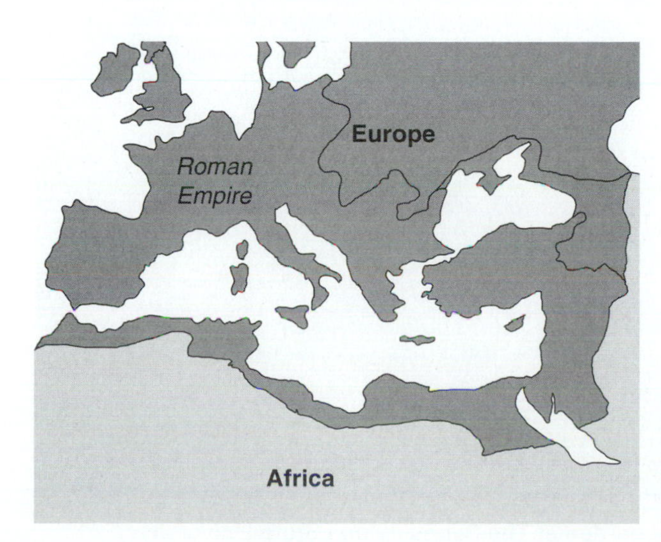

Figure 2-5 The Roman Empire spanned Europe, and parts of Africa and the Middle East.

prosperous Roman had a little piece of land with a house in the country. The estate typically included fruit orchards where apples, pears, figs, olives, pomegranates, and others were grown. In addition, flower gardens contained lilies, roses, violets, pansies, poppies, irises, marigolds, snapdragons, and asters. The wealthy Roman had an estate with a mansion surrounded by frescoed walls containing statues, fountains, trellises, flower boxes, shaded walks, terraces, topiaries (bush sculptures), and, in many cases, even heated swimming pools. The numerous courtyards and symmetric plantings were characteristics typical of Roman cities.

Medieval Horticulture and Agriculture

The time between the fall of the Roman Empire and the Renaissance was the medieval period also known as the **Dark Ages.** During this period, the development of horticulture was static, as was the case with the other arts and sciences. The Dark Ages were restricted to the West, because in the East (India, China, Japan, and others), horticultural and other cultural activities were flourishing. During this period, monasteries became an important way of saving the horticultural and agricultural skills that were perfected by the Romans. Within the monasteries, there were fields of grains, vegetables, and medicinal plants collections together with orchards, located nearby (Figure 2-6). The monks were experts with herbs; they saved many unique plants that would later become very important in the Renaissance.

The **Manorial System,** which was derived from the manor, was an important component of the social structure in the Middle Ages. This system divided plant cultivation practices into the three areas still used today. The first was

Figure 2-6 A Medieval-style garden at The Pennsylvania State University

horticulture, an important component of the gardens close to the manor house, which were enclosed to protect fruits, vegetables, and herbs. The second was **agronomy,** which involved the cultivation of grains and forages in open fields farther away from the manor house. The third was **forestry,** which dealt with the wild lands containing game and forests not maintained to any extent.

Western Europe emerged from the Dark Ages as a result of a series of changes in the political and social structures. The Crusades had a profound influence on horticultural developments because they established routine contact with the ideas and concepts of the East. Merchants such as the famous Italian family, Polo, formed an important link between the Mediterranean coasts and the China Sea by opening trade routes and introducing many goods—including plants—to Western Europe. These were important factors in the cultural enlightenment of Medieval Europe. **Marco Polo** is one of the more well-known Polos; his great contribution was the descriptions of the East in his book *Il Milione*, known in English as the *Travels of Marco Polo*. He provided services to the Great Khan as his administrator to great cities, busy ports, and remote provinces. As part of his services to the Khan, he had to provide detailed written reports on his travels. In addition to his geographical knowledge, Marco Polo reported that the Great Khan had trees planted to provide shade in the summer and to mark the route in winter when the ground was covered with snow.

When meat became an important part of the medieval diet, gardens were used to grow plants as a source of spices and condiments. In the section "The Apple" in the book *Maison Rustique* by Estienne and Liebault (1554), the authors discuss practices of fertilizing, grafting, pruning, breeding, dwarfing, transplanting, controlling insects, girdling to promote flowering, harvesting, processing, and using plants for culinary and medicinal purposes. During medieval times, herbs and herbalists were plentiful; there was little distinction between medicine and botany because plants were used in an attempt to cure all ills. The greatest authority of medicinal plants up to the Renaissance was the Greek physician **Dioscorides,** who wrote the authoritative book *De Materia Medica*.

The Renaissance

The Renaissance has been described as the rebirth of the arts and sciences and is the period in European civilization that immediately followed the Middle Ages. During this period, new continents were discovered and explored, the Copernican was substituted for the Ptolemaic system of astronomy, the feudal system declined, and commerce grew. Numerous items were invented during this period, including paper, printing, the mariner's compass, and gunpowder. The arts were reborn because of painters such as Leonardo da Vinci, Michelangelo, and others.

The Age of Discovery and the New World

During this period, the trading of spices and other horticultural products played a major role in the development of Western Europe and the settlement of the New World. Cinnamon, cardamon, ginger, pepper, and tumeric, plus frankincense, myrrh, and other fragrant resins and gums became very popular in Europe. Although the Arabian traders tried to withhold the true sources of their spices, the Europeans nevertheless discovered their origin. The Europeans built ships and went abroad to the spice-producing countries; these travels became known as the voyages of discovery now found in history books. Prince Henry the

Navigator, Christopher Columbus, and John Cabot were among those who searched for spices in distant lands. For these mariners, the journeys were true adventure, but they were also filled with hard times. For example, Magellan left Spain with five vessels under his command, but only one, the *Victoria*, returned to Spain successfully bringing cloves. However, for European commercial interests, these journeys were highly rewarding because they broke the monopoly held by others in the spice trade and created a wide range of merchandise that was traded between Europe and the Far East, thereby opening up a New World.

The New World had numerous horticultural contributions, including many new vegetables such as maize, potato, tomato, sweet potato, squash, pumpkin, peanut, kidney bean, and lima bean. The New World also contributed fruits and nuts, such as cranberry, avocado, Brazil nut, cashew, black walnut, pecan, and pineapple, as well as other important crops, such as chocolate, vanilla, wild rice, chili, quinine, cocaine, and tobacco. One of the major results of this period was the discovery of America. As time went on, trade routes expanded, which led to further horticultural advances. The exchange between the Old and New World led to the great horticultural industry known today. Two examples of exchanges between the Old and New World include early American gardens reflecting the European origin of the settlers (English influence was predominant, as exemplified by Williamsburg) and horticultural explorers introducing plants from South America, Africa, and the Orient.

BEGINNINGS OF EXPERIMENTAL SCIENCE

Many suggest that the beginnings of experimental science started with the work of da Vinci, Galileo, and Newton in astronomy and the physical sciences. Their innovative findings caused a resurgence in botanical research. Some of the early botanical studies included the following:

- **Marcello Malpighi** (1628–1694) and **Nehemiah Grew** (1641–1712) initiated basic studies in plant anatomy and morphology. Grew wrote *The Anatomy of Plants Begun* and Malpighi wrote *Anatome Plantarum Idea*. Their independent studies were the first important descriptions and statements on the internal structure of plants (plant anatomy). The studies of Malpighi and Grew were so detailed and complete that little new information to the field was added for more than 100 years. In their books, they described the structure of buds, the organization of wood, the characteristics of flowers and their separate parts, the generation of the seed and embryo, and a variety of other topics that had never been explored before.
- **Robert Hooke** (1635–1703) found that living things were made of cells, which led to the future study of cytology.
- **Rudolph Camerarius** (1665–1721) demonstrated sexuality in plants, thereby providing the start of genetics.
- **Linnaeus** (1707–1778) developed a simple yet elegant system for the classification of plants called binomial nomenclature, which is still used today.
- **Stephen Hales** (1677–1761) published the research article "Vegetable Staticks," which was the first significant publication in plant physiology. In this article, he explained some aspects of water uptake by roots, movement of liquid through plants, and evaporation of water from the leaves. His work also advanced the prospect that air provides food for plants and

suggested that light was involved. He was the first scientist to use equipment and methods from the physical sciences to study plants.

- **Joseph Priestley** (1733–1804) showed that burning a candle in a closed container changes the quality of the atmosphere, which results in the flame going out. He also showed that animals placed in that environment would quickly die. However, when a living sprig of mint was placed in the closed container, the candle would burn again and the animal would live in the presence of the burning candle. Today we know that a growing plant takes in carbon dioxide and releases oxygen. Interestingly, upon hearing of this work, Benjamin Franklin sent a letter to Priestly. In this letter he stated, "I hope your findings on the rehabilitation of air by plants will give some check on the rage of destroying trees that grow near houses, which has accompanied our late improvements in gardening from an opinion of their being unwholesome."
- **Charles Darwin** (1809–1882), although better known for his book on the *Origin of Species,* also wrote the *Power of Movement in Plants,* in which he described plants' movement in response to light (phototropism) and gravity (gravitropism). This work laid the foundation for plant hormone research.
- **Gregor Mendel** (1822–1884) was the founder of modern genetics. In 1866, Mendel published a paper representing eight years of his research on inheritance. The original paper was published in German, but the English translation is available.

BEGINNINGS OF AGRICULTURAL RESEARCH IN THE UNITED STATES

In 1862, Congress passed bills constituting the U.S. Morrill Land Grant College Act, which was signed into law by President Abraham Lincoln. That same year, the U.S. Department of Agriculture was created. These events paved the way for the first state **agricultural experiment stations** located in California and Connecticut in 1875. In 1887, the **Hatch Act** was established, which provided yearly support to agricultural experiment stations in each state. Within 10 years, experiment stations across the United States were actively engaged in basic research with the ultimate goal of making agriculture more efficient. More than 13 million acres of federal land was given to states to support the establishment of colleges for the agricultural and mechanical arts. By 1900, there were 60 agricultural experiment stations, and this system became the basis for the U.S. Agricultural Extension service. In 1889, the U.S. Department of Agriculture was elevated to cabinet status, and the newly appointed Secretary of Agriculture had 488 employees and a $1.1 million annual budget. By 1912, the U.S. Department of Agriculture had 13,858 employees and an annual budget of $20.4 million. The overproduction or surplus of goods became a major problem during the Depression of the 1930s. To correct this problem, the Congress passed the **Agricultural Adjustment Act** in 1938. This act was directed at the expansion of utilization research and was accomplished by creating four regional laboratories as part of what is now called the U.S. Agricultural Research Service. Each of these laboratories had specialties in crops grown in a particular region; for example, cotton research was done in the South. In the western United States, wheat, fruits, vegetables, and alfalfa were studied. Animal products, milk, and tobacco were studied in the eastern part of the country, and grain crops, soybeans, and

Figure 2-7 Select frozen vegetables and dehydrated fruits represent significant accomplishments in early agriculture research.

other oil seeds were studied in the northern United States. In 1946, the **Agricultural Marketing Act** was passed in order to correct the imbalance between production and postproduction research. In addition, this act put in place the mechanism for contracting with private research facilities, thereby enabling the government to use the expertise of private-sector scientists. The research addressed problems in targeted areas to obtain quick solutions to problems of national importance. As a result of World War II, regional laboratories quickly redirected their research to respond to the national crisis. USDA scientists discovered methods to produce bulk supplies of penicillin, thereby making this drug available on a wide scale. Other examples of targeting research include establishing ways to produce synthetic rubber, replacements for chemical cellulose, and dehydrated foods, as well as methods for extracting starch from wheat used as a supplement to corn for feeding livestock.

Following World War II, research again focused on dealing with crops in surplus. The development of frozen food technology enabled consumers to have fruits and vegetables year round and to avoid dramatic seasonal changes in prices (Figure 2-7). The ability to concentrate orange juice was an amazing achievement, providing fresh juice all year long. Other accomplishments included instant potato flakes and the development of cotton-blend fabrics that were wrinkle resistant (this led to the resurgence in the use of cotton).

More recently USDA researchers have focused on a number of other areas, such as:

- organic substitutes for petroleum.
- replacing petrochemicals used to manufacture plastics with biodegradable cornstarch derivatives.
- replacing petroleum-based ink products with soy ink, which is made from soybeans.
- replacing gasoline with ethanol or other fuels, which can be made from renewable resources, thereby enabling the United States to cut dependency on foreign sources.
- looking for new and innovative ways to biologically control pests.
- using biotechnology to make foods low in cholesterol and high in vitamins and nutrients.
- searching for new crops worldwide or genetically engineering plants for use as renewable resources for a variety of purposes. Some possibilities include producing a substitute for wood pulp in making paper, producing plants that make oral vaccines to be used for a variety of diseases, and producing plants that produce pharmaceuticals.

From the start of the first U.S. experiment station to the present and even into the future, agricultural research is and will continue to be an important part of our daily lives.

THEORIES OF THE ORIGINS OF CULTIVATED PLANTS

The nineteenth-century French botanist, de Candolle, the twentieth-century Russian agronomist and geneticist, Vavilov, and more recently the American agronomist, Harlan, have published extensively on the geography of plant domestication and crop origins. Both de Candolle and Vavilov concluded that eight centers of plant origin existed (de Candolle, 1959; Vavilov, 1926). Although Harlan (1992) states that just three centers have mostly contributed to the propagation of agriculture around the world, many books have been written on the geographical areas of early agriculture. For more details in this area refer to Harlan (1992).

The current view states that plants were cultivated at approximately the same time over a wide geographical range. The major food crops were domesticated at roughly the same time in the 13 major regions of the world. Today these crops are the staples of our lives. Recall that the 13 major regions where plants were domesticated are North America, Meso-America, Highland South America, Lowland South America, Europe, Africa, Near East, Central Asia, India, Southeast Asia, China, South Pacific, and Australia. The time frame during which the domestication of plants occurred was during the Pleistocene epoch of the Cenozoic era, which was toward the end of the Ice Age or about 11,000 to 15,000 years ago, when early humans began wandering the Earth. Prior to this time, some theories project that tool-using hunter/gatherers had been on Earth for about 4 million years. According to these theories, the domestication of plants began only about 400 generations ago.

Determining where plants were domesticated is complex because some of the plants originally domesticated are much different from the plants used today; in many cases, these plants are the same in name only. For example, three kinds of wheat were originally domesticated, none of which are commercially used today; in fact, these plants are hardly grown at all. The wheats originally domesticated were diploids and tetraploids, known as glume wheats. Today the major wheat species used are hexaploids, known generally as bread wheat, which have very different characteristics than those originally domesticated (Harlan, 1976).

SUMMARY

After reading this chapter, it should be clear that the development of agriculture is one of the most important discoveries affecting the human race, even though its origin is still open for discussion. It was because of agriculture that many other discoveries were made. The Egyptian and Roman civilizations contributed directly to the development and betterment of agriculture, whereas the Greeks contributed in a less direct way. The Egyptians' most significant contribution to agriculture was the development of irrigation systems, drainage methodology, and land-preparation methods using the hoe and plow. The Romans took existing technology and made it better, thereby adding to agriculture in a variety of ways, including:

- grafting and budding.
- legume rotation.
- fertility analysis.
- postharvest storage of fruits and vegetables.
- prototype greenhouse used to force vegetables.

Although the Greeks did not have a direct impact on agriculture, they did provide us with great thinkers such as Plato, Aristotle, and Aristotle's student, Theophrastus. Following the Egyptian, Greek, and Roman civilizations, there was a period of stagnation during medieval times known also as the Middle Ages and the Dark Ages. During this time, horticulture made little strides in the West; however, in the East (India, China, and Japan), major discoveries were being made. After the Dark Ages, the Renaissance occurred and, men such as da Vinci, Galileo, and Newton were key to the surge in botanical research. Key names to be remembered for their contributions to the botanical sciences are Marcello Malpighi, Nehemiah Grew, Robert Hooke, Rudolph Camerarius, Linneaus, Stephen Hales, Joseph Priestly, Charles Darwin, and Gregor Mendel. The following key dates in U.S. agriculture should also be noted.

- In 1862, the U.S. Morrill Land Grant College Act was passed, which led to the establishment of land grant colleges throughout the United States. At the same time, the U.S. Department of Agriculture was formed.
- In 1875, the first state agricultural experiment stations located in California and Connecticut were established.
- In 1887, the Hatch Act was established, which provided yearly support to agricultural experiment stations in each state.

There were 13 major regions of the world where plants were domesticated. You have read that plants were domesticated toward the end of the Ice Age, or 11,000 to 15,000 years ago, and this occurred over a wide geographical range at basically the same time. Considering that theories project man to have been on Earth for approximately 4 million years, the domestication of plants is relatively new. When considering where plants were domesticated, it is important to understand that the answer is much more complex than it appears. (For more details in this area, a good starting point is Harlan's works published in 1976 and 1992.)

Review Questions for Chapter 2

Short Answer

1. What is the definition of agriculture?
2. List five theories on the origin of agriculture and provide a brief explanation of each.
3. What was the major limiting factor in primitive societies?
4. What was the immediate reward of agriculture, which provided an available and dependable supply of food?
5. What are the top four major food crops produced today?
6. Provide examples of food crops domesticated in each of the 13 major regions of the world.
7. What was the time frame during which the domestication of plants occurred?

Define

Define the following terms:

Hatch Act | Agricultural Adjustment Act | Agricultural Marketing Act

True or False

1. The time frame during which domestication of plants occurred was toward the end of the Ice Age.
2. The current view holds that plants were domesticated at different times over a wide geographical range.

3. One of the major accomplishments of the Egyptian civilization was the development of irrigation systems through the use of hydraulic engineering.
4. The Greek civilization was only involved in practical agriculture in a minor way.
5. Although the Greeks were great scientific thinkers, they did not show much interest in the day-to-day problems of agriculture.
6. The Romans produced very little new discoveries, but they did make great improvements on existing technology.

Fill in the Blanks

1. There are ____________________ major regions of the world where major food crops were domesticated.
2. Agriculture is the production of ____________________ and ____________________ to meet basic human needs.
3. The ____________________ was the period of time which immediately followed the Middle Ages.

Matching

1. Hatch Act (1887)
2. Research Marketing Act (1946)
3. Agricultural Adjustment Act (1938)

A. Passed to correct the imbalance between production and postproduction research
B. Initiated to correct the problem with overproduction or surplus of goods resulting from the Depression
C. Provides yearly support to agricultural experiment stations in each state

Matching

1. Stephen Hales (1677–1761)
2. Charles Darwin (1809–1882)
3. Marcello Malpighi (1628–1694) and Nehemiah Grew (1641–1712)
4. Rudolph Camerarius (1665–1721)
5. Linneaus (1707–1778)
6. Gregor Mendel (1822–1884)
7. Joseph Priestly (1773–1804)
8. Robert Hooke (1635–1703)

A. Responsible for the initiation of basic studies in plant anatomy and morphology
B. Found that living things were made of cells leading to the future of cytology
C. Demonstrated sexuality in plants, thereby providing the roots of genetics
D. Developed a simple yet elegant system for the classification of plants called binomial nomenclature
E. Published "Vegetable Staticks," which was the first significant publication in plant physiology
F. Showed that plants purify air
G. First to describe plant movement in response to light and gravity
H. Founder of modern genetics

Activities

1. Contact the U.S. Department of Agriculture in Washington, D.C., and find URLs for research currently being done at experiment stations located in the United States. Select two research projects of interest to you, summarize them, and specify why you feel that they are important.

2. Contact the U.S. National Research Initiative Program in Washington, D.C, and find the URLs summarizing research efforts that are currently being funded by this agency. Select two research projects of interest to you, summarize them, and explain why you feel that they are important.
3. Visit a supermarket and make a list of fresh fruits and vegetables available. Go through the store looking for all the ways in which two crops are processed (for example, apples and potatoes).
4. Visit a processing plant for commodities, such as tomatoes, apples, peaches, potatoes, or cotton, which may be found in your location. Summarize what you saw and why it was interesting or not interesting.
5. Visit a supermarket and look at the nutritional facts on processed fruits and vegetables, and then make a list of crops that you consider to be high, medium, and low in sugar, starch, and cholesterol.

References

Anderson, E. (1954). *Plants, Man and Life.* London: A. Melrose.

Berndt, R. M., & Berndt, C. (1970). *Man, Land, and Myth in North Australia: The Gunwinnggu People.* East Lansing: Michigan State University Press.

Cohen, M. (1977). *The Food Crisis in Prehistory.* New Haven, CT: Yale University Press.

de Candolle, A. (1959). *Origin of Cultivated Plants* (2nd ed.). New York: Hafner. (Original work published 1886)

Harlan, J. R. (1976). The Plants and Animals which Nourish Man. *Scientific American* 235(3), 88–97.

Harlan, J. R. (1992). *Crops and Man* (2nd ed.). Madison, WI: American Society of Agronomy and Crop Science Society of America.

Saur, B. D. (1952). *Agricultural Origins and Dispersals.* Cambridge: M.I.T. Press.

Vavilov, N. I. (1926). *Studies on the Origin of Cultivated Plants.* Leningrad: Inst. Appl. Plant Breeding.

CHAPTER

3

The Horticulture Industry: An Important Part of Agriculture

ABSTRACT

This chapter begins with a discussion of why horticulture has become such a large, diverse industry. The concept of land grant universities is also discussed. The College of Agriculture has traditionally been broken down into 12 departments; however, there are a number land-grant universities that are reducing this number to four major departments. The chapter also covers the three major areas of plant science found in agriculture, with horticulture being one of the areas. Horticulture is broken down into three major areas and a description provided for each. The four basic characteristics of horticulture are also discussed. The chapter concludes with a discussion of significant horticultural scientists and their impact on horticulture today.

Objectives

After reading this chapter, you should be able to

- explain why horticulture is such a diverse industry.
- explain the departmental structure within colleges of agriculture at land-grant universities.
- identify the three important areas of plant science found in agriculture.
- break down horticulture into three major areas and describe each.
- discuss the four basic characteristics of horticulture.
- discuss significant European and U.S. horticulture scientists and their achievements.

Key Terms

agriculture
colere
floriculture
horticulture
hortus
interiorscaping
landscape horticulture
olericulture
ornamental horticulture
pomology

INTRODUCTION

After you have completed this chapter, it will be apparent why the **horticulture** industry is so complex. One reason the industry has become so large is that people like many types of plants at specific times throughout the year. To target special dates, growers must be skilled in the plant sciences and related areas, such as weed control. In addition, the grower must have business management, marketing, computer, soils, engineering, and a variety of other skills. To implement each of these areas successfully, the grower requires facilities, materials, and supplies, thereby broadening each area as well.

This chapter explains how horticulture is a very important part of **agriculture** and how land-grant universities support horticultural research throughout the United States. You will learn the differences among horticulture, agronomy, and forestry, and that these applied plant sciences are closely interrelated. You will explore the background information of the

Figure 3-1 Plants used in ornamental horticulture (magnolia tree), pomology (apple tree), and olericulture (tomato plant)

major areas of horticulture, including **ornamental horticulture (floriculture, landscape horticulture, interiorscaping), olericulture,** and **pomology,** along with more specific characteristics of horticulture (Figure 3-1). You will meet key scientists in the field, such as Mendel (1822–1884), who laid the foundation for the science of genetics. The background information in this chapter will help you with horticulture courses and a variety of other courses, such as landscape architecture.

THE HORTICULTURE INDUSTRY

People like plants, and they like to have them where and when they want them, which requires special skills in a variety of areas in the horticulture industry. Individuals working in the horticulture industry must have skills in the plant sciences and related areas, such as taxonomy, anatomy, plant physiology, entomology, plant pathology, plant breeding, and genetics. In addition, they must be knowledgeable in soil preparation; soil management; basic sciences, such as chemistry, physics, and math; as well as business management, marketing, and computers. For example, knowing how to grow plants properly requires a strong background in cultural methods required to grow the plant, such as plant nutrition, and knowing when to fertilize and when not to. Knowing the light levels required for maximum growth and how to influence flowering by modifying the day length is also important. Knowing when to water and when not to water seems simple, but it involves knowing the plant's physiology and, in many cases, using engineering skills to determine the proper irrigation system to use. Today, growers can use a variety of soil mixtures and media not containing soils, and they must know the media formulation that works best for each particular crop. Understanding the proper methods of soil preparation and soil management has also been the difference between failing or successful farming operations. As another example, selecting the proper greenhouse structure, design, location, and heating/cooling process requires expertise in engineering, plant physiology, knowledge of the environment, and other areas.

The ability to control a variety of pests—including insects, diseases, and weeds—is extremely important in the horticulture industry. Pests must be controlled without adversely affecting the environment while at the same time producing a high-quality crop.

Knowledge of computers is a must because computers are used to control heating, cooling, humidity, light, and other functions in many greenhouses.

People in the horticulture industry must also know the proper way to treat a crop after it has been harvested to prevent postharvest deterioration. In fact, improper treatment of the crop after it has been harvested results in up to 25 percent of losses in developed countries and up to 50 percent in undeveloped countries. Knowing how to handle crops properly after they have been harvested can minimize postharvest losses and increase profits. After successfully producing a crop and minimizing postharvest losses, business and marketing skills are necessary to sell the crop for a profit. For these reasons and many others, a large horticulture industry has emerged.

Flowers have special meanings to people, and they are in high demand on specific days (Figure 3-2). This high demand makes it vital for specific flower types to be available in a narrow window of time, which requires many of the skills described previously. Following are examples of days on which particular plants are in high demand:

- **Valentine's Day.** Red long-stem roses can draw up to 10 times their value on this day as compared to any other day of the year.
- **Thanksgiving.** Chrysanthemums are typically in high demand because of their beautiful fall colors.
- **Christmas.** Poinsettias are in high demand right after Thanksgiving through Christmas day; after this time the value drops substantially. For growers to be successful, they must have their crop come in at a specific time; otherwise, millions of dollars can be lost. Compare prices of plants

Figure 3-2 Bouquets of roses ready for Valentine's Day

prior to and following Christmas day to see just how much money can be lost when a crop is not ready on schedule.

- **Easter.** Lilies are in high demand for one month prior to Easter day; after Easter day, the value of this crop drops substantially.

The pinpoint accuracy required to have a crop come in on time has been one of the reasons leading to the emergence of a large, diverse horticulture industry, which is more than just growing plants or pretty flowers. The following list provides examples of the diversity of the horticulture industry.

- **Supplies.** The supplies required by a grower include everything from pots to chemicals, including tractors for plowing fields, irrigation systems, prefabricated greenhouses or other structures, rooting hormones, chemicals used for pest control or growth retardation, and numerous others.
- **Services for the grower.** The grower requires numerous services, such as pest control, nutrient analysis, computer support, transport of crops, telephone lines, regular maintenance of heating and cooling systems in greenhouses, and a variety of others.
- **Actual production of crops.** Crop production requires skilled and unskilled laborers with expertise in all the areas previously mentioned to actually produce the crop. A variety of plant materials and facilities is required, such as transgenic plants or genetically modified organisms (GMOs), bioreactors, and buildings with aseptic rooms to successfully grow plants in tissue culture (Figure 3-3). In addition, greenhouses (Figure 3-4), fertile land for growing plants in the field, clean water, and many other factors are involved in the actual production of the crop.

Figure 3-3 Bioreactor for growing plants housed in an aseptic room

Figure 3-4 Greenhouse used for growing crops

Figure 3-5 A photograph taken at a farmers' market may be used in a marketing tool. *Courtesy of Getty Images, Inc.*

- **Marketing of crops.** Marketing of crops requires trained professionals who are knowledgeable in the methods to market a particular crop successfully (Figure 3-5).

HORTICULTURE—AN AREA OF AGRICULTURE

Agriculture describes the production of plants and animals to meet basic human needs. Humans need food, clothing, and shelter to survive, and all of these come from plants. Agriculture is manifested in a variety of disciplines in major land-grant

Figure 3-6 Agriculture appears in the form of many disciplines in land-grant universities.

universities today (Figure 3-6). Traditionally, the College of Agriculture has consisted of 12 departments. However, these 12 departments can be broken down into these four groups:

- **Plant Sciences.** Horticulture, agronomy, forest resources, plant pathology, entomology, and food science.
- **Animal Sciences.** Dairy and animal sciences, poultry science, and veterinary science.
- **Engineering.** Agricultural and biological engineering.
- **Social Sciences.** Agricultural economics and rural sociology, agricultural and extension education.

Presently, there are 12 departments at many land-grant universities, but this will probably change in the future. Departments at some land-grant universities are already starting to combine common departments and downsize to overcome duplications.

Agriculture includes three important areas of plant science: horticulture, agronomy, and forestry. Horticulture, agronomy, and forestry are interrelated because they are all applied plant sciences (Figure 3-7); however, they remain distinct from one another. The focus of this textbook is on horticulture, which is the culture of plants for food, comfort, and beauty. The word *horticulture* was initially derived from the Latin words ***hortus,*** meaning "garden," and ***colere,*** meaning "to culture." Horticulture includes a wide range of important and

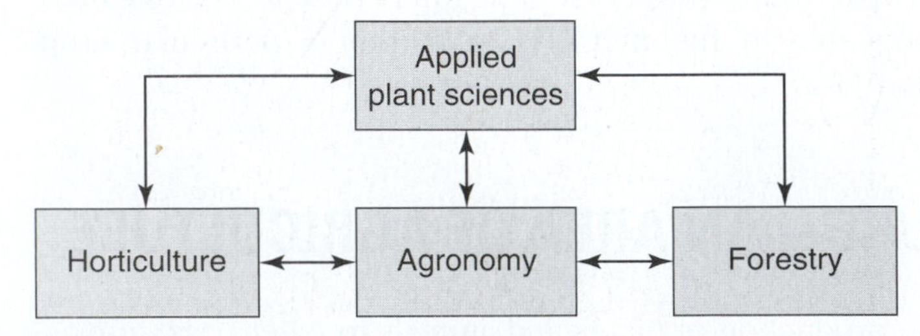

Figure 3-7 The interrelationships among horticulture, agronomy, and forestry, which are all applied plant sciences

useful plants, but it differs from agronomy, which traditionally deals with grain and fiber, and forestry, which focuses on tree production for timber.

AREAS OF HORTICULTURE

Horticulture can be broken down into three major areas: ornamental horticulture, olericulture, and pomology (Figure 3-8).

Ornamental Horticulture

The first major area of horticulture is ornamental horticulture. This involves growing and using plants for their natural beauty and can be further broken down into the following three groups: floriculture, landscape horticulture, and interiorscaping.

Floriculture

Floriculture is the production, transportation, and use of flower and foliage plants. Floriculture is a huge portion of the horticulture industry, especially in Europe, Japan, and the United States (Figure 3-9). In the United States, floriculture

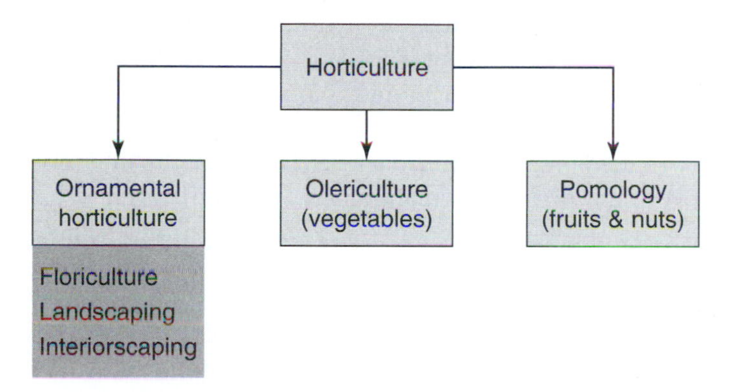

Figure 3-8 The three main areas of horticulture

Figure 3-9 The flowering geranium is an important component of the floriculture industry.

accounts for slightly more than half of the nonfood horticulture industry. The wholesale and retail markets are in the billions of dollars annually. Because most cut flowers and potted plants are grown indoors in temperate climates, floriculture is primarily a greenhouse industry. With the explosion of technology, the floriculture industry has become a highly competitive and highly technical business for the overall horticulture industry. Precise scheduling of floricultural crops is extremely important because the day after a specific holiday, major sales are lost by the grower. To reemphasize this, poinsettias are in high demand from Thanksgiving through Christmas day, after which there is little interest in poinsettias. In floriculture, after producing a high-quality perishable product at a precise time, the product must be handled properly and transported to the retail stores in excellent condition. The floriculture industry can be broken down into the wholesale market and the retail market. Today's high-speed transportation enables flowers to be shipped fresh all over the world in a short period of time, thereby making the floriculture industry very competitive.

Landscape Horticulture

Landscape horticulture deals with producing and using plants to make an outdoor environment more appealing (Figure 3-10). Landscape horticulture can be divided into several components: growing plants to be used in the landscape, designing the landscape, and establishing and maintaining the landscape. Each of these components is key to a successful landscape. If any of these three are unsuccessful, the landscape will fail even though high-quality plants were used.

The growing component of landscape horticulture is called the nursery industry, in which young plants are grown and maintained prior to planting. The first commercial nursery in the United States was established on Long Island, New York by the Prince family around 1730. Since this time, many nurseries have spread throughout the United States; today the top 10 states with respect to production (in terms of cash receipts) are California, Florida, North Carolina, Texas, Ohio, Oregon, Michigan, Pennsylvania, Georgia, and Oklahoma.

The nursery industry involves producing and distributing both woody and herbaceous plants. Although the nursery industry includes bulb crops, fruit trees, some perennial vegetables, Christmas trees, and others, the most important crops are ornamental trees and shrubs followed by fruit trees and bulb crops. The single most important plant grown outdoors is the rose. The types of plants grown in a nursery depend on many factors, which are discussed later in this text. The nursery industry in the United States can be divided into wholesale, retail, and mail-order businesses. There are two general types of nurseries: field nurseries and container nurseries, both of which will be discussed later in this text.

Another growing component of landscape horticulture is the turfgrass industry, which includes the production and maintenance of specialty grasses and other ground covers used for recreation, utility, and overall beautification. Lawn and turf maintenance, which is an important component of the turfgrass industry for residential homes and golf courses, has become a big money-making industry.

Interiorscaping

Interiorscaping is the use of foliage plants to create pleasing and comfortable areas inside buildings. In the United States in the 1970s, interiorscaping began when plants were used in shopping malls to create an outdoor environment during

Figure 3-10 Landscaping has the power to beautify an outdoor environment.

the cold months, when many people were tired of long winters in certain portions of the country. Plants are used in the interior landscape to reduce stress. Scientific studies have shown that interiorscaping increases the efficiency of workers (Figure 3-11). Today, interiorscaping has become a multimillion-dollar industry because big companies are willing to pay to make their workers more productive.

Olericulture

The second major area of horticulture is olericulture. Olericulture is the growing, harvesting, storing, processing, and marketing of vegetables (Figure 3-12). Commercial vegetable production can be broken down into two main categories: fresh market products and processed products, such as canned, dried, or frozen vegetables. The total value of vegetable crops in the United States is estimated to be well over $10 billion annually. The leading states producing vegetables are California, Florida, Arizona, Texas, and Georgia, with California

Figure 3-11 Plants in the interiorscape reduce stress and create comfortable areas.

Figure 3-12 The tomato plant is an important component of the vegetable industry.

alone producing almost half of the vegetables grown in the United States. The commercial production of vegetables has become highly complex, competitive, and technological in an effort to produce high-quality crops at a reasonable price. The increased interest in vegetables is due to known health benefits, which have been scientifically shown in recent years. Both the American Cancer Society and the American Heart Association highly endorse the consumption of vegetables as a means of reducing cancer and heart disease. Because vegetable consumption is now on the rise, the vegetable business has become more complex and has added to the size of the horticulture industry. In addition to commercial vegetable production, home gardening has become popular because large quantities of vegetables can be produced from a very small portion of land. The popularity of home gardening has led to an increased demand for products and services for the homeowner, thereby further increasing the size of the horticulture industry.

Pomology

The third major area of horticulture is pomology, which is the growing, harvesting, storing, processing, and marketing of fruits and nuts (Figure 3-13). Like vegetable crops, the fruit and nut industry can also be broken down into fresh market and processed foods. The total value of the fruit and nut industry in the United States is now in excess of $10 billion annually. This industry is continuing to grow because of the many benefits associated with fruits and nuts. Today, people are starting to heed the USDA food chart recommendation of consuming two to four fruits daily. Although there are many benefits to eating fruits and nuts, one major benefit is that they reduce heart disease. "An apple a day keeps the doctor away" now has some scientific basis, in that fruit pectins play an important role in reducing dietary cholesterol. Because there are numerous scientifically proven benefits of eating fruits and nuts, this industry has become very complex and technology driven to produce high-quality crops. After a high-quality crop is produced, it is extremely important to prevent postharvest spoilage prior to consumption by the consumer.

Figure 3-13 Apple trees are an important part of the pomology industry.

Figure 3-14 A variety of horticultural crops

Specific Characteristics of Horticulture

Horticulture involves cultivating a large number and variety of crops. For example, horticultural crops include everything from petunias to maple trees to broccoli to apples—a wide range of woody and herbaceous crops (Figure 3-14). Specific characteristics of horticulture include the following:

- Horticultural plants and products have a very high water content; generally they contain approximately 90 percent water or greater. In contrast, agronomic products are either dry at harvesting or are allowed to dry before use.
- Horticultural crops are typically intensively cultivated. The plants grown are of high enough value to justify a large input of capital, labor, and technology per unit area of land; for example, strawberries cost a lot more than potatoes.
- Horticultural crops are grown for aesthetic purposes; some common examples are red roses, poinsettias, chrysanthemums, and lilies. This is a unique feature of horticulture, which contributes to its popularity.

KEY HISTORICAL NAMES AND DATES

Early work in horticulture was often by botanists, who called what they were doing "applied botany." The following individuals had a major impact on agriculture and horticulture.

Theophrastus

Theophrastus (377–288 B.C.) was a student of Aristotle and considered to be the most significant Greek horticulturist. Theophrastus made many contributions in the plant sciences; for example, he was the first to suggest that roots were responsible for absorbing nutrients. He also discovered that root pruning encouraged flowering and subsequent fruiting. In addition to these, Theophrastus made numerous other contributions to the plant sciences.

Varro

Varro (116–20 B.C.) developed postharvest storage techniques for fruits. He recommended placing fruits on straw in a cool, dry place, such as a cave. Today, many people are interested in old mines and caves for storing fruits and vegetables in China, the United States, and throughout the world.

Dioscorides

Dioscorides (40–90 A.D.) wrote the authoritative book *De Materia Medica,* which describes roots, stems, leaves, and flowers. For many centuries, no drug plant was considered genuine unless it could be identified by the descriptions found in this book.

Linnaeus

Linnaeus (1701–1778) developed a clear and concise method for classifying plants known as binomial nomenclature, for example, *Solanum tuberosum* L.

Charles Darwin

Darwin (1809–1882) had a profound effect on the plant sciences. He wrote *The Power of Movement in Plants* (1880), in which he described for the first time the ability of plants to move in response to gravity (gravitropism) and light (phototropism).

Gregor Mendel

Mendel (1822–1884) laid the foundation for the science of genetics. He worked with garden peas; for example, when he crossed a tall pea and short pea parental line, the progeny in the F_1 generation were all tall. When he crossed two of the hybrid tall plants, he observed a distinct ratio in the F_2 generation, which was three tall plants and one dwarf plant (Figure 3-15). His work illustrates the characteristics of dominant and recessive traits.

Robert Prince

Robert Prince (1730) established the first commercial nursery in the United States on Long Island, New York.

Andrew J. Downing

Andrew J. Downing (1815–1852) was the first great American landscape gardener.

Frederick Law Olmstead

Frederick Law Olmstead (1822–1903), considered to be the "father of landscape architecture," was the primary architect for Central Park in

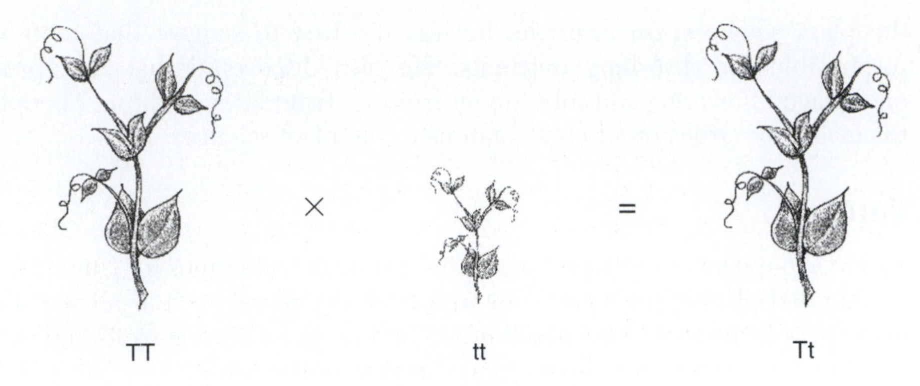

Tall offspring (F_1) (Tt) are crossed.

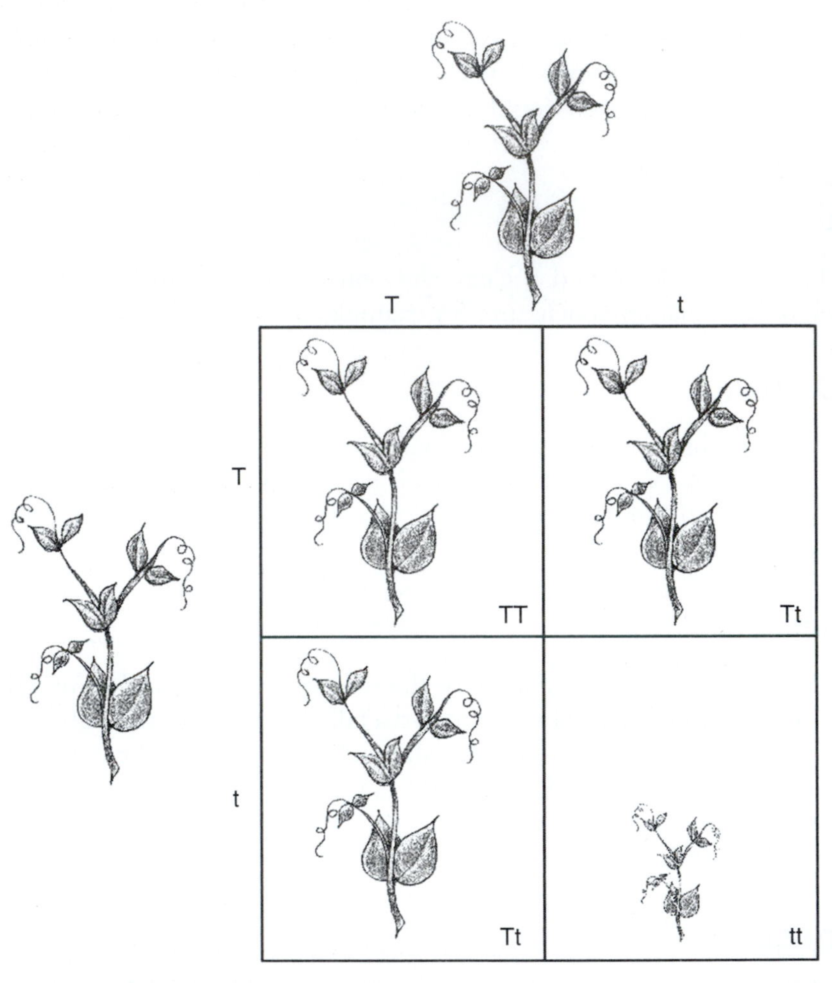

Figure 3-15 Mendel's work with peas showed the characteristics of dominant and recessive traits.

New York City as well as parks in other large cities, such as Buffalo, Chicago, and Detroit.

Liberty H. Bailey

Liberty H. Bailey (1858–1954) is the modern counterpart to Linnaeus. Bailey's books *Manual of Cultivated Plants* and *Hortus Second* are often quoted as the standard authorities on plant nomenclature, taxonomy, pruning, and more.

SUMMARY

The main reason the horticulture industry has grown so large is that people like plants and want specific plants at designated times throughout the year. Because of the seasonal aspect, the grower must be skilled in plant science and related areas together with business management, marketing, computers, engineering, and a variety of other areas. In addition, the grower must have access to a wide range of materials, supplies, and services to produce and market the wide range of plants that are part of the horticulture industry. To strive for continued excellence, the horticulture industry relies on research done both privately and at land-grant universities throughout the United States.

The differences among horticulture, agronomy, and forestry are that horticulture deals with the culture of plants for food, comfort, and beauty while agronomy deals with grains and fibers and forestry deals with tree production for timber. In addition, the major areas of horticulture are ornamental horticulture (floriculture, landscape horticulture, and interiorscaping), olericulture, and pomology. Theophrastus, Dioscorides, Linnaeus, Darwin, Mendel, Prince, Downing, Law Olmstead, and Bailey are all key figures in the history of horticulture.

Review Questions for Chapter 3

Short Answer

1. Why has horticulture become a huge industry?
2. In the College of Agriculture, there are typically 12 departments; break down these departments into four groups.
3. What are three important areas of plant science found in agriculture?
4. Divide horticulture into three major areas.
5. What are four basic characteristics of horticulture?
6. Ornamental horticulture can be broken down into what three groups?

Define

Define the following terms:

agriculture	*colere*	landscape horticulture	pomology
horticulture	ornamantal horticulture	interiorscaping	
hortus	floriculture	olericulture	

True or False

1. Horticulture involves the cultivation of a large number of and variety of crops.
2. The Latin term *hortus* is short for horticulture.
3. The Latin term *colere* means “to culture.”
4. Horticulture includes a wide variety of important and useful plants, including strawberries, tomatoes, and grain crops.
5. Horticultural crops are intensely cultivated.

Fill in the Blanks

1. Floriculture is the production, transportation, and use of ________________ and ________________ plants.
2. Olericulture is the growing, harvesting, storing, processing, and marketing of ________________ crops.
3. Pomology is the growing, harvesting, storing, processing, and marketing of ________________ crops.
4. The word *horticulture* was initially derived from the words *hortus,* meaning ________________ and *colere,* meaning to ________________.
5. Agriculture includes three important areas of plant science: ________________, ________________, and ________________.

Matching

1. Theophrastus (377–288 B.C.)
2. Varro (116–20 B.C.)
3. Dioscorides (40–90 A.D.)
4. Linnaeus (1701–1778)
5. Darwin (1809–1882)
6. Mendel (1822–1884)
7. Robert Prince (1730)
8. Andrew J. Downing (1815–1852)
9. Law Olmstead (1822–1903)
10. Liberty H. Bailey (1858–1954)

A. the modern counterpart to Linnaeus. He wrote the book *Hortus Second.*

B. considered to be the father of landscape architecture. He was the primary landscape architect for Central Park in New York City.

C. wrote *The Power of Movement in Plants,* in which he discussed the capability of plants to respond to the effects of light and gravity.

D. laid the foundation for the science of genetics. His classic experiments illustrate the characteristics of dominant and recessive traits.

E. the first American landscape gardener.

F. a Swedish botanist and physician who developed a method for classifying plants by clear and concise descriptions known as binomial nomenclature.

G. established the first commercial nursery in the United States.

H. student of Aristotle and the most significant Greek horticulturist.

I. wrote the authoritative book, *De Materia Medica.*

J. developed techniques for postharvest storage of fruits.

Activities

1. Go to your local library and find information on land-grant universities in the United States. Pick two universities and provide a summary of how the College of Agriculture is organized, including the number of departments and how you would consolidate them.
2. Select one the main areas of horticulture that you are interested in and visit a nearby location to gain information about it. For example, if you are interested in the area of pomology, visit a local orchard and

find out as much information as possible. Information should include the types of trees, forms of pest control, pruning methods, plant growth regulators, marketed methods (fresh or processed), postharvest storage, and any other information you can find.

3. Pick two key people discussed in this chapter and find out more about the specific accomplishments of those two individuals by using the Internet.

Reference

Darwin, C. (1880). *The Power of Movement in Plants.* New York: Appleton.

CHAPTER

4

Fundamental Steps Leading to Success in a Horticulture Career

ABSTRACT

This chapter presents the proper ways to prepare for, obtain, and succeed in a successful career path in horticulture. Because you will more than likely spend the better part of your life pursuing your career path, it is very important that you choose carefully and work aggressively to be successful.

Objectives

After reading this chapter, you should be able to

- prepare for a career in horticulture.
- obtain and succeed in a job in horticulture.
- choose from selected career areas in horticulture.

Key Terms

arboretum
botanical garden
career goal
goal setting
habitat
horticultural garden
horticulture career
horticulture job
horticulture occupation
horticulture therapist

INTRODUCTION

This chapter shows you how to prepare for a career in horticulture. Because horticulture is more than just growing pretty flowers, it involves art, science, and technology and can lead to numerous career paths, including everything from arranging flowers to gene jockeying. To be successful on your path to a long-lasting **horticulture career,** you must have the proper education and practical experience and be willing to work hard. This all starts with selecting a career path that is right for you. To help you make a good career choice, consider working part time in a given field, career shadowing, talking to school counselors or advisors, and exploring sources of information at your local library or on the Internet. A horticulture career is the path you choose in the field and involves a series of **horticulture occupations** and **jobs.**

Being successful in your horticulture career begins with establishing a **career goal.** The first step in establishing your career goal is through a process known as **goal setting.** Goal setting requires that you decide what your personal interests are and be realistic in establishing those goals. After you establish goals, you prepare an outline showing how the goals will be achieved together with deadline dates and ways and means to accomplish each goal. The amount of education and training

Figure 4–1 Workers (from left to right) in a laboratory, greenhouse, and in the field pursuing horticulture research

necessary will vary depending on your career goals, but you must achieve the proper level of education to be successful in your targeted career goal. In many cases, a high school diploma and a targeted short course will be enough to get a job in the field of horticulture; however, advanced degrees allow more flexibility in obtaining jobs such as managerial, research, and teaching positions. In addition to a formal education, it is extremely important to have part-time or full-time work in the field of horticulture that you intend to pursue. Internships are also available in a variety of areas—from public gardens to Disney World to Bayer Chemical Company—in order to gain practical experience. Individuals with both a formal education and practical experience are very competitive in the job market, whereas those with just one or the other are less competitive.

In today's competitive society, finding the job of your choice is often difficult. Some suggestions on finding the correct position are to contact potential employers in person, look at job advertisements in a variety of sources, and consult teachers, counselors, and others who may provide a lead on a potential job. The job seeker who has networked well is most likely to find the desired job (Figure 4-1).

Filling out a job application and getting together the necessary materials (Figure 4-2) may sound easy, but if not done properly, it could cost you the job interview. Therefore, you should put your best, most professional effort into the job application. In most cases, getting the job interview means that you are among the finalists for the position, which is why you must make your best impression. Be prepared for the interview, take all necessary materials with you, groom and dress appropriately, be on time, and speak clearly and confidently.

After you have a successful interview and take the job, you must do all you can to keep it. Some suggestions for keeping the job are to make every effort to be a team player, work with others in an effective way, take pride in your work, live by the rules, always have a positive attitude, and dress and groom appropriately.

The horticulture industry offers many career paths. The amount of training required ranges from none to postdoctoral experience. Many jobs require

Figure 4–2 Filling out a job application properly is an important first step toward a job in the horticulture industry.

knowledge and training in horticulture, whereas some jobs require no training. Jobs requiring training in horticulture have varying requirements from a high school diploma and a short course in a targeted area to a college education. A college degree in horticulture allows you to pursue a wider range of positions. Although numerous career paths are available in horticulture, the three major areas are ornamental horticulture, olericulture, and pomology. In each of these areas, there are positions for production, managers, workers, designers or architects, researchers, teachers, extension workers, marketing, sales, buyers, and computer specialists. For each of these positions, a variety of subcategories and specialty areas also exist.

HOW TO PREPARE FOR A CAREER IN HORTICULTURE

Horticulture is more than just arranging flowers (Figure 4-3). In addition to being an art, horticulture is a science, which includes genetic engineering (Figure 4-4) and advanced technology, such as the use of computers for hormone analysis in plants (Figure 4-5), making landscape plans or to control environmental conditions in greenhouses. To some, horticulture is a hobby, a way to get outdoors and become one with nature, although to others, it is an exciting profession.

Differences exist, however, among a horticulture career, a horticulture occupation, and a horticulture job. A horticulture career is the path a person

Figure 4–3 The floral arrangement is an example of just one aspect of the horticulture industry.

takes through life relating to work in the field of horticulture; it includes a series of horticulture occupations and jobs (Figure 4-6). A horticulture occupation is specific work that has a title and general duties that a person in the occupation performs. You can change your job and employer while remaining in the same occupation. An example of an occupation is a greenhouse manager. A horticulture job is a specific type of work that a person in a horticulture occupation performs (Schroeder et al., 1999). A job is specific work with specific employers, such as physically removing weeds in a garden.

Numerous career opportunities are available in horticulture because the industry is so diverse. To achieve your career objectives, you need a good education, experience, and hard work. To be successful in your career of choice, you must make good decisions during the selection process. You choose a career that will allow you to do something you like. Some ways to help you make a good choice in your path to a successful career are given in the following sections.

Work Part Time

Part-time work provides you with an opportunity to try an area to determine if a particular type of work is a good fit while getting paid at the same time. For example, you may think that working outdoors (Figure 4-7) on a vegetable farm sounds fun, so experiencing the work on a daily basis can give you a better understanding of what this type of work is like. Or, say you enjoy working with flowers

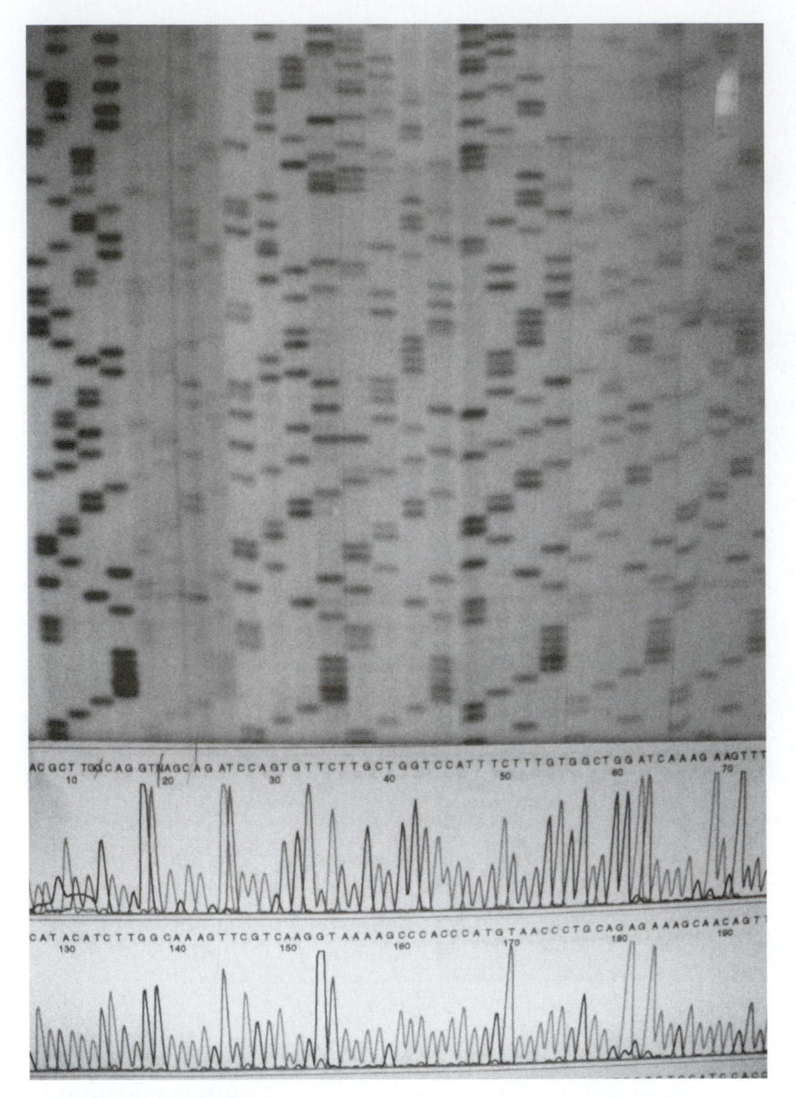

Figure 4–4 DNA sequencing film (top) and results from automated DNA sequencing (bottom)

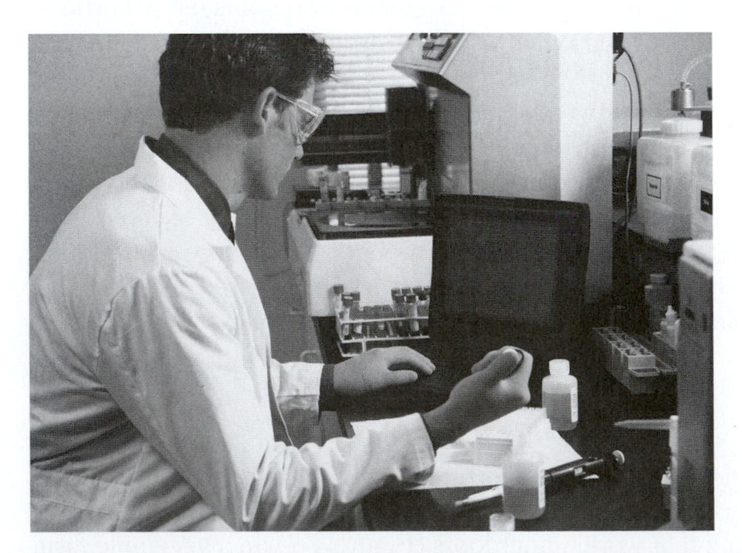

Figure 4–5 Horticulture professionals use a computer to analyze data for plant hormone levels. *Courtesy of Getty Images, Inc.*

Figure 4–6 Caring for flower gardens is one of the many field jobs for horticulture workers.

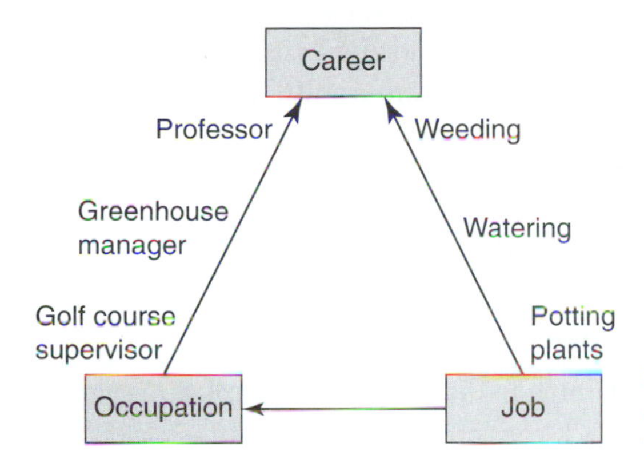

Figure 4–7 The relationship among career, occupation, and job

and creating beautiful designs, so you try working in a flower shop. You find that although the work is enjoyable, you have to work on all holidays. The intensive preholiday and holiday work schedule may be acceptable and exciting to some, but not to others, so working part time brings this aspect to the forefront.

Shadow or Observe Horticulturists

Contact individuals who are working in a field that you are interested in and ask to follow them around for a day, a week, a month, or whatever you feel is necessary. When you find someone, this is a great opportunity to see what actually happens during a typical day without any sugar coating; you will see the good things and the less glamorous parts of the position. For example, you may feel that teaching at a major university simply requires the professor presenting the lecture to show up for class, teach the lecture, and then go back to his/her office and do very little for the remainder of the day. By shadowing the professor for a day or more, you will find that there is preparation prior to class, a variety of meetings to attend throughout the day, and many other duties to perform. In fact, for many professors teaching class is only a small part of their workday.

Consult School Counselors or Advisors

Be sure to consult counselors and advisors and use their knowledge to glean information in a given area of horticulture. They will provide a variety of information, such as potential job leads, networking contacts, and experience with other students. Do not rely on them as your only sources of information, however, because information is readily available in many other places as well.

Research the Library and Internet

Your local library and the Internet have a large amount of information, which you can read and use to make informed decisions. Information obtained includes job openings, application information, and general background information about the different types of jobs available. Again, be sure not to use these resources as your only sources of information.

Seek Other Sources of Information

Interview or shadow horticulture teachers, researchers, extension agents, graduate students, and senior students, who will provide their experience in order to help you to get a clear picture of the particular career path in which you are interested.

TIPS FOR SUCCESS IN A HORTICULTURE CAREER

Setting a career goal (Figure 4-8) that reflects the level of accomplishment you want to attain in your work and career is key to creating a successful career. The first step is to set goals by basically describing what you want to achieve in life. The process of goal setting involves the following steps (as described in the next sections).

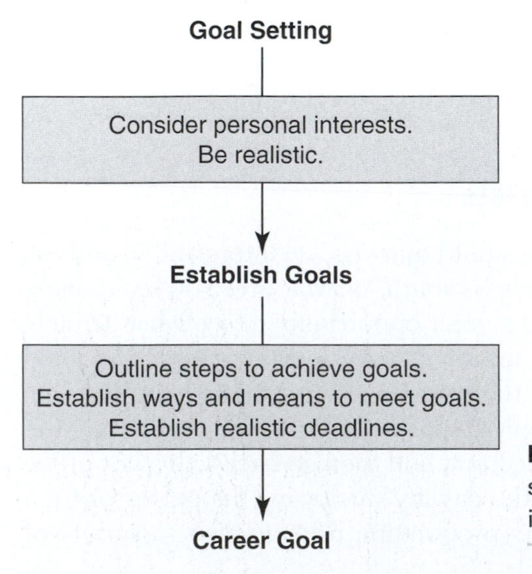

Figure 4–8 Establishing career goals starts with considering your personal interests.

Considering Personal Interests

Ask yourself what you interested in doing for the rest of your life. A relatively simple way to answer this question is to think systematically about what you like and dislike. Ask yourself the following questions:

- Do I prefer working indoors or outdoors?
- Am I the type of person who can conform to regular 9 A.M. to 5 P.M. working hours or is it important for me to have flexible hours?
- Do I like to travel?
- Is job security together with fringe benefits important?
- Do I enjoy working alone or as part of a team?
- What level of education am I willing to pursue?

Along with these suggested questions, there are many others you might need to ask. Take your time, think, and remember to select a career path that leads to work you will enjoy doing.

Being Realistic

Don't set impossible goals; for example, if you have a hard time with science, don't try to become a scientist. If you look at a plant and it dies, you probably do not want to grow plants for a living. Basing your career path on the amount of money you will earn is usually not a good idea. For example, if you are a person who must be outside and working under fluorescent lights drives you crazy, an indoor job, no matter how much it pays, is not the job for you.

After you set your goals, you need to

- carefully outline the steps required to achieve each goal. Make this outline as detailed as possible to make sure that it can be readily followed.
- identify ways and means of accomplishing each goal. Establish logical ways to achieve each goal that are easy to follow. Make sure you have enough money to achieve each goal. For example, if you decide several advanced degrees will be necessary to achieve your career goals, plan how you will have the money available to get these degrees. Take time to evaluate the pros and cons of obtaining several advanced degrees. For example, if you will incur so much debt to get the degrees and will be unable to pay it off for many years because of the low salary you will be making, you should rethink your strategy.
- establish deadlines for reaching each step toward individual goals. Goals can change; therefore, you should regularly assess and modify your goals as necessary. This is not to say when things get tough you should change your goals, but if you have tried your best and the goal cannot be achieved, you may need to modify your goals accordingly. Also, you might proceed through your goals without a problem and see the need to heighten your goals.

Education and Training

The amount of education and training you need depends on what you want to do. Be very careful when determining the level of education to pursue because job requirements may be from a high school diploma through postdoctoral experience—too much or to little may cost you the job of your dreams. For example, a common problem that occurs when an individual does not plan for the future properly is that he or she obtains too many degrees. This trap occurs when the specific jobs are unavailable, so the individual pursues advanced degrees while waiting for a position to open. The person would have been better served by getting practical experience while waiting, because too much education can overqualify for some positions. Knowledge in horticulture can be obtained in two ways, schools and on-the-job experience.

Schools

Education requirements can range from none to some college to postdoctoral experience (Figure 4-9). Your education level may be a high school diploma, B.S. or M.S. degrees, Ph.D. degree, or postdoctorate level. The education to

Figure 4–9 University-level education is necessary for some horticulture careers.

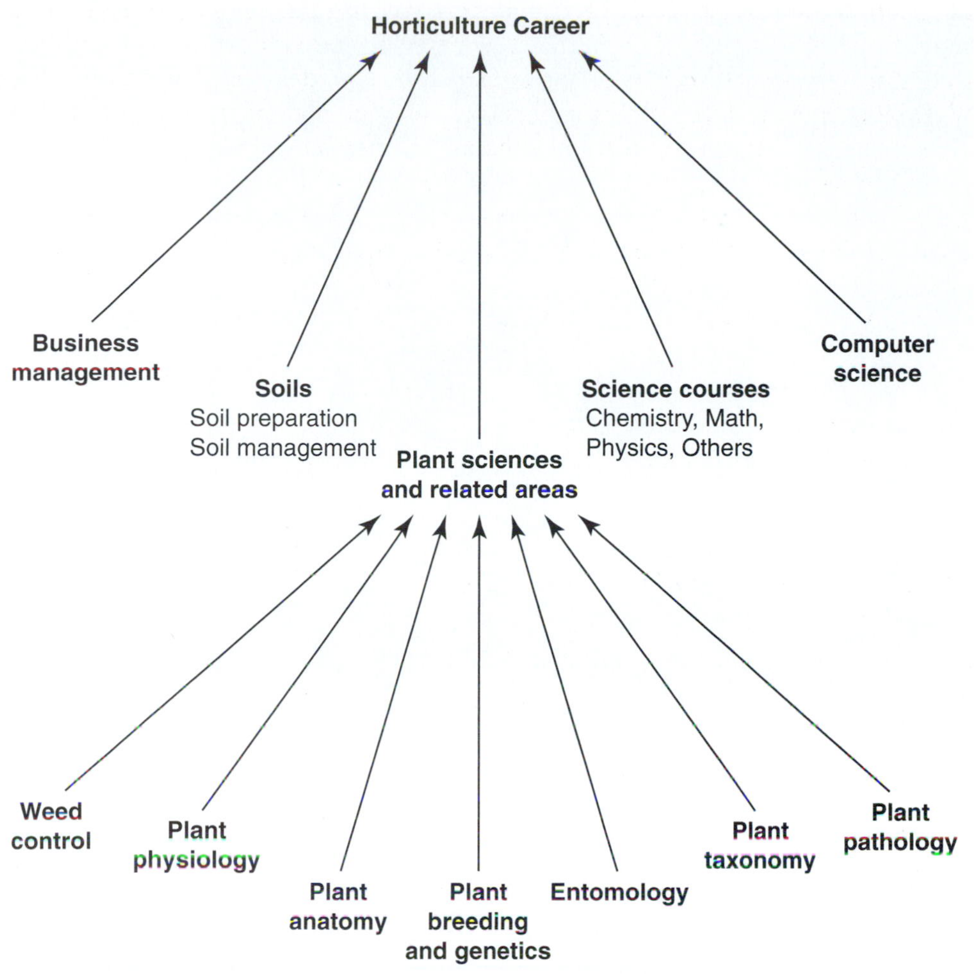

Figure 4–10 The educational areas necessary to pursue a horticulture career

prepare for a horticulture career should provide a good background in a wide variety of areas (Figure 4-10), including:

- Plant sciences and related areas (taxonomy, anatomy, plant physiology, entomology, plant pathology, plant breeding and genetics, weed control, and others).
- Soils (soil preparation and soil management).
- Science courses (chemistry, physics, math, and others).
- Business management.
- Computers.

On-the-Job Experience

Job experience in your field of expertise has become very important. In fact, people with job experience have a greater chance of landing the better jobs than those with only education. Work experience can be in the following forms:

- Full-time jobs in your field, while you are waiting for a job that is more in tune with what you want to do for the rest of your life.
- Part-time jobs in greenhouses, vegetable farms, orchards, flower shops, (Figure 4-11), and a variety of other places.

Figure 4–11 Worker making a floral arrangement.

- Internships in public gardens, flower shops, greenhouses, and companies from Disney to Bayer Chemical Company. Experience can be in turf management, tissue culture, plant breeding, gene jockeying, and a variety of other areas.

SUGGESTIONS ON HOW TO GET A JOB

In today's competitive society, finding the right job can be very difficult. Following are some suggestions for finding job openings:

- Contact a potential employer in person, by telephone, or by e-mail. If you have to leave a message by telephone, be sure to make follow-up calls. The same is true for e-mail.
- Visit as many personnel or placement offices as possible. During your visit, you may establish a contact or at the least let people know who you are and what type of personality you have.
- Look at all forms of job advertisements, including newspapers, magazines, television, radio announcements, and online job postings (Figure 4-12).
- Use contacts with teachers, counselors, and others who may provide a lead on a potential job. Sometimes using contacts is the best way to find a job; networking is a must in our competitive society.

The Job Application

The application may involve filling out one or more forms, providing school transcripts, describing your interests or background, and going for the interview.

Figure 4–12 Search for job advertisements in all media forms.

The Interview

Be prepared.

Be on time.

Groom and dress appropriately.

Bring necessary information with you.

Speak clearly and confidently.

Use good manners.

Figure 4–13 Tips for a successful job interview

Be careful to prepare properly all written materials for the job, because this material often determines whether you get the interview. Your application must be well thought out and professional. In today's job market, literally hundreds of applications are submitted for a limited number of positions. For example, people are quickly deleted from the application process if transcripts or any other piece of requested information is not sent with the application. If your application package is prepared properly, your chances are greatly increased, which may make you one of the select few getting a job interview. Some tips for the job interview (Figure 4-13) are as follows:

- **Be prepared for the interview.** Do your homework, know what the position involves, be familiar with the organization of the company and people working there, and do anything you can think of to show off your confidence and knowledge. For example, you are interviewing at a major commercial greenhouse range. The company is internationally known and has a Web site describing what it does as well as biographical sketches of individuals who will be conducting the interview process. You should study this information and use it to your advantage, because people will be impressed when you actually know what their jobs are and can discuss their jobs with them. Gathering this information is simple to do and could make the difference in getting the job or just being in the top five candidates.

- **Be on time.** Nothing is more annoying than someone who is late for an interview. This starts off the process in a negative way and could mean disaster for the remainder of the interview.
- **Groom and dress appropriately.** Know the company's expectations and meet them; for some companies, men wear a suit and tie and women wear a dress, whereas other companies are more casual. If you are applying to a company that has a strict dress code and you do not like it, you should reconsider whether this job is for you.
- **Take needed information and materials with you.** Bring copies of any written materials required for the interview. For example, if you need to bring a statement of your future goals, be sure to bring a copy for the person conducting the interview; don't say, "This is my only copy, could you make one for me."
- **Speak clearly and confidently and use good manners.** The person conducting the interview will be evaluating your personality to determine whether it is compatible with the position. If you speak clearly, confidently, and use good manners, this will, in some cases, steal a position; in other words, if the job comes down to being between you and another individual, your demeanor might help you get the job.

Keeping the Job and Being Successful

The bottom line is that if you don't fit in with the group where you are working, no matter how good you are, getting fired is a very strong possibility. Following are a few very important skills to have (Figure 4-14):

- **Work well with others.** Research indicates that more people lose their jobs because they can't get along with other people than for any other reason. For example, an individual may be the best at what he/she does, but if that individual is harmful to the overall team's performance, the individual will eventually be fired. If you are fired for not being able to work as a team member, then the chances of obtaining subsequent jobs are greatly decreased.
- **Be honest and live by the rules.** Living by the rules is important and garners respect; sometimes the rules can be bent, but don't break them.

Keeping Your Job and Being a Success

Work well with others.

Be honest and live by the rules.

Groom and dress appropriately.

Develop good work habits. (Take pride in your work.)

Practice a responsible life style. (Don't burn the candle at both ends.)

Figure 4–14 Tips for keeping a job and being a success

- **Groom and dress appropriately.** If there is a dress code, be sure to follow it as closely as possible. If you work at a company that requires a shirt and tie for men and skirts for women, rebelling against this rule may cost you a job.
- **Develop good work habits.** Take pride in what you do and go the extra mile to make a given project work. Be a high-energy individual and always have a positive attitude. You will progress rapidly if you always maintain a positive attitude. Negative attitudes lead to lost jobs.
- **Practice a responsible life style.** If you burn the candle at both ends and are always tired at work, you will not be productive. A wild life style while at the university may lead to problems with grades; however, if these bad habits carry over to your first job, the job you love may be lost forever.

CAREER PATHS IN HORTICULTURE

The horticulture industry is diverse, which leads to a variety of jobs for many types of people both directly and indirectly. The amount of training could be from none to postdoctoral experience as mentioned earlier. Many jobs require knowledge and training in horticulture, although some require no training at all. Jobs requiring training in horticulture may demand only a high school diploma and a short course in a targeted area or a traditional college education. A college degree in horticulture will allow you to pursue a wider range of positions. Although numerous career paths are available in horticulture (Figure 4-15), the three major areas are as follows:

- Ornamental horticulture.
    - Floriculture.
    - Landscape horticulture.
    - Interiorscaping.

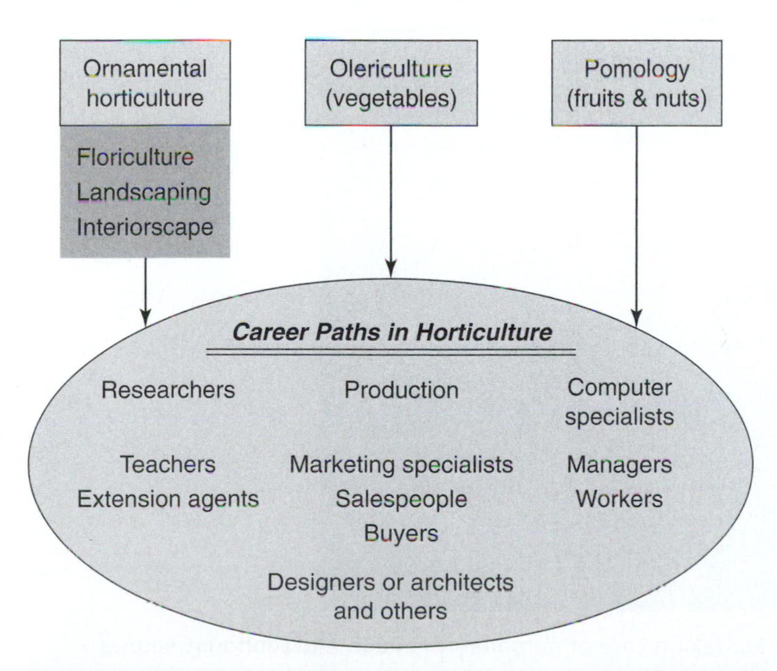

Figure 4–15 Career paths in the three main areas of horticulture

- Olericulture.
- Pomology.

In each of these main areas, there are positions for production, managers, workers, designers or architects, researchers, teachers, extension agents, marketing, sales, buyers, and computer specialists. For each of these positions, there are a variety of subcategories. For example, the subcategories for production could include, but are not limited to, propagation workers, plant breeders, growers, irrigation specialists, inventory managers, and construction supervisors.

Specific Career Areas in Horticulture

There are many careers specific to horticulture requiring varying levels of training and education. Below are several examples.

- **Florist.** There are two types of florists, wholesale and retail. The wholesale florist sells flowers, ornamental plants, and hard goods to retail florists (Figure 4-16). Retail florists sell flowers, ornamental plants, dish gardens, and terrariums. A major part of the retail florist's work deals with corsages and flower arrangements for special occasions such as weddings and holidays (especially Christmas and Easter) (Reiley & Shry, 1997).
- **Golf course superintendent.** This individual is responsible for supervising the construction and maintenance of a golf course (Figure 4-17). The golf course supervisor is responsible for a wide variety of jobs, such as servicing and repairing turf equipment, keeping records for the golf course, and preparing budgets and reports, plus everything else that keeps a golf course running smoothly.
- **Tree surgeon.** This person maintains tree health by proper pruning and care.

Figure 4–16 Worker taking care of geranium plants in a greenhouse setting

Figure 4–17 Golfers at a driving range enjoy the benefits of a golf course superintendent's work

- **Irrigation specialist.** This person is responsible for irrigation systems on golf courses, orchards, vegetable farms, and other areas of horticulture where irrigation is necessary. The individual must understand irrigation installation and maintenance. In addition, the irrigation specialist must have a working knowledge of pumping systems, irrigation lines and nozzles, controllers, hydraulics, and systems-powering method (Rice & Rice, 2000).
- **Horticulture therapist.** This person uses horticulture to help people with health problems or disabilities. This career includes therapy, such as gardening, that involves all aspects of growing plants, from preparing the soil through planting and subsequent growth, making corsages, creating terrariums, and more. Horticulture therapy is a form of stress reduction, and it helps people with disabilities become more mobile, helps terminally ill people cope, and much more.
- **Cooperative extension agent or horticulture specialist.** This person is involved in local activities, such as disseminating the most recent recommendations of researchers to growers and the general public for no charge.
- **Consultant.** This individual provides the cutting-edge recommendations of researchers to people in a specific area of horticulture. You can make a lot of money on this career path if you are skilled in your specific area.
- **Teacher.** This person can teach students at levels beginning with elementary schools, high schools, technical schools, community colleges, two-year colleges, and up to various levels at universities (instructor, assistant, associate, or full professor).
- **Research scientist.** This person performs basic or applied research to improve horticultural crops at universities or companies (Figure 4-18).

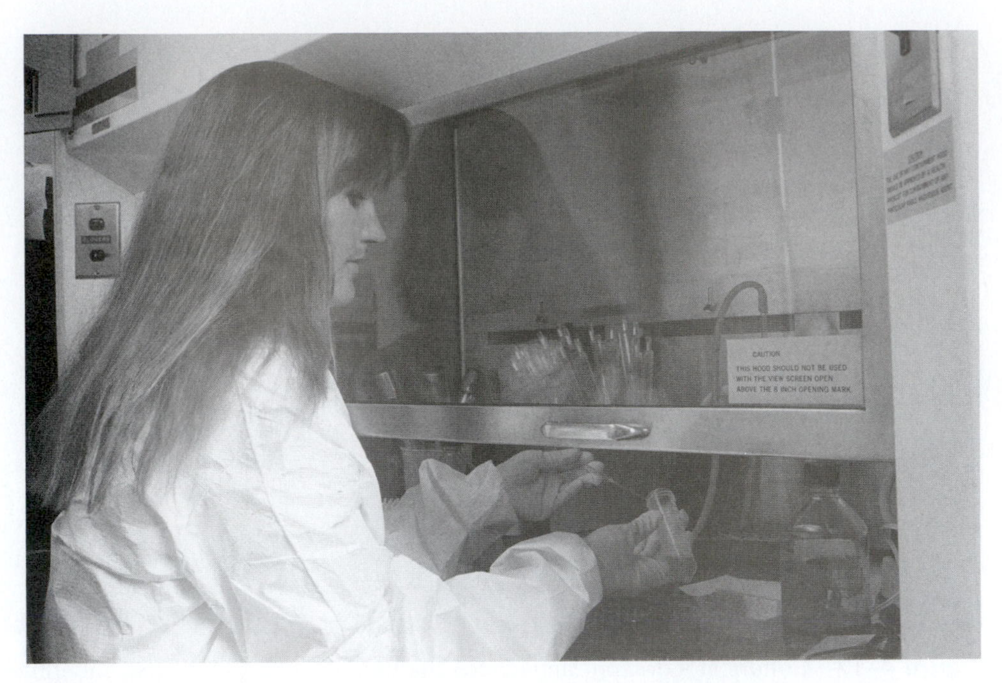

Figure 4–18 Researcher working in a tissue culture hood.
Courtesy of Getty Images, Inc.

- **Arboretum, botanical, and horticultural gardens.** All over the world, many plant collections are available to the general public. Some of the occupations available in these different types of plant collection areas include writers, researchers, propagators, and managers. These collections may be found in each of the following:
  - **Arboretum** is a collection of trees arranged in a naturalized fashion.
  - **Botanical garden** is a plant collection that forms a **habitat** (a place where wildlife live in nature).
  - **Horticultural garden** contains an assortment of plants represented by many horticultural varieties, arranged to achieve a desirable aesthetic effect.
- **Computer specialists in horticulture.** Many areas in the field of horticulture use computers. A variety of positions are available for individuals with computer skills coupled with horticulture experience. Some of these special areas where computers are used in horticulture are listed here:
  - **Horticulture teacher.** This person uses computers as a major teaching aid. In fact, many courses are taught solely via computer.
  - **Computer landscape designer.** This individual uses computers to design landscapes.
  - **Crop modeler.** This person uses computers to predict a crop's response to environmental conditions or its productivity in response to various inputs.
  - **Equipment automater.** This person is a specialist in the use of computers for automatically monitoring and controlling environmental factors, such as light, moisture, and temperature indoors. These positions are available in greenhouses and in the field.

- **Record keeper.** This individual maintains records for a variety of purposes, such as inventory and financial aspects of the business.
- **Publicist.** This individual establishes Web sites and uses computers to get the word out on specific products.

SUMMARY

You should now have a better perspective on how to prepare for a horticultural career and recognize many of the steps required to be successful. The differences among a horticulture career, occupation, and job will become clearer after the completion of this chapter. Proper preparation will help you obtain and succeed in a job leading to a successful career path in horticulture. There are vast career options available in the horticulture field. The general message of this chapter is to make sure you are careful in selecting your career path. Aggressively pursue your career and work hard to be successful in the career of your choice.

Review Questions for Chapter 4

Short Answer

1. Distinguish among career, occupation, and job.
2. What is one of the major factors that should be considered when selecting a career?
3. To be a success in achieving your career objectives, what is required?
4. What are the three major areas of the horticulture industry?
5. What are some examples of careers in horticulture?

Define

Define the following terms:

horticulture career	career goal	arboretum	horticultural garden
horticulture occupation	goal setting	botanical garden	
horticulture job	horticulture therapist	habitat	

True or False

1. A career goal is the level of accomplishment you want to make in your work.
2. A horticulture career is specific work that has a title and general duties that a person performs.
3. A horticulture job is specific work that a person in a horticulture occupation performs.
4. A horticulture occupation is specific work that has a title and general duties that a person in the occupation would perform.
5. Goals do not change; therefore, it is not necessary to assess and modify goals on a regular basis.
6. Research indicates that more people lose their jobs because they can't get along with other people than for any other reason.

7. It is not really important to dress appropriately for a job.
8. Extension agents disseminate the most recent recommendations by researchers to people at no cost.
9. An arboretum is a collection of trees arranged in a naturalized fashion.
10. A person can change jobs and employers while remaining in the same occupation.

Multiple Choice

1. The education to prepare for a horticulture career should provide a good background in which of the following?
   A. Plant science and related courses
   B. Soils and soil preparation and management
   C. Business management
   D. All of the above
2. A horticulture therapist does the following:
   A. Helps plants with their psychological problems
   B. Helps heal people with health problems and disabilities
   C. Genetically engineers plants
   D. None of the above

Activities

In this chapter, we talked about preparing for a career in horticulture and how to achieve success. However, before you can prepare, you must have some idea of what you want to do. In this activity, you will research career opportunities in horticulture. For example, landscape design and horticulture merchandising are two careers within the field of horticulture. Try to find general information, such as the amount of education or training needed for the job, the different occupations available in the career field, and a description of the work. The Internet will be a good starting point for this activity. Another source of information might be horticulture textbooks or books about careers. From the information available, create a list of four horticulture careers along with general information about each.

References

Reiley, H. E., & Shry, C. L., Jr. (1997). *Introductory Horticulture* (5th ed.). Danville, IL: Interstate Publishers.

Rice, L. W., & Rice, R. P., Jr. (2000). *Practical Horticulture* (4th ed.). Upper Saddle River, NJ: Prentice Hall.

Schroeder, C. S., Seagle, E. D., Felton, L. M., Ruter, J. M., Kelley, W. T., & Krewer, G. (1999). *Introduction to Horticulture: Science and Technology* (3rd ed.). Danville, IL: Interstate Publishers.

The Relationship Between Horticulture and the Environment

CHAPTER 5

Objectives

After reading this chapter, you should be able to

- show the benefits and potential dangers horticultural practices present to the environment.
- define pollution and explain the differences between point and nonpoint sources of pollution.
- discuss how the public's demand for high-quality products puts the grower in a difficult position.
- discuss Integrated Pest Management (IPM) strategies as the best ways to control pests.
- discuss the hydrologic cycle and its importance to the environment.
- define eutrophication and the problems associated with it.
- discuss the nitrogen cycle and why it is important not to upset the delicate balance of nitrogen in nature.
- define wildlife and discuss how pesticides and fertilizers can have an adverse effect.
- define wetlands and discuss why they were once thought to have little value and now are protected by law.

Key Terms

abiotic
antisense technology
biological control
bioremediation
biotic
chemical control
chlorosis
cultural control
denitrification
environment
eutrophication
genetic control
genetically modified organisms (GMOs)
habitat
hydrologic cycle
Integrated Pest Management (IPM)
mechanical control
mineralization
nitrogen cycle
nitrogen fixation
nonpoint sources of pollution
pesticides
phytoremediation
plant breeding
point sources of pollution
pollution
sense technology
wetland
wildlife

ABSTRACT

A healthy environment is important for everyone on Earth. The information in this chapter describes how to maintain a healthy environment. Horticulture contributes to the environment because plants add beauty, help prevent pollution, create oxygen for people and animals, and provide other beneficial effects. Unfortunately, some horticulture practices can create environmental problems indirectly as a result of improper practices resulting from human error. For example, the incorrect use of pesticides and fertilizers can cause pollution, or soil erosion can occur if ground covers are not used. This chapter defines pollution and explains point and nonpoint sources of pollution. The public's demand for high-quality products at a low cost puts the grower in a tough place. Integrated Pest Management (IPM) strategies are now being used as an environmentally friendly means of controlling pests. The hydrologic cycle, eutrophication, and the nitrogen cycle are defined and explained. The use of pesticides is very controversial; therefore, the benefits and problems associated with them are presented. The excessive use of fertilizers and pesticides also poses problems for wildlife. Finally, wetlands that were once thought to have little value are protected by law today because of the many benefits they provide.

Figure 5–1 Horticulture practices must consider *potential* benefits without creating environmental problems.

INTRODUCTION

Horticulture helps promote a healthy **environment** but it also harbors potential dangers. Horticulture is beneficial to the environment in many ways because it provides aesthetic beauty (Figure 5-1), controls soil erosion, absorbs atmospheric pollution, absorbs noise, has a cooling effect in hot weather, removes fertilizers and **pesticides** from the environment, and provides many other benefits. Unfortunately, some horticulture practices can create environmental problems, such as the incorrect use of pesticides and fertilizers and the use of horticultural practices that may lead to soil erosion.

Pollution is also a horticultural issue that includes differences between **point** and **nonpoint sources of pollution** (nonpoint sources are the most difficult to control). Both **phytoremediation** and **bioremediation** methods for cleaning up pollution from the environment may offer the only effective means of restoring large areas of polluted land. The public's demand for high-quality products at a low cost places the grower in a difficult position. **Integrated Pest Management** (**IPM**) strategies are now being used as an environmentally friendly way to control pests while being environmentally responsible. The first step of any IPM program is to prevent the pest problem; however, when this is not possible, the pest should be properly identified prior to implementing a control strategy. The five general categories used to control pests include **cultural, biological, mechanical, genetic,** and **chemical control.**

Another important environmental concern is the water required for life. The **hydrologic cycle** is the cycle of water in the environment, which is delicately balanced in nature. If any part of the cycle is disrupted, water resources are put in danger. **Eutrophication** occurs when lakes or streams have too many nutrients in their water due to excessive nutrient runoff. Excessive nutrients in lakes, streams, and rivers create problems with excessive algae growth, oxygen depletion, and reduced water quality. The **nitrogen cycle** is the delicately balanced circulation of nitrogen in nature. When this balance is disrupted, excessive nitrogen gets into the water systems and causes eutrophication.

Also covered in this chapter are the pesticides used to control pests. These chemicals are widely used in agriculture to great benefit although they remain highly controversial. Pesticides are good when used as part of a solid IPM program; however, problems can occur when they are used improperly. Improper use includes applying excessive amounts, using highly volatile compounds, and using pesticides such as DDT that leave behind toxic residues. The proper way to use pesticides to minimize potential error includes reading the label, using the proper equipment, being careful with all calculations, disposing of pesticides and containers properly, and cleaning all equipment after usage in a safe way.

Wildlife (plant or animal organisms that live in the wild) rely on natural food and **habitat** (the place where wildlife live in nature) to survive, so the improper use of pesticides and fertilizers puts wildlife at risk directly by modifying their habitat. Another habitat, **wetlands,** include swamps, ponds, and other areas where water often stands. Although wetlands were once thought to be of little value and agriculture and urban development destroyed many of them, today the federal government protects wetlands.

HOW TO MAINTAIN A HEALTHY ENVIRONMENT

Before a healthy environment can be maintained, it is important to understand the proper definition of the environment, which includes all the factors that affect the life of living organisms. The environment is made up of **biotic** or living factors and **abiotic** or nonliving factors. This definition shows the complexity of the environment and how difficult it is to control all the factors affecting the life of living organisms and solve the problems associated with maintaining a healthy environment. Horticulture is very beneficial to the environment, as shown in the following examples:

- **Aesthetic value.** The use of plants in the landscape increases property value by making the outdoor environment more attractive (Figure 5-2). Plants also relieve stress in people indoors for a variety of reasons, one of which is that plants create a soothing effect by simulating the outdoor environment (Figure 5-3). Horticulture also enhances the outdoor environment by providing a natural environment in public parks, for example, allowing people in large cities to get away from the "asphalt jungle" (Figure 5-4).
- **Soil erosion control.** With proper use of plants, topsoil can be protected from erosion. In addition, controlling erosion minimizes runoff of pesticides and fertilizers, which reduces pollution in water supplies.

Figure 5–2 A nicely landscaped home increases property value.

Figure 5–3 Indoor plants can create a soothing effect by simulating the outdoor environment.

Plants can be used to reduce soil erosion by never leaving the soil exposed; for example, always use cover crops in areas where crops are not being grown. On roadsides in Pennsylvania, crown vetch (Figure 5-5) is used to provide erosion control while enhancing the beauty of the roadsides.

- **Absorption of atmospheric pollutants.** Plants absorb pollutants such as ozone, carbon dioxide, and sulfur dioxide from the atmosphere, thereby purifying the air we breathe. In fact, the problems with increased carbon dioxide levels in the atmosphere are being caused by the removal of natural vegetation worldwide. As you know, plants remove carbon dioxide from the atmosphere and return oxygen; therefore, when plants are indiscriminately removed from the environment, the natural balance is disrupted. Indoor landscaping is also beneficial to the indoor environment by removing indoor pollutants such as carbon dioxide (Figure 5-6) (Acquaah, 2002).
- **Absorption of noise.** When used properly, plants can dramatically reduce noise caused by anything from a noisy neighbor to a nearby highway. For example, a row of dense shrubbery planted around your property line will greatly reduce the noise from a neighbor. The necessary depth, height, and density of the shrubs used will depend on how loud the noise is.
- **Cooling effect in hot weather.** Plants have a dramatic effect on cooling the environment during hot weather. For example, you will always be cooler on a hot day standing in a grass field than standing in the middle of an asphalt parking lot.

Figure 5-4 Children playing in the park and enjoying the natural environment

Figure 5-5 Crown vetch along the roadside helps control erosion.

Figure 5–6 This spider plant helps remove indoor pollutants.

- **Removal of fertilizers and pesticides.** Plants use fertilizers for growth, which keeps fertilizers from getting into the water supply. In addition, plants have been shown to remove pesticides from the environment and convert them into nontoxic forms.

Potential Problems from Horticultural Practices

Many types of production practices are used in horticulture, some of which are potentially dangerous to the environment. The danger is not caused by the practice itself, but rather by human error, which can result from the situations covered in the following sections.

Excessive and/or Improper Use of Fertilizers and Pesticides Creates Pollution

More is not always better when it comes to using pesticides or fertilizers, especially because fertilizers and pesticides are expensive. When they are used at the recommended levels, they are beneficial and cost can be minimized. However, using excessive amounts costs more and pollutes the environment, which is harmful to everyone. Excessive amounts of fertilizer or pesticide get into the environment by miscalculating how much material to apply; for example, simply slipping a decimal place could cause you to apply 10 to 100 times the appropriate amount. Another way in which problems can occur is by using applicator equipment that has not been properly calibrated. A simple mistake can lead to disaster.

Improper Soil Management Leads to Soil Erosion

If soils are left for even brief periods of time without cover crops, the topsoil readily washes away when heavy rains occur. A number of practices are used to minimize soil erosion today. Recently, conservation tillage has become very popular, although it is still not a commercial practice. Tillage is the practice of loosening the soil to maintain its physical condition until the crop is planted; after the crop has been planted, the same process is called cultivation. Conservation tillage is also known as minimum tillage, no till, and reduced tillage, which involve tillage practices that leave 30 percent or more of the crop on top of the soil to prevent erosion. Among the variety of conservation tillage methods, probably the most common one is strip tillage, in which strips of vegetation are left between the rows to prevent erosion. Soil mismanagement is evident when the soil is left bare after harvest. An unexpected heavy rainfall will erode the soil.

Effects of Pollution on the Environment

Pollution comes from a variety of sources, some of which are easy to control or regulate and some of which are subtler and more difficult to control. Pollution occurs when harmful or degrading materials get into the environment. Pollution can come from two main sources: point sources and nonpoint sources.

Point sources of pollution are definite and identifiable. For example, wastewater from a factory is point pollution (Figure 5-7). The factory is highly visible and water coming out of it can be readily monitored. Another example is a nuclear power plant, which is also highly visible and can be readily monitored on a regular basis (Schroeder et al., 1999).

Figure 5–7 Factories are often examples of point sources of pollution.

Nonpoint sources of pollution cannot be specifically identified. For example, a nonpoint source of pollution from horticulture could come from leaching and runoff containing nutrients and pesticides (Figure 5-8). There are so many farms all over the world of varying sizes that it is difficult to monitor all of them. In most cases, many farms have either no problems or minimal problems with pesticides and fertilizers; however, because problems are typically due to human error, it is difficult to determine when the problem happened and to locate the source of contamination. Some situations are complicated further by climatic conditions such as rainfall, which can wash pesticides or fertilizers from the point of origin to a location far away. Therefore, most environmental concerns in horticulture are from nonpoint sources of pollution because they are the most difficult to control.

Figure 5–8 Farmland is often a source of nonpoint pollution.

Two methods are used to clean up the pollution: phytoremediation and bioremediation.

Phytoremediation refers to using plants to remove pesticide spills, heavy metals, and other pollutants from the environment. Phytoremediation occurs, for example, when rapidly growing plants are used to take up large quantities of a given pesticide from the soil and convert it into a form that has no adverse effect on the environment. These plants can then be plowed back into the soil and used as organic matter, posing no harm to the environment. Much research is dedicated to determining which plants are the most efficient for a specific job. Determining which plants are best for the wide range of pesticides, heavy metals, and other pollutants that have been released into our environment is very complicated.

Similar to phytoremediation, **bioremediation** refers to using living organisms to remove pesticide spills, heavy metals, and other pollutants from the environment. Biological materials have been used for many years to clean up environmental hazards, and the method is becoming increasingly more sophisticated. An example of bioremediation is the use of bacteria to clean up oil spills. Bacteria, which feed on oil as a carbon source, are inoculated on waters where an oil spill has occurred. The bacteria feed on the oil until it is gone and then die because their carbon source is gone. In recent years, these bacteria have been genetically engineered to make them even more efficient at cleaning up oil spills. Another use for the bacteria is to inoculate the holding tanks of oil tankers, which have been emptied and require cleaning. The bacteria devour the oil remaining in the tanker, and the remaining water can be flushed into the environment without causing a problem.

In recent years, both phytoremediation and bioremediation have generated great excitement in the scientific community because they may offer the only effective means of restoring thousands of square miles of land that has been polluted by human activities. Currently, many other clean-up methods, such as physically removing contaminated soil from a site and burying it elsewhere, are too costly and destructive to the environment to be applied on the large scale that is now required.

Public Demand Puts the Grower in a Difficult Position

The public demands high-quality plant materials. Supermarket shoppers naturally push any damaged fruits or vegetables to the side to get to an unblemished product (Figure 5-9). This demand challenges the grower to produce a high-quality product, at a low cost, without harming the environment.

Due to economic concerns or availability, most people are unable or unwilling to pay more for a product that is grown without pesticides. However, the environment must be protected as well. To ensure the least amount of damage to the environment and produce high-quality plants while preserving our natural resources, Integrated Pest Management strategies should be used.

Integrated Pest Management (IPM) uses approaches that are environmentally friendly to control pests. The primary goal of IPM is not to eradicate the pest, but to manage it in a way that minimizes economic loss. The first step of any IPM program is prevention; however, when this does not work, the pest must be properly identified in order to implement the proper control strategy, which is an interdisciplinary approach to control pests (Figure 5-10). Five general categories IPM utilizes to control pests are listed next:

- Biological controls use living organisms as predators to control pests. An excellent example is *Bacillus thuringinensis* (BT), which is released into

Figure 5–9 Shoppers demand and look for high-quality, unblemished products, such as the plum and the apple on the right.

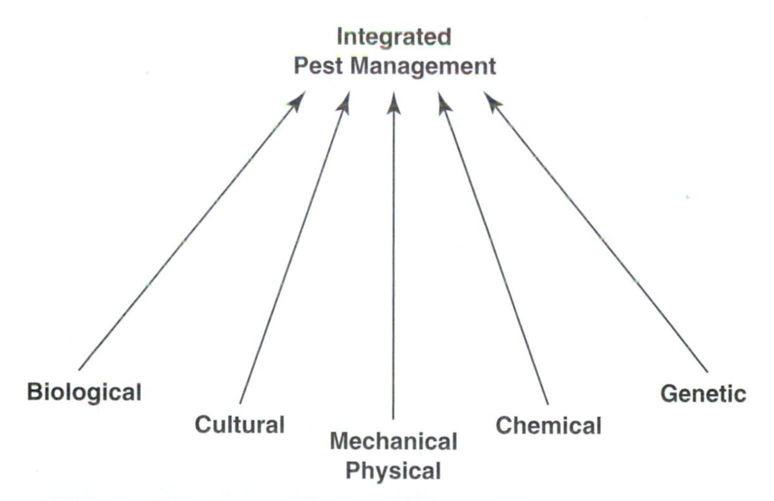

Figure 5–10 Five categories of control measures are used in integrated pest management.

fields and kills various species of worms. Today, plants are being engineered with a gene that enables the plant to produce the BT toxin in order to protect itself. Other examples of beneficial insects include ladybugs, praying mantises, and nematodes. For biological control to be effective, it is important to keep pest populations as low as possible. When pest populations go above a certain threshold, biological control loses its effectiveness, and other forms of control must be used to supplement it.

- Cultural controls use management techniques to control pests. Healthy plants tolerate low pest populations, so if you are a good grower, you can reduce the need for pesticides. However, if good cultural practices are not used to control pests, the only alternative may be to use pesticides. Some examples of cultural pest control include proper irrigation, fertilization, pruning, soil management (aeration and so on), mulching, and sanitation. This may sound like simple common sense, but in many cases all it takes is the mismanagement of any cultural factors to lead to a problem.
- Mechanical controls use tools or equipment to control pests. Examples of commonly used mechanical controls include plowing, cultivation, mowing, and mulching. More specifically, weeds are controlled by physically pulling them out and discarding or by plowing them under. Another example of mechanical control involves the tomato hornworm, which is large enough that it can be physically removed from the plant and destroyed as a means of mechanical control on a small-scale basis.
- Chemical controls use a pesticide to control pests. They are typically very toxic to humans and should be used with caution. It is important to choose the proper pesticide to do the job efficiently and with minimum damage to the environment. The steps involved in selecting the proper pesticide include the following:
  1. Determine if there is a problem.
  2. After a problem has been detected, correctly identify the insect or pathogen.
  3. The cost of control must not be greater than the cost of crop loss, so determine the cost effectiveness.
  4. Determine when the pest is a problem. Determine at what stage of the organism's life cycle it poses a threat and when is it vulnerable to chemical control.
  5. Select a pesticide that is approved and effective for your crop and use the pesticide with the lowest toxicity possible.
  6. Determine the most efficient method of application.
  7. Make sure the pesticide is effective in controlling the pest.
  8. Determine the effectiveness of the application shortly after treatment and decide if a follow-up application is required.
  9. Keep close records of what was done in case problems arise in the future.
- Genetic controls use genetic engineering and plant breeding to modify plants to make them more resistant to specific pests. Plant breeding is the science and art of modifying plants in a way that is advantageous to humans. It involves making crosses between plants and selecting for plants that have the best attributes. General types of breeding objectives include increasing yield and quality and producing disease-, insect-, and salt-resistant plants. More specifically, insect- and disease-resistance goals for cabbage breeders (*Brassica oleracea* L.) are to breed for resistance to cabbage worm, cabbage looper, powdery mildew, downy mildew, and a variety of other pests.

Genetic engineering is not much different from plant breeding, but the consumer must be educated to understand this. Following are the two forms of genetic engineering techniques commonly used:

- **Antisense technology.** Putting a known gene sequence into the plant backward to block a process in plants.
- **Sense technology.** Putting a gene sequence into the plant in the correct orientation to stimulate a process in plants.

- Plants that are genetically modified are called **genetically modified organisms** (**GMOs**). GMOs carry a foreign gene that has been inserted by laboratory techniques into all cells.

Remember that IPM is an interdisciplinary approach; therefore, it is important to optimize each of the five general categories of control to control pests.

WATER RESOURCES

Although Earth is covered with a large amount of water, only a small portion of that water is fit for human use, and problems caused by pollution damage that water supply. Remember that horticultural crops are made of 90 percent or more water. Nature has a variety of ways to purify water; however, if the delicate balance that nature uses to purify water is upset, many problems can arise. The cycle of water in the environment, or the hydrologic cycle, is the way water is purified in nature (Figure 5-11). Evaporation from oceans, lakes, rivers, and streams together with transpiration, which is the loss of water from plants and trees, results in condensation in the form of clouds in the atmosphere. Precipitation comes from the clouds and brings water back to Earth; this water percolates into the soil and enters the groundwater. Surface runoff also goes into our

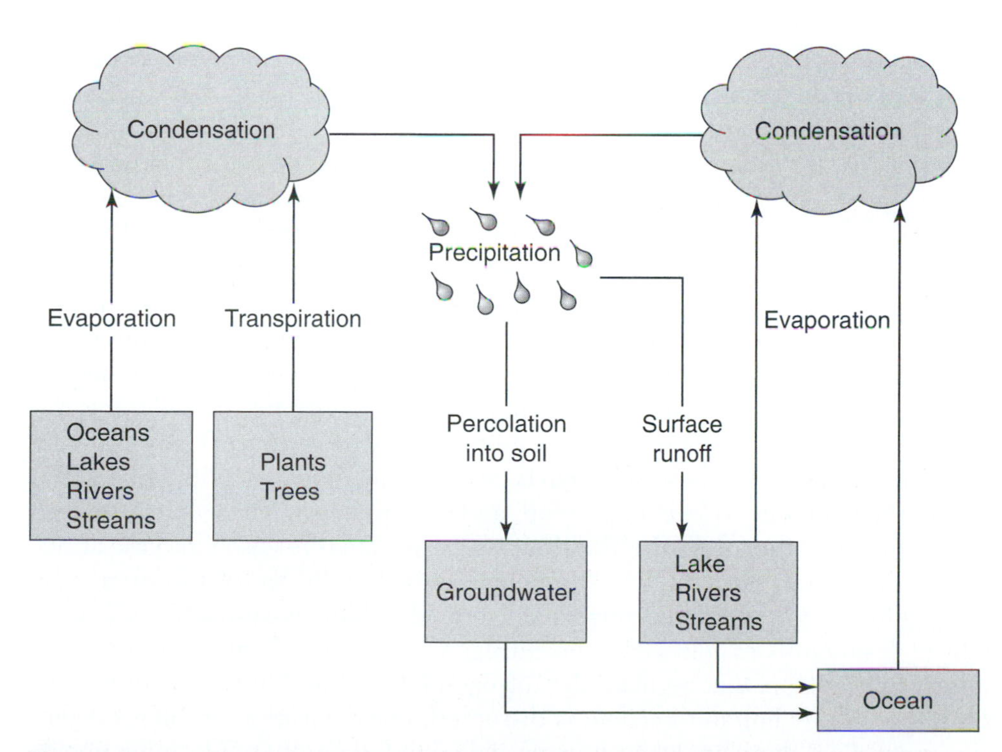

Figure 5–11 The hydrologic cycle refers to how water is purified naturally.

streams, rivers, and lakes, which can lead to the ocean. All parts of the scheme are interrelated; if one part is removed, the system falls apart. For example, if a lake is drained for a parking lot or a shopping mall, less water will be available for evaporation to form clouds. If forests are indiscriminately cut to the ground and grassy fields are covered with large cities, water transpired from these plants to be used for condensation will be lost. In addition, when we pollute the atmosphere and the land, water that has been purified by nature is contaminated as it falls from the sky. You can see that when the delicate balance of nature is disrupted, many problems can and will arise.

EUTROPHICATION

The process of eutrophication occurs when lakes or streams have too many nutrients in their water due to excessive nutrient runoff. Excessive nutrients washing into lakes and streams create problems such as

- excessive algae growth.
- oxygen depletion.
- reduced water quality for aquatic life and for recreational purposes.

Fertilizers are commonly used in horticulture, but they must be used with caution. They are equally as dangerous to the environment as pesticides, but people have a tendency to be less cautious with fertilizers. The proper use of fertilizers will help prevent eutrophication. Excessive use of fertilizers is bad for the environment, and it costs more—two very good reasons to use the correct amount of fertilizer.

Nitrogen is one of the three primary macronutrients used by plants and is the most commonly used element for plant nutrition. Plants absorb nitrate ions (NO_3^-), which are the inorganic form of nitrogen, although in some cases, nitrogen can be absorbed as ammonium (NH_4^+). The nitrogen cycle, which is the circulation of nitrogen in nature (Figure 5-12), is the way nitrogen levels are regulated in nature. After nitrate ions are absorbed by the plant, they are immobilized by their conversion into organic forms, which are used as building blocks in the plants; for example, nitrate is used to produce amino acids and proteins. When the plant dies, it decomposes and the organic form of nitrogen is converted into the inorganic form by a process known as **mineralization.** Microbes can release nitrates from dead tissues, and some microbes can fix atmospheric nitrogen through the process of **nitrogen fixation.** Nitrogen fixation involves a symbiotic relationship between bacteria known as *Rhizobia* and legume roots. The biological fixation of nitrogen involves two chemical reactions, ammonification and nitrification. Nitrogen is very important to plants; for example, it is the key component of chlorophyll and a variety of enzyme systems. Therefore, any deficiency in nitrogen causes **chlorosis** or yellowing of leaves. Nitrogen can be lost by the plant due to leaching, runoff, and crop removal. Nitrogen can also be lost by the plant due to **denitrification,** where nitrate is released into the atmosphere as N_2, N_2O, NO, or NO_2. Ammonia can be lost due to volatilization or by fixation to clay in the soil. Fertilizers and animal or human wastes typically contain high amounts of nitrogen. The nitrogen cycle is made up of a variety of interrelated factors that regulate the nitrogen levels found in the environment; therefore, if one link in the chain is disrupted, eutrophication problems occur. For example, if there are a few animals and a limited number of farming operations and homes or industry along a waterway, the nitrogen cycle typically regulates the level of nitrogen without any problems (within limits). At some point,

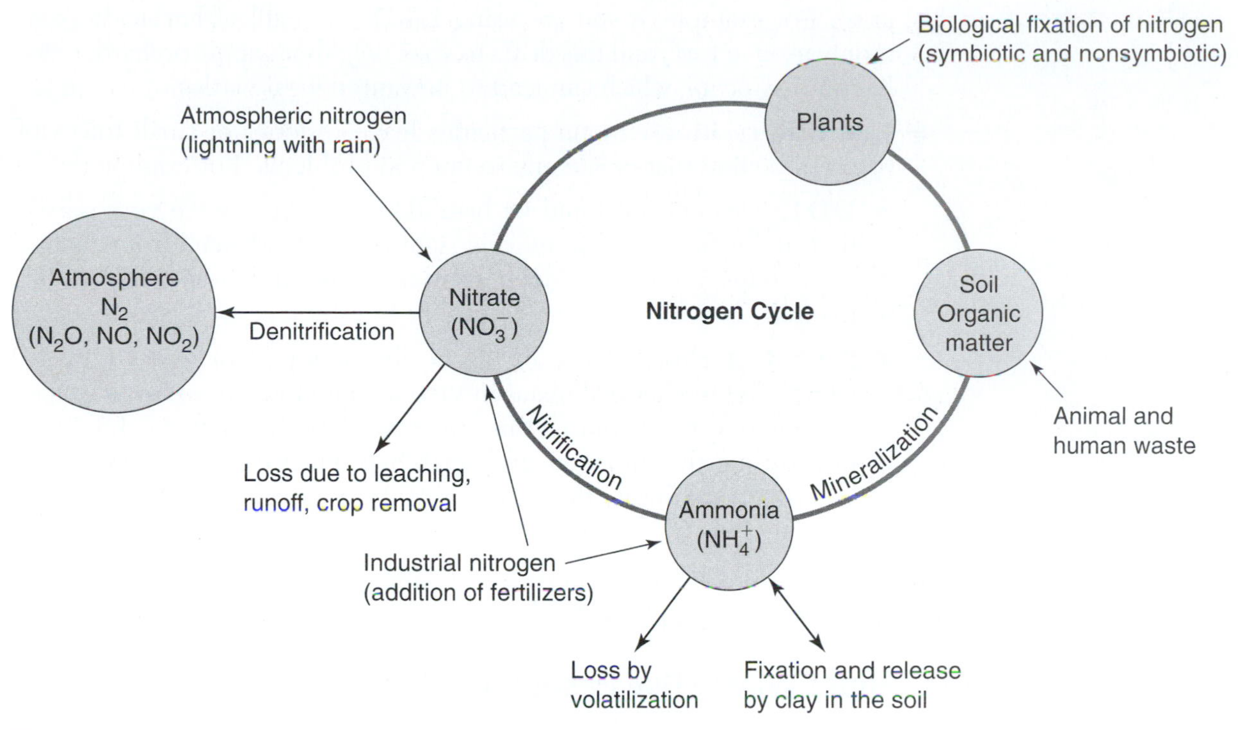

Figure 5–12 Nitrogen levels are regulated via the nitrogen cycle.

as development occurs and population booms, the nitrogen cycle cannot handle the levels of nitrogen it has to deal with; at this point, problems with eutrophication occur.

PESTICIDES

Pesticides are chemical materials used widely in agriculture to help control weeds, insects, diseases, and other organisms that damage plants. Even though they are beneficial, the use of pesticides is still controversial. When pesticides are used as the only source of control, they are bad because the possibility of polluting the environment is increased. However, pesticides are beneficial when used in moderation as part of a solid IPM program. Problems with pesticides occur when they are improperly used, such as in the following examples:

- **Applying excessive amounts.** Carefully reading the label and following the manufacturer's specifications is very important; otherwise, problems with leaching into the groundwater and/or surface runoff into lakes, streams, and rivers will occur. More is not always better, so improvising with pesticides is not recommended. All calculations must be correct prior to pesticide application, because an amount applied that is off by just one decimal point can cause serious problems. Use properly calibrated equipment to ensure that you are applying exactly what is needed.
- **Using highly volatile compounds.** The use of highly volatile compounds, such as esters, causes problems because of drift into areas where they are not wanted. If you are treating for a specific pest on your property and are safely using this material, there will be no problem with the specific plants you are growing. However, if the pesticide you are using is highly volatile and drifts to your neighbor's property, major problems may

arise. For example, if you are using 2,4-D (a broadleaf herbicide commonly used in turf) and this drifts to your neighbor's apple orchard, defoliation may occur, which can lead to unwanted legal battles.

- **Pesticide residues.** Some pesticides leave residues or small traces of material behind that create environmental problems. For example:
  - **DDT.** This chemical and its breakdown products were very persistent in the environment, moving right up the food chain from aquatic ecosystems to humans, which eventually caused its removal from the market.
  - **2,4,5-T.** A close relative of 2,4-D, this was part of Agent Orange, which was used as a defoliant in Vietnam and thought to cause cancer in military veterans during this era. It was later found that 2,4,5-T was not causing the problem; instead, a byproduct called Dioxin, which was produced during the synthesis of 2,4,5-T, was responsible for the cancer.

Note that anyone who uses pesticides commercially must have a Pesticide Applicators License to apply pesticides legally.

The Proper Way to Use Pesticides

Pesticides can be used safely and properly by doing the following:

- Be sure to use chemicals only as they are specified on the label. If you do not follow the manufacturer's specifications, then you are responsible for any problems or litigation that may occur from improper use.
- Correctly operate and regularly service equipment used to apply pesticides.
- Be careful with all calculations so you know how much to apply, and then be sure the equipment is properly calibrated so that the proper amount is actually being applied.
- Dispose of unused chemicals and containers properly.
- Carefully clean all equipment after usage.

WILDLIFE

Figure 5–13 Animals in the wild rely on their environment for food, water, and shelter.

Wildlife refers to plant or animal organisms that live in the wild and rely on natural food and habitat (the place where wildlife live in nature) to survive (Figure 5-13). When used properly, horticultural practices do not adversely affect wildlife. However, when pesticides and/or fertilizers are used improperly, wildlife are exposed to their dangers, which in turn offsets the delicate balance in nature and causes major environmental concerns. For example, exposing wildlife to pollution may cause these problems to move up the food chain to humans (as happened with DDT).

WETLANDS

Wetlands are swamps, ponds, and other places where water often stands (Figure 5-14). Wetlands were once considered of little value, and many were destroyed for agricultural production and urban development. Today, with a

Figure 5–14 Wetlands are important natural rescources.

good understanding of ecological processes and environmental values, wetlands are now thought of as natural resources and are protected by law. In fact, programs have been implemented to restore previously destroyed wetlands. Wetlands are important for the following reasons:

- **Habitat for wildlife.** When wetlands were destroyed, many species of plants and animals became extinct.
- **Aesthetic and recreational purposes.** Because of their beauty, many people visit wetlands to enjoy a fascinating part of nature and to hunt or fish.
- **Flood control.** Wetlands provide an area for water to go during storms, which helps prevent flooding.
- **Improvements in water quality.** Wetlands act as a filter to purify contaminants that find their way into the water supply. Synthetic wetlands are currently being explored as a means of purifying wastewater from greenhouses and other applications (Figure 5-15).

SUMMARY

You should now have a better understanding of how horticulture contributes to the maintenance of a healthy environment and how to avoid potential dangers that can result from horticulture due to human error. Pollution and soil erosion can be controlled properly by using Integrated Pest Management (IPM), which is an environment-friendly method. When the balance of hydrologic cycle and nitrogen cycle is disrupted, eutrophication occurs and adversely affects the environment. Wildlife and wetlands, which provide a habitat for specific life forms, are in danger when fertilizers and pesticides are improperly used. Putting wildlife at risk also puts humans at risk because when the environment is abused and wildlife is harmed, the problem moves up the food chain.

Figure 5–15 Synthetic wetlands are currently being explored as a means of purifying wastewater from greenhouses and other applications. *Courtesy of Dr. Robert Berghage, Department of Horticulture, The Pennsylvania State University.*

Review Questions for Chapter 5

Short Answer

1. List some benefits and potential dangers of horticultural practices.
2. Why does the public's demand for high-quality products put the grower in a tough position?
3. Prior to implementing IPM control strategies, pest prevention should be attempted; if prevention fails, identifying the pest is the next step. After the pest is identified, what are the five general categories that IPM uses to control pests?
4. Define the hydrologic cycle and explain its importance to the environment.
5. Define eutrophication and provide three problems associated with it.
6. Define the nitrogen cycle and explain why it is important not to upset its delicate balance found in nature.

7. Explain how pesticides can be used safely and how problems can arise from their use.
8. Define wildlife and discuss how pesticides and fertilizers can have adverse effects.
9. Define wetlands and provide examples of why they are important.
10. In the past, wetlands were thought to have little value; list two general ways in which wetlands were destroyed.

Define

Define the following terms:

environment
biotic
abiotic
phytoremediation
bioremediation
Integrated Pest Management (IPM)
cultural control
biological control
mechanical control
chemical control
genetic control
antisense technology
sense technology
genetically modified organisms (GMOs)
plant breeding
mineralization
nitrogen fixation
chlorosis
denitrification
pesticides
pollution
point sources of pollution
nonpoint sources of pollution

True or False

1. Nonpoint sources of pollution are the most difficult to control.
2. One of the benefits of horticulture in helping the environment is to absorb noise.
3. Healthy plants can tolerate moderate pest populations.
4. It is important to use cover crops to prevent soil erosion.
5. Antisense technology is putting a gene sequence into the plant in a linear orientation to block a process in plants.
6. Excessive use of fertilizers is not recommended because it is bad for the environment and is more expensive.
7. Pesticides are not good as part of an IPM program.
8. In some cases, it is legal to use chemicals for purposes other than those specified on the label.
9. It is not necessary to have a pesticide applicator's license to apply pesticides in greenhouses.
10. Wetlands are important for flood control.
11. Today wetlands are viewed as a natural resource and are protected by law.

Multiple Choice

1. Antisense technology is a technique that involves
   A. making plants more sensitive to touch.
   B. inserting a gene into a plant in the backward orientation.
   C. making plants more sensitive to light.
   D. None of the above
2. Sense technology is a technique that involves
   A. making plants more sensitive to touch.
   B. putting a gene into a plant in the backward orientation.

C. inserting a gene into a plant in the correct orientation.

D. None of the above

3. The cancer-causing material found in Agent Orange is

A. 2,4,5-T.

B. Dioxin.

C. 2,4-D.

D. All of the above

Fill in the Blanks

1. Many types of production practices are used in horticulture, some of which are potentially dangerous due to human error. Excessive and/or improper use of ____________ and ____________ creates pollution. In addition, ____________ can occur if ground covers are not used.
2. Antisense technology involves putting a known gene sequence into the plant in the ____________ orientation to ____________ a process in plants.
3. Sense technology involves putting a known gene sequence into the plant in the ____________ orientation to ____________ a process in plants.
4. Excessive use of fertilizers should not be done for two very simple reasons: it is ____________ for the environment and more ____________ to do.
5. Eutrophication occurs when lakes and streams have too many ____________ in their water due to excess ____________ runoff.
6. Wetlands were once considered to have little value; therefore, many of them were destroyed for ____________ and ____________.

Activities

Now that you know how horticulture can benefit and harm the environment, you will have the opportunity to explore potential benefits and threats in more detail. In this activity, you will report on specific horticultural practices and how they affect the environment. Your activity is to search the Internet for two sites that present information on the benefits that horticulture provides to the environment and two sites that present information on the dangers horticultural practices pose to the environment. For each site that you find, answer the following questions:

- What is the horticultural practice?
- What is the danger or benefit to the environment associated with the practice?
- If the practice threatens the environment, how can the threat be minimized or avoided?
- What is the URL for the Web site where you found the information?

References

Acquaah, G. (2002). *Horticulture: Principles and Practices* (2nd ed.). Upper Saddle River, NJ: Prentice Hall.

Schroeder, C. S., Seagle, E. D., Felton, L. M., Ruter, J. M., Kelley, W. T., & Krewer G. (1999). *Introduction to Horticulture: Science and Technology* (3rd ed.). Danville, IL: Interstate Publishers.

CHAPTER

6

Classification of Plants and Plant Anatomy

Objectives

After reading this chapter, you should be able to

- explain why we classify plants and describe the two main categories used to classify plants in early systems.
- list and explain the eight ways in which scientists group plants by using similarities between them for classification.
- define scientific classification and morphology and explain why common names and Latin polynomials were never accepted as the standard for classifying plants.
- explain binomial nomenclature and list and explain the nine groups used for botanical classification.
- explain why cultivar identification is important and describe commonly used methods for identifying cultivars today.
- provide the major and minor taxa for the "Delicious" apple as an example for botanical classification.
- discuss the different phases of the plant's life cycle, including juvenile or vegetative phase, reproductive or mature phase, and the senescence phase.
- discuss the basic vegetative parts of the plant, including leaves, stems, and roots.
- discuss the reproductive parts of the plant, including flowers and fruits.
- discuss seeds and seed germination.

Key Terms

alternate leaf arrangement
angiosperm
annual
anther
apical meristem
axillary bud
biennials
bract
bud scale
calyx
cambium
complete flower
compound leaf
cortex
cultivar
dead heading
deciduous perennial
deciduous plants
dicots
dicotyledonae
dry fruit
epidermis
epigeous seed germination
evergreen perennial
evergreen plants
fertilization
fibrous root system
filament
filicinae
fleshy fruit
fruit

(continues)

ABSTRACT

Many benefits are associated with plants; however, to classify them, you must have a better understanding of their anatomy. In this chapter you will learn how to classify plants and recognize their various parts and functions. Classifying plants is an important task that many years ago was accomplished using two main categories. Scientists today group plants by using similarities among them for classification into eight general groupings. Scientific classification and morphology will be defined in this chapter, and you will learn why common names and Latin polynomials were never accepted as the standard for classifying plants. The binomial nomenclature system developed by Linnaeus for classifying plants is also discussed and you will learn about the nine groups used for botanical classification. You will then explore why cultivar identification is important and discover the different ways cultivars are identified today. You will see an example of botanical classification using the "Delicious" apple, including both major and minor taxa. This is followed by a discussion of the different stages of the plant's life cycle and the principal vegetative and reproductive structures. The last portion of this chapter discusses seeds and seed germination.

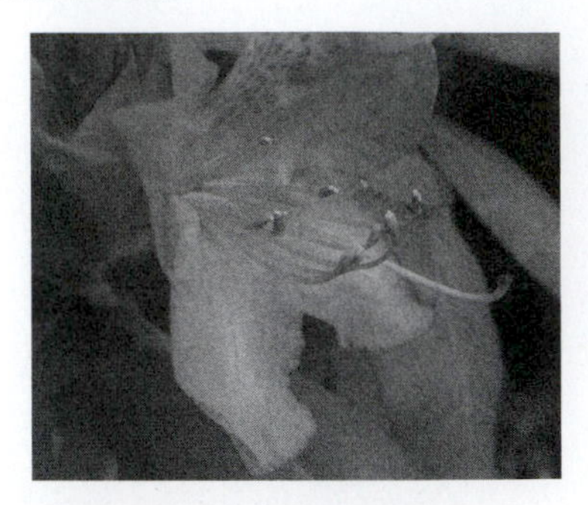

Figure 6–1 Plants are classified for identification and cataloguing, purposes as well as to overcome language barriers, and better understand plant origins and plant relationships.

Key Terms *(continued)*

germination
gymnosperm
hardy plants
herbaceous perennial
herbaceous plant
hypogeous seed germination
incomplete flower
imperfect flower
internode
juvenile or vegetative stage
leaf apex
leaf base
leaf blade
leaf covering
leaf margin
leaf scar
leaflet
lenticel
midrib
modified root
monocots
monocotyledonae
morphology
node
nonwoody plant
opposite leaf arrangement
ovary
perennials
perfect flower
petals
petiole
phloem
photosynthesis
pistil
pith
pollination
primary root
reproductive or mature phase
respiration
root cap
root hairs
scientific classification
scout
secondary root
seed
senescence
sepals
simple leaf
stamens
stigma
stomata
style
taproot system
taxon (taxa)
taxonomy
tender plants
tendril
terminal bud
terminal bud scale scar
transpiration
veins
vernalization
viable
whorled leaf arrangement
woody plant
xylem

INTRODUCTION

In this chapter, you will learn how to classify plants by better understanding their anatomy. Plants are classified for identification and cataloguing purposes as well as to overcome language barriers and better understand plant origins and plant relationships. Early systems of classification simply categorized plants as either harmful or useful. Today, plant classification has become much more complex. Scientists use similarities among plants to classify or break down plants into eight characteristics: kind of stem, size of plant, stem growth form, kind of fruit, life cycle, foliage retention, temperature tolerances, and number of cotyledons. Scientific classification uses the **morphology** of plants as a means of plant classification. In 1753, Linnaeus established a simple yet elegant scientific classification system called binomial nomenclature, which is still used today. Plants can be broken down into major and minor taxa. A **taxa** is a group name applied to organisms that make up a hierarchy within a formal system of classification; for example, kingdom is a taxa. The botanical classification system is commonly used for horticultural crops: the major taxa starts with kingdom followed by division or phylum, class, subclass, order, and family; the minor taxa starts with genus followed by species and **cultivar.**

The plant's life cycle can be broken down into three phases: the **juvenile or vegetative phase, reproductive or mature phase,** and **senescence,** which is ultimately followed by death. The juvenile phase is characterized by exponential increases in size and the inability to flower. The mature phase is characterized by changes that enable the plant or organ to express its full reproductive potential. The senescence phase involves deteriorative changes, which are natural causes of death.

To classify plants properly, you should be familiar with the plant's leaves, stems, and roots, which make up the primary vegetative structures of the plant. The main function of the leaf is to produce food for the entire plant through photosynthesis. Leaves have several basic parts that can help you identify plants: the **petiole, leaf blade, midrib, veins, leaf margin, leaf apex, leaf base, leaf covering,** and **stomata.** The leaf form, apexes, margins, and bases are most commonly used to identify plants. The pattern of leaf attachment to stems—including **opposite, alternate,** and **whorled arrangements**—is another way to classify plants. Modified leaves, which include **bracts** and **tendrils,** have also been used to identify plants.

Stems can also be used to classify plants. The basic parts of the stem are the **terminal bud, bud scale, terminal bud-scale scar, axillary bud** or **lateral bud, node, internode, leaf scar,** and **lenticel.** The internal anatomy of the stem is also used to classify plants by these key anatomical features: **cambium, xylem, phloem,** and **pith.** Modified stems, which include stolons, rhizomes, tubers, corms, and bulbs, are specialized stems used for storing food reserves and for reproducing.

The primary function of the root is to absorb water and nutrients to sustain plant life. The key parts of the root are the **primary root, secondary root, root hairs,** and **root cap.** The two different classes of roots used to classify plants are plants with a **tap root system** and plants with a **fibrous root system.** Some plants also have **modified roots,** which serve as a reserve food system; for example, a sweet potato is a modified root that is vastly different from an Irish potato, which is a modified stem.

The major function of the flower (Figure 6-1) is to attract pollinators and to produce fruit and seed. Flowers are commonly used to classify plants because they come in a variety of sizes, shapes, and colors. A typical flower contains the following parts: **sepals (calyx), petals, stamens (filament, anther),** and **pistils** (**stigma, style,** and **ovary**). The four classes of flowers—**complete, incomplete, perfect,** and **imperfect**—can be used as means of classifying plants. Another means of classification is through the use of **fruits** that can be **fleshy** or **dry.**

At the end of this chapter, you will learn that in order for germination to occur the seed must be **viable,** be exposed to the proper environmental conditions, and be able to overcome primary dormancy. The four stages of seedling germination include the imbibition of water, formation or activation of enzyme systems, breakdown of storage food reserves, and emergence of the radicle from the seed. At the conclusion of this chapter, you will learn the two common forms of seed germination: **epigeous** and **hypogeous.**

HOW PLANTS ARE CLASSIFIED AND NAMED

Before you begin to classify plants, it is important to understand why they are classified. Plants are classified for many reasons, the first being that identifying plants is important. Second, after a plant is identified, it can be catalogued for future reference. Third, the classification of plants overcomes language barriers because when plants are scientifically classified and named, the plant retains that same name all over the world (unlike common names that can be very different within the same country or even the same state). Fourth, by classifying plants, we can better understand their origins. Fifth, classifying plants helps uncover the relationship among plants, which is useful information when growing different types of plants together for the benefit of each.

The early systems used for classifying plants simply determined whether plants were harmful or poisonous and should not be eaten or whether they were useful and could be used for food, spices, medicinal purposes, construction, transportation, and communication (Janick, 1986).

Today, scientists use similarities among plants to break them down into the following eight groups.

Kind of Stem

Stems can be broken down into two basic categories: woody or nonwoody (herbaceous). The **woody plant** (shrubs, trees, and others) produces wood and has buds, which survive above ground level during the winter months. The **nonwoody** or **herbaceous** plant has soft, nonwoody stems. Nonwoody or herbaceous plants typically include turfgrasses, herbs, ground covers, and flowering plants.

Size of the Plant

The overall size of the plant at maturity is commonly used to classify and name plants. Plants may be dwarf, which exhibit shortened internodes, or they can be standard size with normal internode lengths.

Stem Growth Form

The stem growth form is based on how the stem stands in relation to the ground. For example, an erect stem is vertical with respect to gravity, whereas a creeping habit stem grows horizontal with respect to gravity. A plant that grows along the ground and then climbs (such as an ivy plant) when it comes in contact with an obstacle has a climbing habit (Figure 6-2).

Figure 6–2 Ivy plants growing on the side of the building have a climbing habit.

Kind of Fruit

The characteristic of the fruit is another means of classifying plants. For example, fruits are classified as fleshy if they are soft and succulent (such as a peach) or dry if they are hard and have a low water content (such as a walnut).

Life Cycle

As shown in Figure 6-3, plants can be designated as annuals, biennials, or perennials. **Annuals** are plants that germinate from seed, grow to full maturity, flower, and produce seeds in one growing season; in other words, these plants go from seed to seed in one season (for example, a marigold). To extend the flowering period for annuals, a process known as dead heading is commonly used. **Dead heading** refers to removing dead or dying flowers from annuals so the plant continues to live and bloom longer. **Biennials** are plants that complete their life cycle in two growing seasons, such as lettuce. Biennial plants require a vernalization period in order to flower. **Vernalization** refers to using cold treatment to induce flowering. **Perennials** are plants that may be herbaceous or woody and live for more than two growing seasons. There are three types of perennials.

- **Herbaceous perennials.** The tops die and roots live throughout the winter months.
- **Deciduous perennials.** The root systems and stems live throughout the winter months.

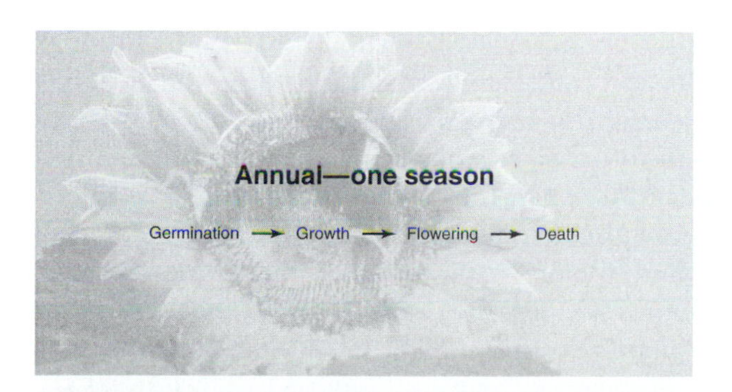

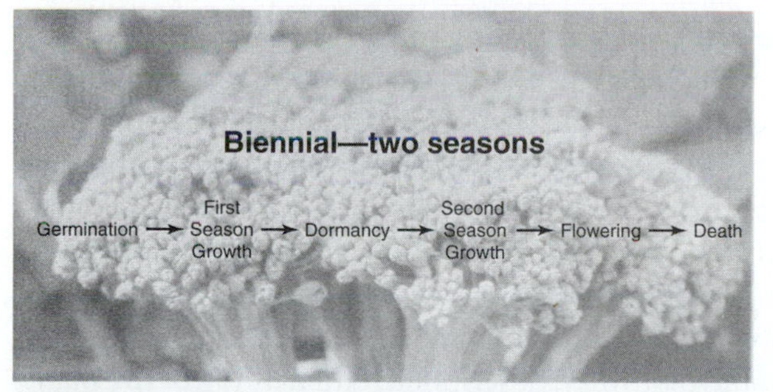

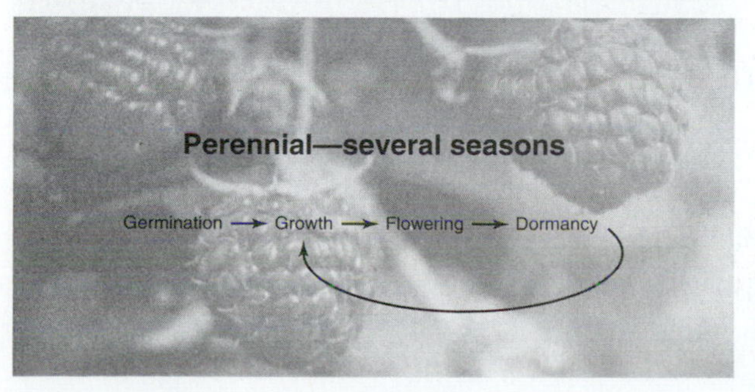

Figure 6–3 The life cycles of annuals (top panel), biennials (middle panel), and perennials (bottom panel)

- **Evergreen perennials.** The root system, stems, and leaves live throughout the winter months.

Foliage Retention

When (or if) a plant loses its leaves is also a characteristic commonly used to classify and name plants. **Deciduous plants** lose their leaves during a portion of the year, usually the winter months. **Evergreen plants** hold their leaves all year.

Temperature Tolerances

How plants handle temperature is also used to classify and name plants. **Tender plants** such as tomatoes and peppers cannot tolerate cool weather. **Hardy plants** such as broccoli and chrysanthemums are less sensitive to temperature extremes than tender plants.

Number of Cotyledons and Venation

The number of cotyledons (seed leaves) and type of leaf venation is commonly used to classify and name plants. **Dicot** plants are characterized by two cotyledons and have reticulate leaf venation. **Monocot** plants are characterized by one cotyledon and have parallel leaf venation.

SCIENTIFIC CLASSIFICATION

Scientific classification uses the morphology of plants (the plant's form and structure) as a means of classification and names plants based on the International Code of Botanical Nomenclature. Using common names for plants was inadequate as a means of classification because of the major problems that arise as a result of language and plant differences throughout the world. Differences in common names can also occur within in the same country and even within the same state. For example, a common name used in upper New York state can have a different meaning at another location such as New York City. Latin polynomials were also found inadequate because they were too long, cumbersome, and difficult to use.

In 1753, Linnaeus published the *Species Plantarum* (a two-volume work), which presented the binomial nomenclature system for plant classification that is commonly used today. An example of the binomial nomenclature system is *Brassica oleracea* L. 'Union'. In this example the first word is the genus (*Brassica*) followed by the species (*oleracea*) (L.), authority, in this case, Linneaus and cultivar (Union). When using the binomial nomenclature system, the genus and species are either underlined or italicized with the first letter in the genus capitalized and the species all lowercase. Many binomials also indicate an authority, which is the person involved in the discovery, description, and naming of the plant.

Botanical Nomenclature

To better understand botanical nomenclature, you should be familiar with the following terms:

- **Taxonomy.** The study of scientific classification and nomenclature.
- **Taxon (taxa).** A group name applied to organisms that make a hierarchy within a formal system of classification.

TABLE 6-1 CLASSIFICATION OF A DELICIOUS APPLE PLANT USING HORTUS THIRD

MAJOR TAXA	
Kingdom	Plant
Division (Phylum)	Tracheophyta
Class	Angiospermae
Subclass	Dicotyledonae
Order	Rosales
Family	Rosaceae (same family as rose or strawberry)
MINOR TAXA	
Genus	*Malus*
Species	*pumila (domestica)*
Cultivar (variety)	Delicious

Botanical Classification

The following is the botanical classification system (Table 6-1) commonly used for horticultural plants.

Major Taxa

1. Kingdom—Plants or animals
2. Division or Phylum
   Tracheophyta includes most horticultural plants. Other phylums include Spermatophyta, Pteridophyta, Thallophyta, and Bryophyta.
3. Class
   Filicinae: Ferns
   Gymnosperms: Primarily evergreen trees, usually have naked seeds born in cones.
   Angiosperms: Flowering plants and seeds develop in fruits.
4. Subclass
   Dicotyledonae (Dicot): Has two cotyledons and reticulate leaf venation, such as a bean plant.
   Monocotyledonae (Monocot): Has only one cotyledon and parallel leaf venation, such as corn.
5. Order
6. Family: Plants are grouped based on natural plant relationships using structural and cultural similarities.

Minor Taxa

7. Genus: Plants are grouped based on morphology and genetic, biochemical, and molecular relationships.
8. Species: Plants are grouped with even greater refinements than the genus but are still grouped based on morphology and genetic, biochemical, and molecular relationships.

9. Cultivar: Horticulturists are generally interested in the botanical variety, form, biotype, and clone; when any of these are intentionally cultivated, they are referred to as a cultivar—cultivated variety (for example, a "Delicious" apple). The term *cultivar* has replaced the older term *variety* to avoid confusion with the taxonomic term. In a taxonomic sense, a variety means a botanical variety.

Cultivars

When a grower puts a lot of time and effort into establishing a cultivar, the grower must establish ownership to prevent patent infringements. Cultivars can be distinguished by the following characteristics:

- **Flower.** Size, shape, and number of the different floral parts.
- **Fruit.** Flesh types, color, size, and shape.
- **Vegetative parts.** Buds, bark color, size, and hardiness.

In today's society, patents and ownership are very important; therefore, using more precise methods of identifying a cultivar are even more critical. Methods used to establish patents and ownership in addition to the preceding are as follows:

- Chemical taxonomy and biochemistry uses proteins, anthocyanins, and other specific chemicals as fingerprints to identify cultivars.
- Molecular evidence uses the Polymerase Chain Reaction (PCR) for DNA analysis.

These tools are necessary to protect patents and resolve infringement disputes. Patent holders may also employ a **scout** to buy plants at various outlets and run analyses of these plants to determine whether the plants are being used legally.

THE PLANT'S LIFE CYCLE

As stated previously, the plant's life cycle has three phases with varying characteristics.

Juvenile or Vegetative Phase

In the initial period of growth, the **apical meristem,** which is the primary growing point of the stem, will not typically respond to internal or external conditions to initiate flowers. The juvenile phase of development exhibits exponential increases in size and the inability to shift from vegetative to reproductive maturity leading to the formation of flowers. In addition, juvenile plants have specific morphological and physiological traits. Examples of morphological traits include the following:

- **Leaf shape.** *Hedra helix,* or ivy, in the immature stage has five lobes, whereas in the mature stage it has none. The ivy commonly seen around college campuses is mostly immature; it is actually fairly hard to find plants in the mature stage (Figure 6-4).
- **Thorniness.** Young growth in *Juniperus virginum* has thorns, whereas older growth does not.

Figure 6–4 Mature ivy is shown on the left and immature ivy is shown on the right.

Examples of physiological traits during the juvenile stage include vigorous plant growth, disease resistance, and a greater ability to regenerate roots and shoots (Arteca, 1996).

Reproductive or Mature Phase

This phase refers to qualitative changes that allow the plant or organ to express its full reproductive potential. Both genetic and environmental conditions determine when plants enter this phase.

Senescence

This phase involves internally controlled deteriorative changes, which are natural causes of death. Changes that occur during senescence include the following:

- Decreases in chlorophyll, protein, nucleic acids (RNA/DNA), and photosynthesis.
- Changes in plant hormones—some increase while some decrease.
- When the senescence phase is complete, abscission occurs.

THE PLANT'S VEGETATIVE STRUCTURES

The leaves, stems, and roots make up the primary vegetative structures of the plant. They take part in growth processes that are essential to the plant's survival.

Leaves

The main function of a leaf is to manufacture food for the plant through **photosynthesis.** Photosynthesis refers to a series of chemical reactions in which carbon dioxide and water are converted in the presence of light to carbohydrates

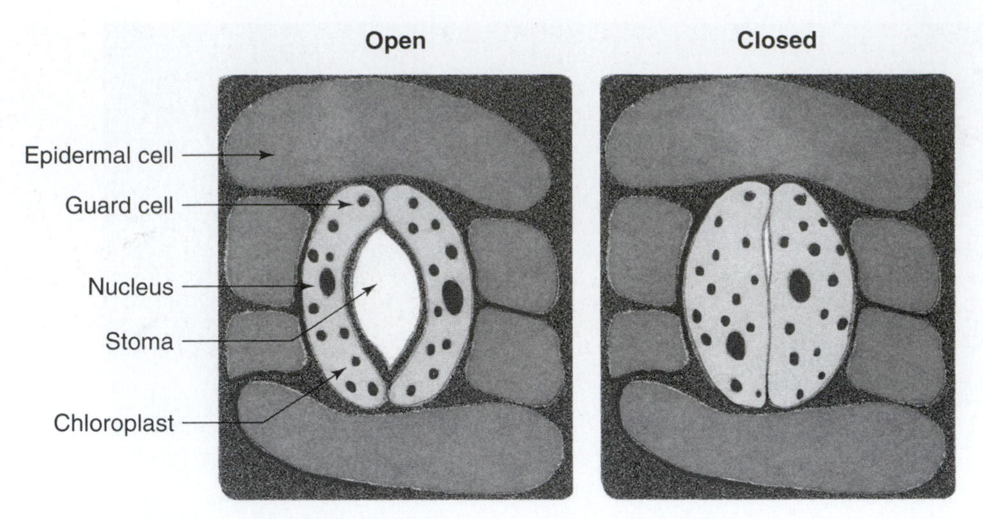

Figure 6–5 Open stomata with associated parts and closed stomata

(sugar) and oxygen. The simplified equation for photosynthesis is as follows:

$$6CO_2 + 6H_2O \rightarrow \rightarrow \rightarrow \rightarrow C_6H_{12}O_6 + 6O_2$$

Both light and chlorophyll are essential to photosynthesis. The two major photosynthetic enzymes found in plants are ribulose-bisphosphate carboxylase (Rubisco) and phosphoenolpyruvate carboxylase (PEP-Carboxylase). Carbon dioxide and other gases enter and exit the leaf through tiny pores in the leaf's surface called stomata (Figure 6-5). Another important function carried out by the leaf is **transpiration,** which is the loss of water from the leaf in the form of water vapor. **Respiration** is another important function carried out by the leaf; this process uses sugars made during photosynthesis and breaks them down into simpler molecules (such as H_2O and CO_2) that are used as energy for plant growth and development.

Parts of Leaves

Leaves consist of several basic parts that help identify them. The major parts of a simple dicot leaf are listed here (Figure 6-6):

- **Petiole.** The leaf stem or stalk that attaches the leaf to the stem.
- **Blade.** The flat thin part of the leaf.
- **Midrib.** The largest vein located in the middle of the leaf.

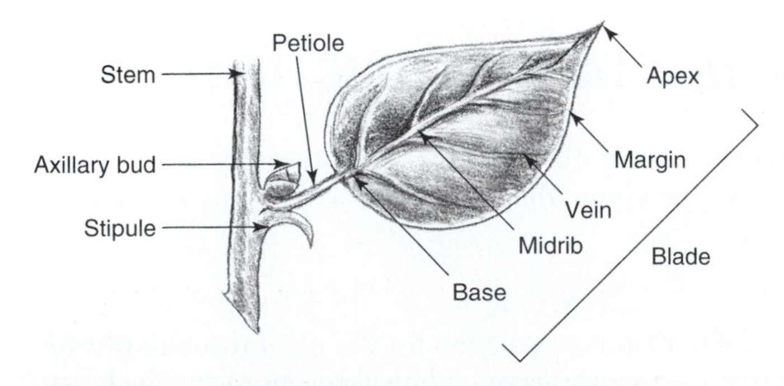

Figure 6–6 Representation of a simple dicot leaf

Figure 6–7 Venation patterns commonly found in plants: parallel, pinnate, and palmate. *Courtesy of Dr. Francis H. Witham, Department of Horticulture, The Pennsylvania State University.*

- **Veins.** Used to transport water and nutrients throughout the plant. The different leaf-venation patterns found in plants can be parallel, pinnate, or palmate (Figure 6-7).
- **Leaf margin.** The outer edge of the leaf blade, which can be lobed, smooth, toothed, or various combinations of the three.
- **Leaf apex.** The tip of the leaf blade, which can be pointed, rounded, or a variety of other shapes.
- **Leaf base.** The bottom of the leaf blade, which can be rounded, pointed, or a variety of other shapes.
- **Leaf covering.** Various leaf coverings include hairy versus not hairy, waxy versus not waxy, and others.
- **Stomata.** The tiny openings in the leaf blade in which gases enter and exit the leaf; they can be located on the leaf's top, bottom, or both.

Leaf forms, apexes, margins, and bases are commonly used to identify plants (Figure 6-8).

Two Basic Types of Leaves

The **simple leaf** consists of one blade per petiole, such as found on an oak leaf (Figure 6-9). The **compound leaf** has two or more leaflets, such as found on a potato leaf (Figure 6-10). The main difference between a leaf and a leaflet is the position of the axillary bud, which is located at the base of the entire leaf; **leaflets** do not have axillary buds.

Patterns of Leaf Attachment to Stems

Differences in how leaves are attached to stems are commonly used as a means of plant identification. Leaves are attached to stems in three major patterns, as shown in Figure 6-11.

- **Opposite.** The opposite pattern occurs when two leaves are directly across from each other.
- **Alternate.** The alternate pattern occurs when leaves are staggered along the length of the stem.
- **Whorled.** The whorled pattern occurs when three or more leaves are attached to the same portion of the stem.

Figure 6–8 Leaf forms, apexes, margins, and bases commonly used to identify plants

Figure 6–9 The oak leaf is an example of a simple leaf.

Figure 6–10 The potato leaf is an example of a compound leaf. *Courtesy of Dr. Francis H. Witham, Department of Horticulture, The Pennsylvania State University.*

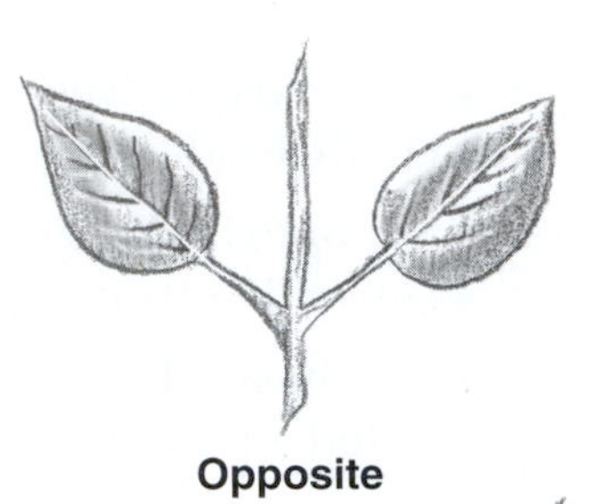

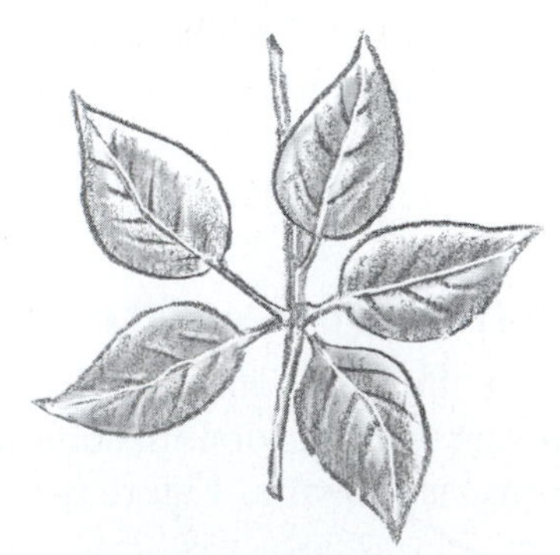

Figure 6–11 Opposite, alternate, and whorled patterns of leaf attachments to stems

Modified Leaves

Modified leaves are commonly mistaken for other plant structures such as flowers or stems. To illustrate this point, examples of modified leaves are provided as follows:

- **Bracts.** Leaves located just below the flower, for example, the poinsettia and dogwood (Figure 6-12).
- **Tendrils.** Appendages produced by certain vines that wrap around a support and allow them to climb, for example, the grape (Figure 6-13).

Figure 6–12 Examples of plants with modified leaves called bracts are the dogwood (top) and the poinsettia (bottom). *Courtesy of Dr. Francis H. Witham, Department of Horticulture, The Pennsylvania State University.*

Stems

Stems can be used to identify plants with some practice. The stem has several important functions:

- **Support.** Stems are used to support leaves, flowers, and fruits.
- **Transport.** Stems contain important transport systems, including the xylem for transporting water and minerals, and the phloem for transporting manufactured food.
- **Photosynthesis.** Stems can be used to manufacture food, but to a lesser extent than the leaves.
- **Storage organ.** Stems can act as a storage organ for food; an example of this is the Irish potato.

Basic Parts of the Stem

The basic parts of the stem are illustrated in Figure 6-14 and described here.

- **Terminal bud.** The bud is positioned at the tip of the stem and contains an undeveloped leaf, stem, flower, or mixture of all.
- **Bud scale.** The bud scale is a tiny leaf-like structure that covers the bud and protects it.
- **Terminal bud-scale scar.** The terminal bud-scale scar is left when the terminal bud begins growth in the spring; it represents one year's growth.

Figure 6–13 The grape tendril wraps around a support.

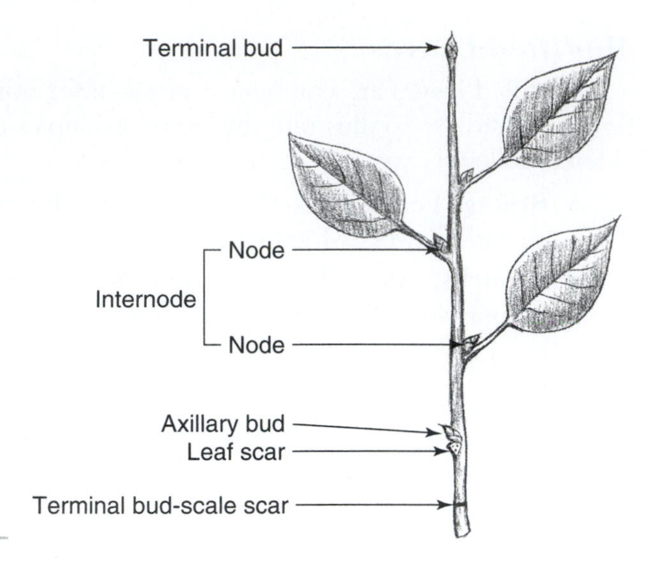

Figure 6–14 The basic parts of a typical stem

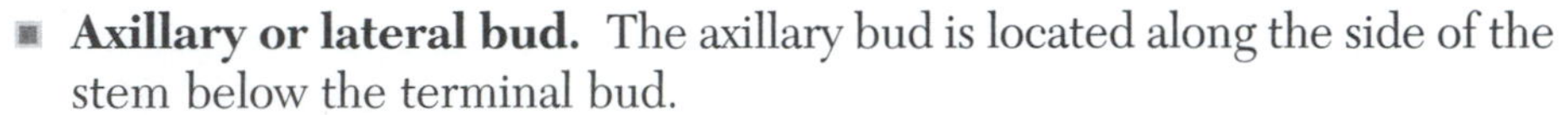

Figure 6–15 Cherry tree lenticels

- **Axillary or lateral bud.** The axillary bud is located along the side of the stem below the terminal bud.
- **Node.** The node is the point along the length of the stem where leaves or stems are attached.
- **Internode.** The internodal region is located between the nodes.
- **Leaf scar.** The leaf scar is left when the leaf drops.
- **Lenticel.** The lenticels are tiny pores located in the stem and are used for gas exchange; an example of a tree with conspicuous lenticels is the cherry tree (Figure 6-15) (Esau, 1965).

Internal Anatomy of Stems

Within the stem, the water, nutrients, and food made during photosynthesis are transported throughout the plant and stored for later use. The internal anatomies of a monocot (has one cotyledon and parallel leaf venation) and dicot stem (has two cotyledons and reticulate leaf venation) are shown in Figure 6-16. The following are key anatomical features found in stems:

- **Epidermis.** The outer layer of plant parts.
- **Cortex.** Primary tissue of the stem or root, which is located between the epidermis and the vascular region.
- **Cambium.** The cambium is an area where new plant cells are formed. When grafting, the cambium layers must match for the graft union to be successful.

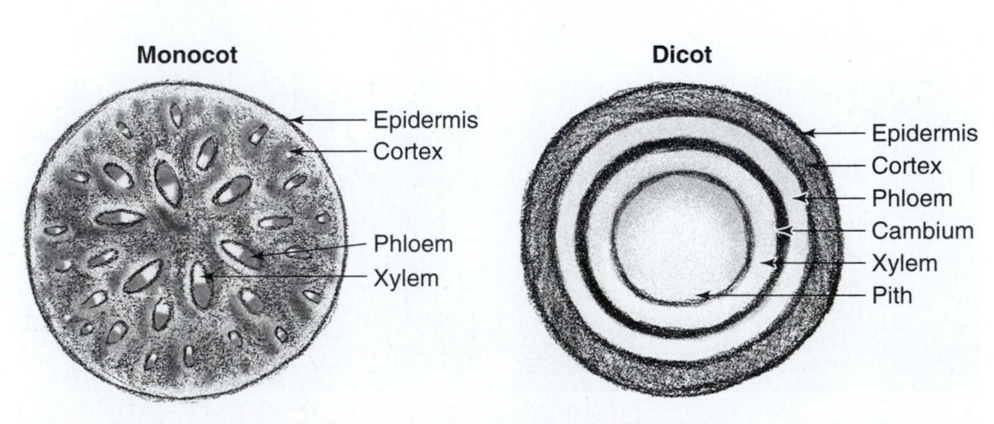

Figure 6–16 Internal anatomy of monocot and dicot stems

- **Xylem.** The xylem is composed of tiny tubes that transport water and nutrients up from the roots to other parts of the plant. The annual rings in a tree are made up of xylem.
- **Phloem.** The phloem is composed of tiny tubes that transport manufactured food and carbohydrates from the leaves down to other parts of the plant, such as the roots and shoots.
- **Pith.** The pith is located in the center portion of the stem where food and moisture are stored.

Modified Stems

In addition to standard stems, there are also modified stems used for storage of reserves and for reproduction. Examples of modified stems are shown in Figure 6-17 and listed here:

- Stolon—strawberry.
- Rhizomes—asparagus or iris.
- Tubers—Irish potato.
- Corms—gladiolus.
- Bulbs—onions, tulips.

Figure 6–17 Modified stems. *Courtesy of Dr. Francis H. Witham, Department of Horticulture, The Pennsylvania State University.*

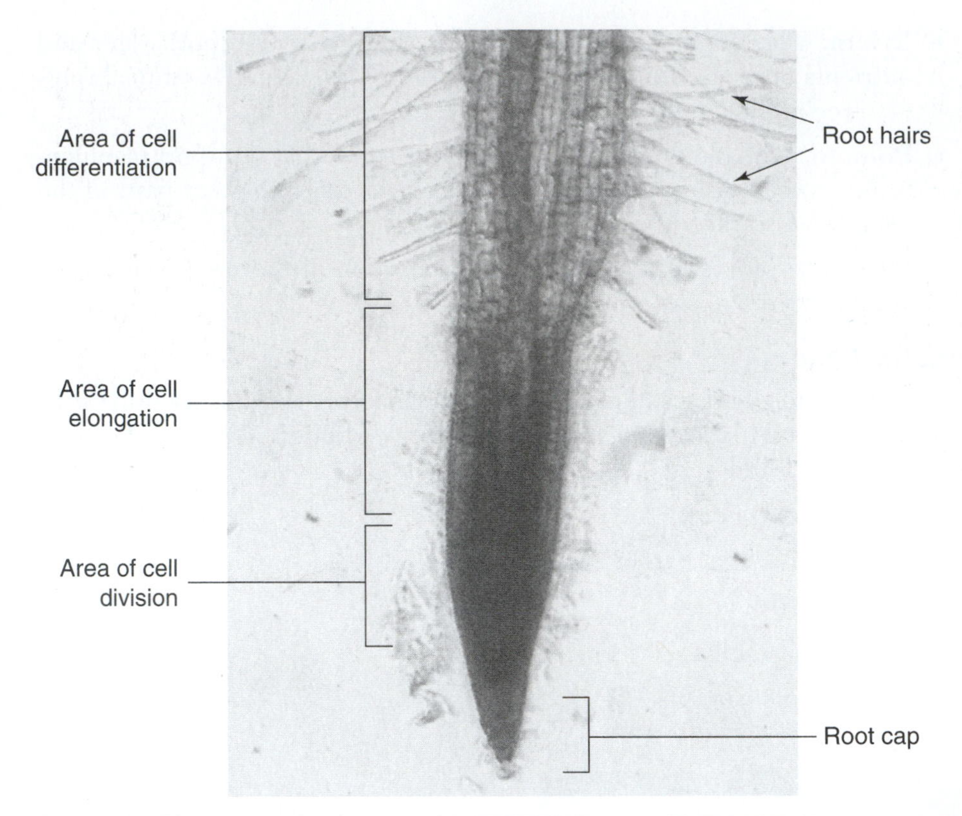

Figure 6–18 The key parts of a root

Roots

The major function of the root is to absorb water and nutrients to sustain plant life. In addition, roots act as storage organs for carbohydrates and provide anchor and support for the top portion of the plant. The first structure to emerge from a germinating seed is the root. The key parts of the root are shown in Figure 6-18 and described here:

- **Primary root.** The primary root is the main root that first emerges from the seed. Starting from the tip of the primary root, there is the root cap, just behind the root cap is the area of cell division, followed by the area of cell elongation, and then the area of cell differentiation.
- **Secondary root.** The secondary root arises from the primary root.
- **Root hairs.** Root hairs are single cells that absorb the greatest amount of water and minerals. Improper handling during transplanting can cause the loss of many root hairs, which decreases the plant's water uptake and results in transplant shock.
- **Root cap.** The root cap is located at the tip of the root and consists of several layers of cells that protect the root as it grows through the soil.

Different Classes of Root Systems

The two major classes of root systems are shown in Figure 6-19 and described here.

- **Taproot system.** The taproot system has a primary root that grows down from the stem with only a few secondary roots; both carrots and dandelions are examples.
- **Fibrous root system.** The fibrous root system has a large number of small primary and secondary roots; potato plants and grasses are examples.

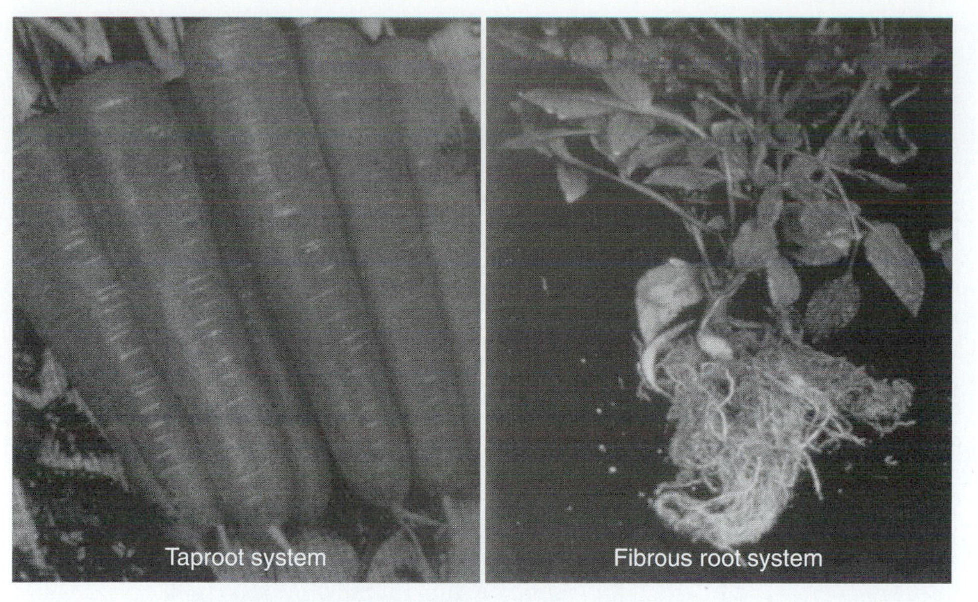

Figure 6–19 The two major classes of root systems. *Courtesy of Dr. Francis H. Witham, Department of Horticulture, The Pennsylvania State University.*

Figure 6–20 Sweet potatoes are modified roots.

Modified Roots

Modified roots serve as a reserve food-storage system; an example of a modified root is a sweet potato (Figure 6-20). Be careful not to be confused the modified-root sweet potato with the modified-stem Irish potato.

REPRODUCTIVE PARTS OF THE PLANT

The reproductive parts of the plant are the flowers, which are pollinated and fertilized to produce the fruits.

Flower

The main function of flowers is to attract pollinators and to produce fruit and seed. Flowers come in a variety of sizes, shapes, and colors to achieve their main function. In addition to their main function, flowers are commonly used for plant identification and produced commercially for their beauty and fragrance.

Parts of the Flower

A typical flower consists of four major parts, as shown in Figure 6-21 and described here:

- **Sepals.** The sepals are green, leaf-like structures located beneath the petals. The calyx is the term used to describe all the sepals on one flower. The calyx is used for protection. For example, some plants have calyx that contain spines, which deter animals from feeding on them.
- **Petals.** The petals are the brightly colored portions of the flower that are used to attract pollinators.
- **Stamens.** The stamens are the male reproductive part of the flower. The stamen consists of a filament that supports the anther, which produces the male sex cells.

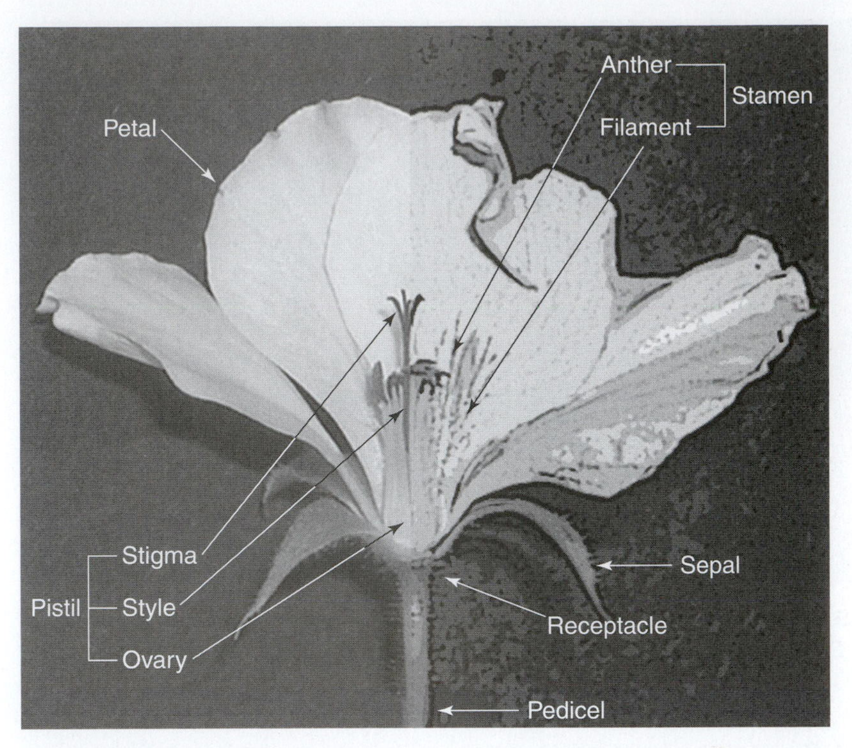

Figure 6–21 A typical flower showing associated flower parts

- **Pistils.** The pistils are the female reproductive part of the flower. The pistil consists of the stigma, which is the sticky surface for collecting pollen; the style, which is the tube that connects the stigma and ovary; and the ovary, which contains ovules or eggs.

Pollination is the transfer of pollen grains from the anther to the stigma. This should not be confused with **fertilization,** which occurs when the male sex cell fuses with the egg cell to form a new plant.

Four Classes of Flowers

Flowers can be broken down into four different classes.

- **Complete flower.** The complete flower contains all four major flower parts: sepals, petals, stamens, and pistils.
- **Incomplete flower.** The incomplete flower lacks one or more of the major flower parts.
- **Perfect flower.** The perfect flower contains both stamens and pistils.
- **Imperfect flower.** The imperfect flower lacks either stamens or pistils; an example of a plant containing this type of flower is corn.

Fruits

Fruits are formed after the flower has been pollinated and fertilized. The definition of a fruit is a mature ovary of a flowering plant. There are two types of fruits:

- **Fleshy fruit.** The fleshy fruit (such as a tomato) has soft fleshy material with or without seeds enclosed.
- **Dry fruit.** The dry fruit (such as a sunflower) has seeds enclosed in a hard fruit wall.

SEEDS AND SEED GERMINATION

A **seed** is a mature fertilized egg that is contained in the fruit. Seeds can be dispersed by the wind (dandelions) (Figure 6-22), stuck to the fur of animals, or spread in other ways. Dicots store their reserve food in cotyledons, whereas monocots store their food in the endosperm.

Seed Germination

The germination process is a series of events whereby the seed embryo goes from a dormant state to an actively growing state. For seed germination to occur, the following criteria must be met:

- The seed must be viable, which means that the embryo is alive and capable of germination.
- The seed must be exposed to appropriate environmental conditions.
- Primary dormancy must be overcome in the seed.

Three stages of seedling germination are as follows:

- **Imbibition of water.** This is the active uptake of water by the seed.
- **Formation or activation of enzyme systems.** After the seed is hydrated, preexisting enzymes are immediately available for breaking down storage reserves, and new enzymes are produced for the breakdown of additional reserves.
- **Breakdown of storage products for use during germination.** Preexisting and newly formed enzymes break down food reserves, which are used for the germination process.

Figure 6–22 Dandelion seeds being dispersed by the wind

The first visible sign of seed germination and growth is the emergence of the radicle followed by growth of the seedling.

The following are two common forms of seed germination.

- **Epigeous seed germination.** The hypocotyl elongates and brings the cotyledons above ground (for example, cherry) (Figure 6-23a).
- **Hypogeous seed germination.** The epicotyl emerges and the cotyledons remain below the soil surface (for example, corn) (Figure 6-23b).

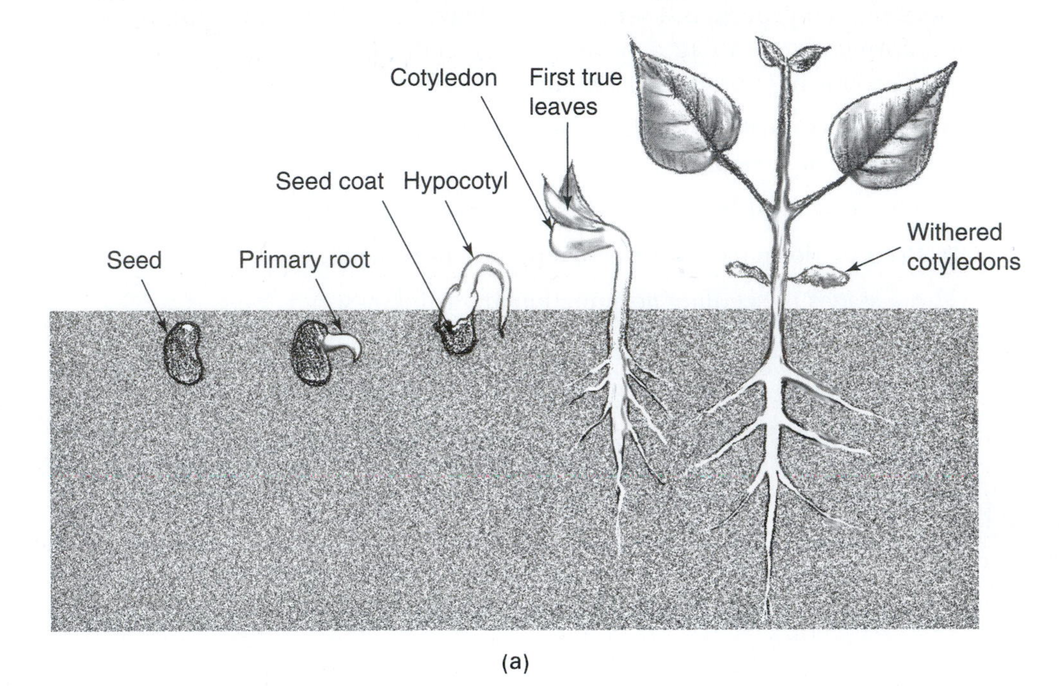

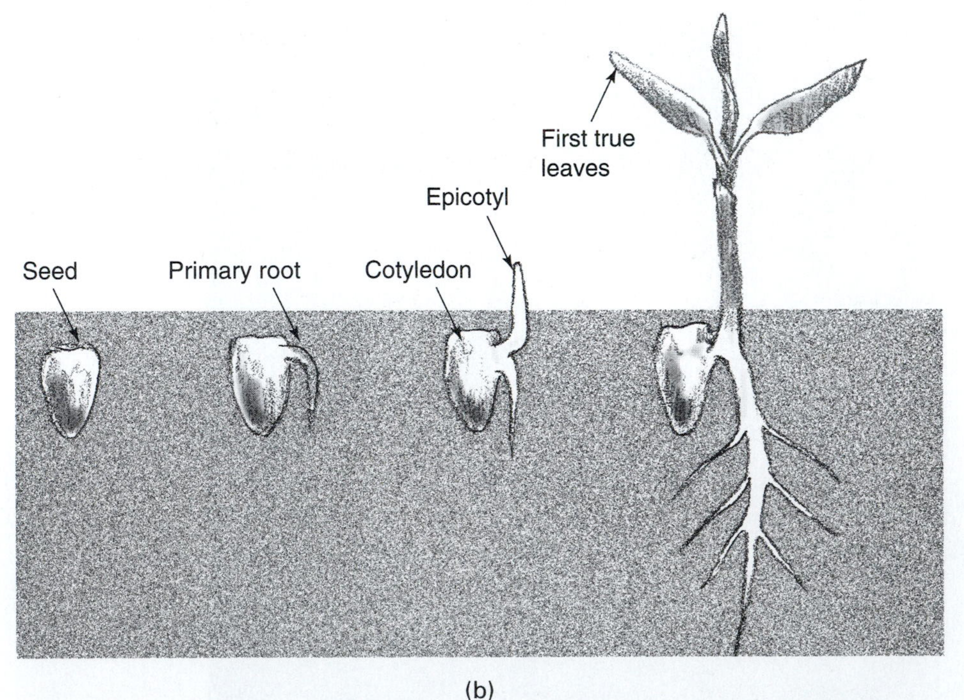

Figure 6–23 Epigeneous germination (top panel) and hypogeous germination (bottom panel)

SUMMARY

The classification of plants has come a long way from the times of our ancient ancestors, who only had two main categories for classifying plants, namely as harmful or useful. Today, scientists group plants in eight ways based on plant similarities for their classification. That botanical classification breaks plants down into nine distinct categories.

Cultivar identification is important, especially when communicating with others. Knowledge of basic plant anatomy, including vegetative (leaves, stems, and roots) and reproductive (flowers and fruits) plant parts, together with understanding the plant's life cycle is important in plant classification as well as in maximizing the plant's potential uses.

Review Questions for Chapter 6

Short Answer

1. What are five reasons for classifying plants?
2. What were two categories used for the classification of plants in early systems?
3. What are eight ways scientists group plants for classification?
4. Why were Latin polynomials and common names found to be inadequate?
5. What group is responsible for the naming of plants today?
6. Give an example of the binomial nomenclature system.
7. List the nine categories used for botanical classification and provide an explanation for each.
8. Why is cultivar identification important, and what are some ways commonly used to distinguish between cultivars?
9. Provide the major and minor taxa of the "Delicious" apple.
10. Name and explain the three major phases of a plant's life cycle.
11. What is the main function of a leaf?
12. What are the two major photosynthetic enzymes found in plants?
13. List the nine major parts of a simple leaf.
14. Distinguish between a leaf and a leaflet.
15. Give two examples of a modified leaf.
16. What are the four major functions of a stem?
17. List the eight parts of the stem.
18. What are four examples of a modified stem?
19. What are the three major functions of the root?
20. Draw a root system and designate the area of differentiation, area of cell elongation, and area of cell division.
21. What are two types of root systems?
22. Provide an example of a modified root.
23. What are the three major functions of flowers?

24. Provide the four major parts of a typical flower and, where applicable, give additional parts associated with each.
25. What are two types of fruits?
26. Where is food stored in monocot and dicot seeds?
27. Provide three factors required for seed germination to occur. After these requirements are met, what are the three phases of seed germination?

Define

Define the following terms:

woody plant	dicots	stomata	petals
herbaceous plant	taxonomy	respiration	stamens
annual	taxon (taxa)	cambium	pistil
biennials	gymnosperm	xylem	pollination
vernalization	angiosperm	phloem	fertilization
perennials	apical meristem	pith	fruit
deciduous plants	senescence	root hairs	seed
evergreen plants	photosynthesis	sepals	epigeous seed germination
monocots	transpiration	calyx	hypogeous seed germination

True or False

1. Removing dead or dying flowers from annuals so the plant will continue to bloom for a longer period of time is known as dead heading.
2. Vernalization is a cold treatment used to make certain plants flower.
3. Angiosperms usually have naked seeds born in cones.
4. Latin polynomial nomenclature is commonly used to classify plants.
5. Gymnosperms are a class of flowering plants that develop their seeds in fruits.
6. CO_2 enters the leaf through tiny pores called stomata.
7. One of the major photosynthetic enzymes in plants is ribulose-diphosphatase.
8. Fertilization is the transfer of pollen grains from the anther to the stigma.
9. Pollination occurs when the male sex cell fuses with the egg cell to form a new plant.
10. A perfect flower contains both stamens and pistils.
11. An imperfect flower lacks either sepals and petals.
12. Dicots store their food in a specialized group of tissues called the endosperm.
13. Epigeous germination occurs when the epicotyl emerges and cotyledons remain below the soil surface.
14. Hypogeous germination occurs when the epicotyl emerges and cotyledons remain below the soil surface.
15. Monocots store their food in the cotyledons.

Multiple Choice

1. Why do we classify plants?
   A. Overcome language barriers
   B. Better understand the relationship between plants
   C. Catalog plants
   D. All of the above
2. Most horticultural crops are in the division (phylum)
   A. Bryophyta.
   B. Pteridophyta.
   C. Thallophyta.
   D. None of the above.
3. The annual rings of a tree are made up of
   A. Xylem.
   B. Phloem.
   C. Pith.
   D. Cambium.
4. Which of the following is the major function of the pith?
   A. Stores food and moisture in the center portion of the stem
   B. Transports water and nutrients
   C. Moves manufactured food and carbohydrates
   D. All of the above
5. Which of the following is the major function of the xylem?
   A. Stores food and moisture in the center portion of the stem
   B. Transports water and nutrients
   C. Moves manufactured food and carbohydrates
   D. All of the above
6. Which of the following is the major function of the phloem?
   A. Stores food and moisture in the center portion of the stem
   B. Transports water and nutrients
   C. Moves manufactured food and carbohydrates
   D. All of the above

Fill in the Blanks

1. The other term for a nonwoody plant is ______________.
2. Horticulturists are generally interested in the botanical variety, form, biotype, and clone. When any of these are intentionally cultivated, they are referred to as a ______________.
3. Linnaeus established a clear and concise method for classifying plants called ______________ ______________.

4. Both ____________ and ____________ conditions determine when plants flower.
5. The ____________ is an area where new cells are formed and is the layer between the xylem and phloem.

Matching Questions

Match the terms with the appropriate label on the diagram (some terms may be used twice).

Terms: Phloem (A), xylem (B), cambium (C), pith (D), monocot (E), dicot (F)

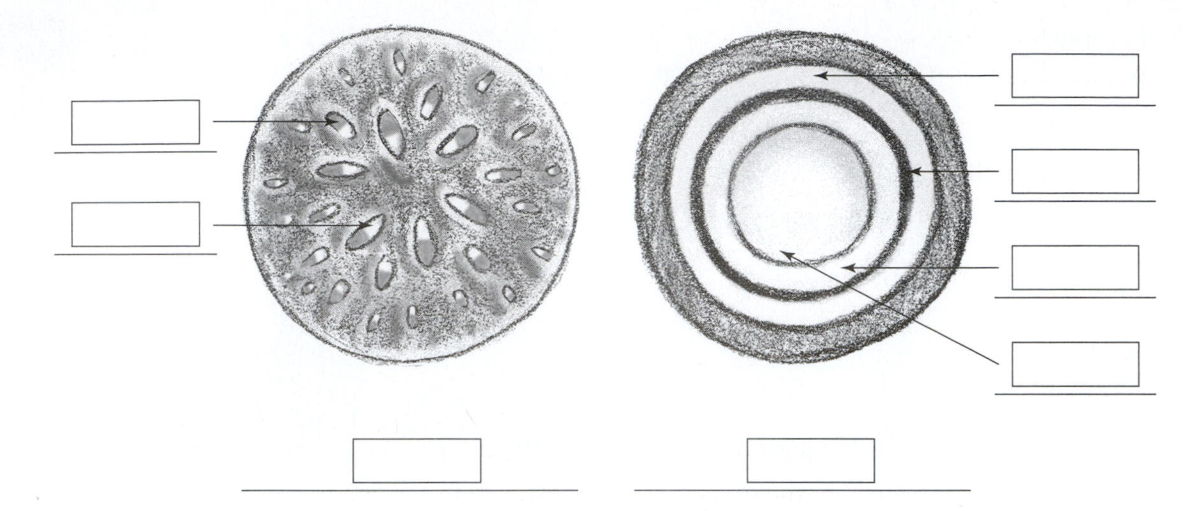

Activities

You now understand basic plant anatomy and how plants are classified. This knowledge enables you to explore additional ways in which plants are classified. In this activity, you will evaluate other ways plants are classified. Your activity is to search the Internet for different sites that present information on the botanical classification of plants. In addition to the Internet, check with your local library for potential reference books that have information on plant classification. From the Internet and/or the library, find the answers to the following questions:

What other forms of botanical classification do these references contain and how similar are they to the ones presented in this chapter?

Which form of botanical classification in your opinion is the easiest to follow?

What is the Web site address or the textbook where you found the information contained in this activity?

References

Arteca, R. N. (1996). *Plant growth substances: Principles and applications.* New York: Chapman & Hall.

Esau, K. (1965). *Plant anatomy.* New York: John Wiley and Sons.

Janick, J. (1986). *Horticultural science* (4th ed.). San Francisco, CA: W.H. Freeman.

CHAPTER

7

Plant Propagation

Objectives

After reading this chapter, you should be able to

- provide background on the factors involved in the production of seeds and the seed-certification process.
- clarify what factors should be considered when planting seeds outdoors (direct seeding) and indoors (indirect seeding).
- list and explain the five categories of primary seed dormancy.
- provide background on the eight types of asexual propagation and factors that should be considered when using each.

Key Terms

air layering
anther culture
apomixis
approach grafting
asexual propagation
bark slips
breeder seed
budding
callus
callus culture
cell suspension culture
certified seed
chemical dormancy
clone
conifer cutting
cuttings
damping-off
deciduous hardwood cutting
detached scion grafting
direct seeding
division
embryo culture
explants
foundation seed
girdling
grafting
hardened-off
herbaceous cutting
heterozygous
homozygous
indirect seeding
layering
leaf-bud cutting
leaf cutting
mechanical dormancy
meristem
meristem culture
morphological dormancy
mound layering
percent germination
physical dormancy
physiological dormancy
plant propagation
registered seed
repair grafting
root cutting
scarification
scion
semihardwood cutting
separation
serpentine layering
sexual propagation
simple layering
stem cutting
stratification
suspension culture
tissue culture
transgenic plant
trench layering
true to type
understock

ABSTRACT

Plant propagation is an exciting field. The two major methods of plant propagation are sexual and asexual propagation. For sexual propagation, important factors must be considered for the production of seeds and the seed-certification process. Both direct and indirect seeding methods may be used. The five major categories of primary seed dormancy are physical, mechanical, chemical, morphological, and physiological; each category of dormancy has important factors to consider. The eight commonly used types of asexual propagation are apomixis, cuttings, grafting, budding, layering, separation, division, and tissue culture.

Figure 7–1 Rooted cuttings and seedlings growing in cell packs

INTRODUCTION

Plant propagation is the reproduction of new plants from seeds and vegetative parts of the plant, such as leaves, stems, and roots (Figure 7-1). The two major methods of plant propagation are sexual and asexual propagation. **Sexual propagation** is the reproduction of plants via seeds. When seeds cannot be used as a means of reproduction, asexual propagation is used as an alternative method. **Asexual propagation** is the reproduction of plants using plant parts.

Sexual propagation is the major means of reproduction in nature and agriculture because it is easy and relatively inexpensive. In a variety of cases, seeds are **homozygous** and produce seeds that are **true to type,** in other words, the same as the parent plant. However, in many more cases, plants are **heterozygous** and are not true to type, so the only way to maintain desirable traits in their progeny is through asexual propagation. By using asexual propagation, the plants produced are always genetically identical **(clones)** to the parent plant.

A variety of factors must be considered for the production and certification of seed. To produce high-quality seed, it is important to choose the proper location, typically an arid climate is best because there are fewer disease problems. After a suitable location is selected, you must know when seeds are ripe to maximize germination rates and produce high-quality seedlings. After the seeds are harvested, they should be sieved to remove debris, cleaned, and properly stored to maintain their high quality.

Plant breeders select plants with desirable characteristics. Prior to commercial distribution of a new cultivar that has been produced by a breeder, four classes of seed production must occur: (1) **breeder seed,** (2) **foundation seed,** (3) **registered seed,** and (4) **certified seed.** After certified seed has been produced, many federal and state laws regulate its packaging and distribution. The law requires that the following information be present on the seed packet: trueness of name, germination percentage, pure seed percentage, percentage of other crop seed and/or weed seed, and percentage of other ingredients.

Seeds can be planted by **direct** or **indirect seeding.** Direct seeding involves planting seeds directly in the soil outdoors. For direct seeding to be successful, a proper site must be selected. After the site is selected, you must properly prepare the seedbed, know the optimum date for planting, use the correct planting depth and spacing, and make sure seeds receive optimal moisture during the germination process. Indirect seeding involves planting seeds indoors and growing seedlings prior to transplanting to a larger container or to a location outdoors. High-quality transplants start with high-quality seeds. After high-quality seeds have been procured, they need a germination medium that has proper drainage and aeration, adequate water and nutrient holding capacity, favorable pH, proper essential elements, and freedom from insects, weeds, and disease organisms. After a high-quality germination medium has been selected, the proper environmental conditions must be provided to ensure good germination and subsequent seedling growth. Prior to transplanting to the field, seedlings should be **hardened-off** to ensure success under the harsh outdoor environment.

For seed germination to occur, the seed must be viable, must be subjected to the proper environmental conditions, and must overcome primary dormancy. The five categories of primary seed dormancy are (1) **physical dormancy,**

(2) **mechanical dormancy,** (3) **chemical dormancy,** (4) **morphological dormancy,** and (5) **physiological dormancy.**

When seeds cannot be used as a means of reproducing plants, asexual propagation is used because it produces a plant that is genetically identical to the parent plant. A variety of types of asexual propagation can be used:

- **Apomixis** is a form of asexual propagation in which seeds are produced without fertilization.
- **Cuttings,** the most common type of asexual propagation, refers to using a detached vegetative part to produce a new plant. The four types of cuttings are (1) **stem,** (2) **leaf,** (3) **leaf-bud,** and (4) **root cuttings.** To ensure optimal rooting of cuttings, start with a high-quality parent plant.
- **Grafting** is the process of connecting two plants or plant parts so they will unite and continue to grow as one plant. For grafting to be successful, the scion and understock must be compatible, and the cambium layers must be in close contact between the scion and understock. Some commonly used grafting methods include the **detached scion** method, **approach grafting,** and **repair grafting.**
- **Budding** is similar to grafting except the scion has been reduced to a single bud with a small portion of bark or wood attached. The three main types of budding are chip budding, T-budding, and patch budding.
- **Layering** is a simple method of asexual propagation in which roots are formed on the stem while still attached to the parent plant. Layering occurs naturally in some species; a good example is the strawberry plant. Five types of layering commonly used in horticulture are (1) **simple layering,** (2) **serpentine layering,** (3) **trench layering,** (4) **mound layering,** and (5) **air layering.**
- **Separation** occurs when a natural structure is removed from the parent plant and grows on its own.
- **Division** relies on cutting plant parts into sections as a means of propagation.
- **Tissue culture** refers to propagating plants from single cells, tissues, or pieces of plant material. Some commonly used methods of tissue culture include **callus culture, suspension culture, embryo culture, meristem,** and **anther culture.**

PLANT PROPAGATION METHODS

The two major methods of plant propagation—sexual and asexual propagation—both are used for duplicating the mother plant with the final goal of producing plants that are exactly the same as the parent plant. Flowering plants produce seeds, some of which can be easily used for propagation purposes and others that are difficult to use. In addition, certain plants do not produce seeds at all. When using seeds for propagation is not a viable way to reproduce plants, asexual propagation is used as an alternative. Asexual propagation involves the use of plant parts for the reproduction of plants. This chapter starts by discussing sexual propagation followed by asexual propagation.

Sexual Propagation

Sexual propagation, as stated previously, is the reproduction of plants by using seeds. Seeds are the major means of reproduction in nature and also in agriculture because they are easy to use and are relatively inexpensive. The genetics of the plant, also known as the genotype, and the mating system used to produce seeds determine whether seeds can be successfully used for propagation of crop plants that are similar to the parents. In homozygous parent plants, similar genes of a Mendelian pair are present, which is the case for dwarf pea plants that contain only the genes for dwarfness. If these plants reproduce by seeds, the seed producer can use the self-fertilization mating system and these seeds will be true to type, that is, identical to the parent plant. However, plants produced by seed are not always genetically identical to the parent plant. When plants produced by seed are genetically different from the parent, they are heterozygous. Heterozygous plants have different genes of a Mendelian pair present in the same organism; such is the case with tall pea plants that contain genes for both tallness and dwarfness, and these plants are not true to type. Many horticultural crops, such as fruits and ornamentals, are heterozygous. The only way to maintain desirable traits in the progeny of these plants is to use asexual propagation because the products are always genetically identical (clones) to the parent plant.

Seed Production and the Certification Process

Those who produce seed range from the home gardener to the commercial seed producer. No matter who is producing the seed, however, the following factors must be considered:

- **Location.** Commercial seed producers are typically located in areas with an arid climate because there are fewer plant disease problems. However, if the individual producing the seed uses good cultural methods, seeds can be produced in a variety of locations.
- **Harvesting and collecting seeds.** One of the most important parts of harvesting seeds is knowing when they are ripe. Immature seeds will never germinate, and if they are harvested too late, the germination rate will be poor. Only through careful observation and accurate record keeping can a grower be confident of when to harvest seeds. Seeds can be harvested either mechanically or by hand. Mechanical methods are faster; however, they may damage seeds or if seeds mature at different intervals, immature seed may be harvested together with the ripe seed. Hand harvesting is slower and more costly but has the advantages of selecting only the ripe seed, causing generally less damage to the seed and collecting less debris during harvesting operations. The grower must weigh the pros and cons of each method and decide what works best in each individual case.
- **Cleaning seeds.** Seeds are typically sieved to remove debris and cleaned. Most ornamental seeds are not treated if the seeds are properly produced. However, in some cases, seeds are coated with a fungicide to protect them from soilborne diseases. Captan is a nonsystemic fungicide that has been used for many years as a seed treatment to protect against soilborne diseases.
- **Storing seeds.** Seed storage is dependent upon the type of seeds. Most seeds can be stored in a cool, dry place. Examples of these are annual

flowers and vegetables, whereas other seeds, such as apple and oak, must be stored in a cool, moist place. It is important to know the best conditions for storing a specific type of seed.

Plants with highly desirable characteristics, such as high yield, disease resistance, and a variety of other traits, are produced for commercial use. Plant breeders use old cultivars and incorporate new genes to improve their characteristics, thereby making a new cultivar. After the plant breeder has extensively tested the new cultivar, the seeds are released to special producers for propagation. Prior to commercial distribution of seed, four classes of seed production occur. First is the production of breeder seed, which is a small amount of seed with desirable traits produced by the breeder and used for the production of foundation seed. Foundation seed is produced under the supervision of agricultural research stations to assure genetic purity and identity. Foundation seed is given to a certified grower for the production of registered seed. The progeny of registered seed is called certified seed, which is sold commercially. Certifying agencies in the region where the seeds are grown regulate the production of certified seed; this ensures that the seeds produced meet the standards set for the crop. Many federal and state laws regulate the shipment and sale of seeds. These laws require that certified seed producers analyze seed and that the seed packet or label (Figure 7-2) contains the following:

- trueness of name, requiring that the name of the plant on the packet be accurate.
- origin, designating where the certified seed was grown.
- germination percentage, indicating the number of seeds that will germinate on the date given on the seed packet or label.
- pure seed percentage, the actual percentage of the seed designated on the label.

Figure 7–2 A typical seed packet contains much information, most of it required by law.

- percentage of other crop seed and/or weed seed.
- percentage of other ingredients, such as inert materials.

Seeds should be purchased from a reputable grower to ensure the highest quality seeds are obtained. High-quality seeds minimize the presence of other crop seed, weed seed, and inert materials mixed in with the desired crop seeds. In addition to purchasing high-quality seeds, you must select seed cultivars that will readily grow in your particular location.

Determining Percentage Seed Germination

After seeds are stored for any length of time, the germination rate should be tested because the germination rate decreases with time. **Percent germination** is the percentage of seeds that will sprout and grow (90 seeds germinate out of 100 seeds tested = 90% germination). If you need 100 plants, you must sow more than 100 seeds. Simply sowing 100 seeds and determining the number of seeds that germinate can easily test the percentage seed germination.

Determining Percentage Living Seeds

A fast and easy way to determine whether seeds are alive is to use the tetrazolium test that utilizes 2,3,5-triphenyltetrazolium chloride. Seeds are soaked in water and imbibed for a brief period of time after which they are cut longitudinally into two sections. One set of halves is placed on a petri plate containing 0.1 percent tetrazolium chloride, while the other set is left untreated. Living seeds will pick up the stain and dead seeds will not. Staining only indicates that the seeds are respiring and does not always translate into germination percentage, however; therefore it should only be used as an indicator.

Planting Seed Outdoors

Direct seeding refers to planting seeds directly into the soil outdoors. Some examples of crops that can be seeded directly are *Zea mays* (corn), *Pisum sativum* (pea), *Lactuca sativa* (lettuce), and *Phaseolus vulgaris* (bean). Direct seeding is very difficult with very small seeds; however, with modern technology, a variety of small bedding plant seeds can be direct seeded. The process simply involves coating the seed with clay to make it larger and easier to handle. Direct seeding is easy and readily adaptable to mechanization. Seeds are easy to transport, handle, and store, unlike transplants, which require much more care during shipping, handling, and storage. In addition, mechanized seed planters are more available than mechanized systems for transplants.

However, there are disadvantages to direct seeding as well, one of which is with species with very small seeds. Very small seeds are much finer than the tilth of the soil and are difficult to plant unless special modifications of the seed are made as described previously. Another disadvantage is that even with a 100-percent germination rate, a 100-percent stand is not always achieved due to a variety of factors. The success of plants grown from the direct seeding method is determined by the following factors:

- **Site selection.** For seeds to germinate properly and the resulting seedlings to grow, the site must be disease free and have adequate environmental conditions such as sunlight, water, temperature, drainage, and aeration. When these conditions are optimal, seeds will germinate and subsequent plant growth will be vigorous.

- **Seedbed preparation.** Remove any large objects and amend the soil with peat moss or sand to create the proper soil texture for optimal moisture retention and aeration.
- **Planting date.** The germination temperature required by the seeds determines the planting date. Soil temperatures required by different crops vary; warm season crops, such as squash or bean, do better at warmer temperatures (between 15 and 25°C), and cool season crops, such as cabbage or cauliflower, do better at cooler temperatures (less than 10°C). Therefore, if seeds are sown too early, they will fail to germinate; for example, if tomato seeds are sown under cool, moist conditions, the seeds will rot and no germination will take place. The planting date is also important for targeting specific harvesting dates to maximize profits.
- **Planting depth and spacing.** Use the recommended depth and spacing from the seed company or your local extension agent to maximize germination and ensure the survival of seedlings. A commonly used rule of thumb for planting depth, if there is no recommendation, is to plant the seed at a depth approximately three to four times the width of the seed. The actual spacing of the seeds is determined by the size of the plant at maturity.
- **Moisture.** Seeds must have sufficient moisture for seed germination to begin. After the germination process has started, the soil must remain moist to permit the germination process to occur normally and to allow for subsequent seedling growth. The soil should not be kept wet, however, because excessive moisture encourages rotting and disease.

Planting Seed Indoors for Transplants

Indirect seeding refers to planting seeds indoors and growing seedlings prior to transplanting them to a larger container or to a permanent location outdoors. Some examples of plants that are commonly indirectly seeded are *Coleus blumei* (coleus), *Pelargonium x hortorum* (geranium), *Zinnia elegans* (zinnia), *Lycopersicon esculentum* (tomato), *Capsicum annuum* (pepper), *Cucumis sativus* (cucumber), and *Allium cepa* (onion) (Figure 7-3).

Some of the advantages of indirect seeding are good establishment and less time to maturity, which reduces the actual field-growing period. Good establishment means that 100 percent of the seedlings become established because only vigorously growing seedlings are planted, unlike direct seeding, in which some seeds do not germinate, resulting in a reduced stand. Transplants put into the field hastens time to maturity by avoiding the adverse conditions that seedlings typically face early in the growing season.

A number of disadvantages are also associated with indirect seeding. The main disadvantage is the higher cost because seedlings must be produced indoors where artificial sources of heat, light, and space are required. Transplants are not as easy as seeds to plant mechanically. In fact, mechanical planters have only been developed for some crops because vegetative planting is more delicate than seed planting. Transplants are more difficult to handle than seeds because the vegetative part of the plant is bulky, and the containers and soil in which they are grown are more cumbersome when transporting. Another problem with indirect seeding is that seedlings must be planted before they get too old and become pot-bound. Immediate postplanting care is also required to avoid transplanting shock. Transplanting shock is a major problem with indirect

Figure 7–3 These transplants in cell packs are the result of indirect seeding.

seeding when seedlings are not properly prepared for transplanting and improper planting techniques are used. Transplanting should not be done during the heat of the day, and newly transplanted seedlings should be watered immediately after planting. These plants must be kept well watered until they have a chance to become established.

Preparing Seedlings for Transplants

Figure 7–4 Different types of plastic cell packs

Starting with high-quality seeds that have a high germination rate is important for producing a high-quality transplant. After high-quality seeds are obtained, choosing the appropriate container to suit specific needs is the next task. Typically, plastic cavity seeding trays that have individual cells varying in size and number are used for the production of transplants (Figure 7-4). Each individual cell has a drainage hole in the bottom. The germination medium is extremely important to ensure seedling success. The best medium to optimize germination provides

- proper aeration and drainage.
- adequate water- and nutrient-holding capacity.
- favorable pH.
- essential nutrients.
- material free of insects, weeds, and disease organisms.

In addition to having a good germination medium, optimal environmental conditions, such as temperature, moisture, and light, are necessary for the highest germination percentages. After seeds germinate, the seedlings should be fertilized on a weekly basis with a complete fertilizer, such as 20-20-20, and kept under optimal environmental conditions to ensure vigorous seedling growth. The tender seedlings must be protected from **damping-off,** which is a fungal disease that causes the stems to rot at the soil line (Figure 7-5). The most effective way to control this disease is to use proper sanitation practices and avoid warm, wet medium. In some cases, when cultural methods are not successful, fungicides are used to control damping-off.

When seedlings are ready for transplanting to larger containers or to the field, they must be hardened-off, or prepared for transplanting to the field. One of the main problems that transplants face is the dramatic change in growing environment from the greenhouse to the field. If seedlings are not properly hardened-off, their chances of survival are greatly reduced and/or the chances of transplanting shock increase. Transplants can be hardened-off by gradually subjecting seedlings to cooler temperatures with less frequent watering.

Figure 7–5 Damping-off fungus attacks seedlings at the soil surface.

Primary Seed Dormancy

For seed germination to occur, the seed must be viable and subjected to the proper environmental conditions. But not all seeds germinate even when the proper environmental conditions are provided. This safety mechanism is called primary dormancy and has five categories:

- **Physical dormancy** (seed coat dormancy). Occurs when the seed covering is impervious to water. Germination can be artificially induced by **scarification,** which is the process of breaking the seed coat. This can be accomplished artificially either by nicking the seed coat or by soaking the seed in dilute acid for a brief period of time. In nature, this type of dormancy can be overcome by microorganisms breaking down the seed coat, by the seed passing through the digestive tracks of birds and animals, by mechanical abrasion, by freezing/thawing, and, in some species, by fire. The *Leguminoseae* and *Solonaceae* are examples of plant families that have physical dormancy.
- **Mechanical dormancy** (hard seed dormancy). Caused by the seed-enclosing structure being too strong to permit expansion of the embryo even though water can penetrate it. Germination can be induced artificially by physically cracking the hard seed coat. In nature, the seed coat can be broken down by microorganisms and/or through the process of freezing and thawing. *Juglans* (walnut) and *Prunus* (stone fruits) are examples of genera that have this type of dormancy.
- **Chemical dormancy** (inhibitor dormancy). Caused by germination inhibitors, which accumulate in the fruit or seed coverings during development. This type of dormancy can be overcome artificially by leaching the seed coat, removing the seed coat, or both. In nature, chemical dormancy is typically overcome by leaching the seed coat. The *Portulaceae* (portulaca) and *Crucifereae* (mustard) are examples of plant families that have this type of dormancy.
- **Morphological dormancy** (rudimentary embryo or undeveloped embryo). Occurs when seeds are shed from the parent plant when their embryos are not fully developed. This type of dormancy can be overcome by treating the seeds with gibberellic acid. In nature, very few seeds will germinate if they are shed from the plant prior to becoming fully developed. The *Ericacea* (rhododendron) and *Umbelliferaceae* (carrot) are examples of plant families that have this type of dormancy.

- **Physiological dormancy.** A general type of dormancy in freshly harvested seeds from herbaceous plants. It is caused by environmental cues, such as temperature and light, which lead to an increase in internal growth inhibitors, spread throughout the seed, and cannot be overcome by removing the seed coat or by leaching, such as chemical dormancy. It can be overcome artificially by stratification, which places seeds on a moist medium at temperatures between 32° and 50°F or by treating the seed with gibberellic acid. In nature, physiological dormancy is overcome by environmental conditions similar to artificially induced stratification (Arteca, 1996).

Asexual Propagation

Asexual propagation is the reproduction of new plants from the stems, leaves, or roots taken from the parent plant. One of the most important benefits of asexual propagation is that plants produced by this method are genetically identical to the parent. A plant that is grown from a piece of another plant and is genetically identical to the parent plant is known as a clone. A variety of species are heterozygous, which means they do not produce offspring that are true to type. Trees and woody perennials are typically propagated by means of asexual propagation to preserve the genotype clonally and reduce breeding time. Asexual propagation has a number of advantages and disadvantages; therefore, when deciding on the most efficient propagation technique to use, you should look at both the pros and cons. The number one advantage is that plants produced are genetically identical to the parent plant, which means heterozygous material can be propagated without any genetic modifications. Other advantages associated with asexual propagation are that plants are established more quickly, they are more uniform, it is sometimes the only means of propagation (for example, seedless fruits such as grapes), and seedborne diseases are not a problem.

Although many advantages are associated with asexual propagation, the number one disadvantage is that typically producing plants vegetatively is more expensive than using seeds. A second problem occurs when plant material becomes infected with a virus that spreads systemically throughout the plant and can thus be transmitted to other plants. Viral infections can be minimized or eliminated by starting with disease-free seedlings and maintaining virus-free stock plants for asexual propagation. Other disadvantages associated with asexual propagation are difficulty with storage, handling, and transport of asexual materials, which are much more cumbersome to handle than seeds. One of the advantages of asexual propagation that can also be a disadvantage is that an entire population of genetically identical plants can be completely wiped out with a single disease infection. Lastly, in some cases, mechanized methods for asexual propagation may not be technically possible with the technology currently available.

Methods of Asexual Propagation

There are a variety of types of asexual propagation. In this section, eight types of asexual propagation will be discussed: apomixis, cuttings, grafting, budding, layering, separation, division, and tissue culture.

Apomixis

Apomixis is a form of asexual propagation in which seeds are produced without fertilization; for example, there is no fusion of the male and female gametes, so these seeds are solely maternal in origin. An excellent example of this type of asexual propagation is the common dandelion, found in most lawns; it is capable of reproducing by seeds, vegetatively, or by apomixis.

Cuttings

Cuttings are detached vegetative portions of the plant that are used to produce a new plant. Propagation by cuttings is the most common method of asexual propagation. Cuttings can be derived from three sources, which include the three primary plant organs: stem, leaf, and root. The time of year, stage of growth, and type of wood will determine if cuttings produce roots. The following sections describe the most common types of cuttings.

Stem cuttings. Stem cuttings use portions of the stem containing terminal or lateral buds (Figure 7-6). Stem cuttings can be made from **herbaceous** or softwood (tissues that are not lignified), **semihardwood** (tissues that contain lignified tissue), **deciduous hardwood** (mature woody stems), and **conifer cuttings** (tissues that are hardwood obtained from conifer plants in early winter). Examples of plants produced by the different types of cutting material are as follows:

- softwood or herbaceous—forsythia, lilac, geranium, or carnation.
- semihardwood—azalea, holly, or rhododendron.
- deciduous hardwood—deciduous trees.
- conifer—juniper, hemlock, or pine.

Leaf cuttings. Leaf cuttings consist of a portion of a leaf blade, a complete leaf blade, or a leaf blade with the petiole attached (Figure 7-7). Leaf cuttings are typically used when plant materials are scarce and when large numbers of plants are needed. Following are examples of popular species, propagated by leaf cuttings:

- *Sansevieria* (snake plant) is very easy to propagate, but the cutting must be placed in the same orientation it is found on the plant (Figure 7-7).

Figure 7–6 Stem cuttings are used to produce a new plant.

Figure 7–7 Leaf cuttings of a snake plant (top) and African violet (bottom)

Figure 7–8 Rex begonia plant (top left), bottom side of the leaf showing venation pattern (top right), and new plant formation from cuttings shown in the lower two panels

- African violets are also easily propagated by detaching the leaf from the plant, inserting the leaf into the soil, and sealing the container with the cutting in a clear plastic bag. New roots and shoots will emerge from the base of the petiole (Figure 7-7).
- Rex begonias can be easily propagated by detaching the leaf, making small cuts on several veins, and pinning the leaf to the soil to assure that the leaf stays in contact with it. The container with the cutting is sealed in a clear plastic bag to hold in moisture, and plants will form roots at the cuts on each vein (Figure 7-8).

Leaf-bud cuttings. Leaf-bud cuttings consist of a leaf, petiole, and a short piece of stem with a lateral bud. This type of propagation is very important when woody plant material is scarce and a large number of new plants are required. Leaf-bud cuttings are taken from plants with fully developed buds and actively growing leaves. Typically, as is the case with most hardwood cuttings, leaf-bud cuttings must be treated with a rooting hormone prior to placement in the rooting medium. The lateral bud should not be completely covered, because the bud will remain too moist and rot (Figure 7-9). To ensure success of rooting, the cuttings should also be placed under high humidity and be supplied with bottom heat. Examples of popular species that are propagated by leaf-bud cuttings are rhododendron, magnolia, camellia, and maple.

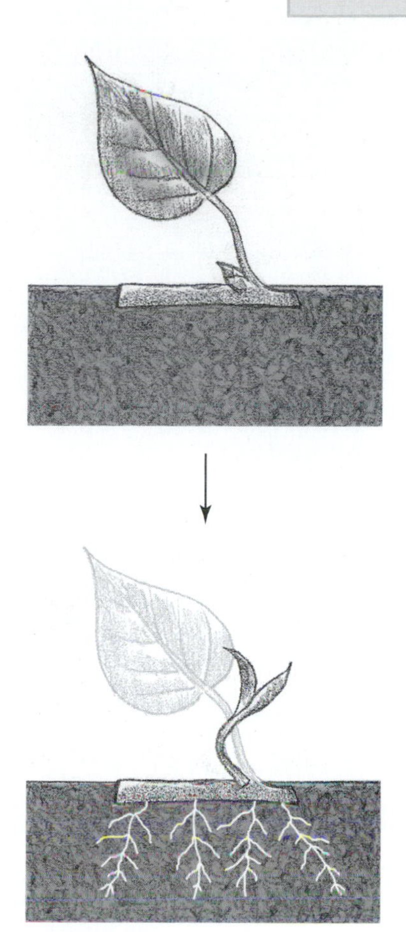

Figure 7–9 The leaf-bud cutting's lateral bud is not completely covered.

Root cuttings. Root cuttings are root pieces taken from young plants as a source of material. Roots are dug in the winter or early spring, cleaned, cut 3 to 6 inches in length, and treated with a fungicide. Root cuttings are typically planted horizontally approximately 2 inches deep. Although the regeneration of plants from root cuttings takes place in different ways, typically the adventitious shoot is produced followed by the production of roots at the base of the newly formed shoot. Raspberries are commonly propagated by using root cuttings. Root cuttings are not commonly done for most types of plants because root cutting is labor intensive and the plant produced can be different from the parent plant (Hartmann, Kester, Davies, & Geneve, 1997).

Factors to ensure optimal rooting of cuttings. To be successful when attempting to root cuttings, five factors should be taken into consideration:

1. **Environmental conditions and physiological status of the parent plant.** The parent plant used for cutting production must be grown under optimal cultural and environmental conditions. When parent plants are well nourished and growing vigorously, they generally produce high-quality cuttings. Consider the following factors when managing parent plants to maximize high-quality cutting production:
   - The nutritional status of the plant should be optimized for the particular plant species being grown. The nitrogen status of the parent plant has an effect on the quality of the cutting.
   - The water status of the plant must be maintained to avoid even brief periods of water stress.
   - The plant should be grown at temperatures and light (intensity, duration, and quality) conditions that have been shown to be optimal for the particular plant species being grown.
   - In some cases, stock plant etiolation, which is the reduction of light levels, has been shown to increase rooting in difficult-to-root species (Hartmann et al., 1997).
   - **Girdling,** which is blocking the phloem's transport by removing a ring of bark from the stem, has been shown to improve rooting in difficult-to-root plants (Hartmann et al., 1997).
   - For many plant species, carbon dioxide enrichment of the stock plant environment has been shown to increase the number of cuttings obtained from the parent plant.
2. **Timing and collection of cuttings.** The time of the year and time of the day to harvest cuttings is very important. Cuttings from herbaceous plants can be taken any time of the year, whereas hardwood cuttings typically root better when materials are collected during the late winter when cuttings are dormant. Softwood and semihardwood cuttings taken from deciduous plants should be taken in the spring and midsummer,

respectively. Determining the best time to harvest cuttings is based on experience and knowledge available in the scientific community; however, when little or no information is available, especially when new crops are being evaluated, a simple way to determine the proper time is to run a small-scale experiment and see what works best. Cuttings should always be harvested during the early morning, when plants are turgid. If cuttings are allowed to wilt, their capability to root will be adversely affected. Clean tools and hands when taking cuttings reduces the chances of disease. In fact, smokers can easily transmit the tobacco mosaic virus if their hands are not properly washed.

3. **Preparation of cuttings.** Some plants are easy to root whereas others are very difficult to root. The plants that are easy to root typically require little or no special treatment (herbaceous plants are generally easy to root). Difficult-to-root plants, which are typically woody, require that the bark is scraped off at the base of the cutting. Both easy-to-root and difficult-to-root cuttings are typically treated with auxins to stimulate rooting. Auxins are especially useful for hard-to-root species. Although auxins are not effective in all plant species, there are several direct benefits (Figure 7-10):
   - A higher percentage of the cuttings produce roots.
   - Root initiation is quicker in most cases.
   - The number and quality of roots per cutting is increased.
   - Uniformity of rooting along the length of the cutting is increased.
4. **Root-inducing environment.** After the cutting is taken from the parent plant and properly prepared, the root-inducing environment must

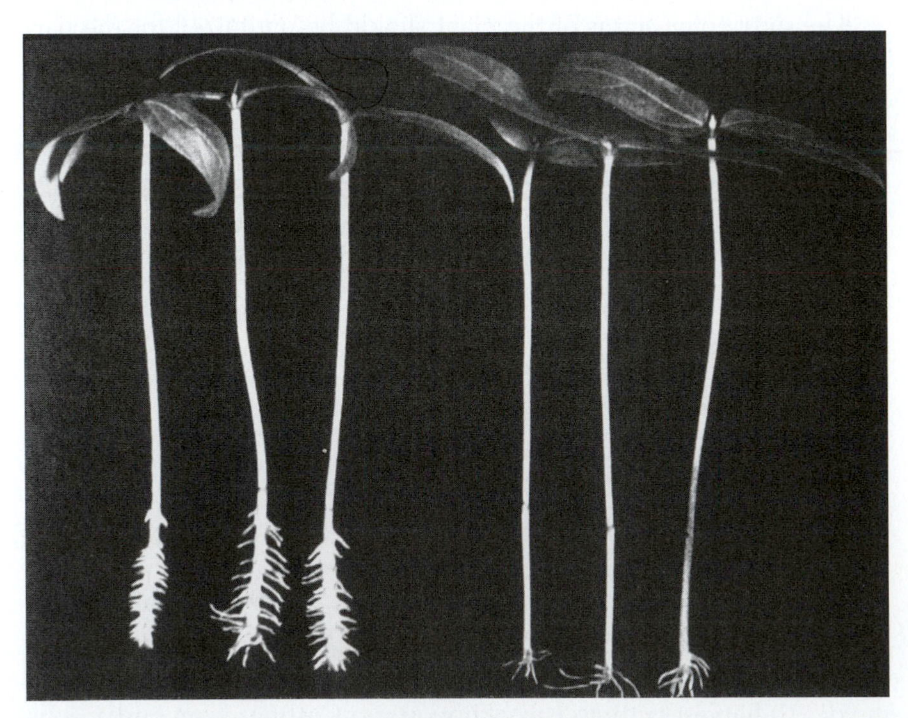

Figure 7–10 Mungbean cuttings treated with auxins on the left and untreated cuttings on the right. *Courtesy of Dr. Charles W. Heuser, Department of Horticulture, The Pennsylvania State University.*

Figure 7–11 A misting bench maintains a high relative humidity. *Courtesy of Dr. E. Jay Holcomb, Department of Horticulture, The Pennsylvania State University.*

be optimized to maximize rooting. The following are important factors to consider for maximizing rooting:

- **High relativity humidity.** Cuttings do not have roots, so they cannot readily absorb moisture from the growing medium. To prevent moisture loss from the cutting, maintain a high relative humidity around the cuttings by misting them (Figure 7-11). Immediately after sticking the cutting, misting is required frequently. As time goes on, the frequency is reduced until finally misting is no longer needed. Because cuttings are grown under high relative humidity, use extreme caution with respect to sanitation to prevent diseases from attacking the cutting.
- **Proper temperatures.** To induce faster rooting, the use of bottom heat is very effective. The temperature used for bottom heat should be about 10°F or 6°C above the ambient air temperature, which is typically maintained between 65 and 75°F. By supplying bottom heat, rooting occurs faster than shoot growth, thereby producing a healthy rooted cutting.
- **High-quality rooting medium.** Many species root very easily and can root in water alone; coleus is an excellent example. Most plants, however, require a high-quality rooting medium for cuttings to root efficiently. A good rooting medium should be disease free and have good drainage and moisture-holding capacity. Perlite and sand are used for drainage and aeration. Vermiculite and peat moss are used for their high moisture- and nutrient-holding capacity. Sand, perlite, peat moss, and vermiculite can be used alone or in various combinations (Figure 7-12).

  Other types of rooting media include rockwool, compressed peat pellets, and a variety of other artificial materials (Figure 7-13).

5. **Fertilization.** Fertilization should occur after the roots have emerged from cuttings. In fact, fertilization prior to root emergence can actually inhibit or delay the rooting process.

Figure 7–12 Peat moss (upper left), perlite (upper right), mixture (center), vermiculite (lower left), and sand (lower right)

The proper time to transplant cuttings differs with the plant species used; however, a general rule of thumb is when an adequate amount of root mass has formed to support plant growth. When transplanting cuttings, carefully remove them from the rooting bed or container and plant no deeper than they originally were planted.

Grafting

Grafting is the process of connecting two plants or plant parts so that they will unite and continue to grow as one plant. For the grafting process, one of the two plants being grafted must serve as the understock, which is the bottom part of the graft union that is in contact with the soil. The other component of the graft is the scion, which is a short piece of stem with one or two buds. For graft unions to be successful, several important factors should be considered:

- The scion and understock must be compatible, which means that the two plants must be closely related genetically. In general, the easiest way to be sure that two plants are compatible is to stay within a species; in other words, you should graft apples with apples not apples with oranges. However, grafting has been shown to be successful between species; for example, some types of plums can be successfully grafted on peach rootstocks.

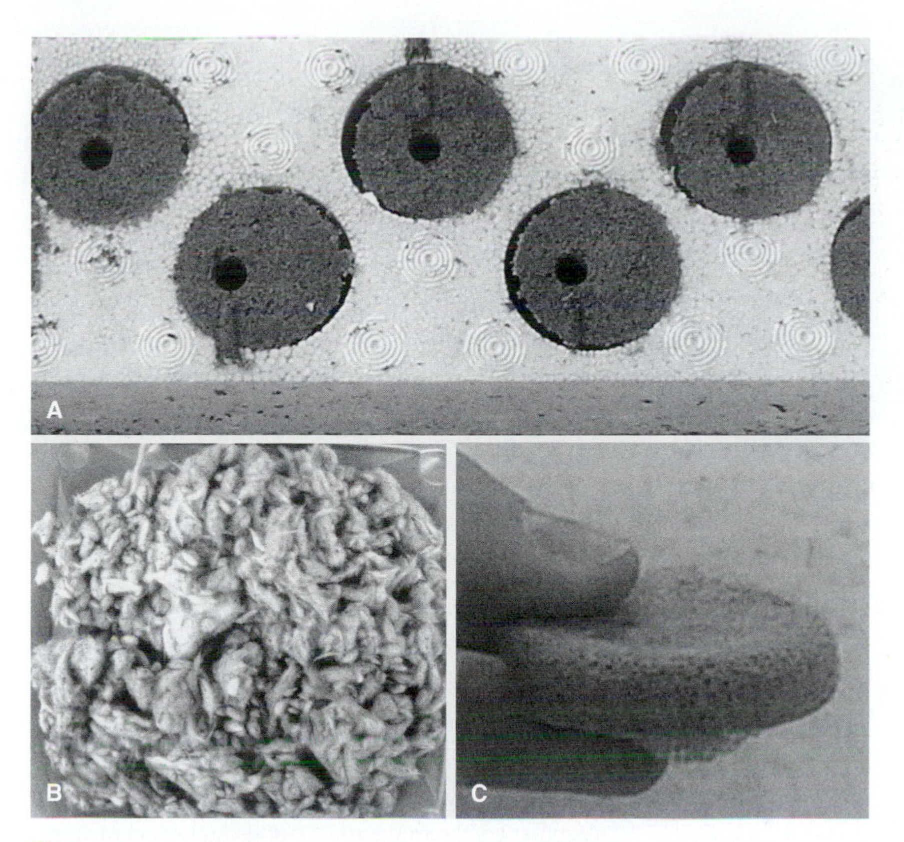

Figure 7–13 (A) Compressed peat pellets, (B) rockwool, and (C) compressed peat pellets are also used as rooting media.

- The diameter of the understock must be equal to or larger than the scion's diameter; typically the scion is much smaller than the understock.
- The cambium of the scion must be in close contact with the cambium of the understock to ensure success of the graft union. To ensure close contact between the scion and understock, use a sharp knife and make clean cuts (Figure 7-14).
- Grafting must be done at the proper time of the year. The understock and scion must be at the correct physiological stage, which means that the scion buds are dormant but they can produce the **callus** tissue required to heal the graft union. The understock may be dormant or actively growing depending on the grafting method used.
- Immediately after grafting, all cut surfaces must be thoroughly covered with grafting wax, which is a water-repellent material composed of beeswax, resin, and tallow, to prevent desiccation.
- The individual who is doing the grafting is critical to the success of the grafting operation; experience in grafting typically leads to success.

Uses of grafting. Grafting can be used

- to maintain clones that cannot be propagated by other asexual methods.
- to gain benefits of certain rootstocks; for example, a variety of disease-resistant and dwarfing rootstocks are currently used.

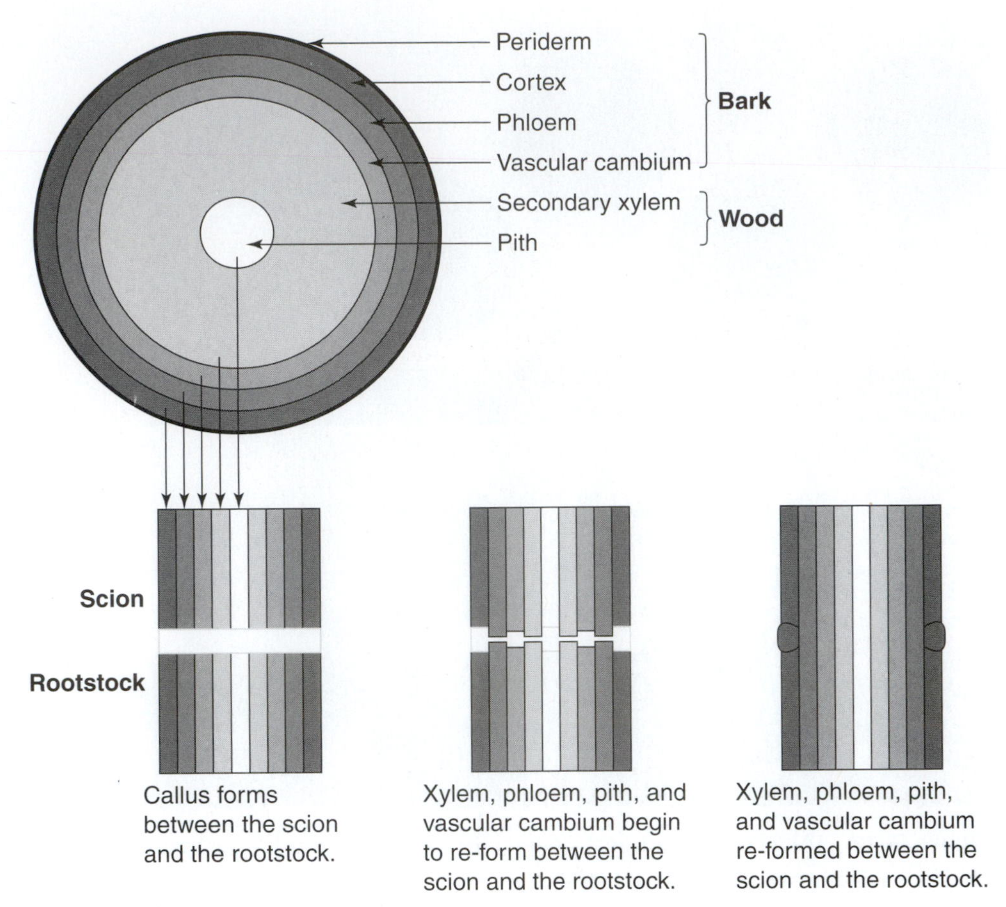

Figure 7–14 The steps involved in the wound repair process are required for the scion and understock to attach. The top shows a cross section of a woody plant stem and the bottom provides a longitudinal section.

- to speed up the time to maturity to promote earlier fruit production.
- to repair damaged parts of trees.

For additional information on grafting and the many reasons for grafting see Hartmann et al. (1997).

Commonly used grafting methods. According to Hartmann et al. (1997), grafting can be classified according to the part of the rootstock on which the scion is placed, such as a root or various places on the top of the plant. The three major categories of grafts are detached scion, approach, and repair grafting:

- Detached scion grafting involves inserting a detached scion a detached scion into the apex, side, bark, or root of the understock; detached scion methods include apical, side, bark, and root grafting. There are six types of apical grafting: (1) whip-and-tongue, (2) splice, (3) saddle (Figure 7-15), (4) cleft, (5) wedge (Figure 7-16), and (6) the four-flap (Figure 7-17). There are three types of side grafting: (1) side-stub, (2) side-tongue, and (3) side-veneer (Figure 7-18). In addition, there are two types of bark grafting: (1) bark and (2) inlay bark (Figure 7-19). There are also two types of root grafting: (1) whole-root or piece-root and (2) nurse root (Figure 7-20).
- Approach grafting occurs when two independent self-sustaining plants are grafted together. This type of grafting can be broken down into three

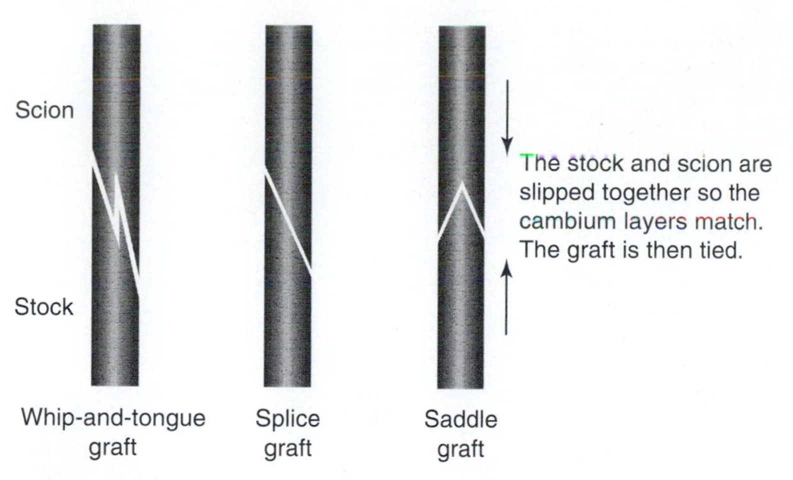

Figure 7–15 Whip-and-tongue, splice, and saddle grafts

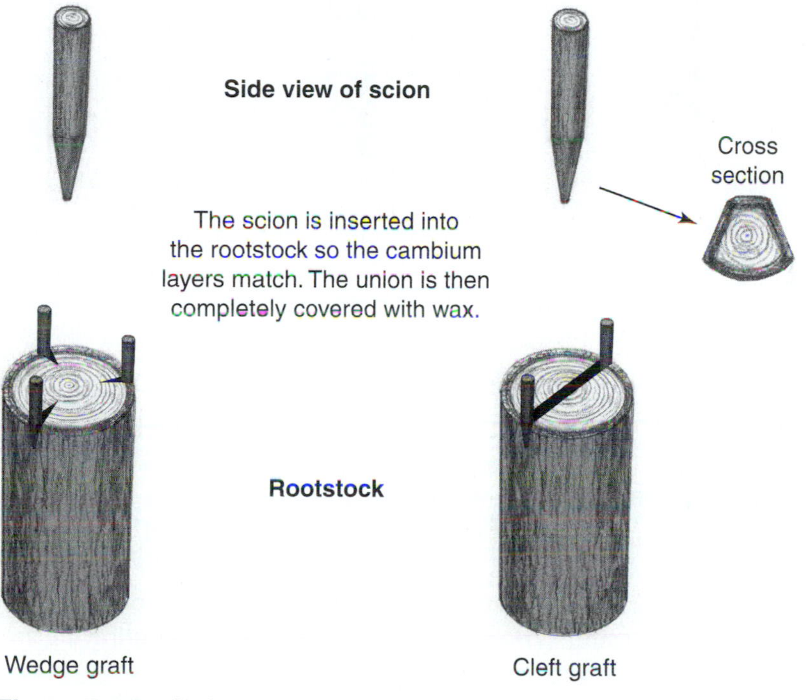

Figure 7–16 Cleft and wedge grafts

categories: spliced approach, tongued approach, and inlay approach (Figure 7-21).

- Repair grafting is used as a specialized way of repairing a plant that has been damaged. Repair grafting can be broken down into three categories: inarching (Figure 7-22), bridge (Figure 7-23), and bracing (Figure 7-24).

Budding

Budding is similar to grafting except that the scion is reduced to a single bud with a small portion of bark or wood attached. Budding methods are typically done in the spring or fall when the **bark slips**—which means that the bark separates easily from the wood (xylem)—and cambium cells are actively dividing. The three main types of budding are chip budding, T-budding, and patch budding (Figure 7-25).

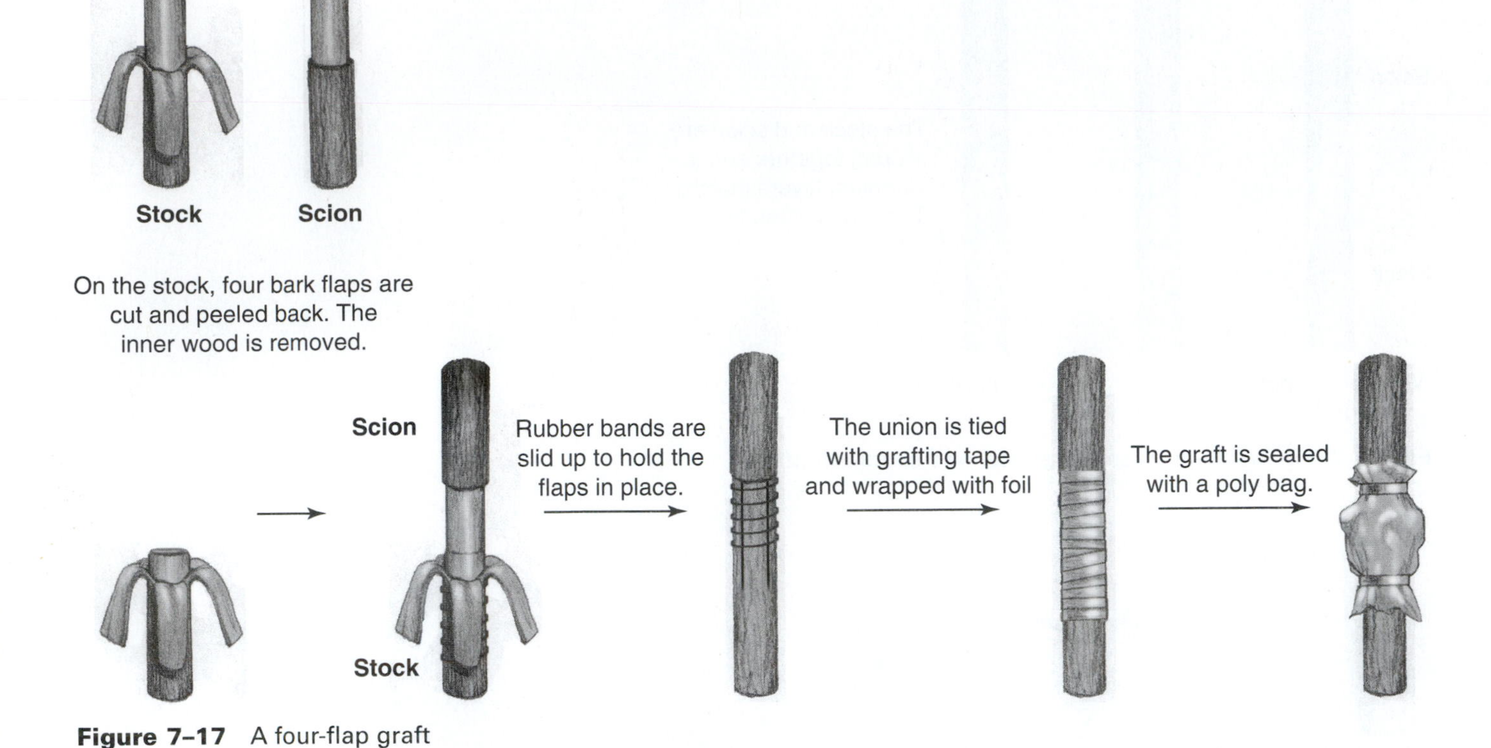

Figure 7–17 A four-flap graft

Layering

Layering is a simple method of asexual propagation in which roots are formed on the stem while it is still attached to the parent plant. The process of layering occurs naturally in some plant species; for example, the strawberry plant produces runners or stolons that elongate. When the node of the runner comes in contact with the soil, it forms roots and then forms shoots, which can be harvested from the plant as a source of a transplant (Figure 7-26). In horticulture, layering is done as a simple means of asexual propagation. Following are the five common types of layering:

- Simple layering occurs when the stem of a plant is gently curved, nicked at the bend, placed in a shallow hole, and the terminal end of the shoot being buried is left exposed. A new plant is formed at the location where the nick was made (Figure 7-27).
- Serpentine layering occurs when the stem of a plant is gently curved at several locations. At each bend, the stem is nicked, placed in a shallow hole, and anchored, and the terminal end of the shoot being buried is left exposed. New roots and shoots arise from each of the locations where nicks were made. After the new plant is formed, it can be detached from the parent plant for subsequent transplanting (Figure 7-27).
- Trench layering occurs when the middle portion of a flexible stem is buried in the soil after nicking it at several locations. New roots and shoots arise from each of the locations where nicks were made. After the new plant is formed, it can be detached from the parent plant for subsequent transplanting (Figure 7-27).

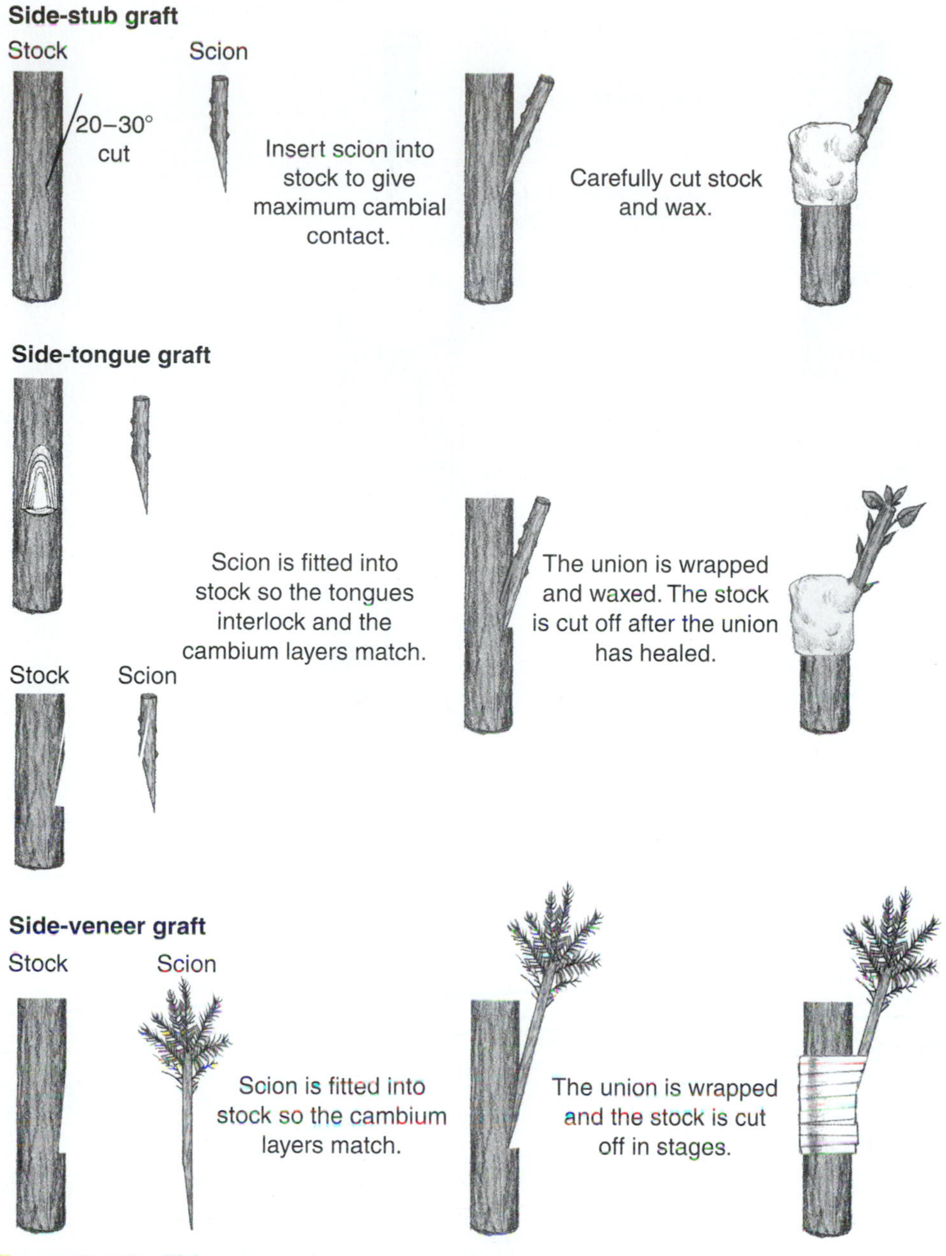

Figure 7–18 Side-stub, side-tongue, and side-veneer grafts

- Mound layering occurs when the parent plant is first cut back to slightly above ground level in late winter and covered with soil. The pruning causes new shoot growth to occur in the spring. After this growth occurs in the spring, soil is mounded around the base of the shoot. As the shoot grows, the additional soil is placed around the shoot and roots then develop around the base of the shoot (Figure 7-27).
- Air layering involves removing a portion of the bark on the stem, placing moist material such as sphagnum moss around this wounded site, wrapping with clear plastic, and sealing both ends to hold in moisture. Common houseplants, such as the Indian rubber plant (*Ficus elastica*), can be propagated by this method (Figure 7-28, Figure 7-29).

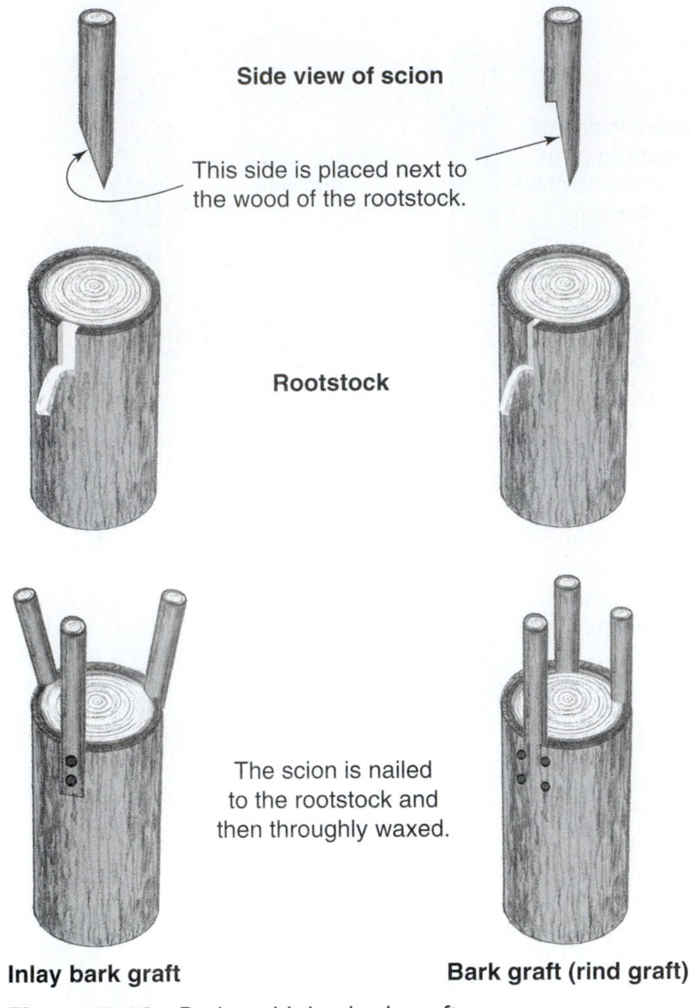

Figure 7–19 Bark and inlay bark grafts

Separation

Separation is a propagation method in which natural structures are removed from the parent plant and planted to grow on their own. Examples are bulbs (tulips or lilies) or corms (gladiolus or crocus) (Figure 7-30).

Division

Division is a method of asexual propagation in which parts of the plant are cut into sections that will grow into new plants naturally. Examples are rhizomes (iris) or tubers (Irish potato) (Figure 7-31).

Tissue Culture

Tissue culture is a method for producing new plants from single cells, tissue, or pieces of plant material called **explants** on artificial medium under sterile conditions. Some commonly used methods of tissue culture are as follows:

- Callus culture is the use of callus, which is an undifferentiated mass of cells that can be induced naturally by wounding or artificially by using

Nurse root graft

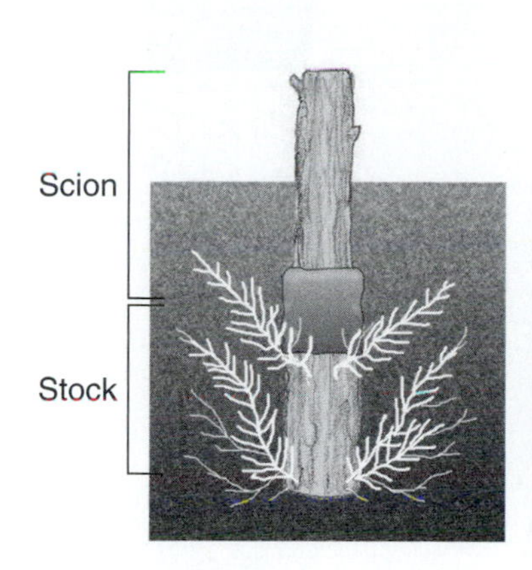

The distal end of the rootstock is joined to the proximal end of the scion.

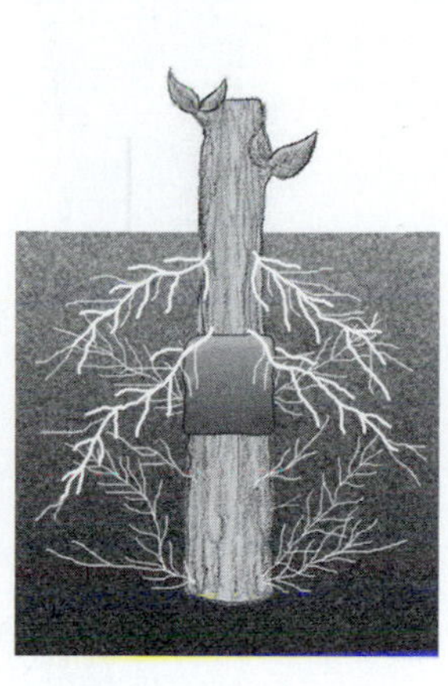

Nurse root graft helps the scion grow until its own roots take over.

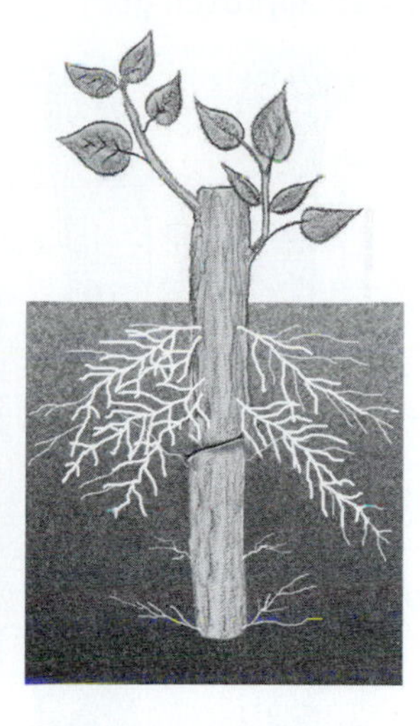

The nurse root dies after the scion is established.

Figure 7–20 Nurse root graft

plant hormones as a means of growing plants or different plant parts on a solid medium (Figure 7-32).

- Cell suspension culture is a method in which plant cells are suspended in a liquid media under continuous agitation to provide aeration. Suspension cultures are initiated by placing callus in a liquid medium and agitating to aerate and to disperse the cells (Figure 7-33).
- Embryo culture (also called embryo rescue) occurs when the embryo is removed from the seed aseptically and grown on a solid gel medium under optimal environmental, nutritional, and hormonal conditions to promote growth of the embryo, which would not germinate within the seed.
- Meristem culture is a technique that uses the smallest part of the shoot tip as an explant, which includes the meristem dome and some leaf primordia. The meristem is a region of the plant consisting of undifferentiated tissue whose cells can divide and differentiate to form specialized tissues.
- Anther culture is a technique in which immature pollen is induced to divide and generate tissue, either on solid media or in liquid culture. Anthers containing pollen are simply removed from the plant and placed on culture medium, where some microspores survive and develop. The generated tissue can be either embryo tissue, which can be transferred to the appropriate medium to allow root and shoot development to take place, or callus tissue, in which case the tissue is placed in an appropriate solution of plant hormones to induce differentiation of shoots and roots.

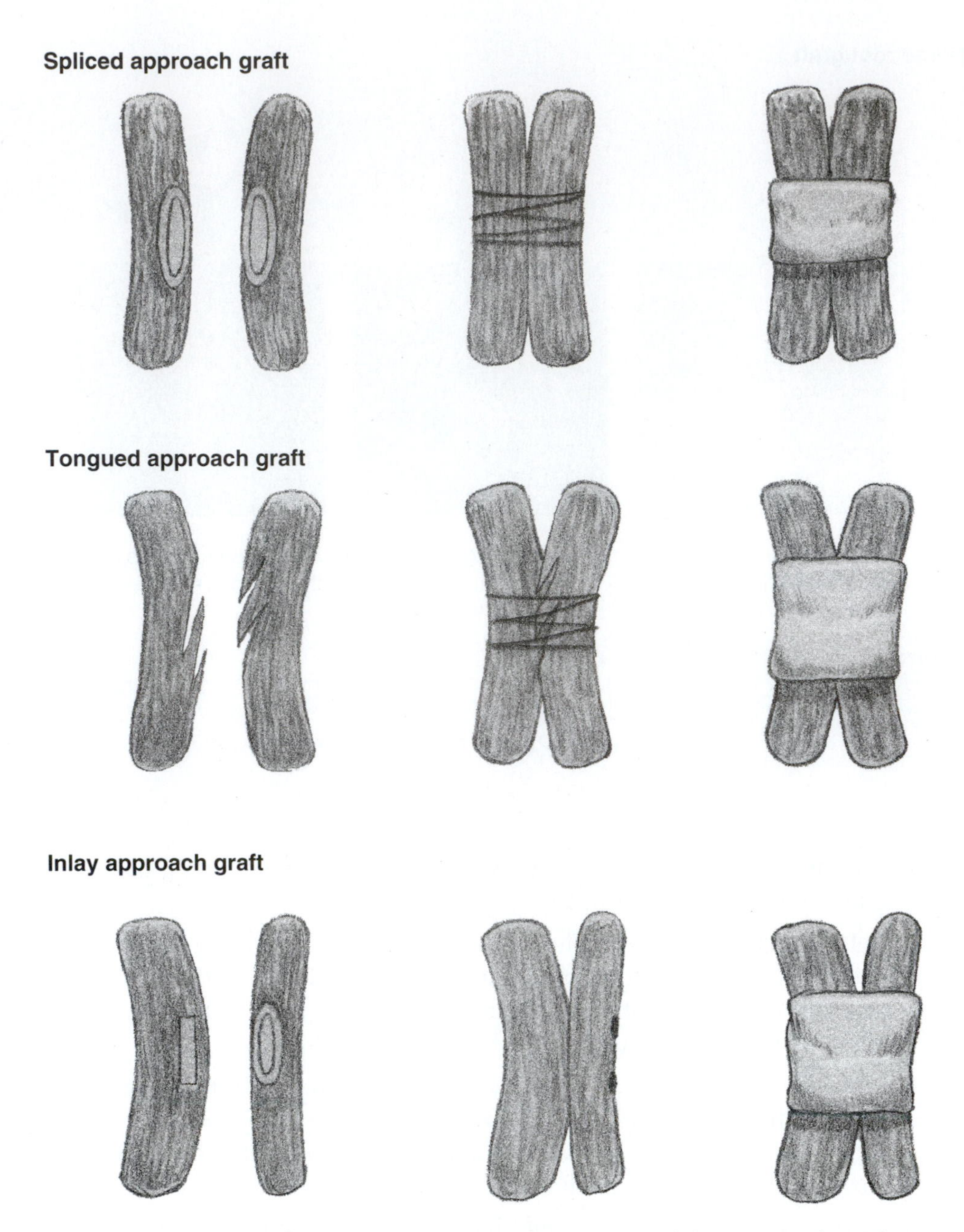

Figure 7–21 Spliced approach, tongued approach, and inlay approach grafts

Inarching graft

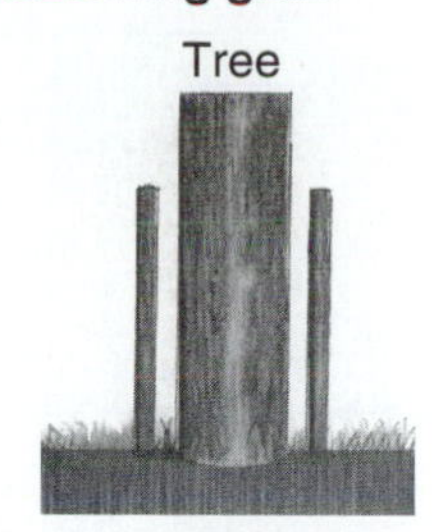

Sucker from the base of the tree, seedlings, or rooted cuttings planted at the base of the tree are used.

Side of the seedling is placed next to the wood.

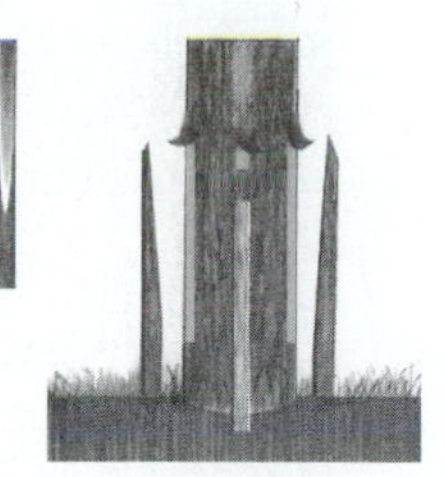

Vertical cuts are made near the base of the tree. Bark is removed except for short flap at top. Suckers, seedlings, or rooted cuttings are cut to fit into the tree.

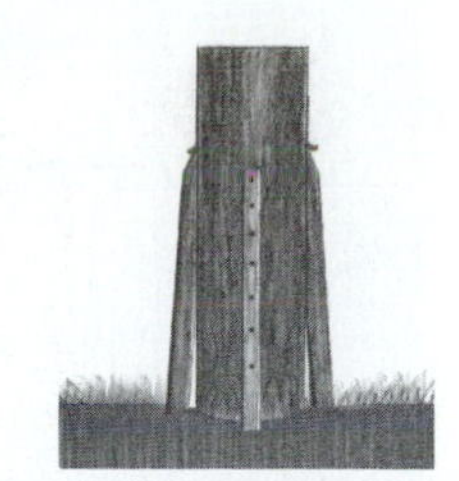

The seedling fits into the slot and is nailed in place and waxed.

Figure 7–22 Inarching graft

Bridge graft

Scion

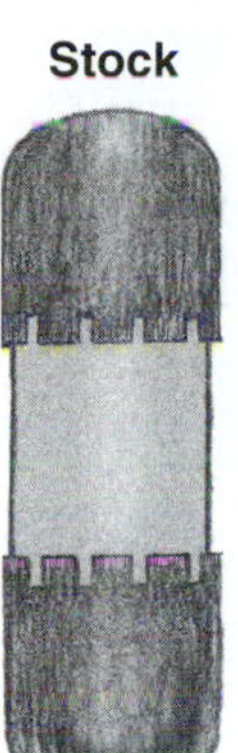

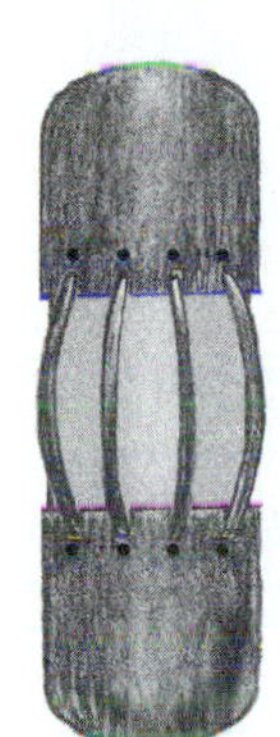

Scions are placed right side up, nailed into place, and thoroughly covered with grafting wax.

Figure 7–23 Bridge graft

Bracing graft

Tree limbs are braced with rope or cord.

Smaller shoots are woven together and secondary growth occurs.

Figure 7–24 Bracing graft

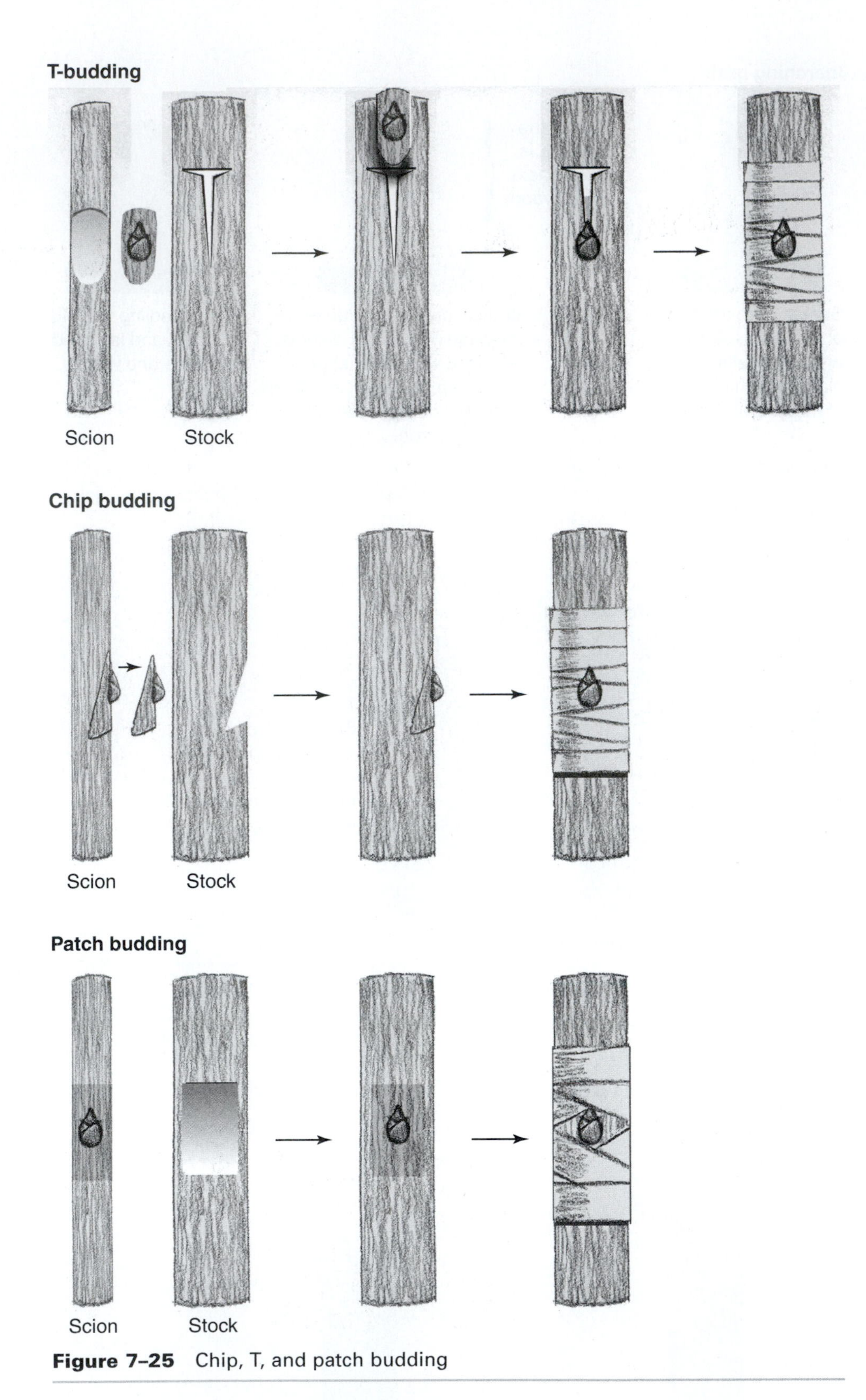

Figure 7–25 Chip, T, and patch budding

Figure 7–26 Strawberry runner showing natural layering

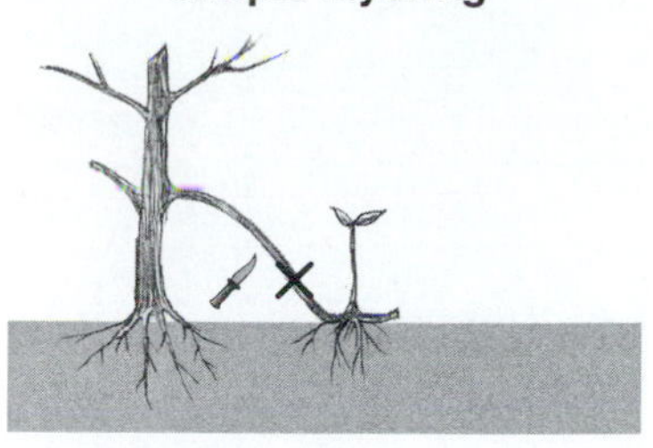

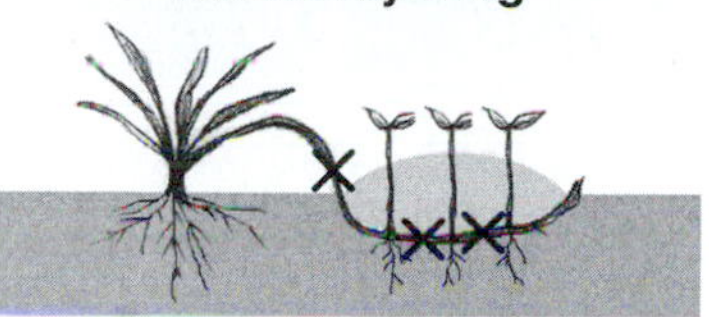

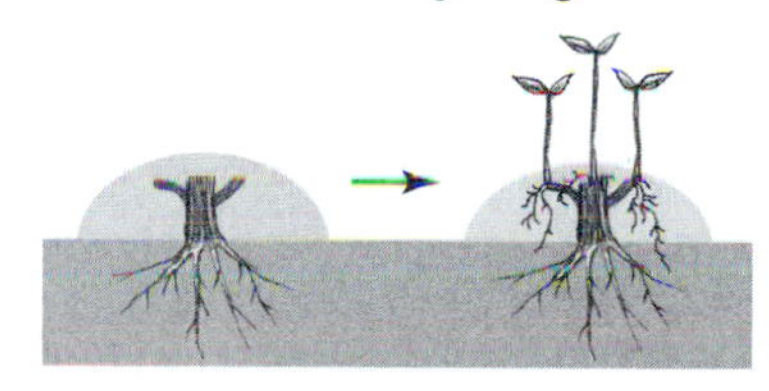

Figure 7–27 Simple, serpentine, trench, and mound layering

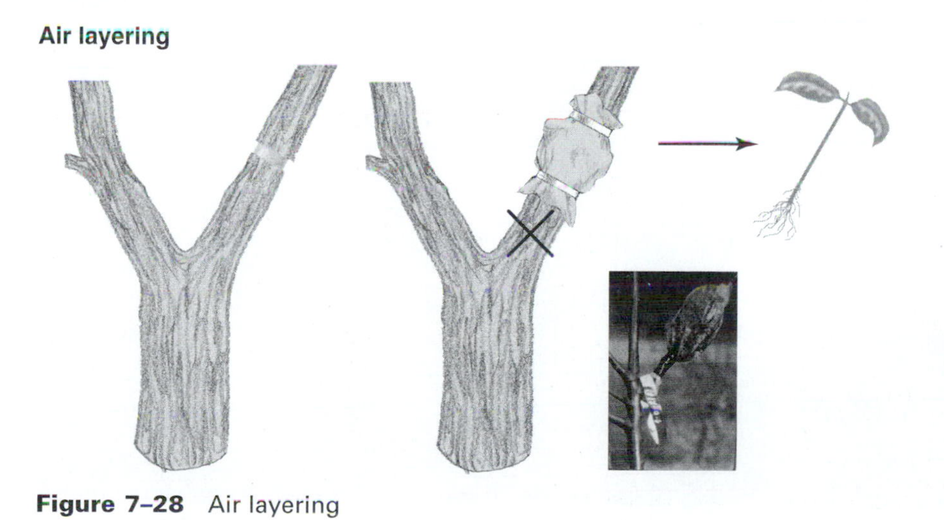

Figure 7–28 Air layering

Figure 7–29 Rubber tree plant

Figure 7–30 Lily bulbs

Figure 7–31 Iris rhizome (left) and Irish potato (right)

Figure 7–32 Callus on agar

Figure 7–33 Cell suspension culture

SUMMARY

You have learned about the two major means of propagation: sexual and asexual propagation. Sexual propagation is the major means of propagation in nature and agriculture today because it's easy and relatively inexpensive. You learned important factors involved in the stages of seed production from producing breeder seed through certified seed. All the factors involved in both direct and indirect seeding are clear by obtaining a better understanding of each. Primary dormancy must be broken for seeds to germinate; you now know the five categories of primary seed dormancy and factors affecting each. Asexual propagation is commonly used when sexual propagation is not an option. The nine commonly used types of asexual propagation and factors that should be considered when using each were discussed.

Review Questions for Chapter 7

Short Answer

1. Why is propagation by seeds the major means of reproduction in agriculture?
2. What are the four classes of seed production that must occur prior to commercial seed distribution?
3. What are four factors that should be considered for the production of seeds?
4. How can you experimentally determine percentage seed germination?
5. Describe direct seeding and list five factors that determine the success of this method.
6. What is indirect seeding?
7. What are four important factors that should be considered when selecting a germination media?
8. What is damping-off, and what are two ways to control it?
9. What are the five major categories of primary seed dormancy? Provide an explanation for each.
10. What is the most important benefit of asexual propagation?
11. What are the eight types of asexual propagation?
12. What are three factors that determine if cuttings will produce roots?
13. List four types of cuttings and provide an explanation for each.
14. What are several factors that should be considered to ensure optimal rooting of cuttings?
15. What are four direct benefits of treating cuttings with auxin?
16. List several important tasks that should be done for successful graft unions.
17. What are four major uses of grafting?
18. List three types of budding and explain what time of the year budding should be done and why.
19. List and explain five types of layering.
20. What are five commonly used methods of tissue culture?

Define

Define the following terms:

plant propagation
sexual propagation
homozygous
true to type
heterozygous
scarification
stratification
hardened-off
asexual propagation
clone
apomixis
girdling
grafting
scion
separation
division
tissue culture
explants
callus
meristem
genetic engineering

True or False

1. The most important factor to be considered when harvesting seeds is knowing when they are ripe.
2. Plants produced by seeds are genetically different from their parents.
3. Mechanical dormancy is caused by the seed covering being impervious to water.
4. Mechanical and physical seed dormancy are both caused by the seed covering being impervious to water.

5. Physical dormancy occurs when the seed covering is impervious to water.
6. Mechanical dormancy is caused by the seed-enclosing structure being too strong to permit expansion of the embryo even though water can penetrate the enclosing structure.
7. Chemical dormancy is caused by germination inhibitors that accumulate in the fruit or seed covering during development.
8. Stratification occurs when seeds are placed in a moist medium at high temperatures to overcome dormancy.
9. Stratification is commonly used to overcome chemical dormancy.
10. Apomixis is a form of incompatibility in plants.
11. One of the major benefits of asexual propagation is its ease.
12. Leaf cuttings are used when plant materials are scarce and when large numbers of plants are needed.
13. Fertilization of cuttings is not necessary until roots have emerged.
14. Leaf cuttings consist of a leaf, petiole, and a short piece of stem with a lateral bud.
15. The phloem of the scion must be in close contact with the cambium of the understock for a graft union to be successful.
16. *Scion* is the term used to describe the short piece of stem with one or two buds used for grafting.
17. Budding should be done during the summer months when the bark separates easily.
18. Budding should be done during the spring or fall when the bark separates easily from the wood.

Multiple Choice

1. Commercial seed producers are located in which of the following areas to minimize disease problems?
   A. Tropical climate
   B. Arid climate
   C. At a location with high humidity
   D. All of the above
2. Many federal and state laws regulate the shipment and sale of seeds. These laws require that seeds be tested and labeled for which of the following?
   A. Trueness of name
   B. Pure seed percentage
   C. Origin
   D. All of the above
3. Scarification is
   A. a method for breaking the seed coat to promote germination.
   B. a scaring process in plants following damage.
   C. a wound-healing process.
   D. All of the above
4. Stratification is
   A. placing seeds in a moist medium at temperatures between 32° and 50°F.
   B. subjecting plants to cold temperatures to induce flowering.
   C. a method to promote rooting.
   D. None of the above

5. What time of the day should cuttings be taken?
   A. Early morning
   B. Early afternoon
   C. The time of day does not matter.
   D. None of the above
6. Simple layering occurs when
   A. a branch is placed in a trench and roots form at the nodes.
   B. the terminal end of the shoot being buried is left exposed.
   C. soil is mounded around new shoots.
   D. None of the above
7. Air layering occurs when
   A. a portion of the bark on the stem is removed, moist material is placed around the wounded site, and this material is wrapped with plastic to hold in the moisture.
   B. the terminal end of the shoot is buried and the shoot tip is left exposed to the air.
   C. a branch is placed under the ground with portions left exposed to the air.
   D. None of the above
8. Which of the following tissue culture methods are commonly used?
   A. Callus culture
   B. Embryo culture
   C. Meristem culture
   D. All of the above
9. Which of the following is a fast and easy way to determine if seeds are alive?
   A. Xenobiotic test
   B. Phenylpropylene test
   C. Tetrazolium test
   D. None of the above

Fill in the Blanks

1. A fungal disease that causes the stems to rot at the soil line is called ______________.
2. The two major methods of propagation are ______________ and ______________ propagation.
3. The three major categories of grafts are ______________, ______________, and ______________.
4. Budding is typically done during the ______________ and ______________ because this is when the bark slips or readily separates from the wood.
5. ______________ is a simple method of asexual propagation in which roots are formed on the stem while it is still attached to the parent plant.
6. ______________ is an undifferentiated mass of cells.
7. The ______________ is a region of the plant consisting of undifferentiated tissue whose cells can divide and differentiate to form specialized tissues.

Activities

1. In this activity, you will search university Web sites throughout the world for information about research currently being conducted pertaining to plant propagation. More specifically, you should explore horticulture programs (some suggestions are provided next) that have faculty doing research in some area of plant propagation. Following are universities with horticulture programs:

Auburn University	Ohio State University	University of Florida
Clemson University	Oregon State University	University of Georgia
Iowa State University	Penn State University	University of Hawaii
Kansas State University	Purdue University	University of Kentucky
Michigan State University	Texas A & M University	University of Maine
New Mexico State University	University of Arkansas	University of Wisconsin
North Carolina State University	University of Connecticut	

   Choose four universities from the list or others that are doing research in the area of plant propagation. Write a summary about the research, including the purpose of the research, what is being evaluated, and why you selected the research. Be sure to include the name of the university and a title for the research topic.

2. Select one of the main areas of plant propagation that you are interested in and visit a nearby location to gain information about it. For example, if you are interested in the area of grafting, visit a local orchard and find out as much information about grafting as possible. Write a summary of your findings.

References

Arteca, R. N. (1996). *Plant growth substances: Principles and applications.* New York, NY: Chapman & Hall.

Hartmann, H. T., Kester, D. E., Davies, F. T., & Geneve, R. L. (1997). *Plant propagation: Principles and practices* 6th ed. Upper Saddle River, NJ: Prentice Hall.

CHAPTER

8

Media, Nutrients, and Fertilizers

ABSTRACT

Media plays an important role in plant life, and the more commonly used kinds of media include soil, soilless, mixtures, and hydroponics. The inorganic and organic components of the soil also affect soil structure and profiles. Most plants require 16 essential elements for normal plant growth and development. For an element to be essential, it must meet two main criteria. Certain factors also affect pH, availability of nutrients, and methods for monitoring nutritional levels in plants. The chapter concludes by discussing fertilizer nomenclature, the different types of fertilizer formulations and methods, and the timing required for fertilizer application to maximize plant growth.

Objectives

After reading this chapter, you should be able to

- provide background information on plant-growing media by discussing the two major roles of media and the four major kinds of media used today.
- provide information on the inorganic and organic components found in the soil and how these components affect the structure of the soil and soil profiles.
- provide the two principal criteria by which an element can be judged essential to any plant and key information on the 16 elements required for normal plant growth and development, factors affecting pH, availability of nutrients, and methods for monitoring nutritional levels.
- generate a better understanding of fertilizer nomenclature and the different types of fertilizer formulations, methods, and timing of fertilizer application to maximize plant growth.

Key Terms

active ingredient
bag culture
bare root system
broadcasting
cation exchange capacity (CEC)
chlorosis
complete fertilizer
controlled release fertilizer
dirt
fallow
fast release fertilizer
fertigation
fertilizer
fertilizer analysis
floating system
foliar analysis
green manure
growing medium
horizons
hydroponics
incomplete fertilizer
inert ingredient
luxury consumption
macronutrients
micronutrients
nutrient film technique (NFT)
nutrient solution
nutrients
organic matter
parent material
peat
peat moss
perlite
side dressing
slow release fertilizer
soil
soil aeration
soil compaction
soil pH
soil profile
soil structure
soil testing
soil triangle
soilless medium
substrate system
top dressing
vermiculite
visual diagnosis

INTRODUCTION

Media, **nutrients,** and **fertilizers** are important for growing high-quality crops with high yields. Plant-**growing medium** does not always refer to soil or soilless medium; rather it refers to the material in which plants grow. A good growing medium meets the specific needs of a plant, which can differ from species to species. In general, a good plant-growing medium must anchor the plant, provide good aeration and drainage, provide good nutrient- and water-holding capacity, be free of pests, be reproducible and available, and should wet easily and not decompose.

Many kinds of plant-growing media are available, including soil, soilless, mixtures, and hydroponics. **Soil** is the top few inches of the Earth's crust consisting of minerals, air, water, and **organic matter** that provide for plant growth. Although soil has a number of advantages, the main disadvantage is that it is difficult to get a consistent, reproducible, and inexpensive source of soil. That is why **soilless media,** which contain no topsoil, have become very popular. Organic sources of a soilless medium are **peat, peat moss, bark,** and other wood products; the inorganic materials used are sand, **perlite,** and **vermiculite.** Mixtures with various combinations of soil, peat, vermiculite, perlite, or sand are used to maximize growth of a variety of plants. **Hydroponics** is an alternative to solid media. Hydroponic systems are classified as **substrate systems,** in which the plant roots are surrounded by inert solid materials that provide support for plants. The second type of hydroponic system is the **bare root system,** in which no physical support is given to the root system that is immersed in the nutrient solution. Hydroponic systems are typically used to grow high-value crops, such as tomatoes.

The inorganic portion of soil is made from weathered rocks by a very slow process involving a variety of physical, chemical, and biological factors. The organic portion of the soil is derived from decaying plant and animal materials. **Soil structure**—which is the arrangement of soil particles—and **soil profile**—which is the vertical section of soil in a particular location—are both affected by how the soil was derived and profoundly affect plant growth and development.

Most plants require 16 essential elements for normal plant growth and development. Those elements are carbon (C), oxygen (O), hydrogen (H), nitrogen (N), phosphorous (P), potassium (K), calcium (Ca), magnesium (Mg), sulfur (S), iron (Fe), manganese (Mn), molybdenon (Mo), copper (Cu), boron (B), zinc (Zn), and chlorine (Cl). Carbon, hydrogen, and oxygen are the main elements found in plants and are derived from carbon dioxide and water, respectively. The remaining 13 elements derived from the soil are broken down into primary **macronutrients,** which are N, P, and K; secondary macronutrients, which are Ca, Mg, and S; and the remaining 7 are **micronutrients.**

Prior to fertilization, **soil testing** should be done. After these results are obtained, the soil temperature, **cation exchange capacity** (**CEC**) of the soil, and the **soil pH** should be examined because each of these has a profound effect on the availability of nutrients. **Foliar analysis** is similar to soil analysis except that foliar analysis tells the grower what nutrients have been taken up and accumulated by the plant, and soil testing provides information on the nutrients available in the soil.

The proper soil fertility is critical for maximizing crop production, so nutrient levels should be closely monitored by soil and/or foliar analysis. Nutrients can be added back to the soil by allowing the land to go **fallow,** which is leaving the land uncultivated for several years to rejuvenate naturally (although this is not a common practice). Nutrients are typically added by fertilizer application organically through the addition of **green manure,** which is the use of a leguminous cover crop for the main purpose of plowing it under to increase soil fertility. Inorganic fertilizers are compounds derived from mineral salts; these are the most commonly used in agriculture today because of the many associated benefits. Inorganic fertilizers come in several formulations; for example, **complete fertilizer** contains all three primary fertilizer nutrients—nitrogen, phosphorous, and potassium—and may have select micronutrients. A second inorganic type is an **incomplete fertilizer,** which lacks one or more of the three primary elements found in a complete fertilizer. Other types of inorganic fertilizers include **fast release, slow release, controlled release,** and fertilizers with herbicide.

You must understand the nomenclature on the fertilizer bag to apply nutrients properly to the soil. The **fertilizer analysis** on the side of the bag has all the information necessary to make an informed decision on the application of fertilizers. The three numbers on the bag are the percentages of nitrogen, phosphorous, and potassium with the total of the three being the total **active ingredient;** the remaining percentage is the **inert ingredient,** which is carrier or filler. After you determine the necessary amount of nutrients, there are three methods of fertilization. The first is dry application, which can be applied by **broadcasting, top dressing,** or **side dressing.** The two other methods are liquid or gas application. After you choose a method of application, you must establish the frequency and rate of fertilizer application, which should be determined by plant genetics, environmental factors, soil conditions, biotic factors, and the stage of plant development.

PLANT-GROWING MEDIA

Under field conditions, plants are grown in soil, which is the top few inches of the Earth's surface, consisting of minerals, air, water, and organic matter that provides for plant growth. Note that loose dirt, of its own, does not constitute soil. **Dirt** is soil that is out of place or, as many a gardener has remarked, dirt is what you find under your fingernails. Good topsoil is becoming limited, so getting a consistent supply at a reasonable cost can be difficult. Soil is also heavy, which translates into more costs when shipping potted plants. Today, a variety of soilless growing media is available that closely mimics soil. Many mixtures also contain a small amount of soil with artificial amendments, making them better than soil alone or soilless media for specific purposes. For that reason, when the term *growing medium* is used, it does not always refer to soil or soilless media, but rather to the material in which the roots of plants grow.

Roles of Media

A good growing medium is formulated to meet a specific need because different types of plants may have different media requirements. Certain media formulations work best for germination and early growth of plants, whereas others are better suited to the long-term growth of the plant. However, all types of growing media must satisfy several basic criteria to maximize plant performance. All growing media must

- provide the plant with nutrients, which are the substances that roots absorb with water from the medium (not carbohydrates).
- anchor the plant to keep it from falling over.
- provide good drainage and aeration.
- provide good nutrient- and water-holding capacity.
- ensure that there are no natural toxins, diseases, insects, or weeds.
- be reproducible and available.
- not decompose and must wet easily.

Kinds of Media

A variety of types of potting media are available including soil, soilless, mixtures, and hydroponic, each of which has its own advantages and disadvantages. Potting mixtures typically contain organic and inorganic components.

Soil

The organic matter portion of soil is derived from decaying plant and animal materials. The proportion of sand, silt, and clay contained within the soil is the inorganic portion that determines its texture. Soil must first be pasteurized to kill harmful bacteria, fungi, nematodes, weeds, and other pests or toxins that may be harmful to plants. This can be accomplished by steam or by chemical pasteurization. Steam pasteurization (the most common method) sterilizes the soil by aerating the soil with steam at a temperature of 140°F for 30 minutes. The soil temperature in turn rises to 180°F, which kills nonbeneficial organisms while leaving the beneficial ones alive. You must be careful not to overheat the soil; otherwise, beneficial organisms will also be destroyed and other problems may arise. The advantages of steam sterilization are that the soil can be used right after it cools down, and, in some cases, **soil aeration** and drainage are improved. Steam sterilization can also be used on pots, tools, and equipment to kill harmful pests.

Plants grow best in soil under field conditions. Soilless mixtures and mixtures of soil and other components are used to mimic nature, which is sometimes difficult to do. As stated previously, the disadvantage of soil is that getting a consistent, reproducible source of soil is expensive. Soil is also heavy, making it more expensive to ship plants that have been potted in only soil.

Soilless Medium

Soilless medium contains no topsoil. The following are common sources of organic matter found in soilless media:

- **Peat and peat moss.** Peat is partially decayed plant material and composed mainly of organic matter, which provides some nutrients (including nitrogen). Many types of peat is available with varying compositions. The most common type of peat used is peat moss. Peat moss is obtained from moss plants growing in bogs; specific species are *Sphagnum* and *Ploytrichum*, with *Sphagnum* being the most commonly used. Peat is used because of its moisture- and nutrient-holding capacity.
- **Bark and other wood products.** Due to the rising costs of peat, bark and other wood products are now being used as alternatives. Bark from both softwoods (conifers) and hardwoods (oak) is used in a variety of

potting mixes as a substitute for peat. The bark from trees consists primarily of lignin, which decomposes slowly. Bark also contains major elements such as nitrogen, phosphorous, and potassium, plus other elements. Wood chips and sawdust are also used in soilless mixtures; these also contain a variety of elements. In some cases, these elements are not a problem, but depending on where the tree was grown, high levels of specific elements can pose potential problems. Nitrogen deficiency can occur when adding sawdust into the growth media because bacteria, which decompose sawdust, first use the existing nitrogen. Some initial nitrogen fertilization should be applied to overcome this problem.

In addition to organic materials in soilless media, there are a number of inorganic materials that are mineral in origin, containing no carbon. Commonly used inorganic materials that are found in potting mixes are as follows:

- Sand is a commonly used component in many potting mixes because it provides good aeration and drainage. Sand is easy to pasteurize and comes in a variety of particle sizes. Larger-particle sand should be avoided because the sand may settle out of the mix and become compacted, which actually reduces the infiltration of water.
- Perlite is made from light rock that is volcanic in origin. Advantages of perlite include its capability to provide aeration and drainage, and it can be pasteurized easily. Perlite is white in potting mixes (Figure 8-1). Perlite also has some disadvantages. Perlite's low weight causes it to float to the surface of the medium when watered. Perlite also produces dust when mixed with other components of the potting mixture; however, this can be overcome easily by slightly wetting the material prior to mixing.
- Vermiculite is heat-treated mica that can be coarse or fine. It has a high moisture- and nutrient-holding capacity. Vermiculite is golden colored in potting mixtures (Figure 8-2).

Figure 8–1 Perlite alone (right) and in a potting mix (left)

Figure 8–2 Vermiculite alone (right) and in a potting mixture (left)

Mixtures

Various combinations of soil, peat, sand, perlite, or vermiculite can be made. Many companies sell a variety of soil and soilless mixtures in various proportions to maximize the growth of a specific plant type. Some are amended with fertilizer, limestone, or wetting agents to produce a better mix.

Soilless mixtures have many advantages and disadvantages. The main advantages of soilless mixtures are as follows:

- Mixtures have good aeration and drainage to enhance plant growth.
- Mixtures have excellent moisture- and nutrient-holding capacity to maximizes plant growth and development.
- The mixture is both physically and chemically uniform and can be obtained reproducibly year after year. The homogenous nature of the mixture makes it possible to standardize a plant's growth and development.
- Shipping costs are typically reduced because of the lightweight nature of soilless mixtures.
- Initially, mixtures are fairly sterile, which maximizes seedling growth and reduces chances of problems with damping-off, diseases, insects, and weed seeds.
- Mixtures can be custom made to meet specific needs.

The disadvantages of soilless mixtures are as follows:

- Some mixes are so light that when dry, plants can easily topple over with even a slight wind or touch.
- Not all soilless mixtures have nutrients added, or they may have nutrients that are not well suited to a specific crop.
- In some cases, mixtures are very different from what the plant experiences under field conditions.

Figure 8–3 This substrate hydroponic system uses gravel. The right side of the container shows the inside with roots being supported by gravel.

Hydroponics

Hydroponics is a method of growing plants in which the nutrients needed by the plant are supplied by a **nutrient solution** that contains nutrient salts dissolved in water. Hydroponic systems have been used for years to study nutrition in plants. In more recent years, a variety of important crops have been grown successfully under hydroponic conditions. A number of advantages are associated with hydroponic growing systems over soil culture:

- Plant nutrition is completely controlled. The nutrients contained in the solution can be monitored carefully and modified as necessary together with the pH, which affects the availability of nutrients to the plant.
- Yield per unit area is greater. Because nutrition and environmental conditions can be regulated carefully, the yield for a given cultivar can be maximized.
- The need for weed-, disease-, and insect control is greatly reduced.

The main disadvantage of using a hydroponic system is the expense, which is not economically feasible for a variety of plants. For example, hydroponically grown tomato plants generate enough revenue to justify this method, whereas growing hydroponic potato plants do not. Another disadvantage associated with plants grown under hydroponic conditions is algae growth, although this can be controlled with good management practices.

Hydroponic systems are classified in two ways: substrate system and bare root system. In a substrate system, plant roots are surrounded by either inert or organic materials that provide support for the plant. Materials commonly used for inert substrate systems include sand, gravel (Figure 8-3), vermiculite, perlite, and horticultural rockwool (mineral wool). Natural organic substrates commonly used are peat, sawdust, wood chips, and bark. Natural organic substrates are typically used in bag culture. **Bag culture** is when plants are grown in bags containing these materials and are watered using a drip irrigation system.

Figure 8–4 Aeroponically grown *Taxus* roots

The second type of hydroponic system is the bare root system, in which no physical support is given to the root system, which is suspended in the nutrient solution. Several types of bare root systems are used. One is the aeroponic system, in which bare roots enclosed in a chamber are suspended in the air, and a fine mist of oxygen-rich nutrient solutions is sprayed on the roots at various time intervals (Figure 8-4). Another type is the continually aerated system, which is similar to the aeroponic system except that the roots are bathed in the aerated nutrient solution instead of being intermittently sprayed with oxygen-rich nutrient solutions (Figure 8-5).

There are also **floating systems,** in which the plant is put on a raft that floats on the nutrient solution with the roots completely immersed below (Figure 8-6).

Finally, the **nutrient film technique (NFT)** system involves recirculating a shallow stream of nutrient solution with no solid rooting medium. The plant is supported above the trough in rockwool or another material that can support the plant, and the roots are placed in troughs with a shallow stream of circulating nutrient solution. A main tank is located at the lowest point of the setup, which contains nutrient solution that is pumped up to the ends of the trough. The solution then flows down a 1 percent slope back to the main tank, where it is recirculated again. The plants produce a root mat, part of which is above the nutrient

Figure 8–5 This bare root hydroponic system is under continuous aeration to grow geranium plants.

Figure 8–6 A bare root floating hydroponic system. The top panel shows a flowering *Arabidopsis* plant floating on a membrane raft, and the bottom panel shows a close-up of the plant in the early stages of development.

Figure 8–7 The nutrient film technique recirculates a shallow stream of nutrient solution. *Courtesy of Dr. E. Jay Holcomb, Department of Horticulture, The Pennsylvania State University.*

solution while the lower part is submerged and bathed in the circulating nutrient solution (Figure 8-7).

Hydroponic systems are typically used in greenhouses in the winter to grow high-value crops, such as tomatoes, lettuce, cucumbers, snapdragons, and a variety of other flowering crops and vegetables. In all cases, aerating nutrient solutions is important to maintain oxygen levels and avoid carbon dioxide buildup, which can be toxic to plants.

SOIL: CONTENTS, STRUCTURE, AND PROFILE

Soil is composed of rocks, minerals, organic matter, air, and water. In addition to its composition, the soil's structure and profile are also key to plant growth and development.

Soil Contents

The top few inches of the Earth's surface consist of minerals obtained from weathered rock, air, water, and organic matter. Basically, five different factors are involved in the formation of soil: **parent material,** climate, types of organisms, slope of the land, and time. The three types of parent materials are igneous (granite), sedimentary (sandstone), and metamorphic (slate) rocks. The inorganic component of soil comes from several kinds of mineral materials, including sand, silt, and clay.

The organic component of soil comes from organic matter, which is the decayed remains of plants and animals. In general, soil is made up of 45 percent minerals, 25 percent water, 25 percent air, and only 5 percent organic matter. Organic matter is very important for soil productivity because it is a source of nutrients when decomposing. Organic matter also promotes the water- and nutrient-holding capacity of the soil. If the soil is not properly managed, erosion can occur, which depletes the soil of organic matter. Soil should be properly managed to maintain a good amount of organic matter and beneficial organisms, such as earthworms, fungi, and others. Proper management of soil includes maintaining physical conditions that promote good aeration and drainage, which allow microbes to do their jobs. The soil pH should be maintained above 4.5 because acidic soils hurt many soil microorganisms. Organic matter can be increased by using green manure, which is a cover crop that is plowed under.

Soil texture is the proportion of sand, silt, and clay present in the soil. A **soil triangle** (Figure 8-8) is used as a method of classifying soil on the basis of

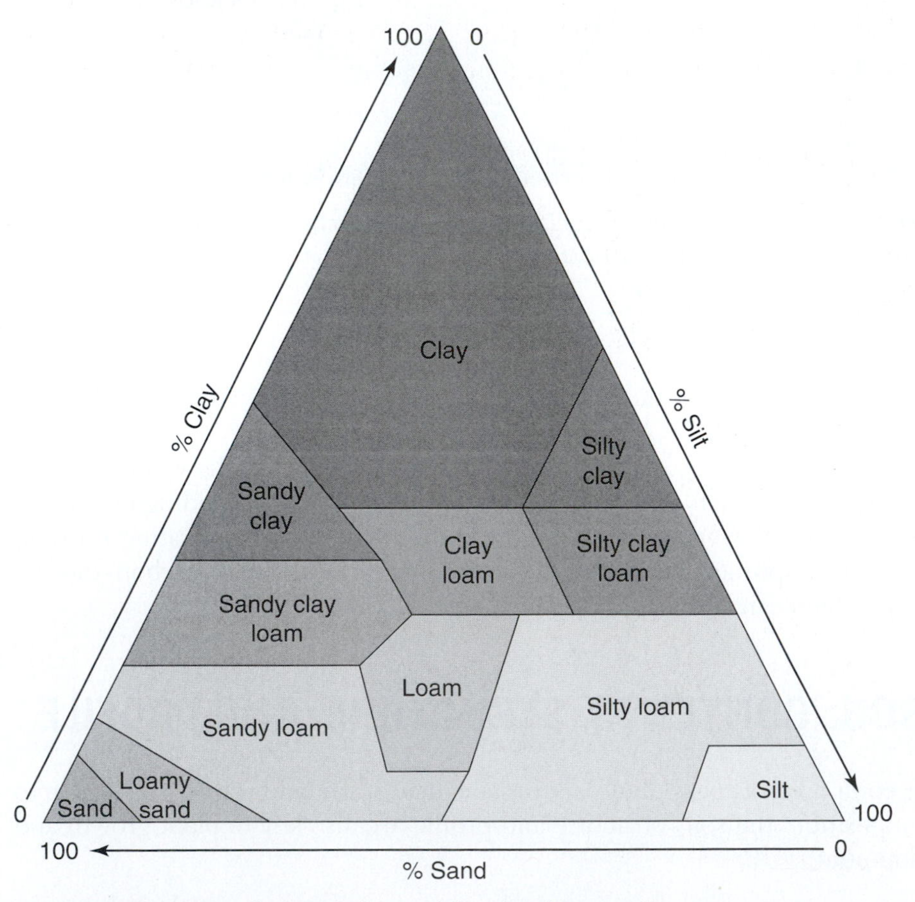

Figure 8–8 The soil triangle is used for classifying soils.

mineral content (texture). The USDA soil textural triangle is easy to use. First, analyze the soil to find out the proportions of sand, silt, and clay. For example, if you have 40 percent sand, 30 percent silt, and 20 percent clay, the first step is to find 40 percent on the sand axis, and then draw a line parallel to the silt axis. The next step is to find either 30 percent silt or 20 percent clay. If you chose clay, look for 20 percent on the clay axis and draw a line parallel to the sand axis to intersect the previous line from the sand axis. The part of the triangle where the two lines intersect indicates the class of the soil texture. Soil texture affects drainage of the soil, with sandier soils having more drainage and soils with more clay having less drainage.

Soil Structure

Soil structure is the physical arrangement of soil particles. Soil structure profoundly affects pore size, water-holding capacity, and the capability of water and air to penetrate the soil. Poor soil structure hinders plant growth and development. Factors that have been shown to affect soil structure include tillage, traffic, and rainfall. For example, bringing heavy machinery into the field when the soil is too wet damages the soil and leads to **soil compaction,** which occurs when soil is compressed into a relatively dense mass. Soil compaction adversely affects pore size, water infiltration, and soil aeration, which is the movement of atmospheric air into the soil. Organic matter or sand can be incorporated to improve soil structure.

Soil Profile

A soil profile is the vertical section of soil at a particular location showing the various layers, also called **horizons,** that have developed over a period of time (Figure 8-9). The development of a soil profile depends upon the age of

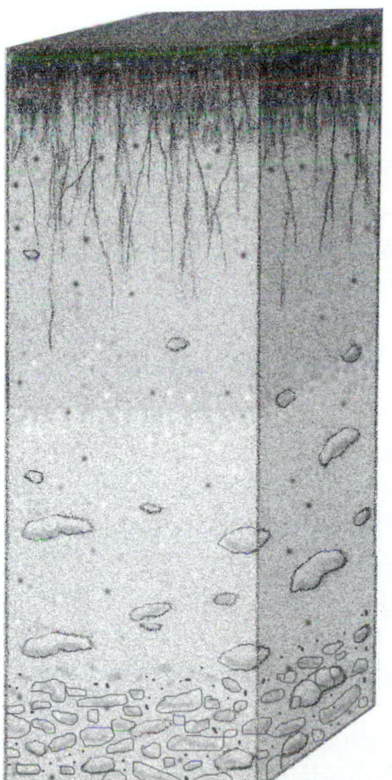

Figure 8-9 Soil profile showing the horizons, or layers

the soil; typically, young soils show less development than older ones. Soil profiles can be complicated; however, only four general horizons are discussed here. The uppermost horizon is the organic residue layer above the soil surface called topsoil or A-horizon. This layer is where most roots are found and experiences the most intense weathering. The next layer is the B-horizon, which has the subsoil layer containing fine particles, leached materials, and some roots. The third layer, the C-horizon, contains parent material, such as weathered bedrock and some leached materials. The fourth horizon is the R-horizon, which is made up of bedrock or parent material, used to form soil.

PLANT NUTRITION

According to Epstein (1972), the two principal criteria by which an element can be judged essential to any plant are as follows:

- An element is essential if the plant cannot complete its life cycle (forming viable seeds) in the absence of that element.
- An element is essential if it forms part of any molecule or constituent of the plant that itself is essential in the plant. For example, nitrogen is an integral part of proteins, and magnesium is in chlorophyll. Just because an element is in the plant does not mean it is essential, because plants indiscriminately take up a variety of materials in water.

There are 16 essential elements required by most plants for normal plant growth and development. The three major elements used by plants are carbon from CO_2, and hydrogen and oxygen from H_2O. In addition to carbon, hydrogen, and oxygen, 13 other essential inorganic nutrients are required for plant growth and are obtained from the soil.

Primary Macronutrients

Macronutrients are chemical elements that are required in large amounts for normal growth and development of plants. The primary macronutrients include nitrogen (N), phosphorous (P), and potassium (K).

Primary macronutrients are required in large amounts, and soils are commonly deficient in one or more of them.

Nitrogen

Beside carbon, hydrogen, and oxygen, nitrogen is the most important element in growing vegetative tissue. Nitrogen is one of the most widely used elements for plant nutrition. Nitrogen is taken up by the plant as nitrate (NO_3^-) or ammonium (NH_4^+) and is used by the plant to make chlorophyll, amino acids, proteins, and nucleic acids (DNA and RNA). Vegetative growth is promoted by nitrogen fertilization; if applied improperly, this can result in delayed maturity leading to a reduction in crop yields. Deficiency symptoms include stunted growth and/or yellowing of leaves, or **chlorosis.** Chlorosis can be due to poor chlorophyll development or destruction of chlorophyll caused by mineral deficiency or pathogen attack. Chlorotic symptoms begin on the older leaves located at the lower portion of the plant and then progressively spread throughout the plant as the deficiency becomes worse. Nitrogen is highly soluble in water and very mobile, so it can be lost from the soil by leaching, by runoff, by being tied up on soil colloids, or by being taken up by the plant.

Phosphorous

Phosphorous is readily taken up by the plant as orthophosphate ions—the main one being $H_2PO_4^-$—and is used for protein and nucleic acid (DNA and RNA) synthesis and energy transfer processes in plants involving adenosine triphosphate (ATP) and adenosine diphosphate (ADP). Deficiency symptoms include dark green leaves with a purplish color on the underside and stunted plants. Phosphorous is mobile in the plant and moves from older to younger leaves when a deficiency exists.

Potassium

Potassium is taken up by the plant in its ionic form, which is K^+. Potassium is used in many enzymatic reactions in the plant and is involved in protein synthesis and growth in meristematic tissues. Potassium is very mobile and thus is subject to leaching in sandy soils. Unlike nitrogen and phosphorous, which are used to produce compounds for plant growth, potassium is found in plant tissues in soluble inorganic salts. Potassium-deficient plants have marginal burning of leaves, speckled leaves, and leaf curling, thereby reducing their size. Potassium deficiency typically occurs on older leaves and progresses to the younger leaves. When plants are deficient in potassium, an accumulation of high levels of water-soluble nitrogen is found in the leaves, which may be responsible for the observed necrosis (Salisbury & Ross, 1991).

Secondary Macronutrients

Secondary macronutrients are needed in much lower amounts than primary macronutrients. In fact, they are typically required in trace amounts and are not generally deficient in soils.

The secondary macronutrients are calcium (Ca), magnesium (Mg), and sulfur (S).

Calcium

Calcium is taken up by the plant in its ionic form, which is Ca^{2+}. Calcium is involved in a number of physiological processes in plants, including cell division, cell growth, cell wall formation, and nitrogen accumulation. Plants that are deficient in calcium have malformed terminal buds and typically have poor root growth. Tomato plants deficient in calcium have fruits with blossom end rot; that is, the bottom portion of the tomato fruit rots and turns brown or black.

Magnesium

Magnesium is also taken up by the plant in its ionic form, which is Mg^{2+}. Magnesium is an integral part of the chlorophyll molecule (Figure 8-10) and is also required for the formation of fats and sugars. It is highly mobile in plants, and deficiency symptoms typically occur in older leaves with yellowing between the veins. Later these deficiency symptoms move to the younger leaves.

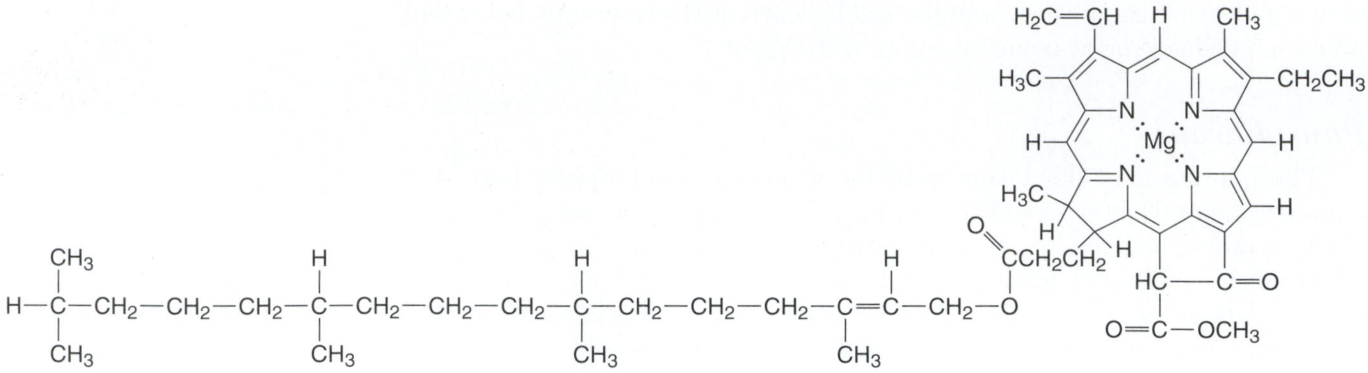

Figure 8–10 The structure of chlorophyll with Mg in the middle

Sulfur

Sulfur is taken up by the plant as SO_4^{-2}. Sulfur is involved in flavors in plants and is also a component in vitamins and amino acids. The main sulfur deficiency symptom is chlorotic foliage.

Micronutrients

Micronutrients are chemical elements that are essential for normal plant growth and development, but they are needed in much smaller amounts than macronutrients. The micronutrients that are essential for normal plant growth and development are iron (Fe), manganese (Mn), molybdenum (Mo), copper (Cu), boron (B), zinc (Zn), and chlorine (Cl).

Boron

Boron is highly mobile in the plant and is taken-up as BO_4^{2+}. Boron is involved in cell division, flowering, fruiting, and a variety of other physiological processes in plants. Deficiency symptoms are death or malformation of the terminal bud, which leads to multiple bud break. The young leaves on these shoots become thick and chlorotic.

Iron

Iron can be absorbed from the leaves or roots mainly as Fe^{2+}. It is a component of a wide range of enzymes and it is involved in the synthesis of chlorophyll. Iron is immobile in the plant and thus deficiency symptoms show up as interveinal chlorosis in younger leaves first.

Molybdenum

Molybdenum is taken up by the plant as MoO_4^{-2} (molybdate). Molybdenum is involved in protein synthesis and is required for nitrogen reduction in some plants. Cereals, grasses, and vegetables exhibit deficiency symptoms in molybdenum-deficient soils. Plants deficient in molybdenum turn pale yellow and the leaves roll up.

Manganese

Manganese is taken up by the plant as Mn^{2+}. It has a pivotal role in chlorophyll synthesis and is important in the activation of enzymes. Manganese deficiency is similar to iron deficiency as interveinal chlorosis is first observed in the younger leaves.

Zinc

Zinc is taken up by the plant as Zn^{2+}. It has a key role as an enzyme activator. Deficiency symptoms typically occur in soils that are high in phosphorous. Typical symptoms of zinc deficiency are reduced leaf size and internode length, resulting in a rosette appearance, and interveinal chlorosis may occur in younger foliage.

Copper

Copper is taken up by the plant as Cu^{2+}. It is important for chlorophyll synthesis and acts as a catalyst for a variety of enzymatic reactions. Younger leaves show interveinal chlorosis; terminal leaves and buds die, resulting in stunting of the plant.

Chlorine

Chlorine is taken up by the plant as Cl^{-}. Although deficiency symptoms are typically rare, the visual effects are stunting, chlorosis, and some necrosis. Typically, there are excessive amounts of chlorine, which cause more problems than a deficiency.

Factors Affecting pH and the Availability of Nutrients

Prior to applying fertilizers, a soil test should be conducted to determine which nutrients are present in the soil. The soil test can be done either by a commercial laboratory or by the grower. To obtain a proper sample for analysis, the tester takes multiple samples over the entire field and combines them. After the results of the soil test are received, the soil temperature, CEC, and pH must be considered because they determine the availability of nutrients in the soil (Figure 8-11).

Soil pH is the measurement of alkalinity or acidity in the soil based on the hydrogen ion concentration. This is based on a scale from 0 to 14; 7 is neutral, values above are alkaline, and values below are acidic. Most plants grow best between pH 5.5 and 8.0 (Figure 8-12). A variety of factors affect pH, such as the amount of rainfall, whether the soil has good or poor drainage, and which crops have been grown at that location over the previous seasons. To raise the pH of a soil, agricultural limestone (contains calcium) or dolomitic limestone (contains calcium and magnesium) is typically used. To reduce the pH of the soil, sulfur compounds are added. The color of flowers on some plants is based on the pH of the soil in which they are grown. The hydrangea has blue flowers in alkaline soils and pink flowers in acidic soils; this plant is sometimes called the “litmus paper” plant because it is so sensitive to pH (Figure 8-13).

Cation exchange capacity (CEC) is the measure of total exchangeable cations a soil can hold. The CEC of a soil is used as an index of soil fertility. Many essential elements carry positively charged ions called cations; they include K^{+}, Ca^{+2}, Mg^{+2}, Cu^{+2}, Fe^{+2}, Mn^{+2}, and Zn^{+2}. When soils are highly fertile, they typically have large surface areas that can attract and hold nutrients. Soils that are rich in organic matter have a high CEC and thus can hold a large amount of cations.

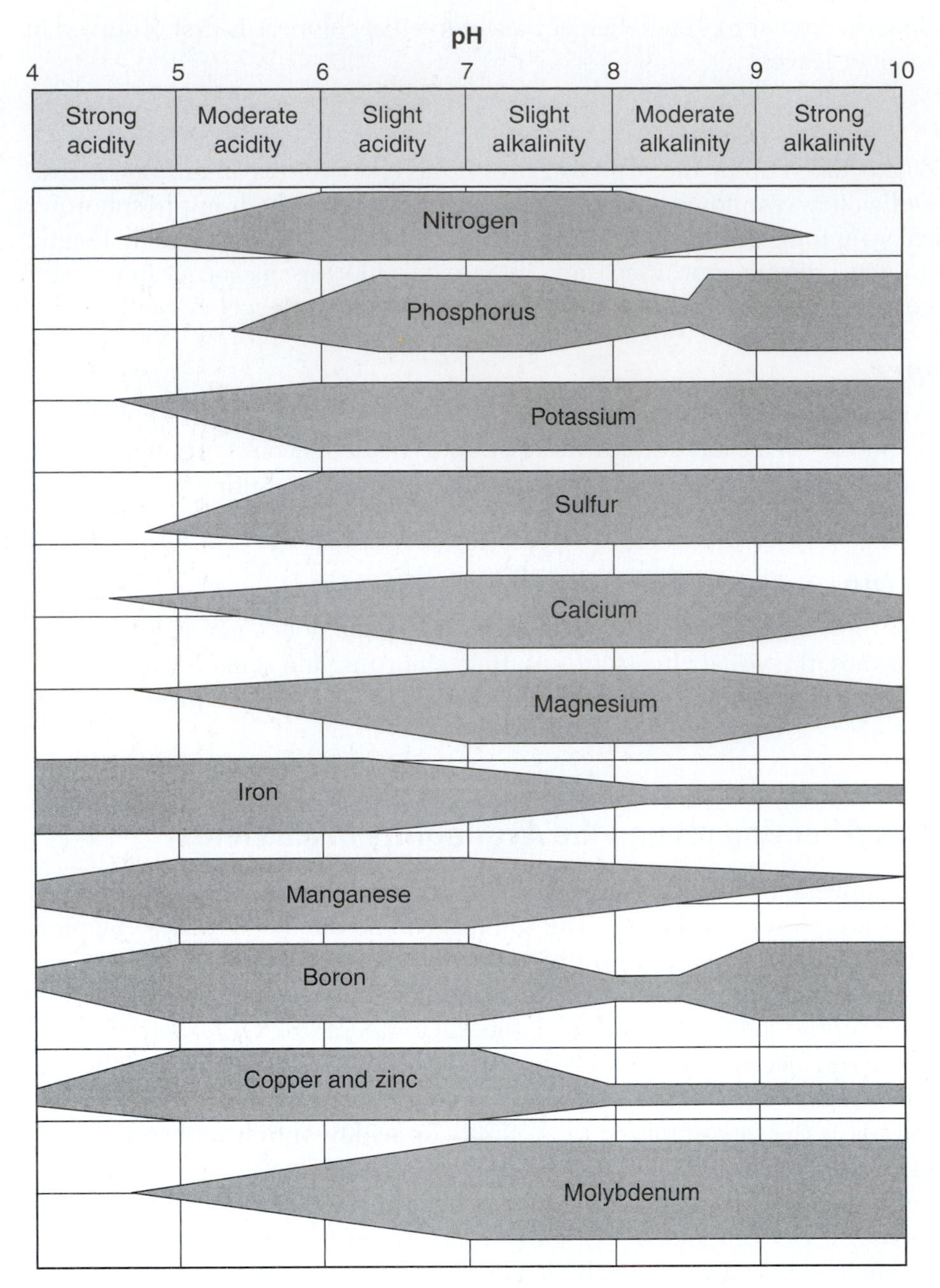

Figure 8–11 The effects of soil pH on nutrient availability

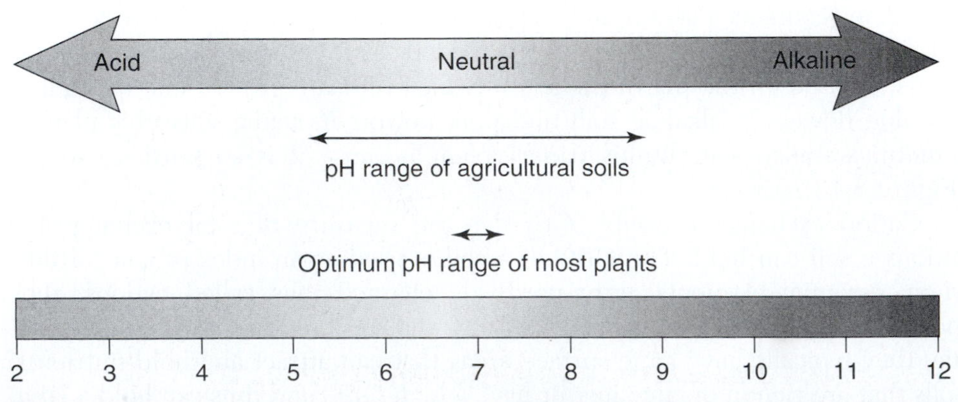

Figure 8–12 The pH range found in agricultural soils and the optimum pH range for most plants

Methods Used to Monitor Nutritional Levels

Three systems are used to determine the nutritional status of the plant: soil testing, foliar analysis, and **visual diagnosis.**

Soil Testing

Soil testing is the most effective way to determine nutritional levels because it provides the pH of the soil, the amount of nutrients found in the soil, and the availability of these nutrients (Figure 8-14). To maximize the effectiveness of soil testing, a proper representative sample for a specific location must be obtained. The land to be analyzed should first be evaluated and divided into sections if necessary. Within each of the sections, multiple samples should be taken and pooled to avoid biasing the sample.

Foliar Analysis

Foliar analysis is similar to soil analysis. The main difference between the two is that soil testing provides information on the availability of nutrients for uptake by the plant, whereas foliar analysis provides information on the amount of nutrients taken up by the plant and accumulated in the leaves. This type of

Figure 8–13 The hydrangea plant is called the "litmus paper" plant. *Courtesy of Dr. Francis H. Witham, Department of Horticulture, The Pennsylvania State University.*

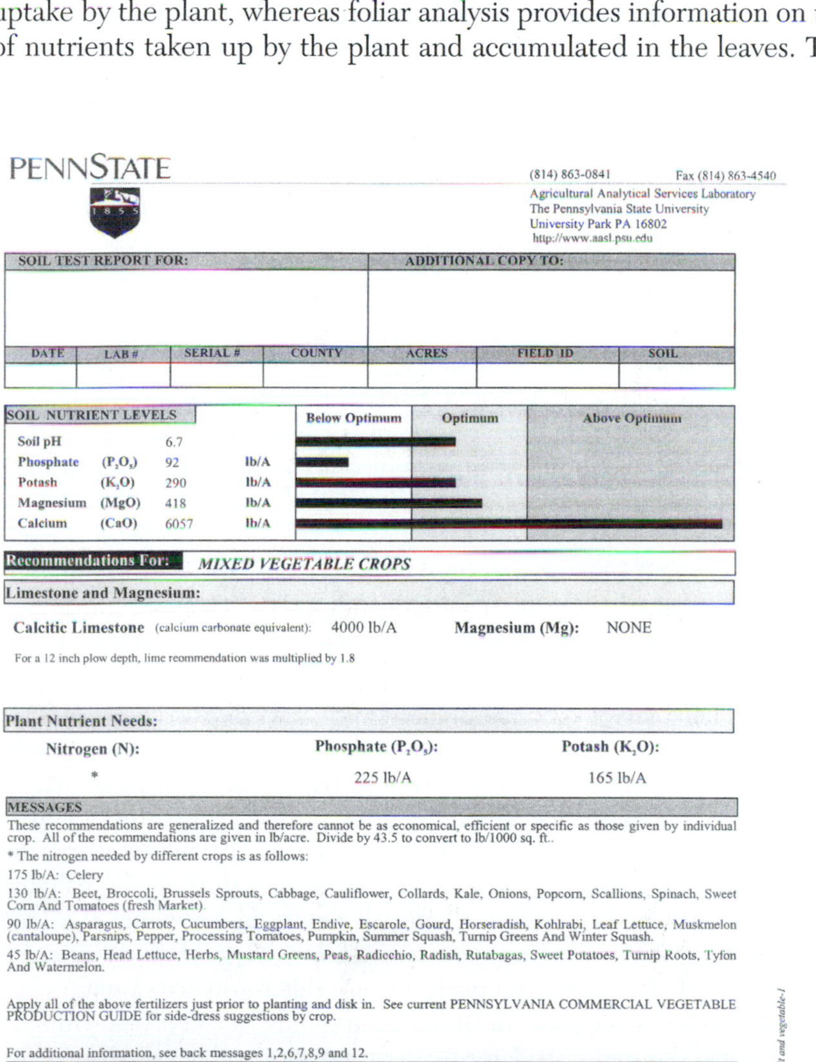

PENNSTATE 1855

(814) 863-0841 Fax (814) 863-4540
Agricultural Analytical Services Laboratory
The Pennsylvania State University
University Park PA 16802
http://www.aasl.psu.edu

SOIL TEST REPORT FOR:	ADDITIONAL COPY TO:

DATE	LAB #	SERIAL #	COUNTY	ACRES	FIELD ID	SOIL

SOIL NUTRIENT LEVELS				Below Optimum	Optimum	Above Optimum
Soil pH		6.7				
Phosphate	(P_2O_5)	92	lb/A			
Potash	(K_2O)	290	lb/A			
Magnesium	(MgO)	418	lb/A			
Calcium	(CaO)	6057	lb/A			

Recommendations For: *MIXED VEGETABLE CROPS*

Limestone and Magnesium:

Calcitic Limestone (calcium carbonate equivalent): 4000 lb/A **Magnesium (Mg):** NONE

For a 12 inch plow depth, lime recommendation was multiplied by 1.8

Plant Nutrient Needs:

Nitrogen (N):	Phosphate (P_2O_5):	Potash (K_2O):
*	225 lb/A	165 lb/A

MESSAGES

These recommendations are generalized and therefore cannot be as economical, efficient or specific as those given by individual crop. All of the recommendations are given in lb/acre. Divide by 43.5 to convert to lb/1000 sq. ft..

* The nitrogen needed by different crops is as follows:

175 lb/A: Celery

130 lb/A: Beet, Broccoli, Brussels Sprouts, Cabbage, Cauliflower, Collards, Kale, Onions, Popcorn, Scallions, Spinach, Sweet Corn And Tomatoes (fresh Market).

90 lb/A: Asparagus, Carrots, Cucumbers, Eggplant, Endive, Escarole, Gourd, Horseradish, Kohlrabi, Leaf Lettuce, Muskmelon (cantaloupe), Parsnips, Pepper, Processing Tomatoes, Pumpkin, Summer Squash, Turnip Greens And Winter Squash.

45 lb/A: Beans, Head Lettuce, Herbs, Mustard Greens, Peas, Radicchio, Radish, Rutabagas, Sweet Potatoes, Turnip Roots, Tyfon And Watermelon.

Apply all of the above fertilizers just prior to planting and disk in. See current PENNSYLVANIA COMMERCIAL VEGETABLE PRODUCTION GUIDE for side-dress suggestions by crop.

For additional information, see back messages 1,2,6,7,8,9 and 12.

LABORATORY RESULTS:											Optional Tests:		
		Exchangeable Cations (meq/100g)					% Saturation of the CEC						
[1]pH	[2]P lb/A	[3]Acidity	[2]K	[2]Mg	[2]Ca	[4]CEC	K	Mg	Ca		Organic Matter %	Nitrate-N ppm	Soluble salts mmhos/cm
6.7	40	2.0	0.3	1.1	10.9	14.3	2.2	7.4	76.4				

Test Methods: [1]1:1 soil:water pH, [2]Mehlich 3 Extractant, [3]SMP Buffer pH, [4]Summation of Cations

6084

Commercial fruit and vegetable-1

Figure 8–14 Soil testing information sheet

analysis is used to determine the level of nutrients actually found in the plant. Plants take up an abundant supply of nutrients from the soil, and, at some point, the uptake of these nutrients does not correspond to an increase in plant growth. This process is called **luxury consumption.** When the amount of nutrients in the plant is greater than normally taken up due to luxury consumption, the nutrients can be toxic to the plant; however, if they are limited, deficiency symptoms can occur. Based on foliar analysis, the proper steps must be taken to overcome the problem found. A representative leaf sample must be harvested to determine accurately the nutrient content of the plant. A wide diversity exists between plant species on the proper leaves to select for analysis; therefore, you should check the current literature or ask your local extension agent as to the proper leaves to use for analysis. In general, samples should be taken at four- to six-week intervals.

Visual Diagnosis

Visual diagnosis is based solely on visual observation and requires a trained eye. The problem with this type of evaluation is that it occurs after there is a problem, which must be corrected quickly to avoid permanent damage. This type of diagnosis is only recommended if no other options are available.

FERTILIZERS

Continual attention to soil fertility maximizes crop production. Soil nutrients have four fates: they can be used by the plant, lost by leaching, lost due to runoff, and/or tied up in the soil by chemical reactions. Lost nutrients must be replaced to ensure that crops have adequate nutrients for maximal plant growth and development. Nutrients can be added back to the soil in a variety of ways, such as use of a fallow period and use of organic or inorganic fertilizers. Leaving the land uncultivated for a period of several years to rejuvenate naturally is called leaving the land fallow. This only works when land is plentiful and not needed for several years. Typically, soil fertility is increased by applying a fertilizer, which is any organic or inorganic material used to provide the nutrients plants need for normal plant growth and development. Organic gardening, which is a system for growing crops without inorganic input, is becoming very popular. Fertilizers used for organic gardening include green manure, which is the use of a leguminous cover crop for the main purpose of plowing it under to increase soil fertility. (This should not be confused with the use of organic wastes from cows, poultry, and others as a source of fertilizer.) Other forms of organic fertilizers are bone meal and dried blood. Organic fertilizers are still only used by homeowners and for specialty crops because of the difficulties associated with them. Problems associated with organic fertilizers are difficulty in calculating the correct amount of nutrients to apply, difficulty in handling, the large amounts of space for storage, and, in many cases, odors. Organic fertilizers typically have a low nutrient content and nutrients are released very slowly.

Inorganic fertilizers, which are nutrient compounds derived from mineral salts, are the most commonly used form in agriculture today because of the many associated benefits. Inorganic fertilizers are easy to apply and store, are commonly available in specific formulations to suit individual needs, can be applied as a foliar spray or to the soil, are available in both liquid and solid forms, and provide many other benefits.

Fertilizer Formulations

A complete fertilizer contains all three primary fertilizer nutrients and may have select micronutrients. An incomplete fertilizer lacks one or more of the three primary elements found in a complete fertilizer. Fertilizers can be obtained as fast release, which means the nutrients are immediately available. These should be used with care because they can cause burning if used improperly. There are also slow or controlled release fertilizers, which means nutrients are available over a long period of time but do not overcome a deficiency quickly. An example of a commonly used slow release fertilizer is Osmocote. There are also fertilizer formulations with herbicide, where the herbicide is 2, 4-dichlorophenoxyacetic acid (2,4-D), which is an auxin compound that selectively kills broadleaf weeds.

Fertilizer Nomenclature

Figure 8-15 shows how to read the label on a fertilizer bag. For example, the fertilizer analysis, which is the proportion of nutrients supplied by a fertilizer formulation for a complete fertilizer containing 16 percent nitrogen, 4 percent phosphate, and 8 percent potash, is marketed as a 16-4-8 analysis or a ratio of 4-1-2. The total active ingredient, which is the total percentage of nutrients being applied, is 28 percent, and the total inert ingredient, which is the carrier or filler ingredient, is 72 percent. The inert ingredient typically makes up the major weight in a bag of fertilizer. Fertilizer can come as either a solid granular or liquid, which can be applied in the irrigation water or a foliar application.

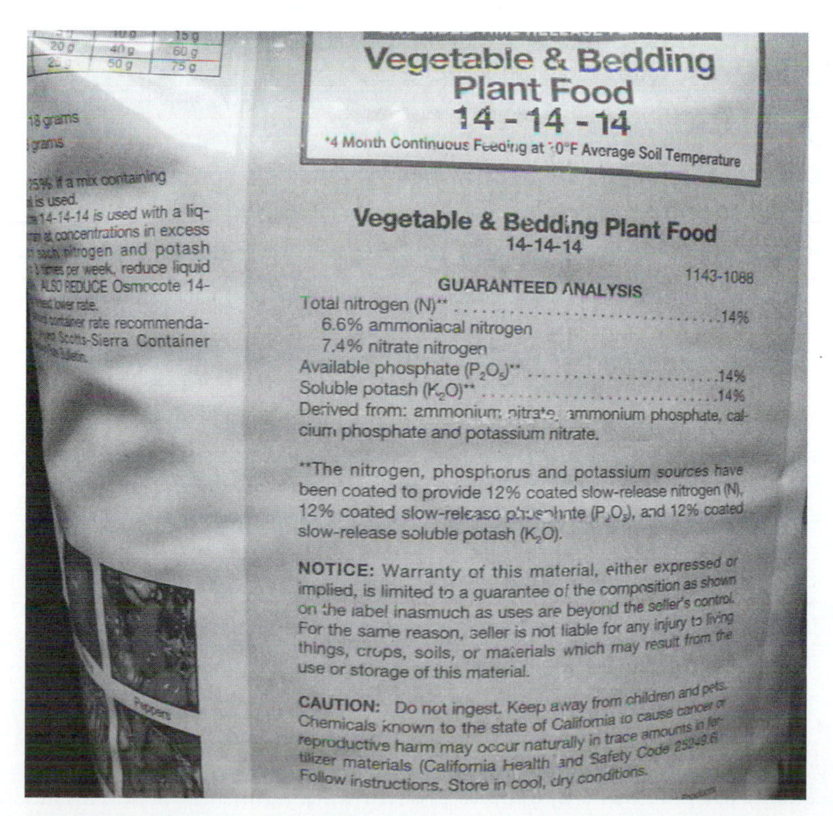

Figure 8–15 Side of fertilizer bag showing information typically present

Fertilizer Application

Prior to fertilizer application, you should run a soil test to know what amount of fertilizer to apply. Fertilizers can be applied prior to planting if necessary, and then at various time intervals during the growing season based on subsequent analysis of nutrients found in the soil or plant by testing.

Methods of Application

Fertilizers can be applied in three ways. The first is dry application, which is applying dry fertilizers to the soil. Dry applications can be made in several different ways. Broadcasting refers to evenly spreading the fertilizer either manually or mechanically on the soil surface. Broadcasting is quick and easy, but fertilizer may be wasted depending upon plant spacing. Top dressing is the placement of fertilizer uniformly around the stem of the plant. Side dressing is when fertilizer is placed in bands along both sides of a row of plants.

The second way to apply fertilizers is through liquid application, which is the application of water-soluble forms of fertilizers to either the soil or leaves. When fertilizer is applied to the soil through the irrigation system, it is referred to as **fertigation.**

A third way to apply fertilizer is by gas application, such as applying nitrogen in the form of ammonia gas.

Timing of Application

After soil tests have been done and the amount of nutrients and method of application has been determined, you must decide on the frequency and rate of fertilizer application. The frequency and rate of fertilizer application depends on several factors, including plant genetics, environmental factors, edaphic or soil conditions, biotic environment, and the plant's stage of development. Plants go through three main stages of plant growth and development. First is the juvenile or vegetative phase, which is characterized by exponential increases in size and the inability to shift from vegetative to reproductive maturity leading to the formation of flowers. The second stage is the reproductive or mature phase, which refers to qualitative changes that allow the plant or organ to express its full reproductive potential. The third stage is senescence, which is internally controlled deteriorative changes, or the natural causes of death. Each of these phases has different nutritional requirements and should be dealt with accordingly. The plant's genetics determines the overall size of the plant and how rapidly the plant goes through each of the stages of growth and therefore should be taken into consideration when deciding fertilization requirements. Environmental factors, soil conditions, and the biotic environment all affect uptake of nutrients by the plant, availability of nutrients to the plant, and leaching of nutrients from the soil. These factors should be taken into consideration when deciding the frequency and rate of fertilizer application.

SUMMARY

You now have a basic knowledge of plant growth media, nutrients, and fertilizers. You know the major roles of media and the four major kinds of media typically used in horticulture today. In addition, you learned how soils are formed and what they are comprised of, and you gained a better understanding of soil structure and soil profile. You learned about the 16 essential plant nutrients required for normal plant growth and development and are knowledgeable on factors affecting pH and availability of nutrients and methods used to successfully monitor nutritional levels in plants. You learned what different fertilizer formulations are available and basic fertilizer nomenclature found on the bag. Also in this chapter you learned commonly used methods of fertilizer application and the proper time for fertilizer application.

Review Questions for Chapter 8

Short Answer

1. What are the seven major roles of media?
2. What are four commonly used kinds of media for growing plants?
3. What are six advantages of using soilless media?
4. What are three disadvantages of using soilless media?
5. What are three advantages and one disadvantage of hydroponic growing systems over soil culture?
6. List two ways to classify hydroponic systems and give two examples for each.
7. What are the three types of parent material used to produce soil?
8. What are the two major reasons for aerating nutrient solutions?
9. What are two principal criteria by which an element can be judged essential to any plant?
10. How many essential elements do most plants require for normal plant growth and development?
11. Where do plants get carbon, hydrogen, and oxygen?
12. List the 16 essential inorganic elements and indicate which are primary macronutrients, secondary macronutrients, and micronutrients.
13. What are two ways nitrogen can be taken up by the plant?
14. What plant constituents have nitrogen as part of their structure?
15. Provide two ways to modify soil pH, one to increase it and one to decrease it.
16. List and explain three systems used to determine the nutritional status of the plant.
17. What are four fates of fertilizers applied to the soil?
18. What are three ways that fertilizers can be applied?
19. What are five factors that must be taken into consideration when deciding the frequency and rate of fertilizer application?

Define

Define the following terms:

soil
dirt
peat moss
perlite
vermiculite
organic matter
soil triangle
soil structure
soil compaction
soil aeration
soil profile
horizons
cation exchange capacity (CEC)
macronutrients
micronutrients
soil pH
soil testing
foliar analysis
luxury consumption
fallow
fertilizer
green manure
complete fertilizer
incomplete fertilizer
fertilizer analysis
active ingredient
inert ingredient

True or False

1. Perlite is made from heat-treated mica and its primary use is for nutrient- and water-holding capacity.
2. Vermiculite is made from volcanic materials and is used in the medium for drainage and aeration.
3. Peat moss is used for its nutrient- and water-holding capacity.
4. Bark is now used as a substitute for peat moss.
5. Wood chips and sawdust are now being used as an alternative to peat moss.
6. Sand is used for aeration and nutrient-holding capacity.
7. Hydroponic methods for growing plants can be broken down into substrate and bare root systems.
8. A soil triangle is used as a method of classifying soils based on the texture of the soil.
9. The CEC of a soil is used as an index of soil fertility.
10. Just because an element is found in the plant does not mean it is essential.
11. The color of flowers on some plants is related to the pH of the soil in which they are grown.
12. Nutrients are carbohydrates produced by the process of photosynthesis in the leaf.
13. Potassium is taken up by the plant as K_2O.
14. Calcium is taken up by the plant in its ionic form, which is Ca^{2+}.

Multiple Choice

1. Cover crops may be used as a fertilizer; a commonly used term for cover crops used in this capacity is
   A. green manure.
   B. supplemental fertilizer.
   C. living fertilizer.
   D. None of the above
2. Some fertilizers contain a herbicide. Which of the following herbicides is commonly found in fertilizers?
   A. 2,4,5-D
   B. 2,4-D
   C. 2,4,5-T
   D. All of the above
3. A complete fertilizer contains which of the following?
   A. All of the essential elements
   B. Both primary and secondary macronutrients
   C. All three primary macronutrients
   D. All of the above
4. An incomplete fertilizer contains which of the following?
   A. All of the essential elements
   B. Both primary and secondary macronutrients
   C. All three primary nutrients and may have some select micronutrients
   D. None of the above

Fill in the Blanks

1. The major disadvantage of hydroponic growing systems over soil culture is ____________________.
2. Aerating the solutions of plants grown hydroponically maintains ____________________ levels and avoids ____________________ buildup.
3. One of the advantages of hydroponic growing systems over soil culture is that the yield per area is ____________________.
4. Soil texture is determined by the proportion of ____________________, ____________________, and ____________________ present in the soil.
5. The CEC is used as an index of ____________________.
6. Nitrogen is taken up by the plant as ____________________ or ____________________.
7. Lime can be used as a source of nutrients. Agricultural limestone contains ____________________, whereas dolomitic limestone contains ____________________ and ____________________.
8. After carbon, hydrogen, and oxygen, ____________________ is the most important element in growing vegetative tissue.
9. Prior to fertilizer application, it is important to run a ____________________ to know what amount of fertilizer to apply.
10. Carbon used by plants comes from ____________________, whereas hydrogen and oxygen come from water.
11. To get accurate results from soil tests, it is important to get the proper sample taken; this is accomplished by taking ____________________ over the entire field.
12. Fertigation is the application of fertilizers through the ____________________ system.

Matching

Some letters may be used more than once.

1. Nitrogen
2. Phosphorous
3. Potassium
4. Sulfur
5. Calcium
6. Magnesium
7. Boron
8. Copper
9. Chlorine
10. Iron
11. Manganese
12. Molybedenon
13. Zinc

A. Primary macronutrient
B. Secondary macronutrient
C. Micronutrient

Activities

Fertilizer is readily available to most consumers in large department stores or small garden shops. Because the proper selection of a fertilizer is essential to the life and growth of the plant, consumers should know what they are buying. In this activity, you will analyze and compare different fertilizers in the same way that a typical consumer should when selecting a fertilizer for a particular use.

1. Visit a local department store and/or a gardening shop and see what types of fertilizer are available.
2. Select one fertilizer type and then evaluate at least three different brands by obtaining the following information:
   - Size of bags and cost per unit
   - Active ingredients
   - Inactive or inert ingredients
   - Crops for which fertilizer is recommended
   - Warnings or cautions
   - Other useful information
3. Compare the information you found for each brand of fertilizer and determine which brand you would recommend to an inquiring consumer. Write a paragraph explaining why you would make that recommendation and what criteria influenced your decision.

References

Epstein, E. (1972). *Mineral nutrition of plants: Principles and perspectives.* New York: Wiley.

Salisbury, F. B. & Ross, C. W. (1991). *Plant physiology* (4th ed.). Belmont, CA: Wadsworth Publishing.

CHAPTER 9

Plants and Their Environment

Objectives

After reading this chapter, you should be able to

- recognize how abiotic and biotic atmospheric factors affect plant growth and development.
- understand how abiotic and biotic edaphic environmental factors influence plant growth and development.

Key Terms

abiotic
arthropods
atmospheric environment
available water
biotic
climate
cuticle
day-neutral plants
ECe number
edaphic environment
epinasty
etiolation
evapotranspiration
facultative fungi
field moisture capacity
fungi
gravitational water
hardened-off
hardpan
hardy plant
humidity
light intensity
light quality
long-day plants
mutualism
nematodes
nonarthropods
obligatory fungi
parasitic fungi
pathogen
pathogenic
photoperiod
plant environment
plant selection
relative humidity
saprophytic fungi
saturated soil
short-day plants
soil aeration
soil pH
soil salinity
tender plant
thatch
transpiration
turgid
water stress
wear
weather
wilting point

ABSTRACT

The abiotic atmospheric environment, including temperature, moisture, light, and wind, affects plant growth and development. The biotic atmospheric environment also affects plants. The abiotic edaphic environment includes water movement into the soil, soil water availability, soil aeration, soil pH, and saline soils. The biotic edaphic environment is filled with a variety of organisms; this chapter covers the four main categories found in this region, including microorganisms, arthropods, nonarthropods, and vertebrate animals, and their influence on plant growth and development.

INTRODUCTION

Plants grow well only when they get what they need from the environment. Horticulture involves growing plants at times and in places that are not natural for the plant; for example, growing poinsettias for Christmas or lilies for Easter in Pennsylvania. Orchids grow outdoors in Hawaii but not in Pennsylvania, unless they are grown in greenhouses. Therefore, understanding the plant responses to the environment enables horticulturists to better create an artificial environment to grow them.

Figure 9–1 Understanding natural plant environments enable horticulturists to create artificial growing environments.

The **plant environment** is the above- and below-ground surroundings of the plant, which has a profound effect on the plant (Figure 9-1). The two main areas of the plant environment are the **atmospheric environment,** which is the above-ground portion of a terrestrial plant's environment, and the **edaphic environment,** which is the soil and area where plant roots are located. Both the atmospheric and edaphic environment can be broken down into the **abiotic** (nonliving factors) and **biotic** (living factors) that affect the environment.

When discussing the atmospheric environment, you must understand the difference between weather and climate. **Weather** is the combined effect of complex interactions among temperature, rainfall, wind, light, and **relative humidity** in a specific location. Weather factors that form patterns on a daily, weekly, monthly, and yearly basis and create the same pattern year after year are the **climate.**

Atmospheric conditions affecting plant growth are both abiotic and biotic. Abiotic factors that affect atmospheric conditions are light, temperature, air, moisture, and wind. The main source of light is from the sun. The three important components of light are **light intensity,** which is the actual quantity of light; **light quality,** which is the actual color or wavelength of light; and **photoperiod,** which is the length of the dark period that influences plant growth. Temperature is also an important abiotic factor because it determines when and where a given type of plant can be grown.

When discussing important abiotic factors that affect plant growth, air does not immediately come to mind; however, without air there would be no plant growth. The main components of air are nitrogen, oxygen, and carbon dioxide. Oxygen is required for almost all normal physiological processes in plants. Carbon dioxide is required for photosynthesis. Good-quality air must be available to maximize growth because pollutants such as ozone, sulfur dioxide, and others cause major reductions in plant growth and development.

Another important abiotic factor to consider is moisture, which is found in the air (in the form of **humidity**) and in the soil or growth medium (in the form of water). Water is used to keep plants **turgid,** which is when cells are full of water; anytime the plant is subjected to **water stress,** the overall potential yield of the plant is reduced. Water is also used by the plant to cool itself through the process of **transpiration,** which is the loss of water from the plant through the leaves in the form of vapor. Plants should be well watered with good-quality water because polluted water can cause a variety of problems leading to a reduction in crop yields.

Wind is another abiotic factor affecting plant growth by cooling the plant, reducing moisture, and reducing disease problems. In addition, wind causes plants to be smaller and more compact with a tougher **cuticle,** which is an impermeable, waxy material on the outside layer of leaves and stems that prevents water loss, thereby making the plant more resistant to numerous stresses. Wind can also be detrimental to plant growth by reducing plant size and subsequent crop yields or by reducing moisture to a point that causes water stress.

Biotic factors also have a profound effect on plant growth and development. Biotic factors affecting the atmospheric environment include insects and related pests; **nematodes;** diseases, such as viruses, **fungi,** and bacteria; weeds; and rodents; and other animals.

The edaphic environment, which is the soil and area where plant roots are located, can also be affected by abiotic and biotic conditions. Abiotic factors discussed in this chapter are water movement into the soil, soil water availability, **soil aeration,** soil temperature, **soil pH,** and **soil salinity.** Following irrigation or rainfall, water must move into the soil or waterlogging or drought can occur. Water movement into the soil is affected by the soil texture, soil structure, **thatch, hardpans,** and the presence of different layers in the soil together with a variety of other factors. Soil water availability is another important factor influencing plant growth and development. The soil must retain an adequate supply of **available water** to prevent water stress. For normal physiological processes to occur in plant roots, soils must be aerated to maintain the proper balance of O_2 and CO_2 in the soil. Soil aeration is also critical to a variety of microorganisms necessary for good soils. Proper soil temperature must also be maintained because it regulates all chemical reactions in the plant and the soil. Soil temperature is affected by the type of soil (sandy or clay) as well as atmospheric conditions such as air temperature, wind, and solar radiation. Because nutrient availability to the plant from the soil is affected by the pH of the soil, the pH range should be between 6.0 and 8.0, which is the optimal range for plant growth. Soil salinity, which is the amount of salt in the soil, is problematic in agriculture because it reduces crop yields in a number of important crops.

Biotic factors also have a profound effect on the edaphic environment. Soil organisms can be broken down into four groups: microorganisms, such as fungi and bacteria; **arthropod** animals, such as centipedes, ants, and grubs; **nonarthropod** animals, such as nematodes and earthworms; and vertebrate animals, such as mice and gophers. More details on biotic factors that affect the edaphic environment will be discussed in Chapter 12.

PLANT ENVIRONMENT

Plant growth and development is first affected by the plant's genetics; however, the plant environment, which is the above- and below-ground surroundings of the plant, also profoundly affects the plant. In fact, genetically identical plants can exhibit a dramatically different appearance when grown under different environmental conditions. Therefore, a grower must be knowledgeable in all aspects of the plant's environment to maximize plant growth and development for high yields and quality crops. The two main areas of the plant environment are the atmospheric environment, which is the above-ground portion of a terrestrial plant's environment, and the edaphic environment, which is the soil and area where plant roots are located. Both the atmospheric and edaphic environment can be broken down into the abiotic (nonliving factors) and biotic (living factors).

Atmospheric Environment

Climate and weather, although linked, are not the same. Weather is the combined effect of complex interactions among temperature, rainfall, wind, light, and relative humidity in a specific location. Weather factors have changing patterns on a daily, weekly, monthly, and yearly basis. The same pattern weather year after year is the climate. The sun influences atmospheric conditions, including temperature, moisture, light, and wind; these conditions, in turn, affect plant growth (Figure 9-2). Atmospheric conditions that have a profound effect on plant growth are discussed next.

Abiotic Factors

Abiotic factors are nonliving factors that affect the atmospheric environment, such as light, temperature, air, moisture, and wind.

Light. The main source of light for all plants is the sun. When discussing light, three components must be taken into consideration: light intensity, light quality, and photoperiod. Light intensity is the actual quantity of light. The highest

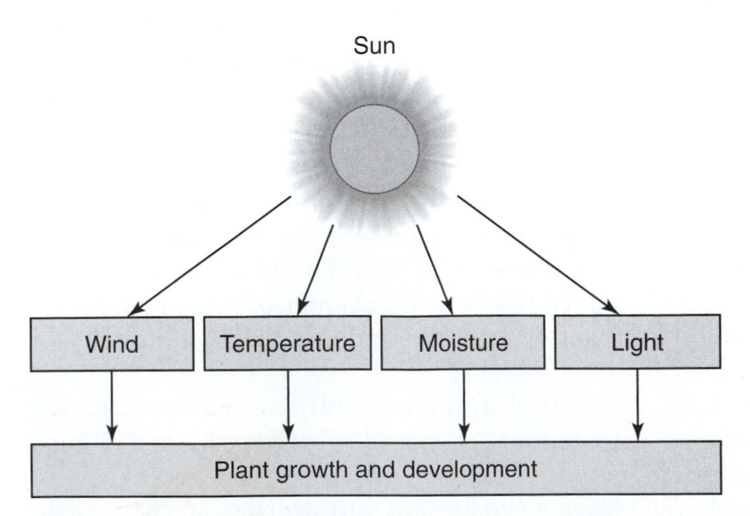

Figure 9–2 The sun affects atmospheric conditions and subsequently plant growth and development.

Figure 9–3 The plant on the left receiving high light levels is short, study, and dark green; the plant on the right is tall receiving low light levels with a spindly stem and has pale green leaves and cotyledons.

light intensity occurs at noon. The maximum amount of light at this time is 2,000 μE m^{-2} s^{-1} or 10,000 foot candles of light. The actual amount of light reaching the plant is affected by a variety of factors, such as clouds, trees, or buildings that cause shading. After the light reaches the leaf surface, it can be absorbed or reflected, so only a small amount of light is used in photosynthesis. Light intensity has a profound effect on plant growth. At high light intensities, plants are typically shorter, appear dark green, and have brittle leaves. Plants grown under low light, however, are tall as a result of **etiolation,** which is exaggerated growth of the stem caused by low light levels (Figure 9-3). The naturally occurring plant hormone indole-3-acetic acid (IAA) plays a key role in etiolation-type growth in plants.

Light quality is the actual color or wavelength of light. Plants absorb light to be used for photosynthesis in the visible portion of the spectrum, which is between 390 and 750 nanometers (Figure 9-4). Most of the light reaching the surface of Earth is in the visible light range. In this range, the three major types of plant pigments—chlorophylls, carotenoids, and phycobilins—absorb specific wavelengths and are involved in photosynthesis. The light range in which a green leaf absorbs light in the visible spectrum is in the blue (approx. 430 nm) and red (approx. 540 nm) regions. The green leaf absorbs light the poorest in the green region (approx. 540 nm) of the spectrum, thereby reflecting mostly green light and making the leaf look green to the naked eye. Light quality helps determine the height of a plant; for example, red/yellow light promotes elongation, whereas green/blue light inhibits elongation and

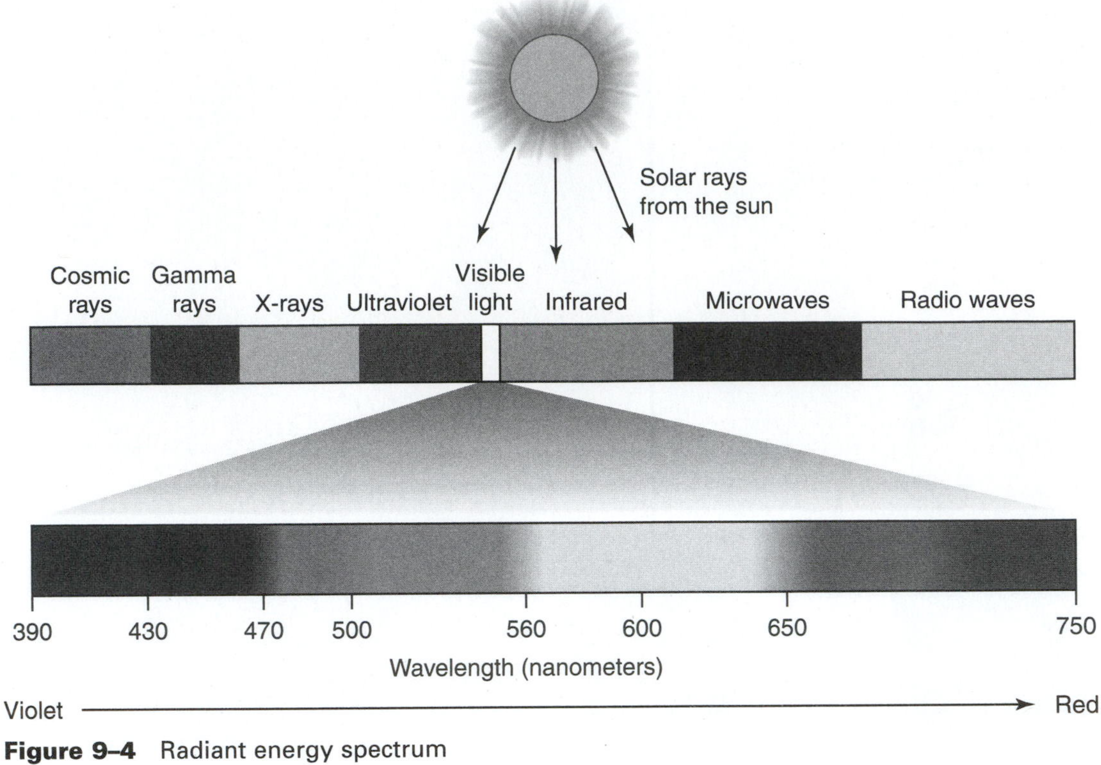

Figure 9–4 Radiant energy spectrum

promotes shorter plants. When supplying supplemental light, you must know the plant's requirement for light quality and quantity due to the wide variety of light sources available.

Photoperiod is the length of the dark period that influences plant growth. The plant photoperiod changes on a seasonal basis; in the fall, photoperiods are shortened, and in the spring, they become longer. Plants can be divided into three groups based on their photoperiod: short-day, long-day, and day-neutral. **Short-day plants** are plants that flower only when the dark period is greater than a certain critical length, because plants measure the length of the dark period in order to flower. Examples of short-day plants are shown in Table 9-1. **Long-day plants** only flower when the dark period is shorter than a certain critical length. Examples of long-day plants are shown in Table 9-2. **Day-neutral plants** determine flower initiation solely by the genotype and have no specific light requirement. Examples of day-neutral plants are shown in Table 9-3.

TABLE 9-1 SHORT-DAY PLANTS

Plant	Scientific Name
Poinsettia	*Euphorbia pulcherrima*
Chrysanthemum	*Chrysanthemum x morifolium*
Kalanchoe	*Kalancho blossfeldiana*
Strawberry	*Fragaria x ananasia*
Violet	*Viola papilionaceae*

TABLE 9-2 LONG-DAY PLANTS

Plant	Scientific Name
Spider plant	*Chlorophytum comosum*
Baby's breath	*Gypsophila paniculata*
Evening primrose	*Oenothera* spp.
Fuchsia	*Fuchsia x hybrida*
Rex begonia	*Begonia rex*

TABLE 9-3 DAY-NEUTRAL PLANTS

Plant	Scientific Name
Cucumber	*Cucumis sativus*
Corn	*Zea mays*
Pea	*Pisum sativum*
Tomato	*Lycopersicon esculentum*
Kidney bean	*Phaseolus vulgaris*

Temperature. Temperature is important for all physiological processes in plants. It determines when and where a given type of plant can be grown. Prior to planting, you should know the temperature extremes in your particular location. The U.S. Department of Agriculture (USDA) Plant Hardiness Map shows regions of the United States where plants can and cannot be grown. This map gives the range of average annual minimum temperatures for 11 zones throughout the United States. Although an excellent starting point for deciding what plants can be planted in your area, this should not be your only source of information. Check with your local weather authorities to determine the temperature extremes in your specific location.

Plants vary in their temperature requirements. **Hardy plants** are less sensitive to temperature extremes, whereas **tender plants** cannot tolerate cool weather. Examples of hardy and tender plants are shown in Table 9-4 and

TABLE 9-4 HARDY PLANTS

Plant	Scientific Name
Geranium	*Pelargonium* spp.
Impatiens	*Impatiens* spp.
Primrose	*Primula* spp.
Hydrangea	*Hydrangea* spp.
Lettuce	*Lactuca sativa*
Carrot	*Daucus carota*
Broccoli	*Brassica* spp.
Radish	*Raphanus sativa*

TABLE 9-5 TENDER PLANTS

Plant	Scientific Name
African violet	*Saintpaulia* spp.
Begonia	*Begonia* spp.
Coleus	*Coleus blumei*
Pineapple	*Ananas comosus*
Petunia	*Petunia* spp.
Cucumber	*Cucurbita sativus*
Tomato	*Lycopersicon esculentum*
Sweet potato	*Ipomea batatas*

Table 9-5, respectively. Damage to the plant by adverse temperatures depends on the physiological state of the plant. For example, actively growing succulent plants with flower buds are more susceptible to frost damage than those in a dormant state. To improve the chances of survival of both hardy and tender plants, the plant should be **hardened-off,** or gradually subjected to cooler temperatures with less frequent watering.

Air. The main components of air are nitrogen, oxygen, and carbon dioxide. For normal physiological processes to occur in the plant, oxygen is required. Carbon dioxide is required for photosynthesis. When either oxygen or carbon dioxide is limited, plant growth is adversely affected. Another important function of the air is to hold water. The water content of the air is called humidity. A psychrometer is an instrument used to measure the relative humidity. Relative humidity is the ratio of the weight of water vapor in a given quantity of air to the total weight of water vapor that a quantity of air can hold at a given temperature (expressed as a percentage).

Good air quality is important to maximize plant growth. Environmental pollutants are becoming a major problem especially in heavily populated or high industry areas. Pollutants that cause significant reductions in plant growth and development are ozone, sulfur dioxide, peroxyacetyl nitrate (PAN) fluoride, and ethylene, plus a variety of pesticides, heavy metals, and others.

Moisture. Water is the most important requirement for plant growth. Water in the atmosphere comes from precipitation and evaporation. Plants use water for all biochemical and physiological processes. Water is required for photosynthesis and normal growth of the plant; when water is limited, photosynthesis is reduced, which leads to a reduction in plant growth and normal development. The plant cools itself during hot weather via transpiration, which is the loss of water from the plant through the leaves in the form of vapor. Nutrients and organic compounds are transported throughout the plant in water; at water deficits, this transport is dramatically reduced. Water keeps cells turgid, which means plant cells are full of water. Water stress occurs when a plant is unable to absorb an adequate amount of water to replace that lost by transpiration. When plants lose turgidity, they wilt; each time the plant wilts, its overall potential yield is decreased. Excessive water also kills the plant because oxygen in the roots is reduced.

Environmental pollutants cause many problems in agriculture. Pesticides and nutrients are pollutants that cause significant reductions in plant growth and

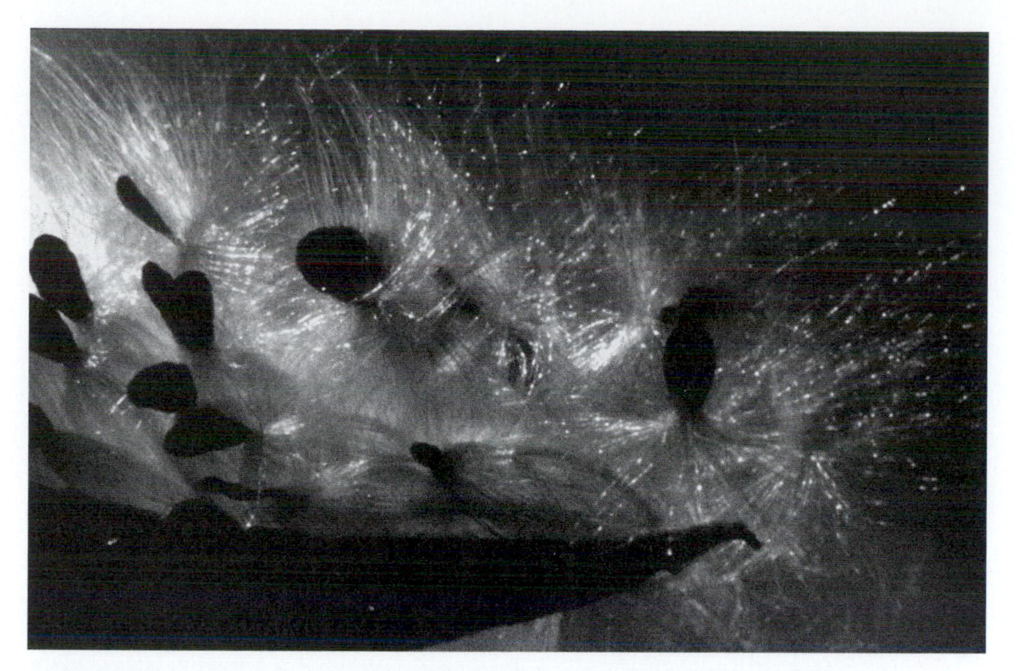

Figure 9–5 These milkweed seeds are ready to be dispersed by the wind. *Courtesy of Dr. Francis H. Witham, Department of Horticulture, The Pennsylvania State University.*

development. Acid rain is caused when sulfur oxide and nitrogen oxide react with water in the atmosphere to form sulfuric acid and nitric acid, which reduces the pH of rain from a normal of pH 6 to as low as 4. The reduction in pH adversely affects the environment in a variety of ways; for example, acidified lakes show dramatically reduced fish populations. Acidification of the soil caused by acid rain alters nutrient availability. This leads to a variety of problems that reduce crop yields.

Wind. Wind is air in motion, which can be beneficial or harmful. Wind cools the plant when temperatures are high and reduces moisture on the plant and soil surfaces, which reduces disease problems. Wind also makes a plant shorter with a tougher cuticle, which is an impermeable, waxy material on the outside layer of leaves and stems that prevents water loss. By shortening the plant and making a tougher cuticle layer, the plant is more resistant to mechanical stress, **pathogen** (a disease-causing organism), or insect attack. Although wind can be beneficial, it can also cause critical problems. Some of the benefits of wind can also be detriments; for example, shorter plants with smaller leaves yield significantly less than normal size plants. Excessive wind can also cause desiccation, which accelerates the senescence process, leading to premature death. In addition, excessive wind can result in physical damage to the plant due to shear force. Wind can also spread fungal spores, weed seeds (Figure 9-5), pollutants, salt, and other pests.

Biotic Factors

Biotic factors are living organisms—ranging from animals and other plants (weeds) to microorganisms—that affect the atmospheric environment. More specifically, biotic factors can be insects and related pests; nematodes; diseases, such as viruses, fungi, and bacteria; weeds; rodents; and other animals. Climate affects the biotic factors; in turn, the biotic factors affect the biotic atmospheric environment. For any plant pest to become a problem, it must have a favorable environment. Both environmental and mechanical stress can adversely affect the biotic atmospheric environment by causing **wear,** which is the physical deterioration of a plant community resulting from excessive stress. Wear can come from a variety of sources in nature, including people, wind, and precipitation. These stresses disrupt the normal atmospheric environment surrounding the plant, resulting in a reduction in normal growth and development. **Plant selection** refers to selecting species that are tolerant to specific environmental stresses, thereby enabling the plant to withstand an adverse biotic atmospheric environment. Plants are being genetically engineered to make them more resistant to wear.

Edaphic Environment

The edaphic environment is the soil and area where plant roots are located. Abiotic and biotic factors affect this environment. Soil provides nutrients, water, gas exchange, and physical support to plant roots. The proper combination of sand, silt, and clay, plus organic matter, is very important for adequate nutrient and water retention, aeration and drainage (gas exchange), and physical support for the plant. Edaphic conditions that have a profound effect on plant growth are described next.

Abiotic Factors

Abiotic factors affecting the edaphic environment include water movement in the soil, water availability, soil aeration, soil temperature, soil pH, and soil salinity.

Water movement into the soil. Water movement into the soil is very important because if water does not penetrate the soil surface, waterlogging or drought can occur. The capability of water from precipitation or irrigation to enter the soil depends on soil texture, soil structure, and the presence of different layers in the soil, together with a variety of other factors. If the soil is subjected to water very rapidly, water will pool on the surface and eventually run off. The pooling of water also decreases the capability of water to penetrate the surface by blocking the surface pores. Water moves into the soil in two scenarios: when the soil is **saturated,** which is when all the pore spaces are filled with water, and when the soil is not saturated. When the soil is saturated, water movement is termed **gravitational water,** because water moves from the large pore spaces due to the pull of gravity. Water in a soil that is not saturated moves in response to a matrix potential gradient; in other words, the water moves from moist areas to drier areas of the soil. Water movement into the soil can be impeded in lawns by thatch, which is the layer of organic residue above the soil surface and just below the green leaves of the host plant (Figure 9-6). When thatch becomes too thick, it causes pooling and water runoff. The downward movement of water can also be impeded due to a hardpan, which occurs when soil is compressed into a very dense mass. This can be caused by the weight of heavy machinery or people. Heavy tractors and harvesting equipment can compress the soil, which causes a hardpan that restricts root growth and water movement, thereby reducing crop yields. Some weeds—such as annual bluegrass and white clover—can tolerate hardpans, which further decreases the plant's quality. In addition to hardpans, the rate at

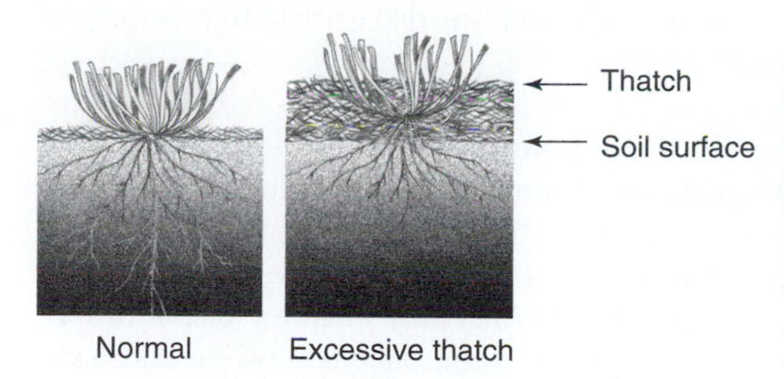

Figure 9–6 The plant on the left has a healthy root system with only a small amount of thatch; the plant on the right has an excessive amount of thatch and a poorly developed root system.

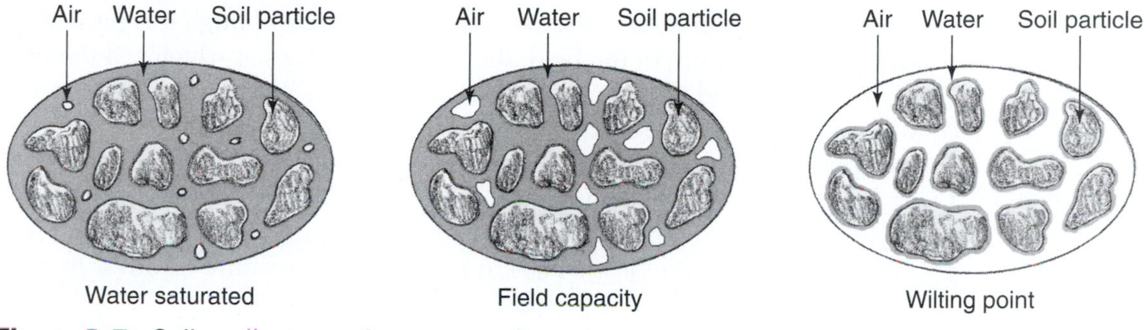

Figure 9–7 Soil at all stages of water retention

which water moves through the soil also depends on the normal composition of the different layers of soil.

Soil water availability. The capability of the soil to retain water is important for normal plant growth and development. When soil receives a large amount of water due to irrigation or a heavy rain, soil becomes saturated, and water freely drains in response to gravity. After all gravitational water has drained out of the large pore spaces leaving only the small pore spaces containing water, the soil is at **field moisture capacity.** This water may be lost from the soil surface by evaporation or from the leaf surface by transpiration; together these are called **evapotranspiration.** When water can no longer be absorbed by the plant, moisture stress occurs, causing the plant to wilt. At this stage, the plant is said to be at its **wilting point.** The difference between soil moisture at field capacity and the wilting point is called the available water (Figure 9-7).

Soil aeration. Soil aeration is the movement of air into the soil. In general, a mineral soil contains about 25 percent air. Keeping soils well aerated is important for maintaining the proper balance of O_2 and CO_2. High levels of CO_2 in the soil are toxic, and low levels of O_2 inhibit root growth. Clay soils are highly susceptible to aeration problems, whereas sandy soils are typically well aerated. Both tomato and pea plants are susceptible to oxygen deficiencies; they will show signs of **epinasty,** which is the downward bending of the petioles (Figure 9-8), wilt, and eventually die if the problem is not corrected. In general, at soil oxygen levels that are less than 10 or 12 percent, plants undergo stress that limits plant growth. Maintaining proper soil aeration and water aeration is important for hydroponically grown plants to maximize plant growth and

Figure 9–8 This hydroponically grown tomato plant has become epinastic due to low oxygen levels in the root zone.

subsequent crop yields. Oxygen levels in the soil are also critical to a variety of microorganisms that are necessary for good soils.

Soil temperature. The proper soil temperature is important for overall plant growth because temperature regulates all chemical reactions in the plant and soil. In general, plant roots stop growth when soil temperatures are 5°C (41°C) or colder. Soil temperature is affected by a variety of factors, one of which is the type of soil. Sandy soils warm up more quickly than clay soils due to the differences in their water content. Sandy soils drain faster, have less water, and have more air spaces than clay soils, which enables them to warm up faster than water-retaining clay soils. In addition, soil temperature is also affected by atmospheric conditions such as air temperature, wind, and solar radiation. Soil temperature can be modified to enhance crop production. Although it is difficult to control soil temperature in the field, there are many ways in which soil temperature can be modified. Applying plastic or mulch to the soil surface is one way to control soil temperature effectively in the field. Covering the soil with different colored plastics and mulches has been shown to be very effective (Figure 9-9). Using light-colored plastic or mulch reflects sunlight and lowers soil temperature; using dark-colored plastic or mulch absorbs light and warms the soil. Some growers use black plastic or mulches to raise soil temperature in order to start crops early.

Another way to modify soil temperature in the field is to promote adequate drainage either by modifying the soil texture by adding sand or by using raised beds to promote soil drainage and soil warming. Controlling soil temperature in a greenhouse is not as hard as in the field. Recent research has shown that supplemental root zone heating can have beneficial effects on plant growth because root zone temperatures in greenhouses fluctuate more than in the field.

Figure 9–9 Different colored plastics cover rows in a field to help control soil temperature. *Courtesy of Dr. Michael D. Orzolek, Department of Horticulture, The Pennsylvania State University.*

TABLE 9-6 PLANTS WITH DIFFERENT pH REQUIREMENTS

pH 4.5 to 5.5	pH 5.5 to 6.5	pH 6.5 to 7.5
Potato	Tomato	Pea
Strawberry	Corn	Broccoli
Blueberry	Watermelon	Cabbage
Dandelion	Zinnia	Poinsettia
Gardenia	Pansy	Gladiolus
Azalea	Boston fern	Crocus
Rhododendron	Dogwood	Apple

Soil pH. Soil pH is the degree of acidity or alkalinity of the soil. The pH of the soil affects nutrient availability and should be maintained in the proper range to maximize crop yields. For example, nitrogen is most available between pH 6.0 and 8.0; as the soil becomes more acidic or more alkaline, the availability of nitrogen is reduced. In addition, nitrogen-fixing bacteria called *Rhizobia* work best in the pH range between 6.0 and 8.0. When the soil experiences low rainfall or it is poorly drained, salts generally accumulate and lead to an increase in soil pH. Soils exposed to heavy rainfall with good drainage typically have a lower soil pH. Examples of pH requirements for selected horticultural crops are shown in Table 9-6. Factors that affect soil pH and plant nutrition are discussed in more detail in Chapter 8, Media, Nutrients, and Fertilizers.

Saline soils. Soil salinity is the amount of salt (NaCl) in the soil. Salt stress is a major problem in agriculture because it reduces crop yields. Problems with salt buildup come from irrigation water and fertilizers. Millions of dollars are lost annually due to salt stress in plants. Selected examples of crop loss with increased salt as shown by **ECe number** (the units of electrical conductivity designated by milliSiemens/cm at 25°C) are shown in Table 9-7. When the salt

TABLE 9-7 CROPS THAT EXHIBIT A 50 PERCENT DECREASE IN YIELD WHEN SOIL SALINITY VALUES ARE BETWEEN 5 AND 10 (EXPRESSED AS [a]ECe)

Plant	Scientific Name
Tomato	*Lycopersicon esculentum*
Cucumber	*Cucumis sativus*
Potato	*Solanum tuberosum*
Bell pepper	*Capsicum annuum*
Chrysanthemun	*Chrysanthemum x morifolium*
Lily	*Lilium longiflorum*
Geranium	*Pelargonium x hortorum*
Grape	*Vitis* spp.
Apple	*Malus pumila*

[a] Units of electrical conductivity (ECe) are milliSiemens/cm at 25°C.

TABLE 9-8 SALT-TOLERANT AND SALT-INTOLERANT PLANTS

GENERALLY SALT TOLERANT

Plant	Scientific Name
Bermudagrass	*Cynodon dactylon*
Beet	*Beta vulgaris*
Broccoli	*Brassica oleraceae*
Muskmelon	*Cucumis melo*
Rose	*Rosa ordorata*

GENERALLY SALT INTOLERANT

Plant	Scientific Name
Kentucky bluegrass	*Poa prantensis*
Strawberry	*Fragaria* spp.
Geranium	*Pelargonium x hortorum*
Azalea	*Rhododendron* spp.
Pepper	*Capsicum annum*

levels in the soil get too high, plants exhibit similar symptoms to water stress, such as wilting, reduced plant growth, and, in some cases, the appearance of a green to bluish-green color. Plants have varying degrees of salt tolerance, as shown in Table 9-8.

Biotic Factors

Biotic factors are living organisms that affect the edaphic environment; they include a variety of living organisms, from animals and other plants (weeds) to microorganisms. Climate affects the biotic factors, which in turn affect the biotic environment. For any plant pest to become a problem, it must have a favorable environment. The edaphic environment is filled with a variety of simple and complex living organisms. These organisms can enhance this environment, thereby increasing crop yields, or they can be pests that damage plants. Soil organisms can be broken down into four groups: microorganisms, arthropod animals, nonarthropod animals, and vertebrate animals.

Microorganisms. The two classes of microorganisms found in the soil are bacteria and fungi. They can either be good or bad for the soil and plant. Bacteria are single-cell organisms that can occur in three different shapes—spherical, rod-like, or spiral—and are widespread in the edaphic environment. Bacteria can have a beneficial effect by causing the breakdown of dead organic material, thereby increasing the soil's organic matter content and the overall physical properties of the soil. Bacteria-plant interactions or symbioses have been shown to have beneficial effects in legumes. The bacteria *Rhizobia* spp. interacts with legume plant roots, enabling the roots to fix nitrogen, thereby reducing the need for nitrogen fertilization. **Pathogenic** (an organism that causes diseases) bacteria have rod-like shapes and thrive in soil with a pH of 6.5 to 7.5, which is the range that many plants also prefer. They enter the host through wounds or natural pores

TABLE 9-9 BACTERIAL DISEASES THAT AFFECT HORTICULTURAL CROPS

Common Name	Pathogen Scientific Name	Host Plant
Crown gall	*Agrobacterium tumifasciens*	Tree fruits and woody plants
Hairy root	*Agrobacterium rhizogenis*	Tree fruits and woody plants
Bacterial wilt	*Erwinia tracheiphila*	*Cucurbits* spp.
Bacterial canker	*Pseudomonas syringae*	Cherry, peach, and plum
Bacterial soft rot	*Erwinia carotovora*	Vegetables, tubers, and fruits
Common blight	*Xanthomonas phaseoli*	Beans

and cause a variety of problems, as shown in Table 9-9. Although bacterial diseases are difficult to control, bacterial disease-resistant cultivars are now available for many species.

Fungi, like bacteria, can also have both good and bad effects. Approximately 75 percent of all seed plants have some form of association with the fungi mychorrizae and its roots; this association is known as **mutualism,** which is when both the host and fungus benefit from the association. This type of association is similar to that observed with the bacteria *Rhizobia* spp. and its symbiotic relationship with legume roots. Much mychorrizal research has been done to understand this relationship better because the relationship enhances phosphorous uptake, which increases plant growth and development. The reproductive structures of fungi are called spores, which come in a variety of sizes, shapes, and colors. Spores are transported by wind, water, insects, birds, and others. Some fungi have coverings that protect them from an adverse external environment. Fungi gain access to the plant by wounds or through natural openings such as plant stomata. Interestingly, most plant diseases are caused by fungi, which are typically multicellular plants that lack chlorophyll. Fungi can be broken down into four classes: **saprophytic fungi,** which live only on dead tissue; **parasitic fungi,** which live on living tissue; **obligatory fungi,** which live on either dead or living tissue; and **facultative fungi,** which live on both living and dead tissues. Even though fungi cause most plant diseases, they are easy to control. Some examples of fungal diseases and their hosts are shown in Table 9-10.

TABLE 9-10 FUNGAL DISEASES THAT AFFECT HORTICULTURAL CROPS

Common Name	Pathogen Scientific Name	Host Plant
Dutch elm	*Ceratocystis ulmi*	Elm trees
Fusarium wilt	*Fusarium oxysporum*	Pea, tomato, and others
Damping-off	*Rhizoctonia* spp.	Commom problem in seedlings
Late blight	*Phytophoria infestans*	Tomato and potato
Rust	*Puccinia graminis*	Turfgrasses

Figure 9–10 Dead patches of grass in the lawn can be caused by grubs.

Arthropods and nonarthropods. Arthropods are animals that have exoskeletons and jointed legs, whereas nonarthropods have neither. Arthropods that have horticultural significance are centipedes, millipedes, termites, ants, grubs, and a variety of other insects. An advantage of termites and ants is that they can increase drainage due to the pore spaces they produce in the soil. However, they both also produce large soil hills in lawns, which are unsightly and cause problems with mowing. Grubs cause major problems in turf because they feed on the roots of grass plants, resulting in dead patches in the lawn (Figure 9-10). Nematodes, also called roundworms and earthworms, are nonarthropods. Nematodes are appendageless, nonsegmented, worm-like invertebrates with a body cavity and complete digestive tract and are the most abundant animal found in the soil. Nematodes are parasites that attack a variety of horticultural crops, including fruit trees, turfgrasses, vegetable crops, and ornamentals. One of the advantages of some types of nematodes is that they feed on the mole cricket, thereby acting as a form of biological control. However, the disadvantage of nematodes is that they feed on the roots of plants and create an entry for disease organisms. In addition, nematodes enter and inhabit the roots, which creates nodules on the roots and reduces the growth of the plant. Earthworms are very beneficial to the soil they inhabit. They do extremely well in areas that are moist and have high organic matter, so they are typically not found in dry, acidic soils. They improve the movement of water into the soil and soil aeration by producing holes in the soil. In addition, earthworms feed on thatch and dead material found in the soil, which increases organic matter and promotes better soil structure.

Vertebrate animals. Vertebrate animals that inhabit the edaphic environment are rodents and small animals such as mice, gophers, chipmunks, ground squirrels, and rabbits. Although the holes produced by these animals can increase soil drainage, these holes are generally too big and unsightly. Turf that the rodents have burrowed under also presents a hazard to people walking on it. In addition, rodents cause heavy economic losses to the grower annually because when they come out of their underground holes, they feed on crops and then retreat to their underground hiding place.

SUMMARY

You now understand how plants are affected by the environment. Abiotic atmospheric environmental factors—including temperature, moisture, light, and wind—together with biotic factors affect plant growth and development significantly. Abiotic edaphic factors include water movement into the soil, soil water availability, soil aeration, soil pH, and saline conditions. The biotic edaphic environment also has a profound effect on the plant. The biotic edaphic environment is rich in a variety of organisms, including microorganisms, arthropods, nonarthropods, and vertebrate animals.

Review Questions for Chapter 9

Short Answer

1. List two main areas of the plant environment and define each.
2. List five abiotic atmospheric conditions that have an effect on plant growth.
3. What are the effects of red/yellow light, green/blue light, high-intensity light, and low-intensity light on plant growth?
4. Why does excessive water applied to the roots kill plants?
5. What are two benefits and two problems associated with wind?
6. What are two problems caused by impeded water movement into the soil?
7. Why is it important to aerate the soil?
8. What are four factors that affect soil temperature?
9. Provide one way to control soil temperature in the field.
10. What are two ways salt builds up in the soil and causes problems with plant growth?
11. List the four main categories of soil organisms and provide examples of each.

Define

Define the following terms:

biotic	etiolation	day-neutral plants	relative humidity
abiotic	light quality	hardy plant	transpiration
weather	photoperiod	tender plant	turgid
climate	short-day plants	hardened-off	water stress
light intensity	long-day plants	humidity	cuticle

pathogen	available water	soil salinity	parasitic fungi
wear	field moisture capacity	ECe number	obligatory fungi
saturated soil	evapotranspiration	pathogenic	facultative fungi
gravitational water	wilting point	mutualism	arthropods
thatch	soil aeration	fungi	nonarthropods
hardpan	epinasty	saprophytic fungi	nematodes

True or False

1. Atmospheric conditions such as temperature, moisture, light, and wind are all influenced by the sun.
2. When plants are grown under high-light intensities, they are generally shorter and darker green than plants grown under low-light intensities.
3. Light color influences plant growth. Red/yellow light promotes elongation growth whereas green/blue promotes shorter plants.
4. Photoperiod is the length of the light period that affects plant growth.
5. One of the benefits of wind is that it promotes shorter plants with tougher cuticles, thereby making them more resistant to stress.
6. The biotic environment is the soil and area where plant roots are located.
7. Soil temperature is not effected by air temperature.

Multiple Choice

1. Excessive watering kills plants
   A. due to elevated levels of oxygen in the root zone.
   B. due to reduced levels of oxygen in the root zone.
   C. due to elevated atmospheric nitrogen levels in the root zone.
   D. None of the above
2. Field moisture capacity is
   A. the amount of water retained by the soil that plants can absorb.
   B. when the water content of the soil fills the small pore spaces.
   C. when the water content of the soil fills the large pore spaces.
   D. None of the above
3. Saline soils are caused by
   A. acid rain.
   B. salt buildup from irrigation water.
   C. improper crop rotations.
   D. All of the above

Fill in the Blanks

1. ____________________ keeps plant cells turgid.
2. Excessive water kills plants due to reduced ____________________ in the roots.
3. Water movement into the soil can be impeded by ____________________, which is the layer of organic residue above the soil surface and just below the green leaves of the host plant.

Activities

Now that you have learned how a plant's environment has a profound effect on its growth, you will have the opportunity to explore in more detail how the environment affects plants. In this activity, you will report on specific abiotic and biotic atmospheric and edaphic environmental conditions and show how they affect plant growth and development. Your activity is to search the Internet for two sites that present information on the effects of different environmental conditions on the plant. For each site that you find, answer the following questions:

1. How do environmental factors affect plant growth and development?
2. Provide potential or existing methods used for modifying environmental factor(s) under field and greenhouse conditions.
3. Provide examples of topics in this area that were exciting and less than exciting; be sure to explain why you came to these conclusions.
4. What is the URL for the Web site where you found your information?

CHAPTER

10

Plant Growth Regulators

ABSTRACT

Plant growth regulators are compounds that regulate plant growth and development. This chapter provides general background information pertaining to plant growth regulators and key definitions. There are six classes of plant hormones: auxins, gibberellins, cytokinins, abscisic acid, ethylene, and brassinosteroids. Each of the six classes of plant hormones is involved in physiological processes in plants, in both synthetic and naturally occurring forms. The chapter concludes with information about the practical uses of plant growth regulators in agriculture today.

Objectives

After reading this chapter, you should be able to

- provide general background information on the six classes of plant hormones, including key definitions.
- discuss each of the six classes of plant hormones, the history involved in their discovery, and their effect on specific physiological processes.
- provide information on the practical uses of plant growth regulators in agriculture.

Key Terms

abscisic acid
abscission
auxein
auxin
Bakanae disease
bolting
brassinolide
brassinosteroids
brassins
callus
cytokinins
dormancy
ethylene
flower initiation
gibberellin
gravitropism
heteroauxin
negative gravitropism
nutrients
parthenocarpy
phototropism
plant growth regulator
plant growth retardant
plant hormone
positive gravitropism
senescence
triple response
typiness

INTRODUCTION

This chapter provides a survey of plant growth regulating compounds and their many effects on plant growth and development, plus their current uses in agriculture. The six classes of **plant hormones** are **auxins, gibberellins, cytokinins, abscisic acid (ABA), ethylene,** and **brassinosteroids.** The first class of plant hormones was discovered more than 60 years ago. Today, plant growth regulating compounds are used for a variety of purposes in agriculture, including promoting or delaying of fruit ripening, inducting rooting, promoting or delaying abscission, controlling fruit development, controlling weeds, controlling size (Figure 10-1), and many others. The main difference between a plant hormone and **plant**

Figure 10–1 The plant growth retardant Sumagic® affects poinsettia height (control plant on left followed by increasing concentrations of Sumagic®). *Courtesy of Dr. E. J. Holcomb, Department of Horticulture, The Pennsylvania State University.*

growth regulator can be summed up by remembering: "All plant hormones are plant growth regulators, but not all plant growth regulators are plant hormones." This means that plant hormones are naturally occurring and so are plant growth regulators, but that plant growth regulators can be both naturally occurring and synthetic and therefore cannot fall into the plant hormone category.

Auxins—the first class of plant hormones discovered—were initially found in human urine. Since this time, they have been isolated from numerous plants and are thought to be ubiquitous in the plant kingdom. The term *auxin* is derived from the Greek word *auxein,* which means "to grow." Indole-3-acetic acid (IAA) is the only active auxin found in plants today and is involved in numerous physiological processes in plants, including cellular elongation, **phototropism, gravitropism,** apical dominance, root initiation, ethylene production, fruit growth, sex expression, and weed control.

Gibberellins were the next class of plant hormones discovered. Japanese farmers observed that their rice plants were taller earlier in the growing season; however, as the season progressed, these plants became spindly, chlorotic, and devoid of fruit. Japanese farmers called these symptoms the **Bakanae** (foolish seedling) **disease.** It was later found that the fungus *Gibberella fujikuroi* was responsible for the Bakanae disease. In 1935, a Japanese scientist named Yabuta isolated and crystallized the active component secreted by the fungus, and he called the compound Gibberellin A. Since this time, more than 100 gibberellins have been identified in plants. Gibberellins are involved in numerous physiological processes in plants: stem growth elongation, bolting and flowering, seed germination, **dormancy,** sex expression, fruit growth, and **parthenocarpy.**

Cytokinins, like auxins and gibberellins, were also not initially isolated from a plant source. In 1955, Miller and coworkers were the first to report the identification and purification of kinetin (6-furfuryl-aminopurine) from aged herring sperm DNA. The first naturally occurring cytokinin found in plants was from *Zea mays* by Miller and coworkers and Letham and coworkers in 1963; this compound was called zeatin. Since this time, more than 30 cytokinins have been identified and purified from plants. Cytokinins are involved in numerous physiological processes in plants, including cell division, organ formation, **senescence,** stomatal opening, lateral bud break, and sex expression (Arteca, 1996).

Abscisic acid was discovered in 1965 simultaneously by two research groups. The Addicott group in the United States used cotton abscission as its experimental system, and the Wareing group in England used dormancy in birch trees. Today, abscisic acid has been shown to be widely distributed within the plant kingdom, and it is the only inhibitor in this category. The major effect of abscisic acid in plants is its capability to signal that a plant is undergoing water stress. In addition to being a plant stress signal, abscisic acid is also involved in dormancy, **abscission,** delaying seed germination, and subsequent plant growth and development (Arteca, 1996).

The ancient Egyptians gashed their figs to stimulate ripening, and the Chinese burned incense in closed rooms to enhance ripening. They unknowingly were treating their fruits with ethylene to promote ripening. In 1935, Crocker and coworkers were the first to report that ethylene was the fruit-ripening hormone. Other roles of ethylene in physiological processes in plants are its effects on seedling growth, senescence, abscission, and sex expression (Arteca, 1996).

In the 1960s, Mitchell and coworkers at the USDA Research Center in Maryland began a screening program in search of a new class of plant hormones. Not until 1979 did Grove and coworkers identify the first brassinosteroid, which was called **brassinolide.** The major role of brassinosteroids in plants is the promotion of stem elongation that is independent of gibberellins. Brassinosteroids also affect crop yields; assimilate uptake; enhance xylem differentiation; enhance resistance to chilling, disease, herbicide, and salt stress; promote germination; and decrease abortion and fruit drop. They also have the potential to be used as an insecticide and in plant tissue culture (Arteca, 1996). More research is necessary to provide further support that these responses to brassinosteroids are widely distributed in the plant kingdom.

A number of plant growth regulators are used in agriculture today for a variety of purposes. The categories of plant growth regulators are **plant growth retardants,** ethylene-releasing compounds, ethylene biosynthesis inhibitors, compounds containing gibberellins, compounds containing auxins, compounds containing gibberellins, and cytokinins. The beneficial effects of plant growth retardants are more uniform and compact plants, better plant appearance, improved plant transplanting, improved shelf life, better capability to withstand stress, and, in the case of turfgrasses, reduced frequency of mowing, and improved turf density, lateral shoot development, and turf color. Plant growth retardants reduce the size of plants by blocking gibberellin biosynthesis and, in some cases, gibberellin action. Ethylene-releasing compounds provide a variety of beneficial effects, including the capability to induce flowering, modify sex expression, promote defoliation, increase lateral branching, and reduce plant height. The ethylene biosynthesis inhibitor Retain® also has a number of other benefits: improved harvest, reduced fruit drop, a wider window of growth to enhance fruit size and color, improved storage potential, and a reduced a variety of physiological disorders.

Compounds that contain gibberellins have a variety of beneficial effects in plants and are used extensively in agriculture today. The beneficial effects include the capability to stimulate uniform germination and seedling emergence, the use as a thinning agent in stone fruits to improve fruit firmness, and the better shelf life and increased size during the season when they are applied. In seedless grapes, the gibberellins loosen the cluster and promote uniform size;

in seeded grapes, they reduce berry shrivel and increase berry size. In addition, compounds containing gibberellins overcome dormancy, delay aging of the rind in citrus thereby preventing physiological disorders, promote **bolting** and uniform seed production, and provide many other beneficial effects.

Compounds containing auxins are used commercially for three main purposes: fruit thinning and sticking of fruits to trees, promoting adventitious rooting, and controlling weeds. Fruit thinning promotes better size, color, yield, and overall fruit quality; it also eliminates alternate bearing of fruits, which ensures that fruit load on trees is uniform year after year. Auxins also prevent fruits from prematurely falling off the tree, thus improving fruit quality. Compounds containing gibberellins and cytokinins are used in agriculture for the many benefits associated with them. These compounds are used for fruit thinning, promoting elongation of apple fruit, and developing more prominent calyx lobes. In addition, they have been shown to increase lateral bud break, increase shoot growth, and improve branch angles on nonbearing trees used for nursery stock, which provides a better tree framework and earlier cropping.

BACKGROUND OF PLANT GROWTH REGULATING SUBSTANCES

In the early 1900s, F. W. Went made the profound statement "Ohne Wuchstoff kein Wachstum," translated, "Without growth substances, no growth." Plant hormones have an important role throughout the plant kingdom. A considerable amount of documentation shows that extremely low concentrations of a plant hormone can regulate many aspects of plant growth and development, from seed germination through senescence and death of the plant. Auxins were the first class of plant hormones discovered more than 60 years ago. Since this time, five additional classes of plant hormones have been recognized: gibberellins, cytokinins, abscisic acid (ABA), ethylene, and, the newest, brassinosteroids. Numerous advances have been made in the use of plant growth regulating substances on a practical scale along with basic research at the biochemical, physiological, and molecular levels. Today, the use of plant growth regulating compounds in agriculture is on the rise and will continue to increase as more scientific information becomes available. Plant growth regulators are used in agriculture for a variety of purposes, which will be discussed in more detail later in this chapter:

- delaying or promoting ripening (ethylene).
- inducting rooting (auxins).
- promoting or delaying abscission (ethylene).
- controlling fruit development (auxins, gibberellins, and cytokinins).
- controlling weeds (auxins).
- controlling size (growth retardants that block gibberellin biosynthesis).

Key definitions in this area are as follows:

Plant hormone or phytohormone is a naturally occurring organic compound that is

- chemically characterized.
- biosynthesized within the plant.
- broadly distributed in the plant kingdom.

- a specific biological activity at extremely low concentrations.
- fundamental to regulating physiological phenomena *in vivo* in a dose-dependent manner and/or due to changes in sensitivity of the tissue during development.

A plant growth regulator is an organic compound other than a **nutrient** (material that supplies either energy or essential mineral elements) that in small amounts promotes, inhibits, or otherwise modifies any physiological process in plants. A plant growth regulator includes both synthetic and naturally occurring regulators.

All plant hormones are plant growth regulators, but not all plant growth regulators are plant hormones. In essence, plant hormones are naturally occurring and can be included with plant growth regulators, but plant growth regulators can be both naturally occurring and synthetic, so they cannot be in the plant-hormone category.

A plant growth retardant is an organic compound that retards cell division and cell elongation in shoot tissues and thus regulates height physiologically without causing malformation of leaves and stems.

Auxins are a class of compounds with activity similar to IAA. The only naturally occurring active auxin is IAA (Figure 10-2).

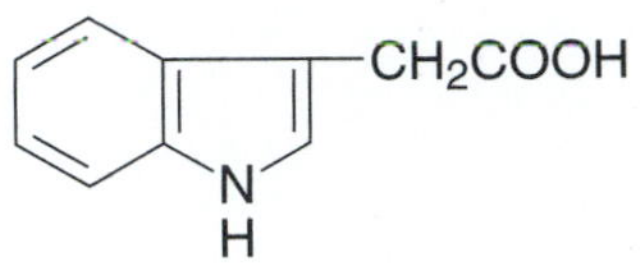

Indole-3-acetic acid (IAA)

Figure 10–2 Structure of the naturally occurring active auxin indole-3-acetic acid (IAA)

Gibberellins are a class of compounds with activity similar to gibberellic acid (GA_3). More than 100 gibberellins are known today; commercially used gibberellins are GA_3 and a mixture of GA_4 and GA_7 (Figure 10-3).

Cytokinins are a class of compounds with activity similar to kinetin, a synthetic cytokinin commonly used today (Figure 10-4); whereas a naturally occurring form of cytokinin is zeatin (Figure 10-5).

The abscisic acid (ABA) compound is produced in response to water stress and is directly involved in stomatal opening and closing. The chemical structure for abscisic acid is shown in Figure 10-6.

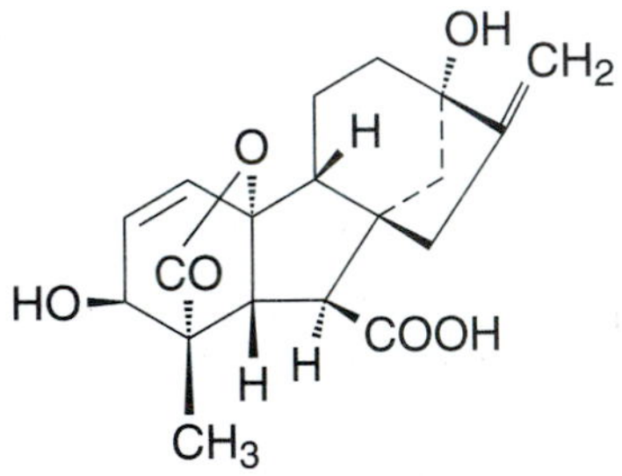

GA_3 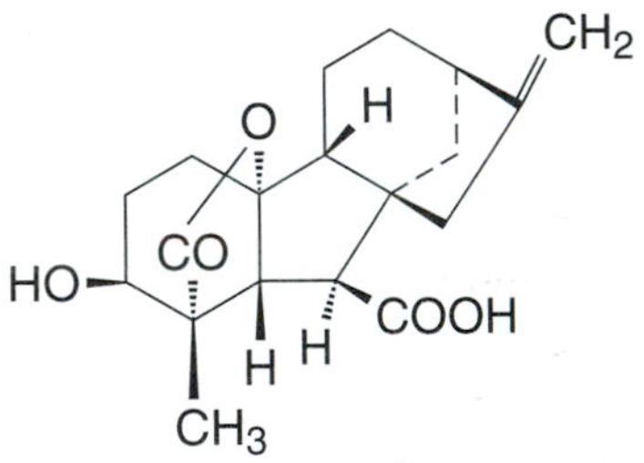 GA_4 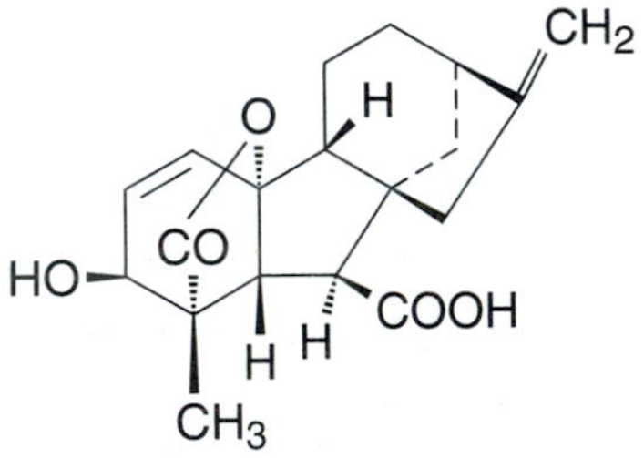 GA_7

Figure 10–3 Structures of the commercially used gibberellins GA_3, GA_4, and GA_7

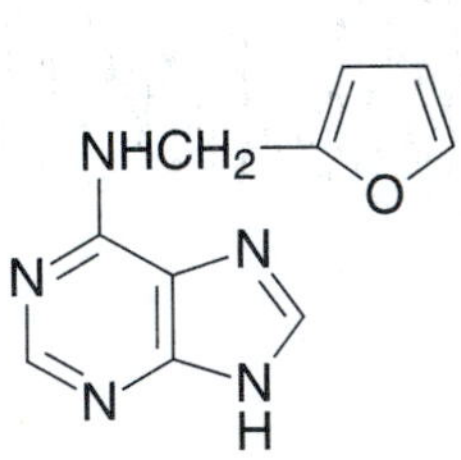

Kinetin
(6-furfurylaminopurine)

Figure 10–4 Structure of the synthetic cytokinin, kinetin

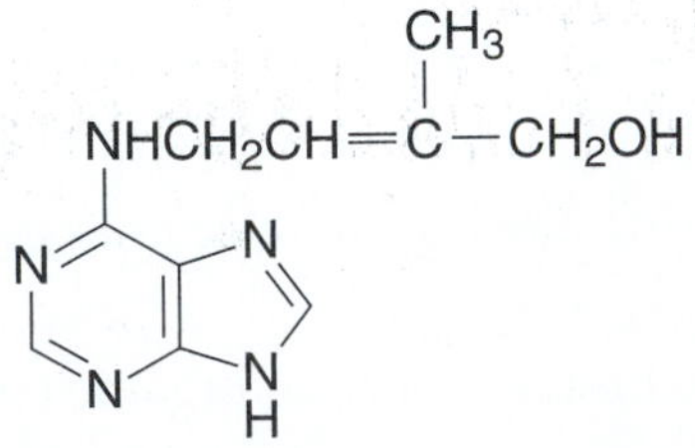

Zeatin
6-(hydroxy-3-methyl-trans-2-butenyl-amino) purine

Figure 10–5 Structure of the naturally occurring cytokinin, zeatin

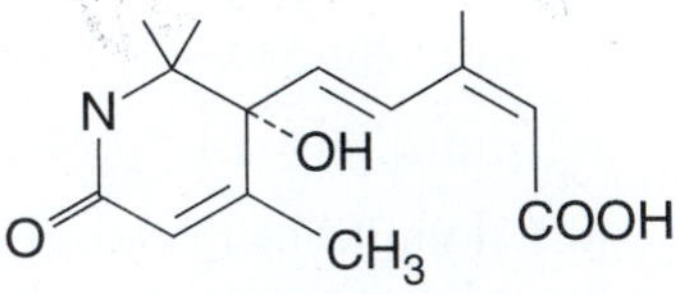

Abscisic acid
(ABA)

Figure 10–6 Structure of the plant-stress hormone abscisic acid

$CH_2{=}CH_2$

Ethylene

Figure 10–7 Structure of the fruit-ripening hormone ethylene

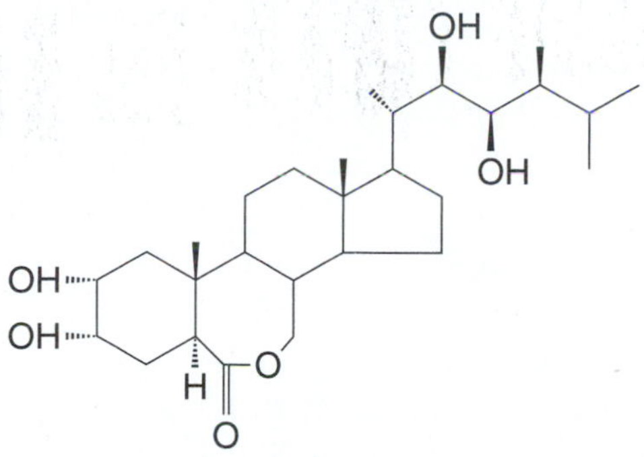

Brassinolide

(2α,3α,22α,23α,-tetrahydroxy-24α-methyl-B-7-oxa-5α-cholestan-6-one)

Figure 10–8 Structure of the naturally occurring brassinosteroid, brassinolide

Ethylene is a simple unsaturated hydrocarbon generally accepted to be the fruit-ripening hormone. The chemical structure for ethylene is shown in Figure 10-7.

Brassinosteroids are a class of compounds having activity similar to brassinolide. The chemical structure of brassinolide is shown in Figure 10-8.

THE SIX CLASSES OF PLANT HORMONES

The following discussion outlines the six classes of plant hormones and provides some historical background and physiological processes affected by each.

Auxins—History and Physiological Processes

Darwin is considered the scientist responsible for initiating modern plant hormone research. In his book *The Power of Movement in Plants* (Darwin, 1880), he describes the effects of light on movement of coleoptiles. In 1928, Went developed a method for the quantification of plant growth substances present in a sample, which was called the *Avena* curvature bioassay and is still commonly used today (Figure 10-9). Kogel and coworkers in 1934 purified **heteroauxin** (other auxin) from human urine; today, heteroauxin is known to be IAA. Since this time, IAA has been isolated in numerous higher plants and is thought to be ubiquitous in the plant kingdom.

The term *auxin* is derived from the Greek word **auxein,** which means "to grow." Today, auxin is defined as a class of compounds with activity similar to IAA. IAA is the only active auxin found in plants today.

Auxins are involved in numerous physiological processes in plants. To cover the key work done with auxins would require many chapters. The following sections, however, summarize some of the physiological processes in which auxins are involved.

Cellular Elongation

Auxins promote a dramatic stimulation in cell elongation in detached plant parts. Typically in these detached tissues, little or no IAA is found; therefore, when these tissues are placed in contact with IAA, there are large increases in

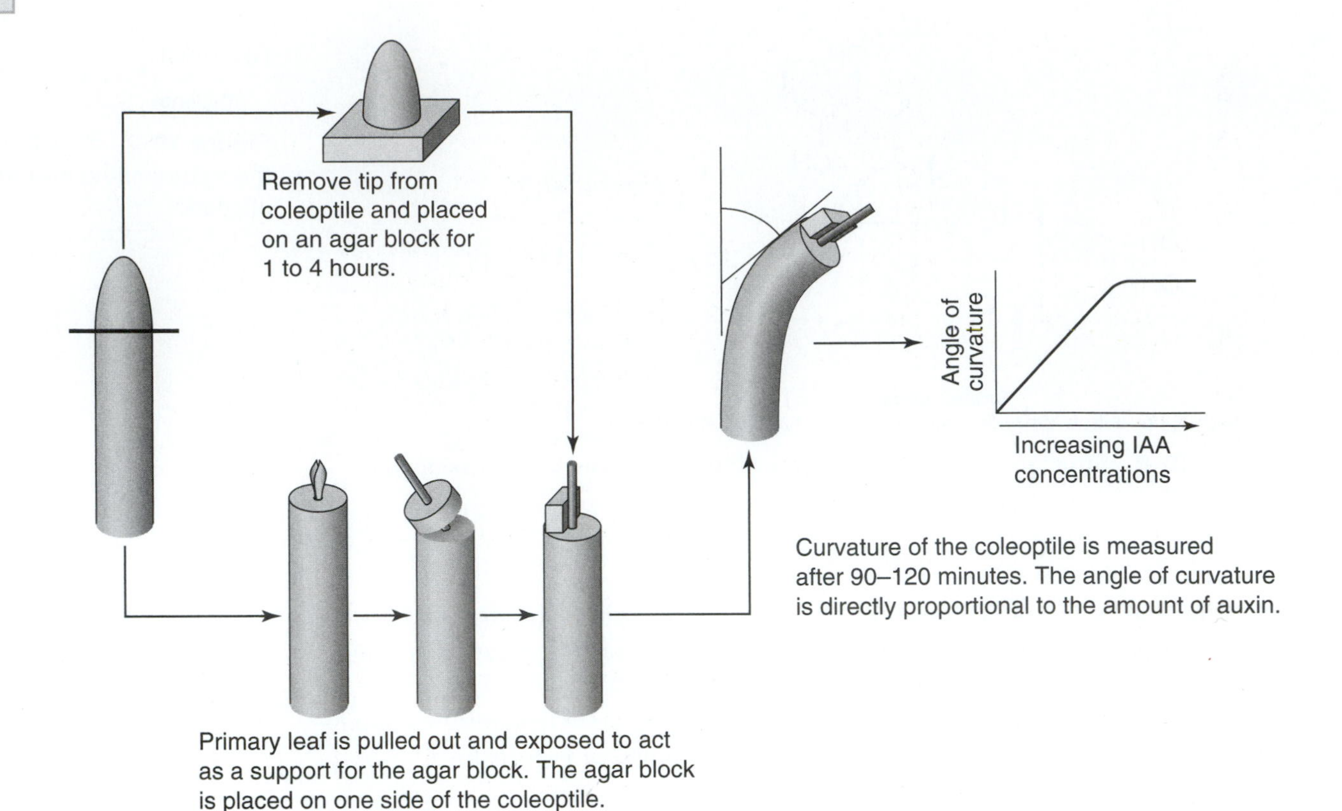

Figure 10–9 Avena curvature bioassay developed by Went

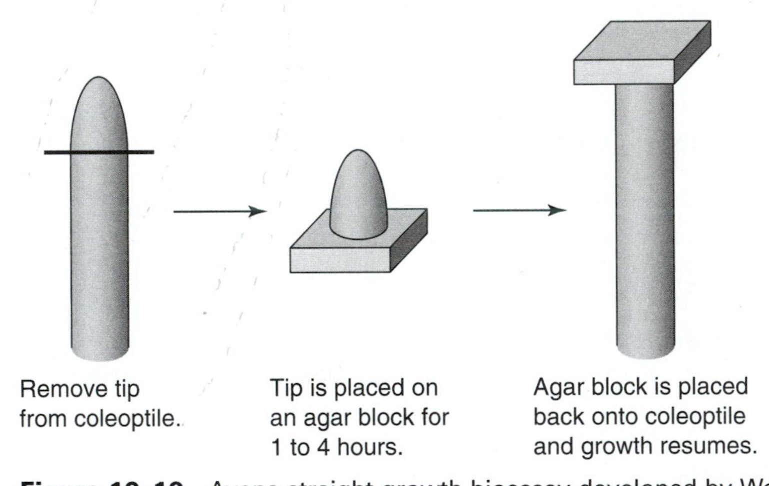

Figure 10–10 Avena straight growth bioassay developed by Went

growth. The dramatic increase in growth promoted by IAA is commonly used in bioassays for quantifying auxins such as the *Avena* curvature and *Avena* straight growth bioassays (Figure 10-10).

Phototropism

Phototropism is the movement of a plant organ in response to directional fluxes or gradients in light. The Cholodny-Went theory provides an explanation of how the phototropism process occurs. This theory states that when higher concentrations of the plant hormone IAA occur on the shaded side of the

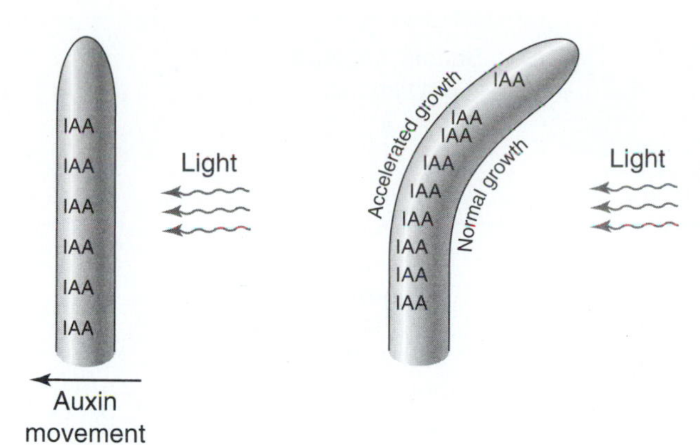

Figure 10–11 Cholodny-Went theory on phototropism. Light from one side of the segment causes the movement of IAA to the shaded side, which results in an increase of growth on the shaded side.

coleoptile in response to light, there is accelerated growth in this region. Growth proceeds normally on the illuminated side, and this differential in growth causes bending toward the light (Figure 10-11).

Gravitropism

Gravitropism is the movement of a plant organ in response to gravity. Research in the area of gravitropism began with the work of Darwin (1880) and is still being pursued by many research groups today. Studying plant growth and development in response to gravity is important for many reasons, such as better understanding how a seed knows which way is up or down during the germination process, understanding how plants grow under weightlessness conditions in an orbiting spacecraft, and considering postharvest problems with flower stalks that exhibit unsightly bending during shipment and storage.

When a seedling is placed horizontally, it responds to Earth's gravitational force with an altered growth pattern. The stem bends upward against gravity, which is called **negative gravitropism,** whereas the roots bend downward with gravity, which is called **positive gravitropism.** The Cholodny-Went theory provides a reasonably sound explanation for gravitropism. It states that an accumulation of the IAA plant hormone on the lower side of the stem and root results in differential growth. The accumulation of IAA on the lower side of the stem accelerates growth on this side, while growth on the upper side proceeds at a normal rate; this differential growth promotes upward bending. The roots are much more sensitive to IAA, so the growth rate on the bottom side of the root is reduced, while the top grows normally; this differential growth results in downward bending (Figure 10-12).

The Statolith theory has been used to describe how plants perceive gravity. A statolith is an object in the plant cell now thought to be an amyloplast, which belongs to a class of colorless plastids called leucoplasts. When the plant is placed in the horizontal position, statoliths fall to the bottom of the cell, which causes changes in IAA gradients resulting in differential growth. This differential growth enables the plant to perceive that it is in the wrong position and reorient its growth pattern (Figure 10-13).

Apical Dominance

Long before regulation of growth by plant hormones was discovered, botanists recognized that the apical bud causes lateral bud suppression. The IAA contained in the apical bud causes lateral bud suppression. A simple experiment

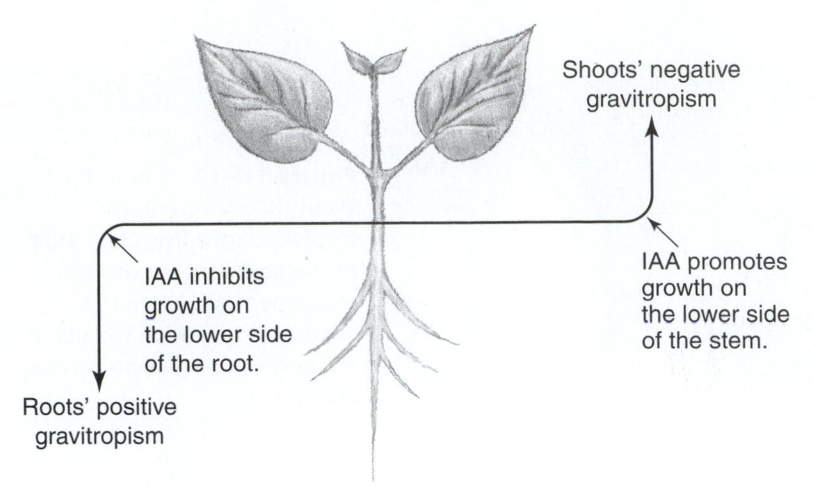

Figure 10–12 Cholodny-Went theory on gravitropism. Gravity causes IAA to accumulate at the bottom of the cell. In stem cells this promotes growth on the lower side of the stem while in roots this inhibits growth on the lower side of the root.

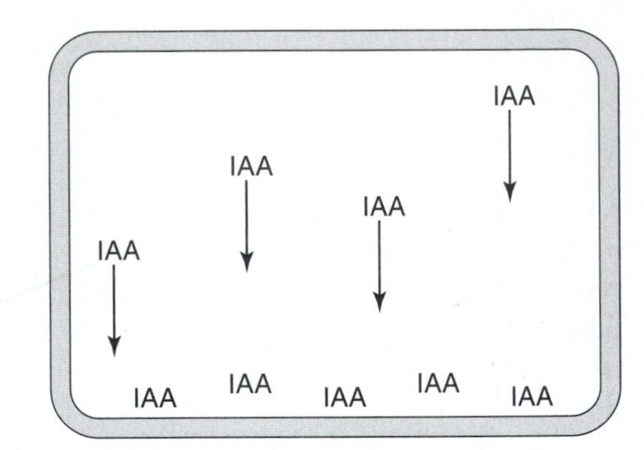

Figure 10–13 According to the Statolith theory, plants perceive gravity due to IAA accumulating at the base of the cell.

to prove that IAA is involved in apical dominance is to remove the apical bud and replace it with gelatin plus or minus IAA. The plant receiving the gelatin block plus IAA should have its lateral buds suppressed, whereas the one receiving no IAA treatment should have lateral bud break (Figure 10-14).

Root Initiation and Cell Elongation

Auxins stimulate root initiation and inhibit root elongation. Auxins are the only class of plant hormones that majorly affect rooting, and they are used commercially to stimulate adventitious rooting (Figure 10-15). The two most commonly used auxins for the promotion of adventitious root formation are indole-3-butyric acid (IBA) and naphthalene acetic acid (NAA).

Ethylene Production

Many factors have been shown to induce ethylene production. In 1935, Zimmerman and Wilcoxon were the first to show that auxin stimulated ethylene production in tomato plants. They treated one tomato plant with auxin, treated the control plant with water, and then placed both plants into a bell jar. Within a brief period of time, both plants became epinastic (downward bending of the petioles and stems), showing that auxin was not responsible for the change in

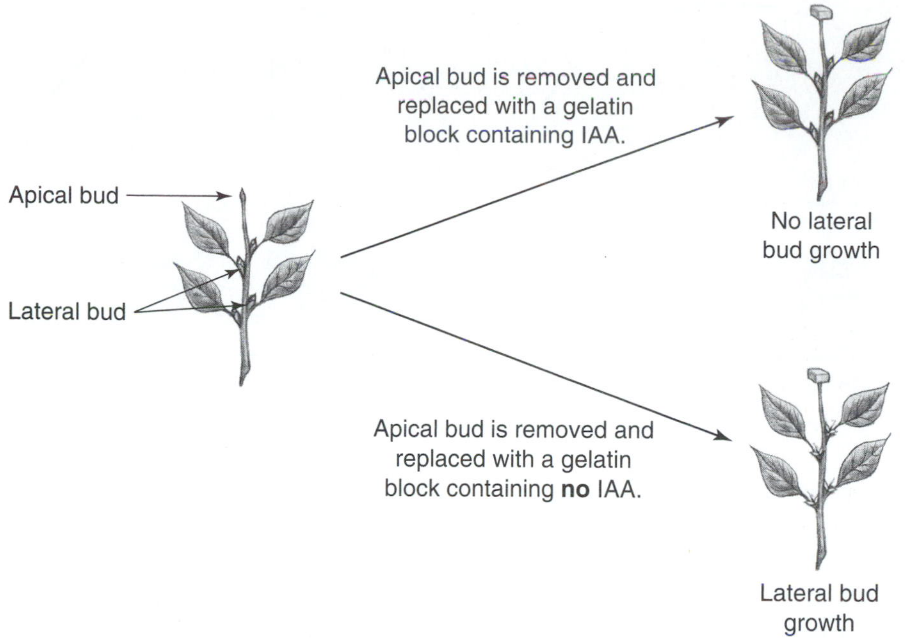

Figure 10–14 Effects of IAA on apical dominance

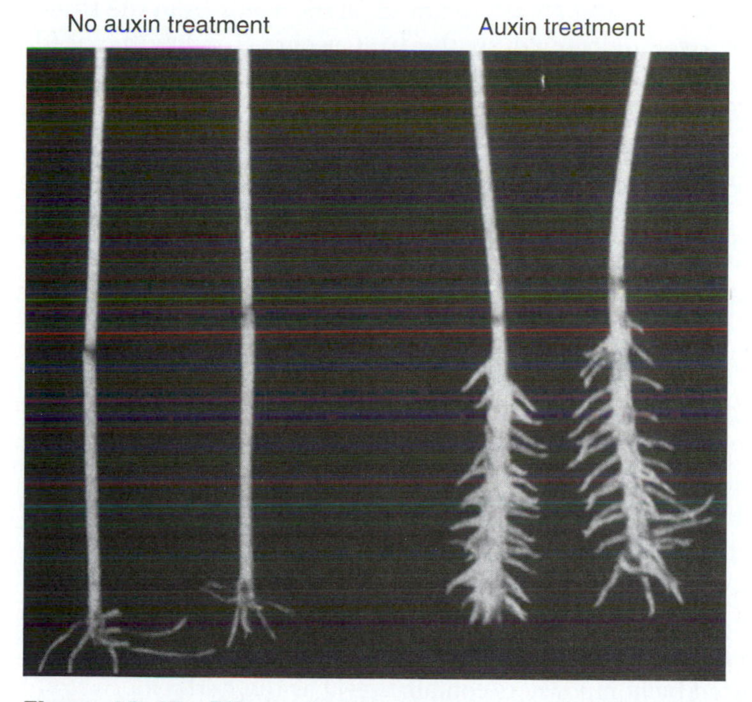

Figure 10–15 Effects of auxin treatment on adventitious rooting in mungbean cuttings. *Courtesy of Dr. Charles W. Heuser, Department of Horticulture, The Pennsylvania State University.*

plant growth and that ethylene was most likely responsible (Figure 10-16). Today, many of the responses once attributed to auxins are now found to be due to ethylene. For example, auxins were sprayed on pineapples to stimulate flowering; now it is known that the ethylene produced in response to auxins is responsible.

Fruit Growth

Auxins promote increases in fruit size, which was shown dramatically by Nitsch's work in 1950 with the strawberry. In this study, no fruit growth was

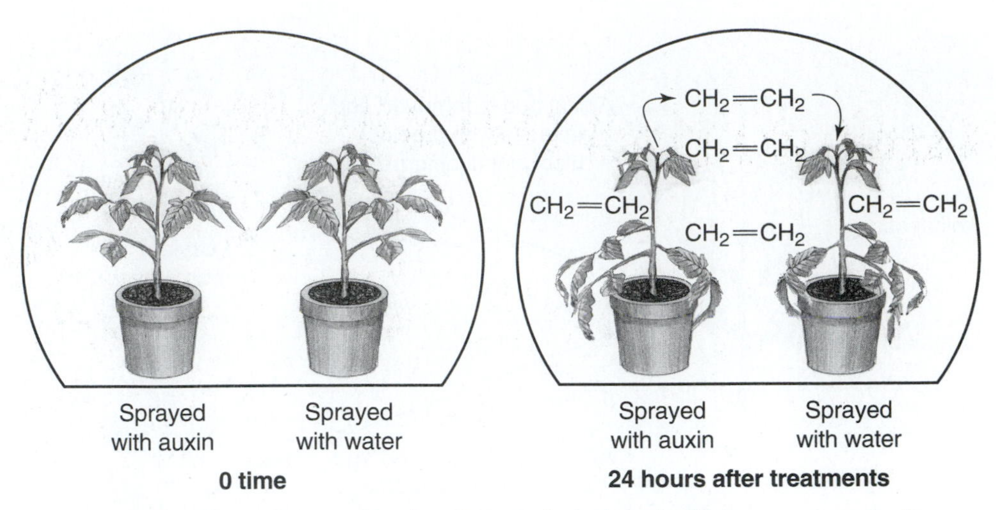

Figure 10–16 Experiment showing that auxin-induced ethylene, not auxin, is responsible for downward bending of petioles in tomato plants

observed when all the seeds (achenes) from the strawberry fruit were removed and lanolin paste was applied to prevent desiccation. When seeds were removed and lanolin paste containing auxin was applied, fruit growth occurred normally. To show the involvement of seeds, Nitsch removed all the seeds from the strawberry fruit except for one, and growth of the fruit occurred below the seed. Nitsch's two lines of evidence showing that auxins were involved in fruit growth were as follows:

- A correlation exists between seed development and the final size and shape of the fruit.
- Exogenous applications of auxins to the fruit induced growth.

Sex Expression

Auxins alter sex expression in plants. When it is sprayed on *Cucumis* or *Cannabis*, there is a dramatic increase in the number of female flowers.

Weed Control

The first selective herbicide was not found until the early 1900s, and today more than 130 selective herbicides are used throughout the world. In the United States alone, 1 billion pounds of pesticides are sold annually and over 65% are herbicides. Selectivity is the key to the widespread use of herbicides. A commonly used selective herbicide is 2,4-dichlorophenoxyacetic acid (2,4-D), which is an auxin that is used to control broadleaf plants in lawns. Most, if not all, weed and feed fertilizers used by homeowners contain 2,4-D as the herbicide.

Gibberellins—History and Physiological Processes

Many years ago, Japanese farmers observed that certain rice plants were taller earlier in the growing season; however, as the growing season progressed, these plants became spindly, chlorotic, and—more importantly—were sterile and devoid of fruit. Japanese farmers called these symptoms the Bakanae (foolish seedling) disease. The fungus called *Gibberella fujikuroi* was shown to cause the Bakanae disease. In 1935, Yabuta isolated and crystallized the active component secreted by the fungus and called this compound Gibberellin A. Since this time, more than 100 gibberellins have been identified in plants.

Figure 10–17 Hydroponically grown tomato plants treated with varying concentrations of gibberellic acid; from left to right is the control and increasing concentrations of gibberellic acid.

Gibberellins are involved in numerous physiological processes in plants, as described in the next sections.

Stimulates Stem Growth in Intact Plants

Numerous reports show that gibberellins promote growth in intact plants (Figure 10-17). The effect of gibberellins on growth of the intact plant is different from auxins, which promote growth only in detached plant tissues. Gibberellins are involved in genetic dwarfism in plants. There are two types of mutants. One involves a single gene mutation in the gibberellin biosynthetic pathway resulting in dramatically shorter plants. Exogenous application of gibberellin promotes rapid growth to produce normal-sized plants. There are also gibberellin sensitivity mutants, which do not respond to exogenous applications of gibberellin.

Bolting and Flowering

Gibberellins have been shown to stimulate bolting, which is rapid stem elongation, and **flower initiation,** which refers to internal physiological changes in the apical meristem that lead to the visible initiation of floral parts. Early researchers working on flowering showed that following the perception of light by the leaf produced a signal that when transmitted to the bud caused flower initiation to occur. Since this pioneering work, there has been a considerable amount of research conducted to identify this signal, which has been called Florigen (flowering hormone); however, there has been little success. Although the flowering hormone has not been identified, it has been shown that gibberellins are involved in promoting bolting and flowering in many plant species.

Seed Germination

A considerable amount of evidence supports that gibberellins are directly involved in controlling and promoting seed germination in a variety of experimental systems. In the barley seed system, which is a popular experimental system to study seed germination, gibberellins have been shown to have a profound

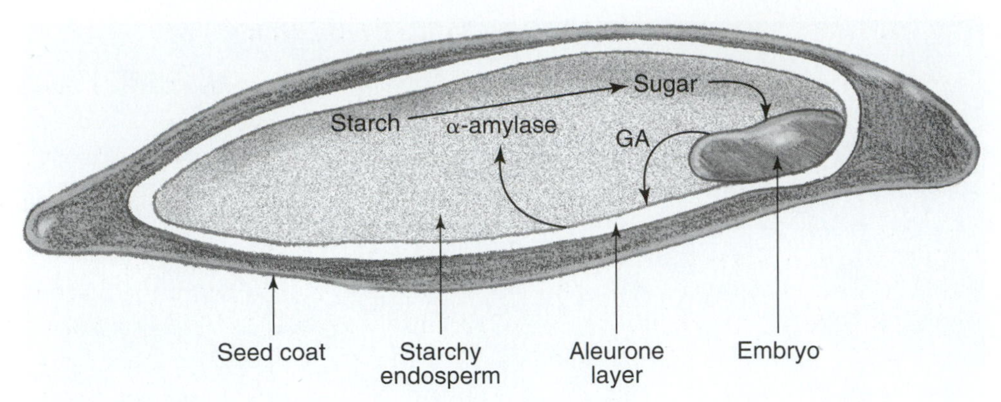

Figure 10–18 The barley endosperm experimental system showing that gibberellins are involved in the germination process

role in the germination process. The series of events that takes place is shown as follows:

1. Gibberellins are produced in the embryo.
2. The gibberellins are then transported to the aleurone layer of cells.
3. Here α-amylase is produced, thereby causing the conversion of starch to sugar in the endosperm. These sugars are used for energy during the germination process (Figure 10-18).

Dormancy

Dormancy is the temporary suspension of visible plant growth. Gibberellins can break dormancy in a variety of crops. There are two lines of evidence showing the involvement of gibberellins in the dormancy process. First, a number of reports show that endogenous levels of gibberellins decrease during the initiation of dormancy and increase dramatically after dormancy is broken. The second line of evidence is that exogenous applications of gibberellins can break certain types of dormancy. A simple experiment to show that exogenous applications of gibberellin can break dormancy is to treat the eyes of one potato with gibberellin and another potato with only water. The gibberellin-treated potato breaks dormancy much more quickly than the water control, thereby showing the ability of gibberellins to break dormancy.

Sex Expression

Gibberellins stimulate male flower production in *Cucumis* and *Cannabis*.

Fruit Growth

Gibberellins have been shown to affect fruit shape. When gibberellins are applied to a Thompson seedless grape bunch, it becomes elongated and the berries become larger and elongated. Gibberellins have also been shown to promote the elongation of apple fruits to make the fruit look more appealing.

Parthenocarpy

Parthenocarpy is the development of fruits without pollination or fertilization, resulting in seedlessness. Gibberellins are commonly used to induce parthenocarpy in grapes.

More than 100 gibberellins are known today. Gibberellic acid, also called GA_3, is the most commonly used, and a mixture of GA_{4+7} is used extensively in apples. There are no synthetic forms of gibberellins.

Cytokinins—History and Physiological Processes

The first step toward the discovery of cytokinins was the observation that a cell division plant growth substance was found to occur in coconut milk (liquid endosperm) (Van Overbeek in 1941). Today coconut milk is still used in tissue culture as source of cytokinins. In 1955, Miller and coworkers were the first to report the identification of kinetin (6-furfurylaminopurine), although it was not found in plants but purified from aged herring sperm DNA. They named it kinetin because of its capability to promote cell division or cytokinesis in tobacco pith tissue. Kinetin has never been found in plants; however, it is still commonly used as a synthetic form of cytokinin in plant tissue culture. The first naturally occurring cytokinin found in plants was from *Zea mays* by Miller and coworkers and Letham and coworkers in 1963. This compound was called zeatin. Since this time, cytokinins have been shown to be ubiquitous in the plant kingdom, and today more than 30 have been identified.

Cytokinin is involved in many physiological processes, as described in the following sections.

Cell Division and Organ Formation

The major function of cytokinins in plants is to promote cell division. Cytokinins promote **callus** formation in most plant tissues. Callus is an undifferentiated mass of cells (Figure 10-19). By manipulating the ratio of auxin to cytokinin in tissue culture, you can obtain a combination of callus, roots, and shoots or a combination of roots and shoots. Typically, higher levels of auxins

Figure 10–19 Petri plate with *Cephalotaxus* callus and plants

Figure 10–20 Geranium plants growing in tissue culture

promote roots, whereas higher levels of cytokinins promote shoots. The capability to regenerate plants from callus is a biotechnological tool commonly used today for selecting plants that are resistant to drought, salt stress, pathogens, herbicides, and others. For example, if a herbicide-resistant plant is desired, callus is grown on toxic levels of a specific herbicide. Large numbers of petri plates containing callus are grown and most will die. However, when a small patch of callus proliferates on a plate containing toxic levels of a specific herbicide, it is selected. This callus cell line is grown and then transferred to a plate containing an auxin-to-cytokinin mixture to induce the formation of plants with roots and shoots. These plants are then grown to determine if they are resistant to the herbicide (Figure 10-20).

Senescence

Cytokinins can delay the senescence process in plants. When cytokinins are sprayed on a single leaf, the plant diverts nutrients to that particular leaf. This is called creating a sink, which delays senescence.

Stomatal Opening

Foliar applications of cytokinins have been reported to stimulate photosynthetic rates in a variety of plants. Some suggest that when cytokinins are sprayed on plant leaves, the stomates will open wider and stay open longer, which leads to an increase in photosynthetic rates.

Lateral Bud Break

When cytokinins are applied to the whole plant they will overcome apical dominance, thereby promoting lateral bud break. There are two lines of evidence that support the involvement of cytokinins in lateral bud break. First, when plants are sprayed with cytokinins, apical dominance is partially overcome. Second, genetic engineering techniques have been used to increase endogenous cytokinin levels in response to heat shock treatment. Transgenic plants that were triggered to overproduce cytokinins exhibited multiple lateral bud breaks and an overall lack of apical dominance, thereby further supporting the involvement of cytokinins in promoting lateral bud break.

Sex Expression

When plants are treated with cytokinins, the number of female flowers increases per plant in *Cannabis*. A commonly used synthetic cytokinin is kinetin, and a naturally occurring form is zeatin.

Abscisic Acid (ABA)—History and Physiological Processes

Abscisic acid (ABA) was discovered in 1965 simultaneously by two research groups: the Addicott group and the Wareing group. The Addicott group in the United States used cotton abscission as its experimental system, and the Wareing group in England used dormancy in birch trees as its experimental system. Today, it is generally accepted that ABA is widely distributed in the plant kingdom and is the only major inhibitor found in this category.

Abscisic acid is involved in a couple of several physiological processes, as described next.

Plant Stress Signal

The major effect of abscisic acid in plants is its capability to signal that a plant is undergoing water stress. When a plant undergoes water stress, ABA levels are increased dramatically. The increase in ABA levels leads to stomatal closure, thereby preventing water loss via transpiration (Figure 10-21).

Normal conditions

K K K K

Potassium enters cells.

K K K K

Water enters and stomates open allowing photosynthesis and respiration to occur.

Stressed conditions

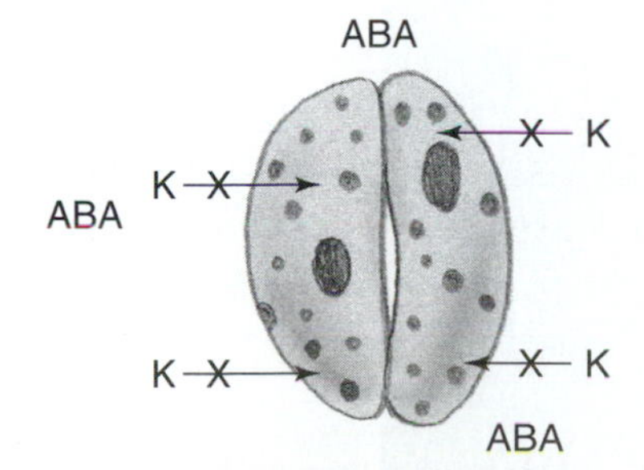

ABA is produced and blocks potassium uptake, so closed stomates remain closed.

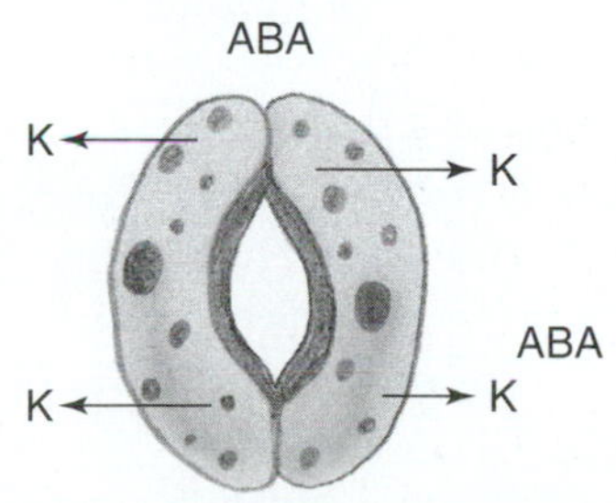

ABA causes potassium efflux thereby facilitating water loss from the stomates.

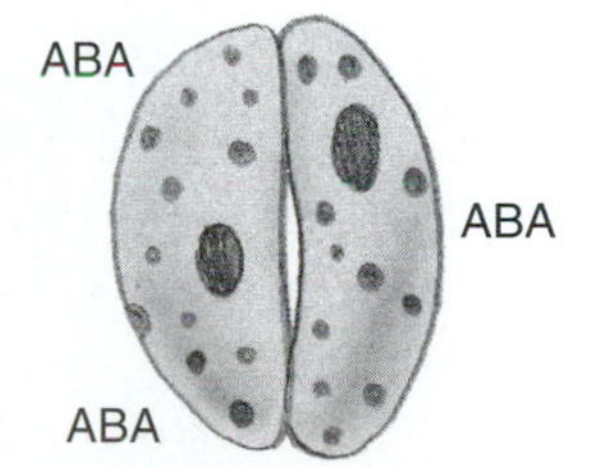

Reduction in turgor causes the opened stomates to close.

Figure 10–21 Effects of abscisic acid (ABA) on stomatal opening and closing

Other Processes

In addition to ABA's capability to act as a plant stress signal, it also promotes dormancy and abscission, delays seed germination, and retards plant growth and development. Currently, no synthetic forms of ABA are available.

Ethylene—History and Physiological Processes

Practical uses for ethylene started many years ago with the ancient Egyptians, who gashed figs to stimulate ripening, and the Chinese, who burned incense in closed rooms to enhance ripening. Neither the Egyptians nor the Chinese were aware that ethylene was causing ripening; in fact, they did not even know what ethylene was. It was not until 1901 that a Russian scientist named Neljubow discovered that ethylene caused the **triple response** in etiolated pea seedlings. Crocker and coworkers in 1935 were the first to propose that ethylene was the fruit-ripening hormone, although today it is generally accepted as fact.

The triple response was the first bioassay used to quantify ethylene and consisted of the following:

- suppression of stem elongation.
- increased radial expansion (lateral expansion).
- promotion of bending or horizontal growth in response to gravity (Figure 10-22).

Ethylene is a gaseous plant hormone involved in many physiological processes, as described next.

Fruit Ripening

Ethylene is the only gaseous plant hormone. When ethylene is applied to fruits it causes them to ripen. For example, green tomatoes are shipped from California to Pennsylvania supermarkets, where they are treated with ethylene to cause them to turn red; however, problems with flavor arise when this is done. In recent years, the ethylene biosynthetic pathway has been completely

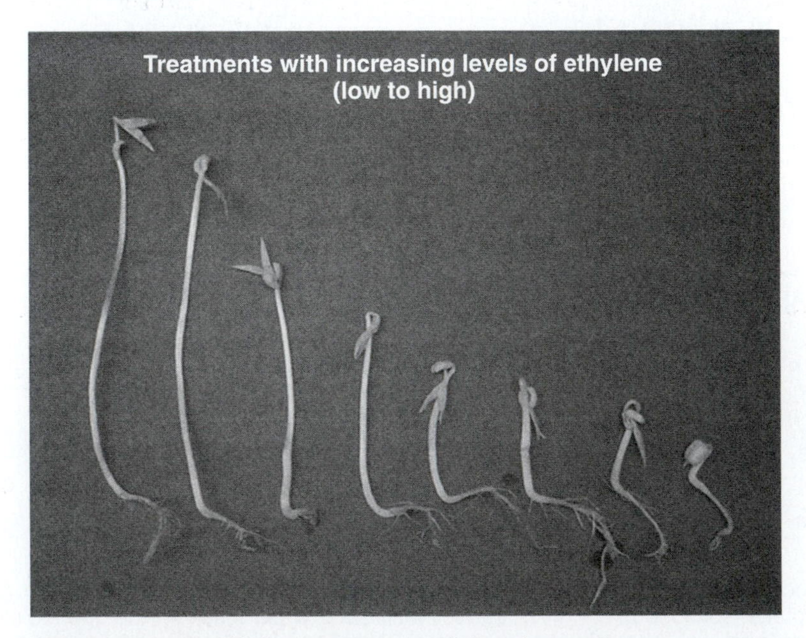

Figure 10–22 Triple response in etiolated mungbeans

characterized and is being exploited to manipulate fruit ripening. The last two steps in the ethylene biosynthetic pathway are currently being used to modify fruit ripening. This will be discussed in more depth in Chapter 11, Postharvest Physiology.

Seedling Growth

In 1901, the Russian scientist Neljubow was the first to show that ethylene has an effect on seedling growth. He demonstrated the triple response showing that ethylene inhibited elongation and promoted lateral expansion and horizontal growth (refer to Figure 10-22). Ethylene can inhibit or promote elongation of stems, roots, or other organs. The inhibition of elongation is rapid and reversible. The triple response promoted by ethylene can act as a survival mechanism in seedlings. For example, when a seedling encounters a barrier such as soil crusting, a rock, or something that restricts its emergence, the seedling responds with the triple response. This allows it either to break through or go around the obstruction, thereby permitting the seedling to reach the soil surface and grow.

Senescence

Senescence is a general failure of many synthetic reactions that precedes cell death and is the phase of plant growth that extends from full maturity to death. Senescence is characterized by chlorophyll, protein, and nucleic acid degradation as well as many other factors. Ethylene promotes senescence (yellowing). As plants age, ethylene is produced, which causes plants to senesce normally; however, when plants are stressed, ethylene is produced, resulting in premature senescence.

Abscission

Abscission is the separation of an organ or plant part from the parent plant. Ethylene promotes abscission in plants. The process of abscission is very important in agriculture because abscission or nonabscission of flowers, fruits, and leaves influences yields, efficiency, and harvesting operations. Ethylene has a natural role in the regulation of the rate of abscission. Three lines of evidence support ethylene's involvement in the abscission process. The first is that ethylene production increases prior to abscission in many abscising plant organs. Second, treatment of a wide range of plant species with ethylene or ethylene-releasing compounds stimulates abscission. Third, inhibitors of ethylene biosynthesis or action inhibit the abscission process. Abscission can be beneficial by facilitating mechanical harvesting or it can be detrimental by resulting in premature abscission of leaves and flowers (Figure 10-23).

Sex Expression

Ethylene promotes female flowers in *Cucumis* and *Cannabis* plants. Many synthetic forms of ethylene-releasing compounds are on the market, including Ethephon, Ethrel®, and Florel®. They are all 2-chloroethylphosphonic acid and release ethylene gas when the pH of the solution is changed. The naturally occurring form of ethylene is the gas C_2H_4.

Brassinosteroids—History and Physiological Processes

In the 1960s, Mitchell and coworkers at the USDA Research Center in Maryland began screening pollen in search of new plant hormones. In 1972, Mitchell and Gregory identified a compound they called **brassins,** which are a crude lipid extract from rape pollen. Mitchell and Gregory showed that brassins

Figure 10–23 Abscission of Regal Pelargonium flowers following a brief exposure to ethylene. *Courtesy of C. F. Deneke.*

can promote rapid growth (Figure 10-24). In 1979, Grove and coworkers identified brassinolide, which is the active component in brassins. Today more than 30 compounds have activity similar to brassinolide. This class of compounds is called brassinosteroids (BR).

Brassinosteroids, the newest class of plant hormones, are involved in many physiological processes.

Shoot Elongation

Numerous reports show that exogenous application of BR promotes elongation in vegetative tissues; however, this did not provide definitive proof that BR was required for normal plant growth and development. Joanne Chory's research group provided solid genetic evidence that BR is required for normal plant growth and development. The group identified Arabidopsis mutants with defects in the BR biosynthetic pathway and showed that these mutants were dwarf and only exogenous applications of BR could restore normal plant growth and development. Steve Clouse's group provided further genetic evidence by identifying a BR-insensitive Arabidopsis mutant, which exhibited a severely stunted appearance that could not be reversed by exogenous BR applications. Since this pioneering research, numerous reports have shown that BR is required for normal plant growth and development that is independent of gibberellins. Therefore, the best-documented effect of BR in plants is the capability to promote dramatic increases in shoot elongation.

Figure 10–24 Brassin-induced splitting of a bean stem caused by rapid growth

Crop Yields

In the early 1980s, USDA scientists showed that BR could increase yields of radishes, lettuce, beans, peppers, and potatoes. However, subsequent results under field conditions were disappointing because of inconsistent results. For this reason, further BR testing for increased crop yield was phased out in the United States. More recently, large-scale field trials in China and Japan over a six-year period have shown that 24-epibrassinolide, an alternative to brassinolide, increased the production of agronomic and horticultural crops, including wheat, corn, tobacco, watermelon, and cucumber. However, once again, depending on cultural conditions, method of application, and other factors, the results sometimes were striking, whereas other times there was no effect. Further improvements in the formulation, method of application, timing, effects of environmental conditions, and other factors need to be investigated to identify the reason for these variable results (Cutler et al., 1991).

Other Effects

Preliminary results indicate that BR has numerous effects in plants, such as uptake of assimilates; enhanced xylem differentiation; enhanced resistance to chilling, disease, herbicide, and salt stress; promotion of germination; and decreased fruit abortion and fruit drop. BR also has potential to be used as an insecticide or in plant tissue culture. Although a number of potential effects of BR have been reported, more research is necessary to validate these preliminary results.

The naturally occurring form of BR commonly used is brassinolide, although more than 30 different BRs have been reported in plants. One of the main synthetic forms of BR used today is 24-epibrassinolide.

PRACTICAL USES OF PLANT GROWTH REGULATORS IN AGRICULTURE

Many plant growth regulators are used in agriculture today, including the following:

- plant growth retardants.
- ethylene-releasing compounds.
- ethylene biosynthesis inhibitors.
- compounds containing gibberellins.
- compounds containing auxins.
- compounds containing gibberellins and cytokinins.

Plant Growth Retardants

Some commonly used plant growth retardants include A-Rest®, Alar® (daminozide), Cycocel® (chlormequat), Sumagic® (uniconazole), Bonzi® (paclobutrazol), Primo® (trinexapac-ethyl), PIX® (mepiquat chloride), Apogee® (prohexadione calcium), MH-30® (maleic hydrazide), and Cutless® (flurpirmidol). Each of these compounds reduces the size of plants by blocking gibberellin synthesis and, in some cases, blocking gibberellin action.

Examples of crops that plant growth retardants are used on include bedding plants (begonia and petunia), bulb crops (Easter lily and tulip), flowering plants (chrysanthemums and poinsettias), foliage plants (dracaena and philodendron), woody plants (azaela and gardenia), and turfgrasses. The major benefits of plant growth retardants are as follows:

- more uniform and compact plants.
- better plant appearance, including better color, more side shoots, and more flowers.
- better plant transplantation because of improved root systems.
- improved shelf life.
- the ability to withstand drought and transport stresses better.
- reduced frequency of mowing, improved turf density, lateral shoot development, and turf color for turfgrasses.

Factors that affect the plant's response to plant growth retardants include the following:

- **Method of application.** Foliar sprays and soil drenches, although each plant growth retardant may work better with one form of application than another, and some work equally well with both.
- **Environmental conditions.** Growers in warmer climates typically need to use higher rates of plant growth retardants and/or more applications than those in cooler climates because rapid stem elongation occurs in warmer temperatures.
- **Cultivars.** When growing cultivars with more vigorous growth rates and rapid stem elongation, higher rates of plant growth retardants need to be applied. The reverse is true for cultivars with slower growth rates.
- **Cultural practices.** Different cultural practices used by different growers, including watering, fertilizing, and spacing, together with environmental conditions such as temperature and light will affect the rate of the plant growth retardant used. The best way to determine the optimal rate to use is to read the label, which should be used as a guideline for small-scale experiments under actual use conditions.

Ethylene-Releasing Compounds

The trade names for three commonly used ethylene-releasing compounds are Proxy®, Florel®, and Ethrel®. The chemical name for the active ingredient in each of these commercially used compounds is 2-chloroethylphosphonic acid; the common name is ethephon.

Proxy

Proxy® (labeled for commercial use only) is used as a foliar spray to slow the growth of turfgrasses, thereby reducing the required frequency of mowing and the volume of clippings collected. Proxy® is also labeled for *Poa annua* or white clover seed head suppression.

Florel

Florel® is used for a variety of purposes, including the following:

- induces flowers in ornamental bromeliads, such as *Ananns*, *Aechmea*, and *Billbergia*.

- modifies sex expression and flowering pattern of cucumbers and squash to facilitate hybrid seed production. Florel® is not labeled for cucumbers or squash to be used for fresh market or processing.
- defoliaties after buds have matured in roses and apple nursery stock to allow digging of stock plants prior to the onset of unfavorable weather.
- increases lateral branching in azalea, begonia, chrysanthemum, geranium, impatiens, and others.
- reduces plant height of potted daffodils and stem topple of potted hyacinths at time of full flower.

Ethrel

Ethrel® is used as a foliar spray to promote earlier coloration and maturity of tomatoes, grapes, apples, and peppers. In addition, Ethrel® promotes fruit abscission (also called slipping) in cantaloupes and accelerates blackberry ripening and loosening. Ethrel® is also used to loosen cherries, walnuts, and apples to facilitate more efficient harvesting. Factors affecting the response of plants to ethylene-releasing compounds are the method of application, timing, environmental conditions, cultivar, and cultural conditions.

Ethylene Biosynthesis Inhibitor

The trade name for the commercially used ethylene biosynthesis inhibitor is Retain®. The active ingredient in Retain® is aminoethoxyvinylglycine, which is a substance isolated from a soil microorganism that specifically inhibits the biosynthesis of ethylene in plant tissues. Ethylene is a plant hormone responsible for the processes of fruit maturation, ripening, and fruit drop, among others. By temporarily turning off ethylene production, the following benefits can be achieved:

- improved harvest.
- reduced fruit drop.
- a wider harvest window that may enhance fruit size and color.
- improved storage potential and better fruit condition after harvest.
- reduction in the incidence of physiological disorders associated with fruit senescence, such as water core, superficial scald, and others. Factors affecting the response to Retain® are the method of application, environmental conditions, cultivars, and cultural practices.

Compounds Containing Gibberellins

There are six commonly used compounds containing gibberellins on the market that are commercially used. The trade names for these compounds are RyzUp®, Release®, Ralex®, ProGibb®, GibGro®, and Provide®. The method of application, environmental conditions, cultivars, and cultural practices have a profound effect on the use of compounds containing gibberellins.

RyzUp

The active ingredient in RyzUp® is gibberellic acid (GA_3), which is used to stimulate vegetative growth in rice, cotton, and other commercial crops. In rice, RyzUp® is applied during the early stages of plant development to stimulate plant growth, resulting in faster, more uniform seedling growth. This enables the grower to flood the field earlier and results in reduced time to harvest and improved grain quality. RyzUp® improves the vigor of cotton crops early in the season, which enables plants to recover from stress caused by adverse growing conditions. In bermudagrass, RyzUp® stimulates growth, which overcomes the effects of cold weather and light frost.

Release

The active ingredient in Release® is gibberellic acid (GA_3). Release® is used as a seed treatment on both semidwarf and tall rice cultivars; it is applied as a dip or fine mist to seeds. Release® promotes uniform germination and emergence, which allows for more accurate and efficient herbicide, fertilizer, fungicide, and insecticide applications, thereby increasing yields and overall quality.

Ralex

The active ingredient in Ralex® is gibberellic acid (GA_3). Ralex® is used as a foliar spray for annually thinning stone fruits (such as plum, peach, cherry, and apricot) to obtain acceptable size fruits and to ensure that the fruit load on trees is uniform year after year. An additional benefit of Ralex® is improved fruit firmness together with better shelf life and increased size during the season when it is applied.

ProGibb and GibGro

The active ingredient in ProGibb® is gibberellic acid (GA_3). ProGibb® is used in seedless grapes to elongate the cluster and promote larger berries with more uniform size. In seeded grapes, ProGibb® is used to reduce berry shrivel and increase berry size. ProGibb® has also been shown to maintain rind integrity and overall fruit quality, thereby allowing for a wider harvest window and better marketing potential for citrus fruit. ProGibb® delays aging of the rind and prevents physiological disorders such as rind staining, water spotting, sticky exudates, and puffiness. ProGibb® is also used to break dormancy in rhubarb, promote uniform bolting and increased seed production in lettuce, and to produce many other beneficial effects.

The active ingredient in GibGro® is also gibberellic acid (GA_3) and has the same uses as ProGibb®.

Provide

The active ingredients in Provide® are Gibberellins 4 and 7. Provide® is used as a foliar spray to apples to increase fruit size, reduce the incidence of the physiological disorder fruit surface russet, reduce fruit cracking, and improve the overall appearance of apples.

Compounds Containing Auxins

Auxins are used commercially for three purposes: fruit thinning and sticking fruits to trees, promoting adventitious rooting, and controlling weeds. The four commonly used compounds containing auxins on the market that are commercially used for fruit thinning and sticking fruits to trees are Amid-Thin W®, Fruitone N®, Fruit Fix®, and Citrus Fix®. The active ingredient in Amid-Thin W® is 1-naphthaleneacetamide. Amid-Thin is used as a foliar spray for thinning apples and pears. The active ingredient in Fruitone N® is 1-naphthaleneacetic acid. Fruitone N® is used to control preharvest fruit drop and thinning apples. The active ingredient in Fruit Fix® is also 1-naphthaleneacetic acid, which is used for thinning apples, citrus, olives, and pears. Fruit Fix® is also used to control preharvest drop of apples and pears. The active ingredient in Citrus Fix® is the highly volatile isopropyl ester of 2,4-dichlorophenoxyacetic acid, which is used to prevent fruit drop in citrus. Because this is a highly volatile compound, extreme caution should be used to prevent drift.

Three commonly used compounds containing auxins are used commercially to promote adventitious rooting: Dip and Grow®, Rhizopon®, and Hormodin®. The active ingredients in Dip and Grow® are indole-3-butyric acid (IBA) and naphthalene acetic acid (NAA) in an isopropanol/ethanol solution. The active ingredient in Rhizopon® and Hormodin® is indole-3-butyric acid (IBA), and they come in powder or tablet formulations.

Many formulations of compounds containing the auxin 2,4-D are commercially available for the control of broadleaf weeds; for example, Weedar64 has the active ingredient 2,4-dichlorophenoxyacetic acid in salt form. The method of application, environmental conditions, cultivars, and cultural practices have a profound effect on the use of compounds containing auxins.

Compounds Containing Gibberellins and Cytokinins

Two commonly used compounds containing both auxins and gibberellins are used commercially: Accel and Promalin®. The active ingredients in Accel are N-(phenylmethyl-1H-purine-6-amine) and gibberellins 4 and 7. Modern apple production requires that apple fruits be thinned early in their development to attain a commercially acceptable final size fruit and to ensure adequate return bloom the following year. Accel contains a cytokinin and gibberellin mixture that promotes fruit thinning and increases the number of even-sized fruits. Promalin contains the same active ingredients as Accel; however, it contains 10 times more gibberellin 4 and 7 than Accel. Promalin improves the shape of Delicious apples through elongation of fruit and development of more prominent calyx lobes. These desirable effects are more evident in years when natural **typiness,** which is elongated fruit and prominent calyx lobes, is limited. In addition, Promalin has been shown to increase the weight of individual fruit and yield per acre. On most apple cultivars, Promalin will increase lateral bud break and shoot growth and improve branch angles on nonbearing trees used for nursery stock. This provides a better tree framework and earlier cropping. The method of application, environmental conditions, cultivars, and cultural practices have a major effect on the performance of these products.

SUMMARY

You now have a better understanding of plant growth regulators and their many uses in agriculture today. You learned the definitions for a plant hormone, plant growth regulator, plant growth retardant, auxin, gibberellin, cytokinin, abscisic acid, ethylene, and brassinosteroids. In addition, you know how each of the six classes of plant hormones was discovered and the key names of scientists associated with their discovery. Each of the six classes of plant hormones is involved in key physiological processes in plants. The major categories of plant growth regulators used in agriculture are plant growth retardants, ethylene-releasing compounds, an ethylene inhibitor, compounds containing gibberellins, compounds containing auxins, compounds containing gibberellins, and cytokinins. Plant growth regulators are used in agriculture today and have potential for use in the future.

Review Questions for Chapter 10

Short Answer

1. Who made the profound statement "Ohne Wuchstoff, kein Wachstum," which when translated means without growth substances no growth?
2. List the six classes of plant hormones.
3. Provide one commercial use for auxins, gibberellins, cytokinins, and ethylene.
4. List three researchers who were pioneers in auxin research and list their contributions.
5. What is the only active auxin found in plants?
6. Provide eight physiological processes in which auxins have a regulatory role.
7. Give two synthetic auxins and a commercial use for each.
8. Who was the scientist that who first crystallized gibberellin A and what material was used for purification?
9. Provide seven physiological processes in which gibberellins have a regulatory role.
10. What is the series of events that takes place during germination in barley seeds starting with where gibberellins are synthesized?
11. Who was the first scientist to purify a cytokinin and what material was used for purification?
12. Who were the two scientists responsible for the purification of zeatin and what source was used to purify this compound?
13. List six physiological processes in which cytokinins have a regulatory role.
14. Who were the first scientists to discover abscisic acid and what material was used for purification?
15. What is the main physiological process affected by abscisic acid?
16. Who discovered that ethylene caused the triple response in pea seedlings?
17. List four physiological processes in which ethylene has a regulatory role.
18. What is the trade name for a synthetic form of ethylene?
19. Who were the first scientists to discover brassins and what material were brassins extracted from?
20. What is the main physiological effect of brassinosteroids?
21. What are four general factors that affect the plant response to plant growth regulators?
22. What are the major benefits of using plant growth retardants in agriculture today?
23. What are five benefits of using the ethylene biosynthesis inhibitor Retain®?
24. What are three commercial uses for auxins today?

Define

Define the following terms:

plant hormone	auxin	Bakanae disease	senescence
plant growth regulator	gibberellin	bolting	abscission
plant growth retardant	cytokinins	dormancy	brassins
nutrients	abscisic acid	parthenocarpy	brassinolide
heteroauxin	ethylene	callus	typiness
auxein	brassinosteroids	triple response	

True or False

1. All plant growth regulators are plant hormones, but not all plant hormones are plant growth regulators.
2. An important component of the definition of a plant hormone is that it must be chemically characterized.
3. The plant hormone indole-3-acetic acid (IAA) was first discovered in human urine.
4. Auxins stimulate root initiation and inhibit root elongation.
5. *Heteroauxin* also means "other auxin," which is known today as indole-3-acetic acid (IAA).
6. Auxins promote cell elongation in intact plants.
7. Negative gravitropism is downward bending.
8. Positive gravitropism is upward bending.
9. IAA accelerates growth on the dark side of the coleoptile, while growth proceeds normally on the illuminated side.
10. Auxins stimulate ethylene production; therefore, many of the responses once attributed to auxins are now found to be due to ethylene.
11. Gibberellins were first discovered in a fungus.
12. GA_1 and GA_2 are the most commonly used forms of gibberellins used today.
13. Kinetin is a naturally occurring form of cytokinin.
14. When a plant undergoes water stress, ABA levels are reduced dramatically to protect the plant.
15. Brassinolide is a crude lipid extract from rape pollen.
16. Brassins are the active component in brassinolide.
17. The ethylene-releasing compound Proxy® is used to reduce the required frequency of mowing and volume of grass clippings collected.
18. Florel® is an ethylene-releasing compound that is used to induce flowering and modify sex expression.

Multiple Choice

1. Which of the following is the naturally occurring auxin found in plants?
   A. Indole-3-butyric acid (IBA)
   B. Indole-3-acetaldehyde (IAH)
   C. Indole-3-acetic acid (IAA)
   D. None of the above
2. Which of the following plant hormones promotes cell elongation in intact plants?
   A. Auxins
   B. Gibberellins
   C. Cytokinins
   D. Brassinosteroids
3. Which of the following is the major physiological effect of abscisic acid?
   A. Gravitropism
   B. Phototropism
   C. Plant stress signal
   D. All of the above

4. Which of the following is the only gaseous plant hormone?
   A. Abscisic acid
   B. Ethylene
   C. Propylene
   D. None of the above
5. Which of the following is the major physiological effect of brassinolide?
   A. Shoot elongation
   B. Root initiation
   C. Plant stress signal
   D. None of the above
6. What is the active ingredient found in Release®?
   A. Gibberellic acid
   B. Indole-3-acetic acid
   C. Zeatin
   D. None of the above
7. What are the active ingredient(s) in Dip and Grow®?
   A. Indole-3-butyric acid and naphthalene acetic acid
   B. Indole-3-acetic acid
   C. 2, 4-Dichlorophenoxyacetic acid
   D. None of the above

Fill in the Blanks

1. Auxins are a class of compounds with activity similar to ________________.
2. During seed germination, gibberellins are produced in the ________________ and transported to the ________________ layer of cells where ________________ is produced, which causes the conversion of ________________ to ________________ in the endosperm. These materials are used for energy during the germination process.
3. Cytokinins promote ________________, which is an undifferentiated mass of cells.
4. ________________ is the only gaseous plant hormone.
5. Brassinosteroids are a class of compounds with activity similar to ________________.
6. The active ingredient in Rhizopon® and Hormodin® is ________________.

Matching

Insert the proper term in the space provided.

Terms: (A) ACC synthase (B) AdoMet (C) ACC oxidase (D) ACC

Ethylene Biosynthetic Pathway

________________ → (________________) ________________ → (________________) **Ethylene**

Activities

Select a specific plant growth regulator and search for Web sites that present information on commercial uses of your selected plant growth regulator. Collect the following facts about the plant growth regulator:

- proper concentration to use.
- methods of application.
- when to apply.
- safety precautions.
- other facts that you feel are important.

After you have compiled this information, create a fact sheet that puts everything important about the selected plant growth regulator in general consumer terms.

References

Arteca, R. N. (1996). *Plant growth substances: Principles and applications*. New York: Chapman & Hall.

Cutler, H. G., Yokota, T., & Adam, G. (1991). *Brassino steroids: Chemistry, bioactivity and applications*. Washington, DC: American Chemical Society.

Darwin, C. (1880). *The power of movement in plants*. New York: Appleton.

Postharvest Physiology

ABSTRACT

Postharvest physiology refers to handling horticultural commodities following harvest. Knowing when to harvest, the harvesting options available, and the importance of proper postharvest handling techniques for horticultural commodities is vital. Seven biological factors—respiration, ethylene production, compositional changes, growth and development, transpiration, physiological breakdown, and physical and pathological breakdown—influence the quality of perishable horticultural commodities. Five environmental factors—temperature, relative humidity, atmospheric composition, ethylene, and light—also influence this quality level. Different postharvest technology procedures are used to maximize the storage life of horticultural commodities and to explore future trends in handling perishable horticultural commodities.

Objectives

After reading this chapter, you should be able to

- provide general background information on when to harvest and the options available for harvesting.
- understand the importance of studying better ways to prevent postharvest losses of horticultural commodities.
- identify the seven biological factors involved in the deterioration of horticultural commodities.
- identify and discuss the five environmental factors influencing the deterioration of horticultural commodities.
- list and discuss postharvest technology procedures to maximize the storage life of horticultural commodities.
- recognize future trends in handling perishable horticultural commodities.

Key Terms

chilling injury
climacteric fruits
controlled atmosphere (CA) storage
freezing injury
modified atmosphere (MA) storage
nonclimacteric fruits
respiration
vine-ripe

INTRODUCTION

This chapter provides some key information for reducing postharvest losses by better understanding biological and environmental factors involved in the deterioration of horticultural commodities and by using postharvest techniques to delay the deterioration process. Knowing the proper time to harvest and the options available for harvesting crops is important. This information is based on whether the crop is to be harvested for fresh market or processing. When crops are intended for the fresh market, they can be harvested **vine-ripe** or they can be harvested prior to maturity and allowed to ripen in storage. The benefit of harvesting prior to maturity is that the postharvest life is longer; however, the disadvantage is that the flavor is not as good as the vine-ripened fruits and vegetables. Vine-ripened fruits and vegetables have a much shorter postharvest life, however. Fruits or vegetables that are used for processing are canned, dried, or frozen. Fruits and vegetables to be used for processing must meet specific guidelines for texture, sugar content, color, and other characteristics. Special instruments are used to determine precisely all the quality parameters used for processed fruits and vegetables. Growers must be knowledgeable in all the quality parameters for particular crops to be successful.

Most horticultural crops are perishable and must be harvested rapidly and used fairly quickly. In fact, the number one factor that determines the postharvest storage life of a given commodity is the ability to harvest quickly and immediately remove field heat, thereby slowing the deterioration process. Harvesting can be done by hand or mechanically. Both have a number of advantages and disadvantages. Hand harvesting is selective, which reduces the need for cleaning and sorting; however, it is also labor intensive, expensive, and slow. Mechanical harvesting is faster than hand harvesting, so many acres can be harvested rapidly, and the commodity can be taken out of the field more quickly, which facilitates removal of field heat. However, the initial outlay of money is high for the harvesting machine and for upkeep.

Between harvest and consumption of horticultural crops, dramatic losses in quality and quantity occur. In developed countries, postharvest losses can be between 5 and 25 percent, whereas in undeveloped countries, postharvest losses are typically between 20 and 50 percent (Kader, 1992). To reduce postharvest losses, growers must understand the biological and environmental factors involved in the deterioration of horticultural commodities and the proper postharvest techniques that delay the deterioration process.

Seven biological factors are involved in the deterioration of harvested horticultural commodities: **respiration;** ethylene production; compositional changes in organic acids, proteins, amino acids, lipids, vitamins, and flavor volatiles; growth and development; transpiration; physiological breakdown; and physical and pathological breakdown. The five environmental factors involved in the deterioration of harvested horticultural commodities are temperature, relative humidity, atmospheric composition, ethylene, and light. Although all five environmental factors are important, temperature has the most profound affect on horticultural commodities placed in storage. For each increase in 10°C above the optimum storage temperature, there is a two- to threefold increase in the rate of deterioration.

Three main groupings of postharvest technology procedures are used to extend the storage life of horticultural commodities: temperature management, control of relative humidity, and supplements to temperature and humidity. Temperature management is the most effective tool for extending the shelf life of fresh horticultural commodities (Figure 11-1). In fact, the first step in effective temperature management is the rapid removal of field heat. Relative humidity affects water loss, decay development, a variety of physiological disorders, and uniformity of ripening. By properly managing the relative humidity in storage, problems can be minimized. There are a variety of supplements to temperature and humidity management, including controlled or modified atmospheric storage, crop curing, damaged fruits removal, waxing or other surface coverings, heat treatments, fungicide treatments, insecticides, sprout inhibitors, calcium, gibberellins, ethylene or antiethylene compounds, and sanitation. None of these supplemental procedures alone or in combination can be used as a substitute for temperature or relative humidity management, but they are very

Figure 11–1 Flowers are stored in a large cooler to maintain their quality.

effective in extending the storage life of harvested commodities beyond temperature or relative humidity alone.

Because billions of dollars are lost worldwide annually due to postharvest losses, research and development efforts are always underway to improve existing technology and to study new and innovative methods to improve the quality of perishable horticultural commodities. Some trends in handling perishables are developing better means of controlling and monitoring temperature and relative humidity, expanding film wrapping and other technology, improving controlled and modified atmospheric storage technology, and genetically engineering horticultural crops to enhance their storage life.

GENERAL BACKGROUND INFORMATION ON POSTHARVEST HANDLING

Producing a high-quality crop requires using good cultural practices with environmentally sound pest management. When a grower produces a high-quality crop, the postharvest performance is much better than that of a poor-quality crop. Previous chapters outlined ways to produce high-quality crops, but prior to beginning this section on postharvest physiology, it is important to start with a brief discussion about when to harvest and the options available for harvesting crops.

Figure 11–2 Vine-ripened tomatoes with stems still attached

Horticultural crops used for consumption can be harvested for fresh market or processing. A crop that is to be used for fresh market can be harvested when vine-ripe, which is when fruits or vegetables are picked after they are ripe and ready for immediate use (Figure 11-2). Crops to be used for fresh market also can be harvested prior to maturity and ripened in storage naturally, or the process can be accelerated by treatment with the plant hormone ethylene in certain cases. For example, bananas are harvested when they are green, and then shipped to distant places where they ripen naturally. The benefit of harvesting prior to maturity is that the postharvest life is much longer, although the flavor is not as good as vine-ripened fruits or vegetables. However, vine-ripened fruits have a much shorter postharvest life.

Fruits or vegetables to be used for processing are canned, dried, or frozen. Fruits and vegetables for processing have specific guidelines for texture, sugar content, color, and more. Special instruments are used to determine precisely a variety of quality parameters. For example, the percentage soluble solids, which are primarily sugars, is important in fruits such as apples and pears. In certain crops, color is very important, whereas in others, the degree of firmness is critical. The grower must understand what quality parameters are necessary for a particular crop to produce successfully and to market fruits and vegetables for processing.

Most horticultural crops are perishable and must be harvested rapidly and used fairly quickly. One of the most critical factors in deciding the postharvest life of a given horticultural commodity is how quickly it can be harvested and immediately removed from field heat to reduce its deterioration. Harvesting can be done by hand or mechanically, and each method has advantages and disadvantages. Harvesting fruits and vegetables by hand has several advantages over mechanical harvesting mainly because they are picked selectively.

This reduces the number of low-quality fruits or vegetables and debris harvested. Hand picking also saves time required for cleaning and sorting after harvesting. In some cases, hand harvesting may be the only choice because specialized machinery is not available for a particular crop. The main disadvantages of hand harvesting are that it is labor intensive, expensive, and slow. Examples of crops that are generally harvested by hand are cabbage, broccoli, pepper, and cauliflower.

The main advantage of mechanized harvesting is speed. Many acres can be harvested very rapidly, thereby getting the commodity out of the field and cooled down quickly and efficiently. The main disadvantage is the high cost of the machines to do the harvest and their upkeep. Products harvested mechanically are more likely to be damaged and contain poor-quality products and debris that requires cleaning and sorting. Examples of products that are commonly mechanically harvested include cherries, tomatoes, corn, potatoes, beets, and carrots.

After products are harvested, either by hand or mechanically, they are typically washed, sorted, and graded prior to placement into storage. Losses in quality and quantity affect horticultural crops between harvest and consumption. In developed countries, postharvest losses are estimated at 5 to 25 percent, whereas in undeveloped countries, postharvest losses are estimated at 20 to 50 percent (Kader, 1992). To reduce postharvest losses, producers and handlers must understand the biological and environmental factors involved in the deterioration of horticultural commodities and the proper postharvest techniques that delay the deterioration process.

Biological Factors Involved in the Deterioration of Harvested Commodities

Fresh fruits, vegetables, and other horticultural crops are living tissues that begin to deteriorate following harvest. All horticultural crops are high in water content and thus subject to desiccation and to mechanical injury, which leads to attack by bacteria and fungi, and other pathological breakdowns. Although postharvest breakdown cannot be stopped, it can be slowed down dramatically.

Respiration

Respiration is the process by which stored organic materials (carbohydrate, protein, fat) are broken down into simple end products with a release of energy. Oxygen is used in this process and carbon dioxide is liberated, which is the opposite of photosynthesis. Therefore, increased respiration means a loss of stored food reserves leading to

- an acceleration of the senescence process.
- a reduced food value for the consumer.
- a loss of flavor, especially sweetness.
- a loss of salable dry weight.

The rate of deterioration (perishability) of harvested commodities is generally proportional to the respiration rate. This is clear in Table 11-1, which shows that horticultural commodities with very low respiration rates, such as dried fruits and vegetables, have a long shelf life, whereas commodities with extremely high rates of respiration, such as sweet corn, have a very short shelf life.

TABLE 11-1 PERISHABLE HORTICULTURAL COMMODITIES AS RELATED TO RESPIRATION RATES

Commodities	Perishability	Respiration Rate
Nuts, dates, dried fruits and vegetables	Very low	Very low
Apple, citrus fruits, garlic, mature potato, onion, watermelon	Low	Low
Banana, blueberry, cabbage, cherry, cucumber, head lettuce, peach, tomato	Moderate	Moderate
Avocado, blackberry, leaf lettuce, lima beans, raspberry	High	High
Bean sprouts, broccoli, cut flowers, snap beans, watercress	Very high	Very high
Sweet corn, asparagus, mushroom, spinach, parsley, peas	Extremely high	Extremely high

Ethylene Production

Ethylene is a gaseous plant hormone involved in the regulation of many aspects of growth, development, and senescence. Ethylene is physiologically active at levels as low as less than 0.1 ppm (parts per million). The regulation of ethylene production and its action is very important in agriculture today. The key steps in the ethylene biosynthetic pathway starts by S-adenosylmethionine (AdoMet) being converted to 1-aminocyclopropane-1-carboxylic acid (ACC) via the enzyme ACC synthase. ACC has two fates:

- conversion to malonyl ACC (which is an inactive endproduct) via the enzyme ACC N-malonyltransferase (ACC-NMTase).
- conversion to ethylene via the enzyme ACC oxidase (Figure 11-3).

Generally, ethylene production rates are increased as the commodity matures and in response to physical injuries, diseases, temperature extremes (either too high or low), water stress, and a variety of other factors (Arteca, 1996).

The production of ethylene can be reduced in horticultural commodities by

- decreasing the storage temperature.
- decreasing O_2 levels to less than 8 percent.
- treating with the inhibitors aminoethoxyvinylglycine (AVG), aminooxyacetic acid (AOA), or cobalt chloride.
- genetically engineering plants to reduce the production of ethylene (through the use of antisense technology and the cloned genes for ACC synthase and ACC oxidase).

The effects of ethylene action on horticultural commodities can be blocked by

- treating flowers with silver thiosulphate (STS) (trade name Floralife). The ethylene receptor is blocked, thereby stopping ethylene's capability to function as a plant hormone.

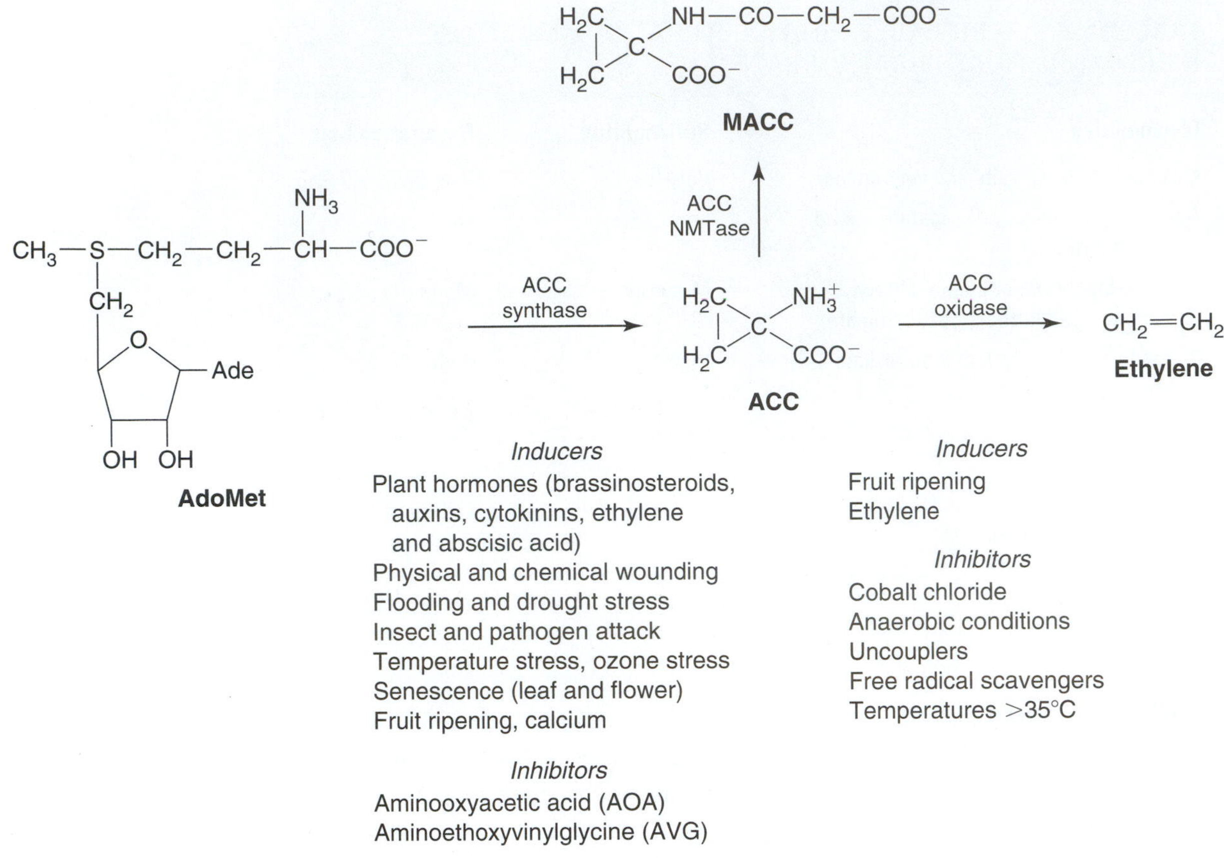

Figure 11–3 Biosynthetic reactions for the production of ethylene from S-adenosylmethionine (AdoMet) to ethylene plus inducers and inhibitors

- placing horticultural commodities in hypobaric storage, which is keeping the commodity under vacuum, thereby not enabling ethylene to stay in contact with or act upon the commodity.
- treating horticultural commodities with levels of CO_2 greater than 2 percent reduce to the capability of the ethylene receptor to accept ethylene, which in turn blocks ethylene action.
- genetically engineering plants that have a modified ethylene receptor that does not accept ethylene, thereby blocking its action.

Based on their respiration (CO_2 evolution) and ethylene production patterns during maturation and ripening, fruits are classified as either climacteric or nonclimacteric. **Climacteric fruits** show large increases in CO_2 and ethylene (C_2H_4) production rates coincident with ripening and ripen in response to exogenous applications of ethylene. **Nonclimacteric fruits** show no change in their generally low CO_2 and C_2H_4 production rates during ripening and do not ripen in response to ethylene. The relationship between ethylene production and respiration during a climacteric rise in bananas is shown in Figure 11-4; nonclimacteric fruits show a flat line with respect to CO_2 and C_2H_4. Selected examples of both climacteric and nonclimacteric fruits are shown in Table 11-2. Horticultural commodities can be classified according to ethylene production

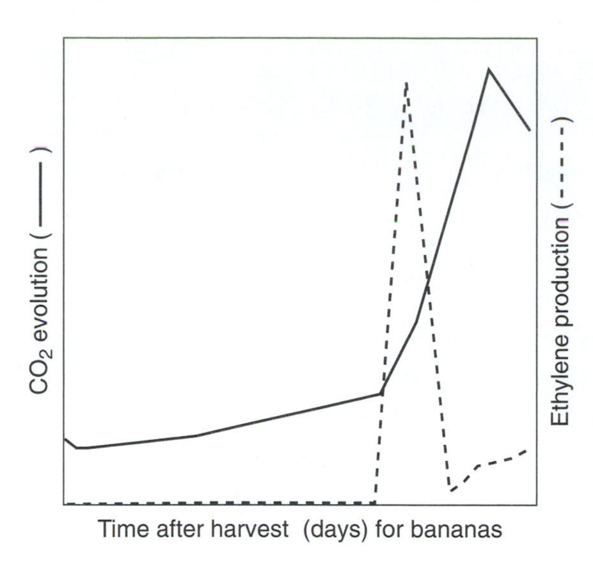

Figure 11–4 Climacteric rise in carbon dioxide and ethylene following harvest of bananas

TABLE 11-2 CLASSIFICATION OF FRUITS BASED ON RESPIRATION AND ETHYLENE PRODUCTION RATES DURING RIPENING

Climacteric Fruits	Nonclimacteric Fruits
Apple	Blackberry
Avocado	Cherry
Banana	Cucumber
Blueberry	Pepper
Fig	Strawberry
Mango	Lemon
Peach	Orange
Pear	Peas
Plum	Watermelon
Tomato	Grape

rates; however, there is no consistent relationship between ethylene production rates and perishability, as shown in Table 11-3.

Compositional Changes

Changes in composition are another biological factor involved in the deterioration of crops.

Pigments. Many changes in pigments take place during development and maturation of a given commodity, some of which may continue following harvest. These changes may either be good or bad. Following are some pigments that change:

- **Chlorophyll (green color).** A loss of chlorophyll in tomatoes is good, but a loss of chlorophyll in broccoli is bad.
- **Carotinoids (yellow, orange, and red colors).** Carotinoids are desirable in fruits such as apricots, peaches, and citrus, giving them their

TABLE 11-3 THE PERISHABILITY OF HORTICULTURAL COMMODITIES AS RELATED TO ETHYLENE PRODUCTION RATES

Commodities	Perishability	Ethylene Production Rate
Strawberry	Very high	Very low
Cherry	Very high	Very low
Grape	Moderate	Very low
Leafy vegetables	Very high	Very low
Raspberry	Very high	Low
Watermelon	Low	Low
Banana	High	Moderate
Fig	Very high	Moderate
Tomato	Moderate	High
Apple	Low	High
Avocado	High	High
Cut flowers	Very high	Very low

yellow and orange color. In tomatoes and pink grapefruit, a specific carotinoid called lycopene gives them their red color.

- **Anthocyanins (red and blue colors).** Anthocyanins give red and blue color to apples, berries, and cherries.

Phenolic compounds. These compounds are responsible for tissue browning; for example, when an apple, banana, or potato is cut, it browns due to phenolic compounds.

Carbohydrates. These can be in the form of starch and sugar.

- **Conversion of starch to sugar.** The conversion of starch to sugar during storage is not desirable in potatoes because when a potato has a high sugar content, it fries dark brown or, in some cases, when sugar levels are very high, the potato turns black. However, a high sugar content is very desirable in apples or bananas.
- **Conversion of sugar to starch.** The conversion of sugar to starch is not desirable in sweet corn because of the reduction in flavor, but it is very desirable in potatoes.
- **Conversion of starch and sugars to carbon dioxide and water during respiration.** The conversion of starch and sugars to carbon dioxide and water is not desirable. The breakdown of pectins and polysaccharides results in fruit softening, which increases susceptibility to mechanical injuries.

Organic acids, proteins, amino acids, lipids, vitamins, and flavor volatiles. Changes in organic acids, proteins, amino acids, and lipids can influence the flavor of horticultural commodities. During storage, vitamin content can decrease—especially ascorbic acid (vitamin C)—lessening the nutritional quality of the product. Many flavor volatiles are produced during the ripening processes that are extremely important to the eating quality of a number of fruits and vegetables. Modifying these volatiles in any way may result in off flavors or smells, which dramatically reduces the quality of the fruit or vegetable.

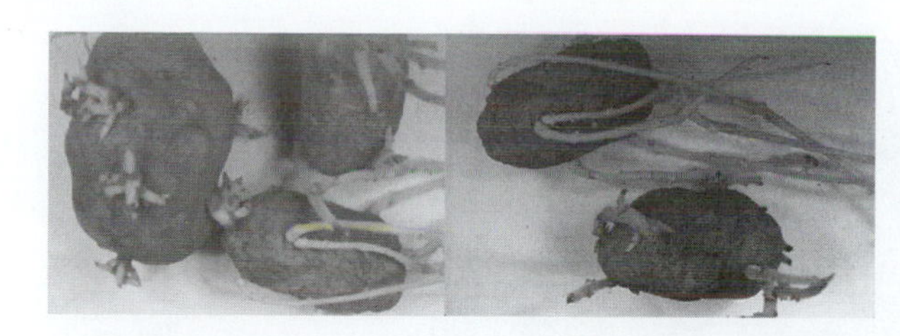

Figure 11–5 Excessive sprouting in potato tubers. *Courtesy of William Lamont, The Pennsylvania State University.*

Figure 11–6 Onions with roots starting to grow

Growth and Development

After a horticultural crop is harvested, retarding any further growth and development reduces loss of postharvest quality. The following are examples of types of growth and development that reduce postharvest quality:

- **Sprouting.** When sprouting occurs in potatoes, onions, garlic, and root crops after harvesting, stored food reserves are diverted to the sprout and away from the vegetable itself, resulting in an acceleration of the deterioration process (Figure 11-5).
- **Rooting.** When onions and root crops form roots, stored food reserves are diverted to the newly formed root, causing a reduction in quality (Figure 11-6).
- **Elongation and curvature.** Asparagus spears continue to grow after harvest if handled correctly. For example, after asparagus are harvested and put in water upright, they will grow another 3 to 4 inches (Figure 11-7). However, if they are placed horizontally they will exhibit gravitropic curvature, which leads to an increase in toughness due to lignification. Increased lignin content is also responsible for toughening in a number of root crops. Other gravitropic responses affect the quality of gladiolus and snapdragons by causing an unsightly appearance.
- **Seed germination.** Seeds germinate in fruits such as tomatoes, peppers, and lemons, producing an undesirable quality.

Figure 11–7 Asparagus stored upright to prevent gravitropic bending

Transpiration

Water loss is the main cause of deterioration because it results in a loss of salable weight, unsightly appearance such as wilting or shriveling, changes in texture such as loss of crispness and juiciness, and reduced overall nutritional quality. The outer protective covering surrounding the commodity has a direct effect on water loss. The rate of transpiration is influenced by morphological and anatomical factors such as the surface-to-volume ratio, surface injuries, and the stage of maturity. In addition, transpiration is influenced by external or environmental conditions, such as temperature, relative humidity, air movement, and atmospheric pressure. Transpiration is a physical process that can be reduced by treating the commodity with waxes and/or other surface coverings or by manipulating environmental conditions, such as maintaining a high relative humidity and properly controlling air circulation.

Physiological Breakdown

Exposure to undesirable temperatures can result in problems with physiological disorders, such as the following:

- **Freezing injury.** Freezing injury occurs in some commodities that are held at temperatures below freezing. The result of freezing damage is a collapse in tissues, resulting in total loss. Selected examples of fruits and vegetables with varying degrees of sensitivity to freezing injury are shown in Table 11-4.
- **Chilling injury.** Chilling injury occurs mainly in tropical and subtropical crops that are held at temperatures above their freezing point. This type of injury exhibits its symptoms upon transfer to higher nonchilling temperatures. The most common symptoms are browning and pitting, water-soaked areas, uneven ripening or failure to ripen, off flavors, molds, and decay. Selected examples of fruits and vegetables classified according to sensitivity to chilling injury are shown in Table 11-5.

TABLE 11-4 FRUITS AND VEGETABLES AND THEIR SUSCEPTIBILITY TO INJURY FROM COLD TEMPERATURE WHEN STORED BELOW FREEZING

Commodities	Susceptibility
Fruits	
Blackberry	Very susceptible
Apricot	Very susceptible
Peach	Very susceptible
Apple	Moderately susceptible
Grape	Moderately susceptible
Pear	Moderately susceptible
Vegetables	
Cucumber	Very susceptible
Asparagus	Very susceptible
Carrot	Moderately susceptible
Celery	Moderately susceptible
Brussels sprouts	Generally resistant
Rutabaga	Generally resistant

TABLE 11-5 FRUITS AND VEGETABLES AND THEIR SENSITIVITY TO CHILLING INJURY

Sensitivity to Chilling Injury	Fruits	Vegetables
Nonchilling sensitive		
	Apricots	Artichokes
	Blackberries	Asparagus
	Blueberries	Beets
	Cherries	Broccoli
	Grapes	Celery
	Pears	Sweet corn
	Raspberries	Garlic
	Strawberries	Onions
Chilling sensitive		
	Avocados	Beans
	Bananas	Cassava
	Citrus	Cucumbers
	Mangoes	Peppers
	Olives	Potatoes
	Papayas	Sweet potatoes
	Pineapples	Tomatoes
	Lychees	Watermelons

- **Heat injury.** Heat injury is caused by exposure to direct sunlight or excessively high temperatures. The symptoms include bleaching, surface burning, uneven ripening, and desiccation.

Preharvest nutritional imbalances. Certain types of physiological disorders are caused by improper nutrition prior to harvest. Examples of this are blossom end rot in tomatoes and bitter pit in apple, which are both caused by calcium deficiency. Increasing the calcium content either by preharvest or postharvest treatments can reduce the susceptibility to physiological disorders. The calcium content also has a major effect on the textural quality and senescence rate of fruits and vegetables. Increasing the calcium content in fruits and vegetables improves firmness, reduces respiration and ethylene production rates, and reduces the rate of decay.

Physical and Pathological Damage

Physical damage can occur due to a variety of factors, such as mechanical damage caused by impact bruising and vibration during harvesting operations. Immediately following bruising, browning occurs in the damaged tissues due to the disruption of membranes, which activates the polyphenol oxidase enzyme. Mechanical damage not only reduces the appearance of fruits and vegetables, but it also accelerates water loss, stimulates carbon dioxide and ethylene production by the commodity, and provides an opening for fungi and bacteria to enter.

Environmental Factors Involved in the Deterioration of Harvested Commodities

Environmental factors can have an effect on the deterioration of harvested crops. These factors include temperature, relative humidity, atmospheric composition, ethylene, and light.

Temperature

The environmental factor that most influences the deterioration rate of harvested commodities is temperature. For each increase of 10°C above optimum storage temperature, a two- to threefold increase occurs in the rate of deterioration. When horticultural commodities are subjected to undesirable temperatures, a wide range of physiological disorders occur. Temperature also has a profound effect on respiration and ethylene-production rates. In addition, temperature has a dramatic effect on spore germination and the growth of a wide range of pathogens. If chilling-sensitive crops are exposed to temperatures that are too cool, there will be a marked reduction in their quality.

Relative Humidity

The rate of water loss from fruits and vegetables depends on the relative humidity; therefore, if the temperature is reduced and the air is not humidified, the crops will lose water, resulting in a loss in quality.

Atmospheric Composition

The amount of O_2 or CO_2 in the storage atmosphere has a dramatic effect on deterioration rates of stored fruits and vegetables. By modifying the atmospheric composition, ethylene effects can be minimized. The effects of atmospheric composition on the deterioration rate of fruits and vegetables depend on the cultivar, physiological age, O_2 and CO_2 levels, temperature, and duration in storage.

Ethylene

The plant hormone ethylene has a dramatic effect on harvested horticultural commodities, which can be desirable or undesirable; therefore, all postharvest handlers must understand completely ethylene and its many effects. For example, fruits can be treated with ethylene to promote faster, more uniform ripening or, in the case of tomato fruits, can be harvested when they are mature (green) and then treated with ethylene to turn them red. However, fruits and vegetables produce ethylene during storage, which can have adverse effects on their quality and should be carefully monitored and regulated.

Light

In certain crops, light should be avoided because it reduces quality. For example, when potatoes are exposed to light, they turn green due to the formation of chlorophyll, which is not harmful to humans. However, in addition to chlorophyll, a compound called solanine, which is colorless and very toxic to humans, is produced simultaneously. Calcium infiltration into potatoes has been shown to decrease greening potatoes; however, this method is not used commercially.

Postharvest Technology Procedures

Various techniques can be used to prolong the storage of harvested crops while maintaining their quality. These include temperature management procedures, control of relative humidity, and supplements to temperature and humidity management.

Temperature Management Procedures

Temperature management is the most effective tool for extending the shelf life of fresh horticultural commodities. The first step in effective temperature management is rapidly removing field heat by using one of the following techniques: hydrocooling, in-package icing, evaporative cooling, refrigeration, vacuum cooling, or hydro-vacuum cooling. Vehicles used to transport horticultural commodities should be precooled prior to loading the commodity. After the commodity has been cooled, it should be stored in a well-engineered and adequately equipped cold-storage facility. All commodities should be stacked in the cold room with air spaces between pallets and room walls to ensure good air circulation. To ensure proper cooling, storage rooms must not be loaded beyond their limit. When monitoring temperatures, the temperature of the commodity should be monitored rather than the temperature of the air. The bottom line of good temperature management is to be sure that proper temperature is maintained from harvest through storage, starting with the rapid removal of field heat.

Control of Relative Humidity

Relative humidity can have a profound effect on water loss, decay development, the incidence of physiological disorders, and nonuniform ripening. By maintaining the proper relative humidity, these problems can be minimized. One or more of the following procedures can be used to control relative humidity: adding moisture (water mist or spray) to the air with humidifiers, sprinkling the horticultural commodity with water, and carefully regulating air movement and ventilation. The proper relative humidity for fruits is 85 to 95 percent; for vegetables it is 90 to 98 percent, except for pumpkins and dry onions, which should be kept at 70 to 75 percent; and for some root vegetables it is 95 to 100 percent.

Supplements to Temperature and Humidity Management

Many procedures can be used to supplement temperature and humidity management. None of these procedures alone or in combination can be used as a substitute for temperature and relative humidity management, but they can help to extend the storage life of harvested commodities. These supplemental treatments include the following:

- using of **controlled or modified atmospheres (CA or MA) storage,** which is the regulation of CO_2 and O_2 levels.
- curing root, bulb, and tuber crops.
- removing damaged fruits, (remember, one rotten apple spoils the whole bushel).
- waxing and other surface coatings.
- applying heat treatment.
- treating with fungicides and insecticides.
- treating with sprout inhibitors.
- treating with calcium, gibberellins, or antiethylene compounds.
- applying ethylene treatment (degreening or ripening).
- using sanitation techniques.

Future Trends in Handling Perishables

Research and development efforts are always underway to improve existing technology and to test new and innovative ideas to improve the quality of perishable items. Some trends are as follows:

- using better and more economical methods for controlling and monitoring temperature and relative humidity.
- reducing losses due to chilling injury by avoiding temperatures that cause it.
- expanding film wrapping and other treatments to replace waxing and reduce water loss.
- increasing mechanization in all aspects of the harvesting and storage operation.
- developing improved CA and MA technology.
- genetically engineering crops to enhance their storage life.

SUMMARY

Proper postharvest handling techniques are critical for horticultural commodities. Seven biological factors are involved in the deterioration of harvested horticultural commodities: respiration; ethylene production; compositional changes in organic acids, proteins, amino acids, lipids, vitamins, and flavor volatiles; growth and development; transpiration; physiological breakdown; and physical and pathological breakdown. In addition, five environmental factors are involved in the deterioration of harvested horticultural commodities: temperature, relative humidity, atmospheric composition, ethylene, and light. Key postharvest technology procedures for extending the shelf life of horticultural commodities include temperature management practices, regulation of relative humidity, and supplements to temperature and relative humidity management. These supplements include using controlled or modified atmospheric storage; curing crops; removing damaged fruits; using waxing or other surface coverings; applying heat treatments; treating with fungicides, insecticides, sprout inhibitors, calcium, gibberellins, ethylene or antiethylene compounds; and using sanitation techniques. Future trends in handling perishable horticultural commodities include genetically engineering crops and increased mechanization.

Review Questions for Chapter 11

Short Answer

1. Describe the main advantage and disadvantage of harvesting horticultural crops to be used for consumption prior to maturity.
2. Describe the main advantage and disadvantage of harvesting horticultural commodities to be used for consumption when they are vine-ripe.
3. What are the main advantages and disadvantages of hand- and mechanical harvesting?
4. What are the estimated percentage postharvest losses in developed and undeveloped countries?
5. What are two factors that producers and handlers must consider to reduce postharvest losses?
6. What are the seven biological factors involved in the postharvest deterioration of horticultural commodities?
7. What are four factors that occur as a result of increased respiration?
8. Provide the last two steps in the ethylene biosynthetic pathway; be sure to include the intermediates and enzymes associated with each of these steps.
9. List five causes of increased ethylene production rates.
10. What are four ways to block ethylene production?
11. What are four ways to block ethylene action?
12. What are four general categories of compositional changes that occur during postharvest deterioration of horticultural commodities?
13. What colors do carotinoids promote in fruits? What is the specific carotinoid responsible for red color in tomatoes?
14. What colors do anthocyanins promote in fruits?
15. What are four types of growth and development that adversely affect horticultural commodities?
16. What are four factors that lead to physiological breakdown?
17. What are five common symptoms of chilling injury?
18. Distinguish between chilling-sensitive and chilling-insensitive plants.
19. What is the first step in effective temperature management to extend the shelf life of horticultural commodities?

20. List six ways to cool down commodities removed from the field.
21. What are three ways to control relative humidity during storage of horticultural commodities?
22. What are 10 supplements to temperature and humidity management that can be used to extend the storage life of horticultural commodities?
23. List five future trends for handling horticultural commodities.

Define

Define the following terms:

vine-ripe
respiration
climacteric fruits
nonclimacteric fruits
controlled atmosphere (CA) storage
modified atmosphere (MA) storage

True or False

1. Postharvest breakdown of fruits and vegetables cannot be stopped, but it can be slowed down dramatically.
2. Climacteric fruits ripen in response to ethylene.
3. Nonclimacteric fruits ripen in response to treatment with brassinosteroids.
4. No consistent relationship exists between ethylene production rates and a commodity's perishability.
5. The rate of deterioration of harvested commodities is proportional to the respiration rate.
6. Phenolic compounds are responsible for changes in color pigments.
7. Apples are climacteric fruits.
8. Strawberries are nonclimacteric fruits.
9. Lycopene is a carotinoid giving tomatoes their red color.
10. The green pigment found in potatoes stored in light is very toxic to humans.
11. Rapid removal of field heat is the first step in effective temperature management to improve postharvest quality of fruits and vegetables.

Multiple Choice

1. Carotinoids are desirable in fruits because they give them which of the following colors?
   A. Green, blue, and red
   B. Yellow, orange, and red
   C. Red, blue, and purple
   D. None of the above
2. Anthocyanins are desirable in fruits because they give them which of the following colors?
   A. Red and yellow
   B. Green and blue
   C. Red and blue
   D. Orange and red

Fill in the Blanks

1. Horticultural crops used for consumption can be harvested for ______________________ or ______________________.
2. Harvesting can be done by ______________________ or ______________________.
3. During respiration ______________________ is used and ______________________ is liberated.
4. Climacteric fruits show a large increase in ______________________ and ______________________ during ripening, whereas nonclimacteric fruits show no change.

Matching

Match the nutrient with the appropriate term. Some letters may be used more than once.

1. ________ Silver nitrate
2. ________ Hypobaric storage
3. ________ Increase CO_2
4. ________ Decrease O_2
5. ________ AOA

A. Blocks ethylene action
B. Blocks ethylene production

Activities

Now that you know about postharvest physiology and its importance, you will have the opportunity to explore this exciting field of research in more detail. In this activity, you will report on specific postharvest practices currently used and cutting-edge research in this area. You will then show how such practices affect horticultural commodities by enabling them to be stored for longer periods of time. Your activity is to search for two Web sites that present information on postharvest practices used and current research in this area. For each site, answer the following questions:

1. How do postharvest handling factors affect(s) the shelf life of horticultural commodities?
2. What potential and/or existing methods are used for modifying environmental factor(s) to be used for the storage of horticultural commodities?
3. What topics in this area were exciting and less than exciting? Be sure to explain why you came to these conclusions.
4. What is the URL for the Web site where you found the information?

After you have completed the Internet section of this activity, go to your local supermarket and see what postharvest practices are used there. In most cases, the supermarket manager is always willing to answer questions about the store and should be used as a resource.

References

Kader, A. (1992). *Postharvest technology of horticultural crops* (2nd ed.). Oakland, CA: University of California, Division of Agriculture and Natural Resources.

Arteca, R. N. (1996). *Plant growth substances: Principles and applications*. New York: Chapman & Hall.

CHAPTER

12

Pest Management

ABSTRACT

Pests must be properly controlled while minimizing any adverse effects on the environment. The five major categories of pests that attack plants are insects and related pests, nematodes, weeds, diseases, and rodents and other animals. The series of events that should take place when controlling pests starts with prevention, if possible. If prevention is not possible, the pest must be properly identified. After the pest has been identified, Integrated Pest Management (IPM) should be employed. The general categories of pest control used as part of a responsible IPM program are cultural, biological, mechanical, chemical, and genetic pest controls.

Objectives

After reading this chapter, you should be able to

- provide background information on the five categories of pests.
- discuss the series of events that should take place when controlling pests starting with prevention, followed by identification, and finally implementation of IPM.

Key Terms

abiotic diseases
active ingredient (a.i.)
band treatments
biological pest control
biotic diseases
broadcast treatments
chemical pest control
chewing insect
complete metamorphosis
cultural pest control
facultative parasites
facultative saprophytes
genetic pest control
genetically modified organism (GMO)
herbicides
host plant
incomplete metamorphosis
insects
mechanical pest control
metamorphosis
mutualism
mycoplasma-like organisms
nematodes
no metamorphosis
nonselective herbicide
obligate parasites
obligate saprophytes
pathogens
pest
pesticides
pheromones
piercing and sucking insect
plant breeding
plant diseases
postemergence treatment
preemergence treatment
preplant treatment
selective herbicide
spot treatments
surfactant
symbiosis
weeds

INTRODUCTION

This chapter identifies the five categories of **pests** that attack plants and provides environmentally sound practices to control these pests. Plants are subject to damage by a variety of pests, which are anything that causes injury or loss to a plant (Figure 12-1). Pests can be broken down into the five categories of **insects** and related pests, **weeds, nematodes, plant diseases,** and rodents and other animals. Insects are animals with three distinct body parts (head, thorax, and abdomen), three

Figure 12–1 The top split panel shows insect damage, the middle panel shows powdery mildew on lilac leaves, and the bottom panel shows a weed infestation.

pairs of legs, and one, two, or no pairs of wings. Any deviation from this definition is called an insect-related pest. Insects can be classified in two different ways; one way is based on the feeding habits of the insect, which can be **chewing** (caterpillars) or **piercing and sucking** (aphids). The second way that insects can be classified is based on their life cycle. The three types of **metamorphosis,** or gradual development of the insect, are **incomplete**

metamorphosis (aphid), **complete metamorphosis** (butterfly), and **no metamorphosis** (silverfish).

Not all insects are harmful. A variety of insects are very useful for biological control, including praying mantis, ladybird beetles, larvae of green lacewings, nematodes, *Aphytus* wasps, Cryptolaemus, Delphatis, and predatory mites. Other beneficial uses of insects are for pollination and for the products they make, such as honey, beeswax, and silk. Even though insects have many beneficial effects, many types of insects can cause dramatic losses in crop yields and overall plant quality if left uncontrolled. Insects that are harmful to horticultural crops include aphids, white flies, ants, codling moths, spider mites, corn ear worms, Colorado potato beetles, cabbage loopers, leaf miners, squash vine borers, thrips, peach twig borers, Mediterranean fruit flies, and a variety of others. Insect-related pests include mites, spiders, millipedes, centipedes, toothed grain beetles, dried fruit beetles, and a variety of others.

Nematodes are worm-like invertebrates that lack appendages, are nonsegmented, and are found in the soil. Some nematodes are beneficial, whereas others cause major problems in a wide range of horticultural crops. Nematodes penetrate root cells causing small wounds in the root, which alone does not create a problem; however, this wound allows fungi and bacteria to enter the root and cause damage.

Weeds are any plants growing out of place or any unwanted plants. They compete with crop plants for water, nutrients, and light, plus they harbor diseases and insects. Weeds cost growers billions of dollars each year due to crop losses and weed control expenses. Important weeds that pose a threat to horticultural crops include dandelion, Canadian thistle, field bindweed, common lamb's-quarter, Johnsongrass, quackgrass, common cocklebur, and large crabgrass.

Plant diseases are abnormal conditions in plants that interfere with their normal appearance, growth, structure, or function. The two main groups are **abiotic diseases,** which are noninfectious or disorders such as a nutrient deficiency, or **biotic diseases,** which are caused by parasites or **pathogens** that are infectious and transmissible, such as Dutch elm disease. Classification of plant diseases is by the causal organism. The four main groups of disease-causing organisms are fungi, bacteria, viruses, and **mycoplasma-like organisms.**

Fungi can be beneficial as well as pathogenic. For example, numerous plant species grow in association with a fungi known as mychorrizae; this association is known as **mutualism,** which is when both the host and fungi benefit. However, fungi have been shown to cause most infectious diseases. Fungi can be broken down into four main categories: **obligate saprophytes, obligate parasites, facultative saprophytes,** and **facultative parasites.** Some common fungal diseases that attack horticultural crops include Dutch elm disease, Fusarium wilt, damping-off, downy mildew, rust, powdery mildew, and late blight.

Bacteria do not cause many economically important diseases in horticultural crops. In some cases, they can even be beneficial; for example, the symbiosis of Rhizobia with legumes promotes the fixation of nitrogen in **host plant** roots. However, in a number of cases, bacteria causes problems. Common blight, fireblight, crown gall, bacterial wilt, bacterial canker, and bacterial soft rot are common examples of bacterial diseases.

Viruses typically consist of a core of RNA or DNA. Viruses are microscopic organisms that can be transmitted by insects such as aphids and leafhoppers. The disease caused by the virus is named after the plant on which it was first studied, but it can also occur in other plants. Viruses that affect horticultural crops include tobacco mosaic virus, tomato spotted wilt virus, tomato ring spot virus, and bean common mosaic virus. At present there is no way to eliminate viruses, but they can be controlled by using resistant cultivars and programs that control vectors that spread viruses.

Mycoplasma-like organisms are small parasitic organisms lacking constant shape and are intermediate in size between viruses and bacteria. These organisms are found in the phloem and disrupt phloem transport, leading to plant yellowing, wilting, distorted plants, and overall plant stunting. Common diseases caused by mycoplasma-like organisms are aster yellows, pear decline, mulberry dwarf disease, and corn stunt. Like viruses, there is no way to stop mycoplasma-like organisms, however, they can be controlled by using resistant cultivars and effectively controlling insect vectors.

Both small and large animal pests cause crop damage by feeding on leaves, stems, fruits, and roots of plants. For some types of animals (rodents), the use of **pesticides** is legal; however, it is illegal to use pesticides on some other animals, such as deer. Understanding the law is important for deciding whether alternatives must be sought. For each of the five categories of pests previously mentioned, billions of dollars are spent annually for the cost of control and lost annually due to crop losses.

To avoid polluting the environment, a simple series of events should be followed when controlling pests, starting with prevention. If prevention is not possible, the problem should be identified and Integrated Pest Management practices (IPM) should be implemented. Five categories of pest control are used as part of a responsible IPM program: **cultural pest control** (crop rotation, irrigation, fertilization, and sanitation), **biological pest control** (beneficial bacteria; natural chemicals, toxins, and repellents; and beneficial insects), **mechanical pest control** (plowing, mowing, mulching, pruning, tilling, hand picking, traps, and temperature and radiation treatment), **genetic pest control** (plant breeding and genetic engineering), and **chemical pest control** (pesticides). By using a sound IPM program, high-quality horticultural crops can be produced with little or no adverse effects on the environment.

Since the inception of agriculture, farmers have continuously battled plant pests. Although there has been a considerable amount of research on improving ways to control plant pests, it is estimated that pests destroy nearly half of the food crops produced worldwide during crop growth, through transport, and into storage.

FIVE MAJOR CATEGORIES OF PESTS

Plants are subject to damage by pests, and a host plant provides a pest with food. Some plants are more likely to be attacked than others; for example, white flies love both tobacco and tomato plants. Most pests can be controlled and losses reduced or eliminated. A pest is anything that causes injury or loss to a plant. Pests can be broken down into five major categories:

- insects and related pests.
- nematodes.
- weeds.
- diseases.
- rodents and other animals.

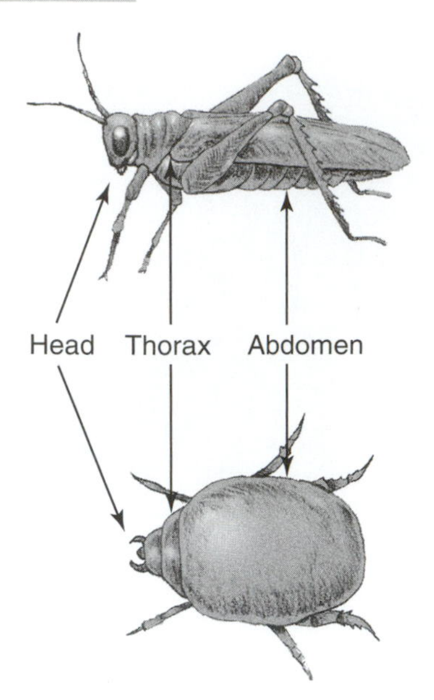

Figure 12–2 Three distinct body parts found in insects

Insects and Related Pests

Insects are organisms with three distinct body parts (head, thorax, and abdomen) (Figure 12-2), three pairs of legs, and one, two, or no pairs of wings. Any deviation from the definition for an insect is an insect-related pest. Insects are in the phylum Arthropoda. True insects belong to the class Insecta, whereas spiders and mites are in the class Arachnida—both account for most insect pests. Pests are economically important because they cause physical damage and spread disease in food crops all over the world. Insects make up more than 80 percent of all animal life. In horticulture, a wide range of insect pests cause a reduction in crop yields and quality. Not all insects are harmful, however; some are very useful in biological control programs aimed at controlling insects that attack horticultural crops, and they can be used on a variety of crops for pollination (Figure 12-3).

Classification of Insects

Insects can be classified by their feeding habits or by their life cycles.

Feeding habits. How insects feed is determined by the structure of the insect's mouth (Figure 12-4). Insects are divided into chewing insects and piercing/sucking insects:

- **Chewing insect.** Uses mandibles for chewing plant parts such as leaves, stems, roots, fruits, flowers, and petals. Examples of chewing insects

Figure 12–3 A bee pollinating a flower

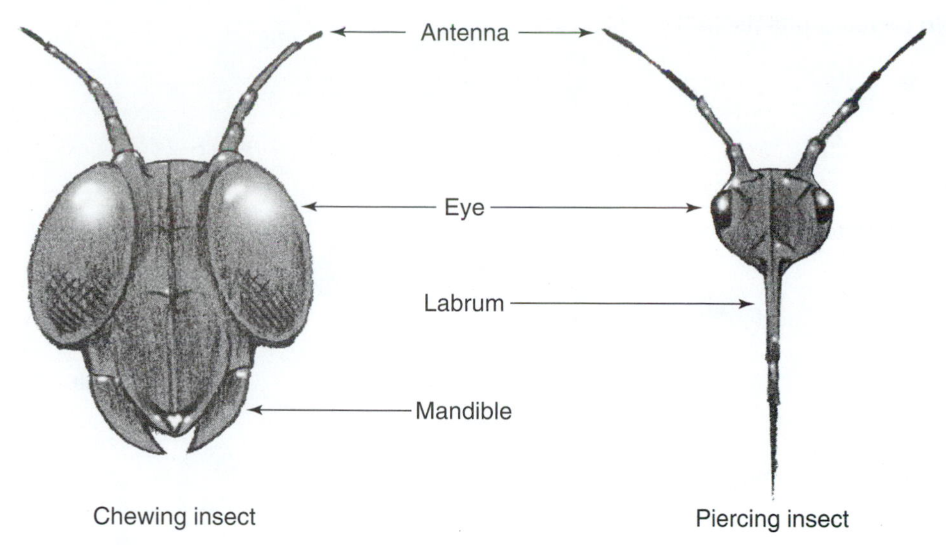

Figure 12–4 How insects feed is determined by the structure of the insect's mouth.

are grubs and caterpillar larvae, adult beetles, and adult grasshoppers. Symptoms of damage caused by chewing insects include defoliation (caterpillars), root damage (wireworms), and boring, which is when the insect makes channels in succulent tissues (corn borers).

- **Piercing and sucking insect.** Uses the labrum to pierce the leaf and suck out the juices. Examples are aphids, leafhoppers, scales, mealy bugs, thrips, and, of course, mosquitoes.

Life cycle. Insects, like all living organisms, have a specific life cycle that consists of various stages of growth and development, starting from an egg and progressing on to an adult. Typically, not all of these stages are harmful to plants. The gradual development of the insect is called metamorphosis. Three types of metamorphosis are used for classifying insects (Figure 12-5):

- **Incomplete metamorphosis.** A gradual change in the insect's size (for example, aphid leafhopper).
- **Complete metamorphosis.** A complete change in the insect. The insect has four distinct stages: egg, larvae, pupae, and the adult stage (for example, butterfly, fly).
- **No metamorphosis.** The insect exhibits no change; it goes from egg to full size (for example, silverfish, springtails).

Insects are most harmful during specific stages of development. In addition, some insects, such as aphids, can have 10 or more generations per year or can take many years to complete one generation. For example, cicadas take 13, 17, or 18 years to complete one generation. Knowing the life cycle of the insect helps determine at what stage control measures should be used. Knowing the number of generations that occur per year is also helpful in preparing a treatment schedule for the year.

The insect orders with the most species that cause problems in horticultural crops are Coleoptera (beetles), Lepidoptera (cutworms), Hymenoptera (ants), and Diptera (flies). Other economically important orders include Homoptera (aphids, mealybugs), Orthoptera (grasshoppers, crickets), and Thysanoptera (thrips).

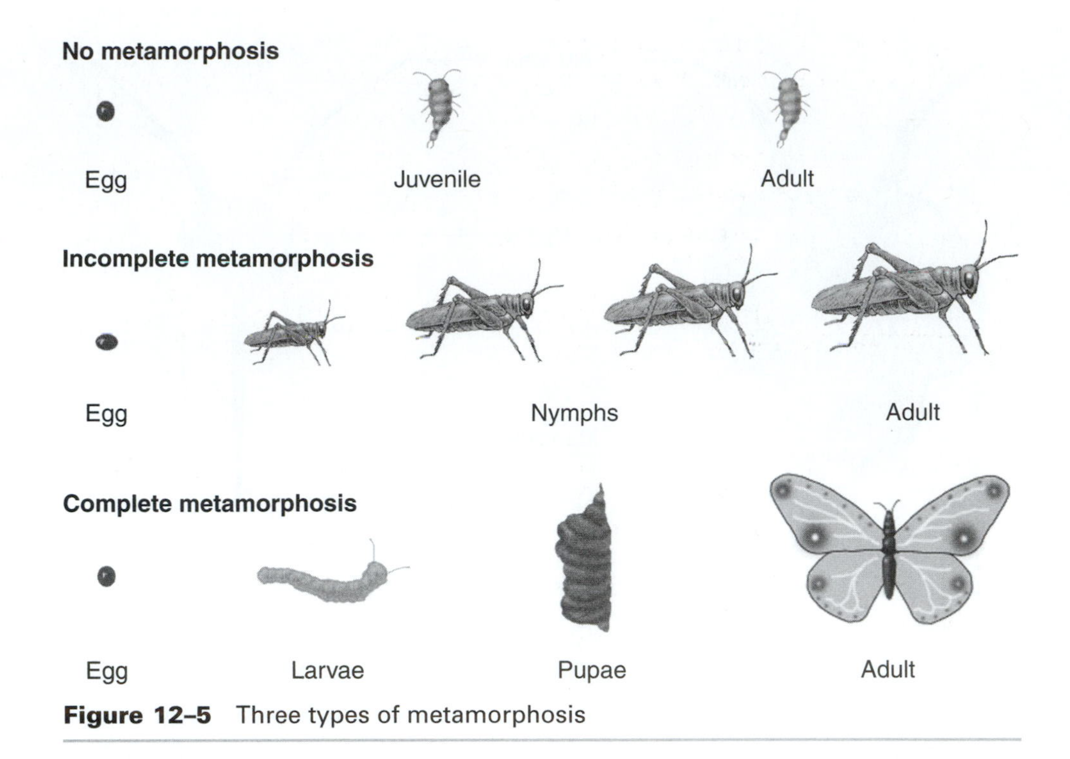

Figure 12–5 Three types of metamorphosis

Beneficial Effects of Insects

Many beneficial insects are used as forms of biological control for harmful insect species. The praying mantis is referred to as an all-purpose predator that attacks a wide variety of insects, including aphids, beetles, caterpillars, grubs, grasshoppers, and virtually anything that moves. The ladybird beetle is an effective predator of aphids, mealy bugs, scales, leaf hoppers, and a variety of other destructive pests. The larvae of the green lacewing feeds on mealy bugs, scale, aphids, white fly larvae, mites, thrips, and many other insects. Green lacewing larvae have been called the "aphid lion" because they can eat up to 1,000 insects per day. Beneficial nematodes are microscopic organisms that destroy more than 200 kinds of soil-dwelling and wood-boring insects, including cutworms, armyworms, rootworms, weevils, grubs, fungus gnat larvae, and many other insects. Other beneficial insects include the predatory *Aphytis* wasp that destroys red and black scale, *Cryptolaemus* for mealybugs, *Delphatis* for white fly larvae, and predatory mites for mites.

In addition to their use in controlling other insects, beneficial insects can be used for pollination. The honeybee and other insects are extremely important for pollinating fruit trees and other crops. Some insects also produce useful products, such as honey and beeswax from bees and silk from the silkworm. Today beneficial insects can be purchased for commercial or home use.

Insect-Related Pests

Mites are insect-related pests in the Tetranychidae family and the order Acrina. They present a major problem to horticultural crops because they are among the most widely distributed pests, affecting many different plant species, including ornamentals, fruits, vegetables, and field crops. Mites are sucking insects that reproduce rapidly and have many generations in one year. They are problems in the field as well as the greenhouse. Fine webs are noticeable on plant parts when there is a bad infestation of mites. Other examples of insect-related pests include spiders, millipedes, centipedes, toothed grain beetles, and dried fruit beetles.

Insects and Mite Pests that Pose a Threat to Horticultural Crops

Numerous insects and mites pose threats to the quality and quantity of horticultural crops produced each year. Following are examples of some key insect and mite pests:

- **Aphids.** Aphids attack the young leaves and the growing tips of stems and young leaves. The most common aphid species is the green peach aphid (*Myzus persicae*), which attacks many ornamental, orchard, and vegetable crops. Aphids have sucking mouthparts, which cause physical damage to plants and also spread a number of viruses such as cucumber mosaic and bean mosaic.
- **White fly.** White fly (*Aleuocanthus woglumi*) was an uncommon pest prior to 1970; however, since then, it has become resistant to existing control measures and is now a serious pest both indoors and outdoors. White flies attack a variety of garden plants, including tomatoes, potatoes, geraniums, fuchsia, gardenias, and many others.
- **Ants.** Ants (*Iridomyrmex humilis*) cause numerous problems in the garden and landscape. Wet weather conditions promote the ant population.
- **Codling moth.** The codling moth (*Laspeyresia pomonella*) attacks apples, pears, and walnuts. The larvae cause an unsightly appearance because they tunnel into the fruits.
- **Spider mites.** Spider mites (*Tetranychus urticae, Panonychus ulmi,* and *Bryobia praetiosa*) are not true insects. They have piercing and sucking mouthparts and live on the underside of leaves. Spider mites attack a wide variety of crops, including vegetables, fruits, nuts, and ornamentals. They attack plants in the field, greenhouse, and growth chamber.
- **Corn ear worm.** Corn ear worm (*Heliothis zea*) caterpillars feed on corn, beans, lettuce, tomato, cotton, alfalfa, clover, peanuts, and tobacco. They get their name because they attack corn silks and kernels, which creates an unsightly wormy appearance.
- **Colorado potato beetle.** The Colorado potato beetle (*Lepinotarsa decimilneata*) larvae and adult are pests that attack tomato, eggplant, and a variety of other crops.
- **Cabbage looper.** The cabbage looper (*Trichoplusia ni)* is one of the most common garden caterpillars. The larvae make holes or completely devour plants in the cabbage family and many other garden species.
- **Leaf miners.** Leaf miners (*Liriomyza* spp.) create visible tunnels under the leaf surface of plants, which causes leaf yellowing and, in severe cases, the leaf abscises. Leaf miners attack a variety of garden plants, including lettuce, melon, tomato, and eggplant.
- **Squash vine borer.** The squash vine borer (*Melittia calabaza*) tunnels its way through the vines of plants, causing wilting and eventually death of squash, cucumber, melon, and a variety of other species.

- **Thrips.** Thrips (*Thrips tabaci* [onion thrips] or *Frankliniella occidentalis* [flower thrips]) are sucking-type insects that attack plants and cause a distorted appearance. Thrips are very common in the garden and attack onions, a variety of flower crops, and others.
- **Peach twig borer.** The peach twig borer (*Anarsia lineatella*) attacks peaches and a variety of other stone fruits. The larvae bore into buds and shoots as they begin to grow in the spring.
- **Mediterranean fruit fly.** The mediterranean fruit fly (*Ceratitis capitata*) attacks peaches, apricots, apples, citrus, bananas, and a variety of other fruits and vegetables.

Nematodes

Nematodes are worm-like invertebrates that lack appendages, are nonsegmented, and are found in the soil. They are microscopic nonarthropods that attack above- and below-ground plant parts; however, most nematodes are soilborne. Some nematodes are beneficial, as mentioned earlier in this chapter; however, many cause major problems to vegetables, fruit trees, ornamentals, and foliage plants. The most common nematodes are the root knot (*Meloidogyne* spp.) and cyst (*Heterodera* spp.). Nematodes feed by penetrating root cells with a hollow-style mouth structure and injecting enzymes into the cells, causing digestion of the root cell. The wound made by nematodes allows fungi and bacteria to enter the root, which causes damage. The major damage is not caused by the nematode itself. The root knot nematode produces irregular-shaped knots that disrupt nutrient flow in plants and lead to yellowing and stunted growth in plants.

Nematode movement in the soil is very slow; they only move 12 to 30 inches per year. In a natural setting, they pose little problem of moving from one area to another; however, nematodes are disseminated by wind, air, tools, and equipment. Accurately diagnosing nematodes requires laboratory testing. When purchasing land for an orchard, for instance, it is crucial to make sure there is not a problem with nematodes. After land has been purchased, measures must be taken to prevent an infestation of nematodes.

Weeds

Weeds are any unwanted plants or any plants growing out of place (Figure 12-6). Weeds cause many problems, which lead to billions of dollars in annual crop losses and control costs. Weeds are problematic because they

- compete with the crop for water, nutrients, and light.
- harbor diseases and insects.

Weeds tend to grow better than crop plants because they typically have characteristics that allow them to grow and survive under harsh conditions. Weeds usually can do extremely well on poor to marginal soils. They have seeds that retain viability for many years and can be dormant until an opportunity exists for success. A variety of weeds have underground structures, such as roots, rhizomes, and others, which are used for reproduction. Common household weeds include dandelion, plantain, and field bindweed (Figure 12-7).

Many good cultural practices that are used to control weeds actually promote the weed's reproductive process. For example, cultivation breaks underground

Figure 12–6 Weeds growing in a flower bed

Figure 12–7 Field bindweed in a flower bed

structures into small pieces and spreads them over a wider area, creating an even larger problem. Accurately identifying the weed and knowing its life cycle (annual, biennial, or perennial) is vital to implementing the proper control measures.

Figure 12–8 The top panel shows the yellow dandelion flower, and the bottom panel shows the mature flower ready to disperse seeds.

Weeds are excellent indicators of soil fertility. Thistle, chickweed, and yarrow thrive in soils that are highly fertile. Sandy dry soils provide excellent conditions for dandelion, stinging nettle, and shepherd's purse; whereas wet, heavy clay soils are excellent conditions for thistle, plantain, and creeping buttercup. In highly acidic soils, foxglove and daisy thrive, whereas in alkaline soils white mustard and wild carrot thrive. Moss and algae also fall into the weed category and have a dramatic effect on turf. Both thrive in soils that have been neglected (low pH, low nutrients, and so on).

Weeds that pose a threat to horticultural crops include the following:

- **Dandelion.** The dandelion (*Taraxacum officinale*) is a perennial plant that has a deep taproot system, which must be completely removed or destroyed to prevent its spread. The bright yellow dandelion flower is common in many lawns where control measures have not been implemented and is easily dispersed by the wind (Figure 12-8).
- **Canadian thistle.** The Canadian thistle (*Cirsium arvense*) is a perennial plant that has deep roots and can reproduce both sexually by seeds or asexually by rhizomes. Canadian thistle is a problem in the northern United States.
- **Field bindweed.** The field bindweed (*Convolvulus arvensis*) is a perennial plant that has deep roots and can reproduce both sexually by seeds or asexually by rhizomes. It has a twinning habit, or it may spread on the ground, making it difficult to control.
- **Common lamb's-quarter.** Common lamb's-quarter (*Chenopodium album*) is an annual weed that reproduces by seed and has a deep taproot.
- **Johnsongrass.** Johnsongrass (*Sorghum halepense*) is a perennial weed that reproduces sexually by seed and asexually by rhizomes. This weed presents its greatest problem in the southern United States.
- **Quackgrass.** Quackgrass (*Agropyron repens*) is a perennial weed that reproduces by seed or rhizomes. It has a fibrous root system and presents its greatest problem in the southern United States.
- **Common cocklebur.** Common cocklebur (*Xanthium pensylvanicum*) is an annual weed that reproduces sexually by seed. It has a hairy stem, which can get as tall as 90 centimeters, and has a deep taproot.
- **Large and small crabgrass.** Large crabgrass (*Digitaria sanguinalis*) and small crabgrass (*Digitaria ischaemum*) are annual weeds with fibrous root systems and spreading habits. Crabgrass is a problem in lawns all over the United States.

Diseases

Plant diseases are abnormal conditions in plants that interfere with the plants' normal appearance, growth, structure, or function. There are two principal groups of diseases:

- **Abiotic (environmental) diseases.** Noninfectious disorders caused by nutrient deficiencies, damage to plant parts, chemical injuries, pollution injuries, and environmental conditions.
- **Biotic (pathogenic) diseases.** Caused by parasites or pathogens (organisms that cause disease) that are infectious and transmissible. The

Figure 12–9 Monoculture cropping systems facilitate the spread of disease.

pathogens responsible for causing most biotic diseases are fungi (Dutch elm disease), bacteria (Fireblight), virus (Tobacco mosaic), and mycoplasma-like organisms.

Horticultural crops are typically more susceptible to diseases than their wild relatives. This susceptibility is due in part to breeding for a variety of desirable traits at the expense of disease resistance. Probably the main reason is due to modern cultural practices, which use dense cropping monoculture systems (Figure 12-9) and facilitate the spread of disease. Disease can be identified by sending samples away to a laboratory. By identifying the disease, the chances of controlling it are much better.

Plant diseases are classified by the causal organism. The four main groupings of disease-causing organisms that attack horticultural crops are fungi, bacteria, viruses, and mycoplasma-like organisms.

Fungi

Fungi can be beneficial as well as pathogenic. Numerous plant species grow in association with fungi and are known as mychorrizae. This association is called mutualism because both the host and fungi benefit. This is similar to **symbiosis,** which is when a mutually beneficial plant bacteria association occurs, as is the case in legumes. Mychorrhizal fungi enhance the capability of plants to absorb phosphorous from the soil. Not all fungi are pathogenic. Fungi also serve many useful purposes; for example, penicillin comes from *Penicillium,* mushrooms are used for food, and fermented beverages and foods—bread, wine, cheese, and beer—use fungi for their production.

Spores are the reproductive structures for fungi that come in a variety of sizes, shapes, and colors. Fungal spores are spread by wind, water, insects, birds, and a variety of other ways. Fungi typically have protective coverings that enable them to withstand adverse conditions. For fungi to infect a plant, the hyphae must enter through plant wounds caused by pests, adverse weather conditions, and many other ways. Fungi have been shown to cause most infectious diseases. Fungi can be grouped into four categories:

- **Obligate saprophytes.** Live only on dead organic matter and inorganic materials.
- **Obligate parasites.** Attack only living tissues.
- **Facultative saprophytes.** Parasitic and attack living tissues but can also live on dead tissues when the proper conditions exist.
- **Facultative parasites.** Normally live as saprophytes but can live as parasites under the proper conditions.

Some Common Fungal Diseases that Effect Horticultural Crops

The following common and scientific names are for fungal diseases that commonly affect horticultural crops:

- Dutch elm (*Ceratocystis ulmi*) attacks elm trees.
- Fusarium wilt (*Fusarium oxysporum*) attacks a variety of crops, including tomatoes.
- Damping-off (*Pythium* spp., *Rhizocotonia* spp.) attacks seedlings of a wide range of crops.
- Downy mildew (*Plasmopara viticola*) attacks grapes.
- Rust (*Puccinia striformis*) is a problem in turfgrasses.
- Powdery mildew (*Erysiphe polygoni*) is a common disease in a variety of horticultural crops, such as lilac (Figure 12-10).
- Late blight (*Phytopheria infestans*) attacks tomatoes, potatoes, and other crops.

Figure 12–10 Powdery mildew on lilac leaves

Bacteria

Bacteria do not cause many economically important diseases in horticultural crops. These unicellular organisms are widespread in the environment. They are found in three shapes: spherical, rod-like, and spiral. Bacteria can be beneficial in plants such as the symbiosis of *Rhizobia* bacteria with legumes to promote the fixation of nitrogen in the host plant's roots. Bacterial diseases can cause problems in horticultural crops:

- **Common blight.** Common blight (*Xanthomonas phaseoli*) attacks field, snap, and lima beans. This type of bacteria can be controlled by using sanitation, disease-free seed, disposal of crop residues, and crop rotations.
- **Fireblight.** Fireblight (*Erwinia amylovora*) attacks pome fruits such as apples, pears, and some ornamentals. This type of bacteria can be controlled by using resistant cultivars, using streptomycin sprays at the bloom stage, and pruning out diseased portions of the plant and ensuring proper disposal of this material.
- **Crown gall.** Crown gall (*Agrobacterium tumefasciens*) causes galls in woody ornamentals and fruit trees. This type of bacteria can be controlled by starting with disease-free stock and maintaining a high level of sanitation.
- **Bacterial wilt.** Bacterial wilt (*Erwinia tracheiphila*) attacks cucumbers and a variety of other curcurbits. This bacteria can be controlled by destroying insects such as spotted and striped cucumber beetles that are responsible for the spread of this disease.
- **Bacterial wilt of corn.** Bacterial wilt of corn (*Erwinia stewartii*) attacks corn, and resistant cultivars are available for its control.
- **Bacterial canker.** Bacterial canker (*Pseudomonas syringae*) attacks a variety of pitted fruit, such as apricots, avocados, cherries, peaches, plums, and almonds. This bacteria can be controlled by using resistant cultivars, spraying with Bordeaux mix, and pruning out diseased areas and properly disposing of the diseased materials.
- **Bacterial soft rot.** Bacterial soft rot (*Erwinia carotovora*) attacks vegetables with fleshy storage organs. This bacteria can be controlled by avoiding damage during harvest, maintaining good ventilation during storage, and using the proper storage temperature.

Viruses

Viruses typically consist of a core of RNA or DNA enclosed in a protein coat. They are not capable of digestion or respiration. Viruses are obligate parasites that get into the host cell and take over the machinery for producing viral DNA or RNA. Viruses are microscopic organisms that can be transmitted by insects such as aphids, leafhoppers, and mites.

After a plant is infected, the virus alters the plant's metabolism and is extremely difficult to treat. Common symptoms of a viral infection are leaf yellowing, loss of plant vigor, and reduced growth that results in stunted plant growth. The disease caused by a virus is named after the plant on which it was first studied, but it can also occur in other plants. Examples of viruses that affect horticultural crops include tobacco mosaic virus, tomato spotted wilt virus, tomato ring spot virus, potato virus, potato leaf roll virus, cucumber mosaic virus, prunus necrotic ring spot virus, and bean common mosaic virus. Although there is no way to eliminate a virus, some control measures can be taken. Using resistant cultivars and programs to control vectors that spread viruses have been shown to be effective for virus control.

Mycoplasma-Like Organisms

Mycoplasma-like organisms are small parasitic organisms lacking constant shape and are intermediate in size between viruses and bacteria. Mycoplasma-like organisms are found in the phloem and rapidly multiply, which causes a disruption in phloem transport. This disruption in transport causes yellowing, wilting, distorted plants, reduced leaf size, and overall plant stunting. Diseases caused by mycoplasma-like organisms are aster yellows, pear decline, mulberry dwarf disease, and corn stunt. As is the case with viruses, the disease caused by the mycoplasma-like organism is named after the first plant it was studied in. For example, aster yellows can also be found in phlox, tomato, lettuce, strawberry, and a variety of weed species. Resistant cultivars and an effective spray program that eliminates insect vectors have been shown to be effective in controlling mycoplasma-like organisms.

Rodents and Other Animals

Small and large animal pests can cause damage to crops because they eat leaves, stems, fruits, and roots of plants. Animal pest control may require the use of pesticides (rodents). However, you must be careful because laws protect some animals, such as deer, from the use of chemical control. The following are ways to control pests without using chemicals:

- Destroy the animal's habitat.
- Capture the animal and move it to another location.
- Use sound to scare animals away.
- Use fencing to keep animals out.

CONTROLLING PESTS

For all five categories of pests previously mentioned, the degree of damage depends on the pest and the specific plant being affected. When using any of the following forms of pest control, keep in mind that the amount of plant loss should be at least equal to the cost of control. Remember, you do not want to spend $1,000 in control costs if the dollar value put on plant loss is only $50. Pests cost growers billions of dollars annually because of control costs and crop losses.

Pest Control Procedure

The series of events that should take place when controlling pests is as follows:

1. Prevent, if possible.
2. If prevention is not possible, identify the problem.
3. Implement an IPM program (Figure 12-11). There are five general categories of pest control: cultural, biological, mechanical, genetic, and chemical.

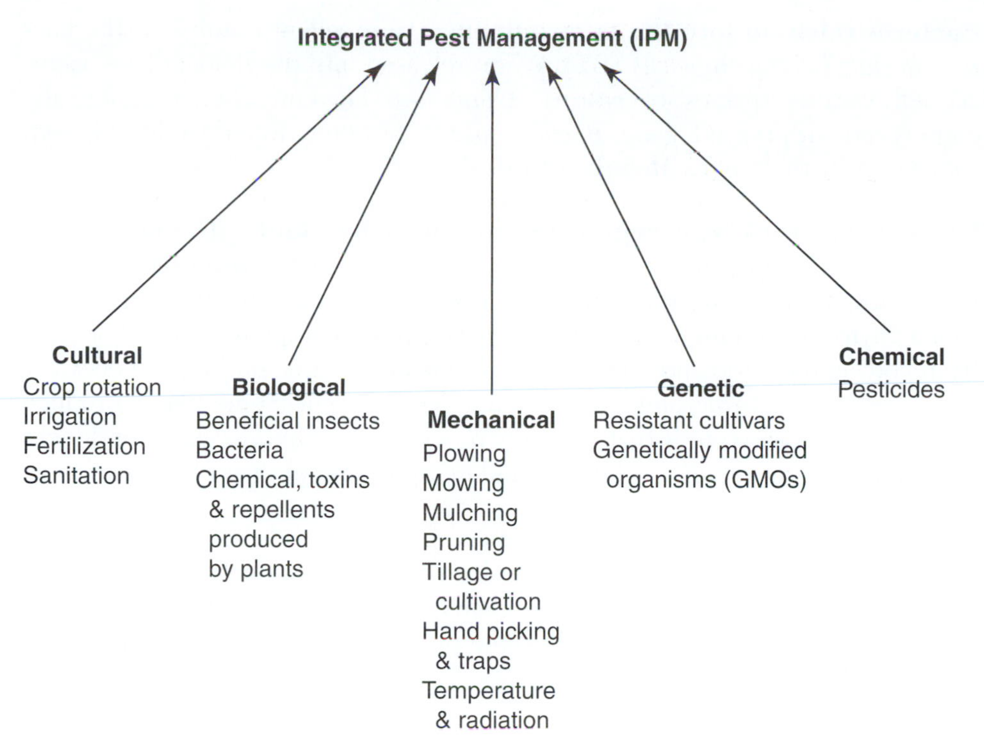

Figure 12–11 IPM control measures

Cultural Pest Control

Cultural pest control uses management techniques to control pests. Cultural techniques include the following:

- **Crop rotation.** Many diseases and other pests that affect horticultural crops are a result of the cropping system used. Monoculture is a system where the same species of plants are grown in a given plot year after year. This encourages a large buildup of diseases and insects. Crop rotation simply means that no single crop is grown on a given plot year after year. For example, if corn is planted in year one, the next year beans are planted, followed by sweet potatoes and cabbage. A variety of other rotations have also been shown to be effective. Disease-causing organisms need the host plant in order to survive; however, they cannot survive if the host is not present for a two- to three-year period.
- **Irrigation.** Irrigation is important to keep crop plants well hydrated to avoid weakening the plant, which makes the plant more accessible to insects and diseases.
- **Fertilization.** Plants that are properly fertilized grow vigorously and have a well-established cuticle layer, thereby reducing access by diseases and insects.
- **Sanitation.** Sanitation is the removal of dead and diseased plant debris that harbors insects and disease organisms. In some cases, this debris must be incinerated to kill the pathogens.

Biological Pest Control

Biological pest control uses living organisms that are predators to control pests or uses naturally occurring chemicals extracted from plants.

Bacteria released into the environment. An excellent example is the bacterium *Bacillus thuringiensis* (BT). When released into the fields, these bacteria kill various species of worms. Plants are currently being genetically engineered with the BT gene. *Bacillus subtilis* has been found to delay brown rot caused by the fungus *Monilinia fruticola*.

Chemicals, toxins, and repellents produced by plants. The best-known chemical insecticides from plants are pyrethrum extracted from plants in the chrysanthemum family, rotenone, and nicotine. Certain plants exude toxins called phytoallexins into the soil in their own defense. This prevents other species from getting too close and competing for nutrients, water, and other necessary resources needed to maximize growth and development. Some plants produce strong volatile compounds that naturally repel insects. Onions, garlic, and leeks repel aphids, mint repels flea beetles, and marigolds repel a variety of insects.

Beneficial insects. A variety of beneficial insects provide excellent forms of biological control for insect species that have an adverse affect on horticultural crops. Examples include the praying mantis, ladybird beetle, the larvae of the green lacewing, beneficial nematodes, the predatory *Aphytus* wasp, and a variety of others.

Mechanical Pest Control

Mechanical pest control uses tools, equipment, or other physical means to control pests. Examples of mechanical control are as follows:

- **Plowing.** The structure of the soil can be improved by plowing and not allowing heavy machinery in the field under wet conditions to avoid producing hardpans.
- **Mowing.** Mowing reduces plant height and keeps plants in their vegetative state to avoid the production of reproductive structures, which could result in the dissemination of unwanted weed seed.
- **Mulching.** Plastic mulching can prevent weed seeds from germinating and growing. In addition, plastic mulches can trap heat, which raises soil temperatures and kills certain soilborne pathogens.
- **Pruning.** Pruning and physically removing dead or diseased materials and discarding them prevents the spread of disease throughout the plant and surrounding plants.
- **Tillage.** Tillage or cultivation is used as a physical means of removing weeds from a given area.
- **Hand picking and traps.** Hand picking and traps are used on a small scale to remove and destroy caterpillars and other large insects by hand. Another mechanical means of control is to utilize traps. A strip of paper or polyethylene (typically yellow) that has been coated on both sides with a sticky material where the insect is trapped and eventually dies is used for aphids, flies, and other insects. For larger pests, such as rodents, mechanical traps are commonly used. In certain cases, light attracts pests, which results in their death.
- **Temperature and radiation.** Temperature and radiation treatments can be used to sterilize soil and kill disease organisms. Cold temperatures do not kill pathogens but greatly reduce their activity in stored produce. Radiation has been used to kill disease organisms to prolong shelf life.

Genetic Pest Control

Genetic pest control uses plant breeding and genetic engineering to manipulate plants to make them more resistant to specific pests. **Plant breeding** is the science of controlled pollination of plants to develop new cultivars. Plant breeders have been manipulating the genotype of many plants over a long period of time to create the wide range of cultivars available today. Through breeding efforts, researchers transfer the capability to resist diseases and other pests from the wild into the cultivated species and from cultivar to cultivar within the species. In some cases, they can transfer resistance across species and genus boundaries. Today, resistant cultivars exist for most of the major pests of horticultural plants.

Genetically modified organisms (GMOs) carry a foreign gene that was inserted by laboratory techniques into all of its cells. The genetic modification of plants is similar to plant breeding except that it is generally faster. Genetic engineering is typically done for traits that are controlled by one gene, only a few genes, or a small cluster of genes. Traits that can be manipulated include resistance to herbicides, resistance to adverse environmental conditions, improved nutritional quality, flower color, keeping quality of horticultural crops, and resistance to viral infection. First the gene is identified, then cloned, and finally transformed into plants using one of the following three insertion methods: *Agrobacterium,* biolistics (particle bombardment), or electroporation, which are discussed in greater detail in Chapter 13 Plant Biotechnology and Genetically Modified Organisms: An Overview.

Chemical Pest Control

Chemical pest control uses a pesticide, which is a chemical used to control pests. Pesticides that are used to control insects are called insecticides whereas chemicals used to control weed are **herbicides.** Pesticide usage must be approved by various government agencies, such as the Environmental Protection Agency (EPA). Getting a pesticide approved and labeled for a specific purpose is very costly, so many minor crops do not have pesticides labeled for them. All chemicals are hazardous to humans and should be used properly. All individuals working for a commercial enterprise and using pesticides must take a test to obtain a pesticide applicator license.

Follow these general safety rules when using pesticides:

- Use only approved pesticides.
- Read the label before application.
- Use the pesticide with the lowest toxicity.
- Use the right equipment.
- Use pesticides only when needed.
- Wear protective clothing.
- Dispose of empty containers properly.
- Apply in good weather.
- Know the proper emergency measures.
- Properly store pesticides.

Classification of Insecticides Based on Their Mode of Action

Pesticides are sold in two general types of formulations or forms: dry and liquid. Dry formulations are available as dusts, wettable powders, pellets, and granules. Liquid formulations include emulsifible concentrates, fumigants, solutions, and aerosols.

The following sections show these insecticides classified by their mode of action.

Systemic poisons. This type of insecticide may be applied as a foliar spray or as a soil drench because it is readily translocated throughout the plant. When the insects chew or suck on the plant, they ingest the poison. When using systemic pesticides on food crops, growers must be cautious to wait for the toxins to break down to an acceptable level prior to use. Two examples of systemic poisons are orthene and glyphosate (not for use on food crops).

Contact poisons. This type of insecticide is absorbed through the insect's exoskeleton (skin). Once absorbed, it attacks the respiratory and nervous systems, resulting in death. Insects that live on the lower side of the leaf are difficult to control with this method. An example of this type of insecticide is malathion.

Fumigants. These are volatile compounds that enter the insect's body through the respiratory system. This type of insecticide is very effective in greenhouses and storage facilities.

Stomach poisons. This type of poison must be ingested by the insect to be effective. Chewing insects such as beetles, caterpillars, and grasshoppers are readily controlled by this type of insecticide. An example of this type of insecticide is rotenone.

Suffocation agents. Some insects, such as scale, can be readily controlled by spraying the leaf surface with oil-based compounds. This plugs up the insect's breathing holes and suffocates the insect.

Repellents. This type of compound is designed to repel, not kill. An example of a repellent is the Bordeaux mix, which prevents powdery mildew from attacking lilacs.

Fatal attraction action. This type of compound utilizes **pheromones,** which are chemicals that many female insects secrete to attract male partners. These chemicals have been successfully identified and used for traps to catch and kill male insects such as the Japanese beetle. By reducing the number of males available for mating, most females are left unfertilized, thereby reducing the population of insects.

Classification of Insecticides Based on Their Active Ingredient

Another way to classify an insecticide is based on its **active ingredient (a.i.),** which is the actual amount of pesticide in a formulation that is responsible for killing the pest. The two general classes of insecticides are inorganic and organic:

- Inorganic insecticides are made from minerals such as arsenic and sulfur. These are becoming less commonly used.
- Organic insecticides can be broken down into natural and synthetic insecticides.

Naturally occurring insecticides. These insecticides are made from plants that have insecticide effects. These are becoming popular because they are typically nontoxic to humans. Examples of naturally occurring insecticides are pyrethrum (from chrysanthemum), nicotine (from tobacco), and a variety of others. Most naturally occurring insecticides act as stomach and contact poisons.

Synthetic organic insecticides. These insecticides are artificially produced and are effective on a wide range of pests. Several classes of synthetic organic insecticides are used today:

- **Chlorinated hydrocarbons**—chlorodane, lindane, methoxychlor, heptachlor, and others.
- **Organophosphates**—malathion, diazinon, phorate, and others.
- **Carbamates**—carbaryl, aldicarb, furan, and others.
- **Pyrethroids**—the synthetic counterpart to pyrethrins, which are naturally occurring in chrysanthemum. These are very effective on a wide range of pests and have low toxicity in humans.

Specific Insects and Chemical Control

Following are selected insect pests, the plants they attack, and the chemical control used:

- **Aphids.** Sucking insects found on the underside of leaves that attack a wide range of houseplants, annuals, perennials, trees, and shrubs. Aphids cause leaf curling and secrete honeydew that attracts other pests such as mites, flies, and ants. Chemicals commonly used to control aphids are diazinon, malathion, and orthene.
- **Mites.** Sucking insects found on the underside of the leaf that cause discoloration patches resulting in a reduction in plant growth and eventually death in a wide range of plants, including houseplants, annuals, perennials, trees, and shrubs. Chemicals commonly used to control mites are malathion, kelthane, and dicofol.
- **Scale.** Piercing and sucking insects that overwinter on the bottom side of the leaf as eggs or young scale, which can be armored or unarmored. Scale attacks houseplants, annuals, perennials, trees, and shrubs. Chemicals commonly used to control scale are carbaryl, acphate, and nicotine sulfate; scale can also be removed by hand.
- **Mealybugs.** Found on the underside of the leaf and have the appearance of white cotton-like material. They attack a wide range of houseplants, including African violet, begonia, gardenia, gloxinia, and many others. The home remedy for this insect is to wipe it off with a cotton swab that had been dipped in alcohol. Chemicals that can be used are orthene, diazinon, and malathion.
- **Whitefly.** As the name implies, these are tiny white flies that attack a wide range of houseplants such as geraniums, begonia, and coleus. They also attack a wide range of succulent annual and perennial plants. Chemicals used are rotenone, orthene, and malathion.
- **Fungus gnats.** These insects cause problems on a wide range of houseplants such as ferns and philodendron. They can be controlled by nicotine sulfate.

- **Fruit worm** or **ear worm.** Large insects that attack tomato, corn, and a variety of others. They can be controlled by simply removing them by hand; however, on a larger scale, carbaryl or diazinon are very effective.
- **Japanese beetles.** These insects are very destructive both in their larval and adult stages. They present major problems on trees, fruits, and a variety of ornamentals. Diazinon has been shown to be an effective control.

Classification of Herbicides Based on Their Mode of Action

Herbicides are chemicals used to control weeds and are the most commonly used method of control on a large scale. The disadvantages associated with the use of herbicides are damage to the environment together with direct health hazards to humans, both of which detract from their role in agricultural production. Herbicides may be classified in the following ways:

- selectivity.
- timing of application.
- method of application.
- chemistry.
- formulation.

Selectivity. **Selective herbicides** are effective in controlling a limited number of plant species. For example, 2,4-dichlorophenoxyacetic acid (2,4-D) kills broadleaf weeds and does not affect grasses.

Nonselective herbicides destroy all vegetation. Examples of nonselective herbicides are Roundup® and atrazine. This category can be broken down further into the following:

- Contact herbicides kill only the portions of the plant that they come in contact with. For this type of herbicide application, the plant must be completely covered to be effective.
- Systemic or translocated herbicides are absorbed and transported throughout the plant and kill the entire plant, not just a portion; an example of this type of herbicide is Roundup®. This type of herbicide does not require complete coverage.

Timing of application. There are three stages in the weed's life cycle to which herbicide can be applied:

- **Preplant treatments** are made to the soil prior to planting the crop. The crops being planted must be able to tolerate the herbicide. This type of herbicide kills weeds at the germination and/or seedling stage of growth.
- **Preemergence treatments** are applied to the soil surface after the crop is planted but before the emergence of the weed seedlings, crop seedlings, or both. They are designed to kill weeds as they germinate and/or as seedlings grow.
- **Postemergence treatments** are made after the emergence of crop plant seedlings, weed seedlings, or both.

Method of application. Another way to classify herbicides is by its method of application. Three methods of application are commonly used:

- **Broadcast treatments** cover the entire area uniformly.
- **Band treatments** apply herbicide in a narrow band around the crop row.
- **Spot treatments** apply herbicide to a specific location, for example, to weeds growing through a crack in the cement.

Depending on a variety of factors—including the type of weed, area to be treated, density of the weed population, and stage of growth—broadcast, band, or spot application can be used. Broadcast application covers the entire area and is typically used when there is no fear of damaging other plants. Both liquid and granular formulations can be used in the broadcast method. Band application is typically used in orchards between the rows. Spot applications are used to get those weeds that break through small gaps in walkways or driveways.

Chemistry. Herbicides may also be classified based on whether they are organic or inorganic. Examples of organic herbicides include phenoxy (2,4-D), dipehenyl ethers (Fusilade®), organic arsenicals (arsenic and derivatives of arsenic), substituted amide (Diphenamid), substituted ureas (Siduron), carbamates (EPTC), triazines (Simazine®), aliphatic acids (Dalapon), and bipyridyliums (paraquat).

Formulations. Herbicides can be formulated in two ways: liquids or granules. Most herbicides are applied as sprays, making it important to use the proper spray equipment. Granular formulations are more expensive because of their bulk and shipping costs; plus, they do not provide uniform application.

The Performance of Herbicides

Many factors can affect the performance of herbicides:

- **Proper identification of the weed.** Knowing the correct name for the weed is very important to choosing the correct herbicide to use.
- **Environmental conditions.** Temperature, wind, rainfall, humidity, and light all affect how well a herbicide will perform.
- **Maturity of the crop and weeds.** Herbicides are typically more effective on younger plants than on older ones, so weeds should be dealt with before they become mature.
- **Soil characteristics.** Some herbicides are absorbed by soils that are high in organic matter, therefore higher concentrations must be used.
- **Chemical concentration.** Reading the label and using the proper concentration is important. If the chemical is overdiluted, the application will be ineffective, which wastes time and money. If the chemical is used at concentrations that are too high, it may injure desired crops and again waste money. **Surfactants** are used to enhance the uptake of herbicides. A surfactant is a material that helps disperse, spread, wet, or emulsify a pesticide formulation (Figure 12-12). An example of a surfactant is soap.
- **Time of application.** Herbicides can be used as preplant, preemergence, or postemergence herbicides, but it is important that they are applied at the correct time.
- **Complete ground coverage.** The ground must be completely covered with the herbicide because incomplete coverage enables weeds to grow in untreated spots, which increases production costs.

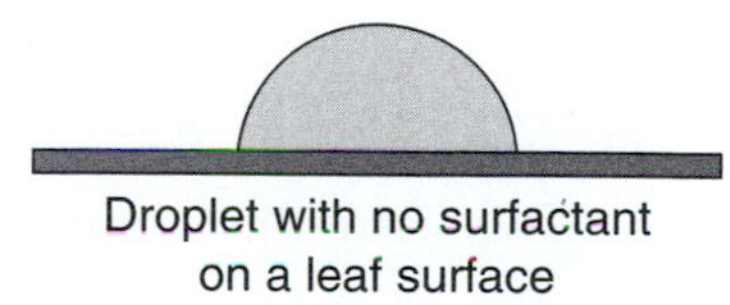

Figure 12–12 Effects of a surfactant on droplet size

SUMMARY

You now have a better understanding of the importance of pest management. The five major categories of pests include insects and related pests, weeds, nematodes, diseases, and rodents and other animals. Insects and related pests are classified based on feeding habits (type of mouthparts) or their life cycle (incomplete, complete, and no metamorphosis). A wide variety of beneficial insects are also used for biological control, although many insect and mite pests pose a threat to horticultural crops. Nematodes can also be beneficial or cause damage to horticultural crops. In addition, weeds are very costly to growers because they compete with the crop for water, nutrients, and light; they also harbor diseases and insects. You learned about both abiotic and biotic diseases, together with the main groupings of disease-causing organisms that attack horticultural crops, including fungi, bacteria, viruses, and mycoplasma-like organisms. The series of events that take place when controlling pests starts with prevention. If prevention is not possible, the pest must be properly identified and IPM should be employed. The general categories of pest control used as part of a responsible IPM program are cultural, biological, mechanical, chemical, and genetic.

Review Questions for Chapter 12

Short Answer

1. What are the five major categories of pests?
2. The way an insect feeds is based upon the structure of its mouth. What are two types of feeding habits commonly found in insects? Provide two examples for each of these types and explain the specific structure used to feed.
3. Give two examples each of insects having incomplete metamorphosis, complete metamorphosis, and no metamorphosis.
4. List two beneficial effects of insects and give two examples for each.
5. List the major adverse effect that insects have on plants and give two examples of insects causing these adverse effects.
6. Do nematodes cause major damage to roots?
7. Provide three ways nematodes are disseminated.
8. What are four problems associated with weeds?
9. What are the most common factors leading to problems with moss and algae in turf?
10. Provide four examples of weeds that pose a threat to horticultural crops.
11. What are five causes of abiotic and three causes of biotic diseases?
12. What are four common fungal diseases that affect horticultural crops?
13. What are four common bacterial diseases that affect horticultural crops?
14. What are four examples of viruses that affect horticultural crops?
15. What are four examples of mycoplasma-like organisms that affect horticultural crops?
16. What are four nonchemical ways to control animals?
17. What are two major reasons pests cost the grower billions of dollars annually?
18. List the series of events that should take place when controlling pests; be sure to list the five categories of IPM control.

19. What are four examples of cultural techniques used to control pests?
20. What are three forms of biological control?
21. What are six examples of mechanical pest control?
22. Why are there no pesticides labeled for many minor crops?
23. Provide 10 safety rules that should be followed when using pesticides to control pests.
24. What are four ways that insecticides can be classified based on their mode of action?
25. What are two ways that insecticides can be classified based on their active ingredient?
26. List five selected pests that attack horticultural crops and the chemical control used for each of these pests.
27. What are three ways that herbicides can be classified based on their mode of action?
28. What are five major factors that affect the performance of herbicides?
29. What are three ways that genes can be inserted into plants?

Define

Define the following terms:

pest
insects
metamorphosis
nematodes
weeds
abiotic disease
biotic disease
pathogens
mutualism
symbiosis
obligate parasites
facultative saprophytes
facultative parasites
mycoplasma-like organisms
pesticides
pheromones
active ingredient (a.i.)
herbicides
selective herbicide
nonselective herbicide
preplant treatment
preemergence treatment
postemergence treatment
broadcast treatments
band treatments
spot treatments
surfactant
plant breeding
genetically modified organism (GMO)

True or False

1. Insects with piercing and sucking mouthparts use mandibles to pierce the leaf and suck out the juices.
2. The type of insect that uses mandibles is said to be a chewing insect.
3. An example of an insect order that contains many species that cause problems in horticultural crops is Coleoptera.
4. An example of an insect order that contains many species that cause problems in horticultural crops is Homoptera.
5. Examples of beneficial insects are ladybugs and grubs.
6. Nematode movement in the soil is very slow; they move 12 to 30 inches per year.
7. Moss and algae are pests in turf that has been neglected.
8. Biotic diseases are often caused by nutrient deficiencies.
9. Abiotic diseases are caused by environmental factors.
10. Abiotic diseases are often caused by nutrient deficiencies and pollution.
11. Biotic diseases are caused by parasites or pathogens that are infectious and transmissible.
12. Fungi and bacteria are organisms that cause biotic diseases.
13. A commonly used method to control animals protected by law is to use sound to scare them away.
14. The goal of IPM is to reduce pest populations to a point where plant losses are at least equal to the cost of the control.

15. An excellent example of a bacteria used for biological control is *Bacillus thuringiesis*.
16. Fertilization is a cultural method of pest control.
17. Mowing, fertilizing, and irrigating are all cultural techniques used for pest control.
18. *Agrobacterium* and electroporation are two methods used to insert genes into plants.
19. Soap is a commonly used surfactant.

Multiple Choice

1. The major problem caused by nematodes is:
   A. They feed on foliage of plants.
   B. They wound the roots of plants, which allows fungi and bacteria to enter the root.
   C. They enter the xylem causing the plant to wilt and die.
   D. None of the above
2. Abiotic diseases are caused by which of the following?
   A. Parasites
   B. Bacteria
   C. Pathogens
   D. None of the above
3. Which of the following is an excellent control measure for animals protected by law that are causing damage to crops?
   A. Destroy the animals' habitat.
   B. Capture the animal and move it to another location.
   C. Use sound to scare away the animal.
   D. All of the above

Fill in the Blanks

1. The four distinct stages involved in complete metamorphosis are ____________________, ____________________, ____________________, and ____________________.
2. Insects have three distinct body parts, which include the ____________________, ____________________, and ____________________.
3. Weeds cause many problems that cost billions of dollars annually due to ____________________ and ____________________.

Activities

Now that you understand pest management and its importance, you will have the opportunity to explore this exciting field of research in more detail. In this activity, you will search Web sites that contain information about IPM strategies, including pros and cons of the strategies. Select three sites and write a description for each site that includes the following information:

- the Web site address.
- the purpose of the site.
- an explanation of each IPM strategy used.
- the pros and cons of each IPM strategy plus any other information that you want to share on IPM.

CHAPTER

13

Plant Biotechnology and Genetically Modified Organisms: An Overview

ABSTRACT

Biotechnology is used today in a variety of areas including hydroponics, tissue culture used for micropropagation, the production of specialty chemicals, and the use of single cells as a source of genetic variability for plant improvement. In addition, foreign genes are transferred via different methods to produce genetically engineered plants. Through genetic engineering, crops are improved and biopesticides are produced. Common questions about genetically engineered plants include the following: Will genetically engineered plants be safe to eat? Is it possible for genetically engineered plants to become weeds or can they transfer their genes to other plants and thereby make them weeds? What impact do genetically engineered plants have on agricultural practices? Opposition to genetically engineered foods comes from ethical considerations, safety considerations, anticorporate arguments, sustainability considerations, and philosophical considerations. Many genetically modified crops are produced in the United States and the world. The United States regulates genetically modified food and agricultural biotechnology products through several agencies.

Objectives

After reading this chapter, you should be able to

- provide background information on biotechnology.
- discuss several examples of biotechnology currently used today.
- provide background information on genetically engineered plants.
- provide answers to commonly asked questions about genetically engineered plants.
- provide reasons why there is opposition to genetically engineered foods.
- provide a description of genetically modified crops produced in the United States and the world and the way in which the United States regulates genetically modified food and agricultural biotechnology products.

Key Terms

biotechnology
callus
clone
electroporation
explants
genetic engineering
genetically modified organism (GMO)
hydroponics
meristem
protoplasts
somaclonal variation
somatic embryogenesis
tissue culture
totipotency
transformed
transgenic plant

INTRODUCTION

This chapter provides an overview of plant **biotechnology** and **genetically modified organisms.** By the year 2100, some estimate that Earth's population will reach 12 billion. The main goal of agricultural research is to maximize crop yields while minimizing adverse effects on the environment. Biotechnology is the manipulation of living organisms, or substances from organisms, to make products that benefit humanity. Plant biotechnology is changing the ways in which food crops, pharmaceuticals, and chemicals are currently produced.

The simplest form of plant biotechnology is **hydroponics,** which is a method of growing plants that provides nutrients needed by the plant

Figure 13–1 Potato tubers (top) and geranium plants (bottom) grown on agar aseptically

via a nutrient solution. Work in this area began in the late 1800s. When hydroponics was first used for commercial applications, plants were grown in large containers with aerated nutrient solution that was regularly flushed with fresh solution. This method has since been replaced by the nutrient film technique, which uses a thin film of nutrients dissolved in water that continually flows past the root system. Hydroponics require much capital and energy, so only high-value crops such as tomatoes are grown using this method.

Another form of biotechnology is **tissue culture,** which is a method of growing new plants from single cells and plant parts on artificial media under sterile conditions (Figure 13-1). Some commonly used methods of tissue culture include **callus** culture, cell suspension culture, embryo culture, **meristem** culture, and anther culture. These methods can be used for micropropagation, for production of specialty chemicals, as a source of variability for plant improvement, and for genetically engineering plants.

The capability to regenerate plants via tissue culture is an important step for producing genetically engineered plants because after the gene is inserted into a cell or tissue, it must be regenerated into a plant for future propagation. Genetic engineering involves the transfer of genes between related and unrelated organisms. Gene transfer into plants can be done in three ways: via *Agrobacterium tumefaciens,* particle bombardment, or **electroporation.**

Transformation via *Agrobacterium* occurs when *A. tumefaciens* attaches to a wound site and transfers its genes into the plant cell. When genes are transferred from a single bacterium to a single plant cell and are integrated into the chromosome of the plant cell, the cells are defined as being **transformed.** Although *A. tumefaciens* is useful in a wide range of plants, many plant species still cannot be transformed using this method. To provide an alternative to *A. tumefaciens,* researchers developed a method called particle bombardment as a means of introducing genes into plant cells.

The particle bombardment method uses a particle gun to shoot DNA-coated particles with enough speed to penetrate the first cell layer of plant tissue. After insertion of DNA into plant cells, the DNA is transcribed into RNA, which is translated into protein. When the transformed cells divide, they pass on the DNA that had been introduced to its progeny.

Another commonly used method to introduce DNA into plant cells is electroporation. In electroporation, plant **protoplasts** are exposed to a sudden electrical shock that opens up pores in the plant cell and enables DNA to enter. The DNA that enters the cell is then incorporated into the plant's chromosome and can be passed on to its progeny. One of the drawbacks of this method is that many plant species cannot be regenerated into whole plants via protoplasts.

Researchers are currently looking for new and improved methods for inserting DNA into plants; however, at present, the previously mentioned three methods are commonly used.

The main goal of genetic engineering is to improve crop plants by introducing foreign genes into them. Many desirable characteristics can be put into plants by inserting a single gene with currently available technology. In theory, inserting a gene into a plant is not very different from what plant breeders have been doing for years. One of the main problems with improving plants via gene transfer is that many useful genes have not been identified precisely. Researchers are currently looking for useful genes that can be used for crop improvement. Another major technical problem with using gene transfer for crop improvement is the capability to regenerate plants from transformed cells. Scientists worldwide are currently developing better methods for regenerating plants from transformed cells.

Although many advantages are associated with genetically engineering plants, there are many concerns with their use for human consumption. Some commonly asked questions include the following:

- Will genetically engineered plants be safe to eat?
- Is it possible for genetically engineered plants to turn into weeds or transfer their genes to other plants thereby making them weeds?
- What effect will genetically engineered plants have on the environmental impact of agricultural practices?

Many countries, groups, and individuals oppose the use of genetically modified organisms. The reasons for this opposition are broken down into five categories: ethical considerations, safety considerations, anticorporate arguments, sustainability considerations, and philosophical considerations.

The use of genetically modified organisms for food and in agriculture has generated a considerable amount of interest and controversy in the United States and around the world. Many people are strong supporters of gene transfer technology, although others raise questions about environmental and safety issues. Approximately 670 million acres of land are under cultivation worldwide, of which 16 percent was used for GMOs in 2000. Four countries that grew 99 percent of the global GMO crop as of the year 2000 included the United States with 74.9 million acres, Argentina with 24.7 million acres, Canada with 7.4 million acres, and China with 1.2 million acres. South Africa, Australia, Mexico, Romania, Bulgaria, Spain, Germany, France, Uruguay, and Indonesia also had significant acreage of GMOs, although much less than the four major countries.

Biotechnology products in the United States are regulated under the same laws that govern the health, safety, efficacy, and environmental impacts of similar products derived by more traditional methods such as plant breeding. The three major agencies that have primary responsibilities for regulating biotechnology products in the United States are the Food and Drug Administration (FDA), the United States Department of Agriculture (USDA), and the Environmental Protection Agency (EPA). Currently the system regulating biotechnology products appears to be working; however, only time will tell whether additional regulations will be necessary.

GENERAL BACKGROUND ON BIOTECHNOLOGY

The production of food over the past several decades has kept pace with the increases in population thus far. The main goal of all agricultural research continues to be to increase crop yields with minimal adverse effects on the environment. Plant biotechnology is changing the ways in which food crops, pharmaceuticals, and plant derived chemicals are currently being produced.

Biotechnology is described as the manipulation of living organisms, or substances from these organisms, to make products that benefit humanity. Although biotechnology is a fairly new term, its origins go back to the beginning of human civilization. Instead of gathering food from the wild, people started domesticating animals and plants for food. The grower selected for beneficial characteristics year after year and unknowingly modified them. Over the past 100 years, the modification of useful organisms has increased dramatically. Plant breeders have selected crops that can grow under a wide variety of conditions while maximizing their yields and quality characteristics. In addition, plants that produce valuable chemicals, such as *Taxus* plants for the production of taxol (an antitumor compound) and peppermint plants for mint oil, have been modified to maximize the production of these important compounds. Other biotechnological advances have been made by industrial microbiologists who have selected for microorganisms that produce drugs such as penicillin or enzymes that are added to laundry detergents for use in the food industry (Sonnewald, 2003).

Today with the advent of genetic engineering technology, dramatic modifications of living organisms is possible. Prior to the introduction of this technology, plant breeders were generally limited to genetic exchanges within and between species. Now genes can be transferred between very different organisms. The ability of scientists to transfer genes among humans, plants, and bacteria has revolutionized biotechnology. Although genetically modifying organisms is an important tool for biotechnological advancement, other technologies—such as plant tissue and organ culture for the mass production of crops—are vital components to the excitement in the field of biotechnology.

HYDROPONICS

Probably the simplest form of biotechnology is hydroponics, which is a method of growing plants that provides nutrients needed by the plant via a nutrient solution. Research on hydroponics began in 1860 when two German plant physiologists found that many plants could be grown in aqueous solutions containing four salts, which were calcium nitrate, potassium dihydrogen phosphate, magnesium sulfate, and a small amount of iron sulfate. Interestingly, nearly 50 years later, it was found that the salts used for these experiments were impure and supplied other trace elements that are required for plant growth. Today, we know that 16 elements are essential for normal plant growth and development.

Early research in the field of hydroponics has led to biotechnological applications today. When hydroponics was first used for commercial applications, plants were grown in large containers of aerated nutrient solutions or in gravel beds, which were regularly flushed with nutrient solution. Today, these methods have been replaced by the nutrient film technique (Figure 13-2). This technique

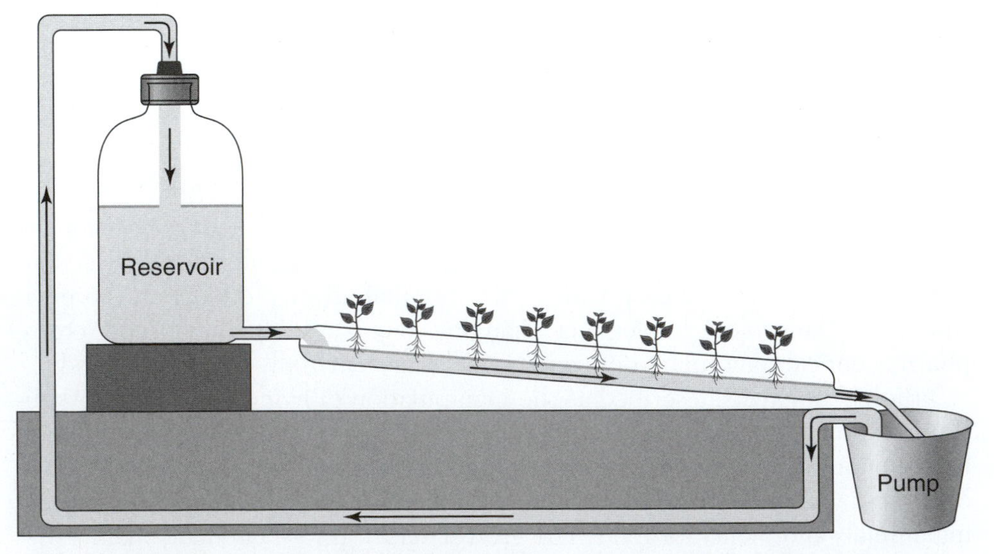

Figure 13–2 The nutrient film technique for growing plants hydroponically features a continuous flow of nutrients.

uses a thin film of nutrients dissolved in water that continually flows past a root system, which is small as compared to plants grown in soil. The continuous flow of nutrients ensures the proper concentration and availability to the plant. The use of hydroponics is expensive and requires a high amount of energy, so only high-value crops such as tomatoes, peppers, and lettuce are grown using this method.

TISSUE CULTURE

Tissue culture is a method for producing new plants from single cells, tissue, or pieces of plant material called **explants** on artificial media under sterile conditions. Some commonly used methods of tissue culture are callus culture, cell suspension culture, embryo culture, meristem culture, and anther culture, which are described in Chapter 7 of this text. These methods can be used for micropropagation, for production of specialty chemicals, as a source of genetic variability for plant improvement, and for genetically engineering plants.

Micropropagation

Micropropagation of plants in tissue culture has become a large industry. Meristem culture is used to produce large numbers of plants. This method uses a small shoot apex consisting of a meristem (a mass of dividing cells) with a few embryonic leaves. When put under the proper conditions, the meristem can be used to grow a whole plant in tissue culture. If a small amount of the plant hormones cytokinin and auxin is added at the proper concentration, multiple shoots will emerge from a single shoot apex. This method of multiplying plants is known as clonal propagation. Each of the plants produced by this method is called a **clone,** which is a plant that is genetically identical to its parent. Within a year, one plant can produce millions of cloned plants identical to the parent plant using this method, which has led to the emergence of a new worldwide micropropagation industry.

More than a thousand different plant species have been propagated by tissue culture. Initially most of the work in this area was with ornamental plants, many of which are sold in supermarkets or other retail outlets that sell in quantity. However, this method has now been extended to strawberries, potatoes, medicinal plants, trees, and others because of the many associated benefits. One of the major benefits of tissue culture as a method of propagating plants is that elite specimens, especially in trees, can be reproduced quickly and efficiently without fear of any changes in the plant's genetic makeup. Obtaining disease-free plants is another major advantage of using tissue culture as a means of propagating plants. Viruses live in plant cells and can be transmitted from one generation to another, causing major problems. Interestingly, generally no viruses are found in plant meristems; therefore, starting plants from a meristem ensures that all derived plants will lack viruses. Commercial strawberry and potato growers start every year with new virus-free planting material using meristem culture. The use of tissue culture is now a major method of micropropagating plants.

Propagating plants via seeds is the major means of reproduction found in nature and agriculture. For some agricultural uses, it would be advantageous if the ease of handling seeds could be combined with the benefits of clonal propagation using *in vitro* somatic embryogenesis. **Somatic embryogenesis** is a pathway to differentiation in plants that are induced in undifferentiated cell, tissue, or organ cultures by appropriately controlling nutritional and hormonal conditions, which results in the formation of organized embryo-like (embryoid) structures. Under appropriate cultural conditions, these organized structures can develop to form plantlets and eventually whole plants. Somatic embryos can be produced from callus growing on a solid medium or in a liquid suspension culture. The use of callus for the widespread production of somatic embryos is not economically feasible due to the high cost. However, the production of somatic embryos in liquid culture can be scaled up in bioreactors, and millions of embryos can be produced at one time. A variety of food and ornamental crops can now be propagated quickly and efficiently. In addition, the transfer of plantlets from liquid solution culture to trays containing a sterile soil mix can now be done using machines at a rate of 8,000 plantlets per hour (Chrispeels & Sadava, 1994).

Figure 13–3 Callus grown in darkness (top) and callus grown in light (bottom)

To use embryos as functional seeds, they would have to be individually packaged to facilitate their handling, transport, storage, and dispersal into the field for growth into plants. At present, embryos are encapsulated in a protective hydrated gel. The way an embryo is encapsulated depends on the species, agricultural application, and the physiological state of the embryo (hydrated or dehydrated, dormant or nondormant). The material used to encapsulate the embryo must provide physical protection and can carry nutrients, growth regulators, and fungicides or bactericides, thereby helping the seedling become established in the field. Currently, synthetic seeds are not used widely due to the technical problems of establishing seedlings in the field.

Production of Specialty Chemicals by Plants

When small sections of leaves, stems, or roots are excised from the plant, surface sterilized, and placed on agar under aseptic conditions with the proper ratio of auxins and cytokinins, they will form callus. Callus is an undifferentiated mass of cells (Figure 13-3). The cells in callus proliferate and grow very rapidly in tissue culture and can be used for the production of specialty chemicals. Rapidly growing callus can be used to produce a cell suspension culture, which is a group of single cells or small clumps of cells grown in liquid culture (Figure 13-4). Plants are a source of specialty chemicals for application in the food, pharmaceutical,

Figure 13–4 Cell suspension culture produced by rapidly growing callus

flavor, and fragrance industries. Some compounds from plants are used in pure form whereas others are used as mixtures. Sometimes plant compounds are further processed chemically to produce new structures, for example, steroid hormones. Approximately 119 chemical substances used as drugs are extracted from plants. So far, only a small percentage of all plants have been screened for useful compounds with the aid of modern scientific tools. Some estimate that only 5 to 10 percent of all plant species have been partially screened for some form of biological activity. The plant kingdom has been called the sleeping giant for drug development.

The great potential has been confirmed in recent years by finding totally new drugs such as taxol (Figure 13-5), a compound that is produced by the yew plant (Figure 13-6). Taxol is a novel antitumor drug that is highly effective in treating breast cancer and a variety of other types of cancer. Taxol, like many other important compounds produced by plants, is typically found in low amounts in plants. Because the chemical synthesis of taxol is complicated and not cost effective, large-scale extraction from wild or cultivated plants is required. However, because the concentrations of these important specialty compounds are very low in plants, difficult, labor-intensive extraction procedures, as well as dependence

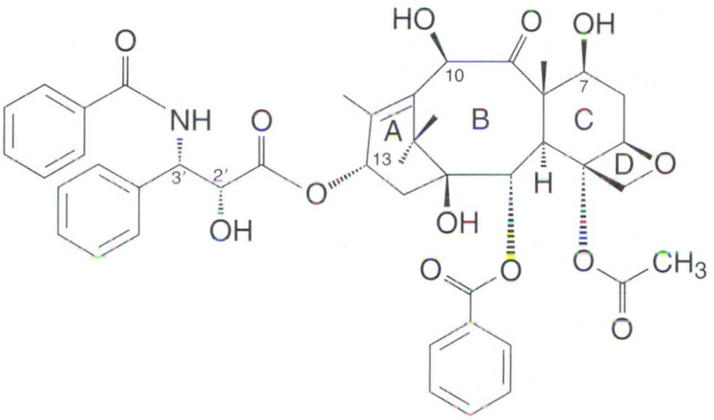

Figure 13–5 Structure of taxol, a compound produced by the yew plant

Figure 13–6 Yew plant

Figure 13–7 Airlift bioreactor

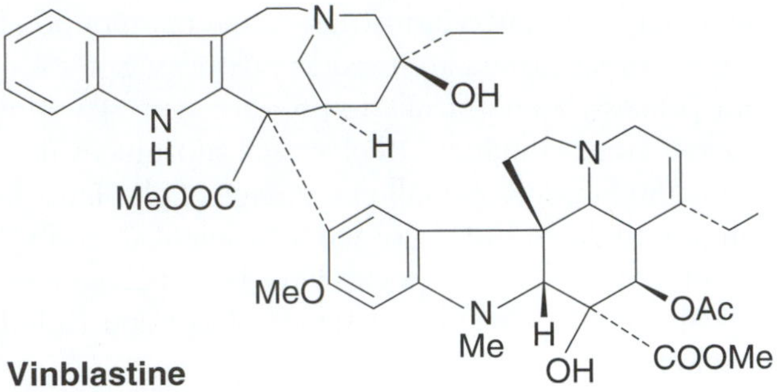

Figure 13–8 Structure of vinblastine, an alkaloid found in periwinkle and used in treatment of various cancers

Vinblastine

on seasonal and geographical factors, make it difficult to obtain large quantities of a given compound. The use of plant tissue culture offers a viable alternative for the efficient production of specialty compounds using bioreactor technology (Figure 13-7). Although bioreactor technology for plant cells is not used widely on a commercial scale, its use for taxol and vinblastine (Figure 13-8) is expected to be on a major scale in the future.

In addition to taxol, many plant species—especially tropical plants—produce unusual chemicals that can be used for a variety of drugs, cosmetics, flavorings, or agrichemicals. Until now, these compounds have been extracted from plants grown in open fields, often in tropical or semitropical regions. The development of new bioreactor technology for the growth of cells in tissue culture may in the near future replace farms with plant-cell factories. This process with plants imitates the advances in microbial biotechnology that took place over the past 50 years.

Use of Single Cells as a Source of Genetic Variability for Plant Improvement

Small pieces of tissue put under the proper conditions (aseptic and others) can be used to produce callus. After callus is formed, manipulating the ratio of the plant hormones auxin and cytokinin promotes callus, roots, shoots or roots, and shoots. With this in mind, scientists from Germany, Japan, and the United States discovered that a whole plant can be grown from a single cell given the correct nutritional and hormonal environment. Making an entire organism requires the correct expression of at least 30,000 genes. The capability of a single mature plant cell to produce an entire organism is called **totipotency.** Because plant cells are all interconnected by cell walls, isolating a single cell is difficult. Isolating a single cell is also tedious with variable yields, so protoplasts, which are plant cells without cell walls, are commonly used.

Protoplasts are obtained by digesting the cell walls of plant tissue with enzymes. Although protoplasts are very fragile, they can be used to produce entire plants. Protoplasts are cultured in liquid nutrient media. After a protoplast is produced, the following series of events takes place: first a complete cell wall is produced, followed by cell division, which results in a small clump of cells. From the small clumps of cells, there are two ways to regenerate a complete plant; the best way usually depends on the species. One way is to transfer these small clumps of cells to a solid medium where they will grow into callus that can form shoots. The shoots are cut off and transferred to a medium that induces root formation, thereby regenerating a new plant. A second way to regenerate a plant is to use these small clumps to produce somatic embryos. Somatic embryos resemble the embryos that normally arise as a result of the

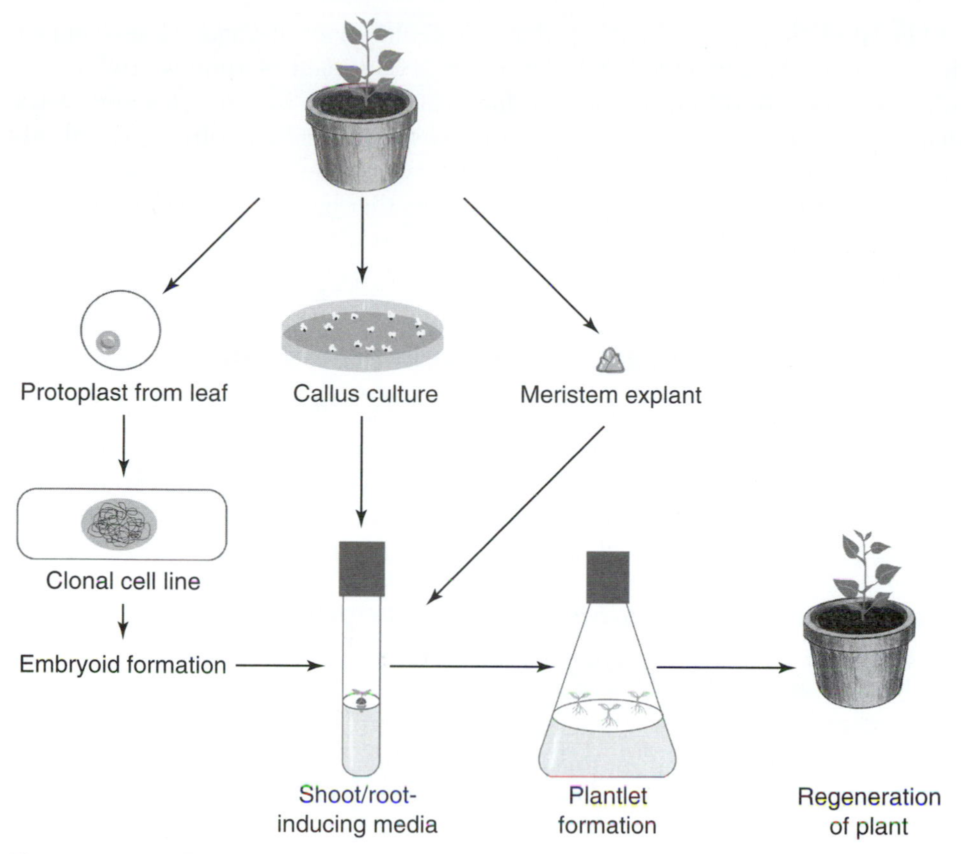

Figure 13–9 Three different routes for regenerating plants in tissue culture

growth of a fertilized egg cell. With certain systems, thousands of embryos can be produced in liquid culture, whereas in other systems, embryos or embryo-like structures are formed at the surface of the embryogenic callus growing on a solid medium. Such embryogenic callus are formed when tissues are obtained from developing seeds used to start the new tissue culture. By removing the embryos from the surface of the callus, they can be grown into complete plants. The three different routes for regenerating plants in tissue culture are from meristems, mature plant parts, and single cells, as shown in Figure 13-9.

When somatic embryos are derived from single cells and are grown into mature plants, the plant's characteristics exhibit some variability called **somaclonal variation.** When scientists first observed this variation, they were puzzled because plants regenerated from single cells were thought to be exactly alike. However, rearrangements of genetic material and mutations occur as plant cells divide before giving rise to an entire plant. When plant breeders discovered that somaclonal variation existed, they exploited it as a new source of genetic variation for crop improvements. This method is still used today for this purpose.

GENETICALLY ENGINEERING PLANTS

Genetic engineering is the isolation, introduction, and expression of foreign DNA in the plant. The first and most critical step involved in genetically engineering a plant is to identify and isolate a specific gene or genes responsible for a given trait or desirable plant attribute. After the gene or genes are identified and isolated, they must be packaged into a recombinant plasmid, which can produce more of the desirable DNA. After the recombinant plasmid is produced, it must be transferred into the plant, which can be done in three different ways. Plants produced by any of these three processes are called **transgenic plants** or genetically modified organisms (GMOs), which are organisms containing a foreign gene or genes. Some examples of traits that have been genetically modified in plants include insect resistance, disease resistance, herbicide resistance, modified fruit quality, and enhanced nutritional content.

Methods of Transferring Foreign Genes into Plants

The ability to regenerate plants from protoplasts, cells, and pieces of plant tissue is an important technology for producing genetically engineered plants. A designated crop can often be improved by introducing a single gene or multiple genes into a plant. Gene transfer into plants can be done via *Agrobacterium tumefaciens,* particle bombardment (biolistics), or electroporation.

Transferring via Agrobacterium tumefaciens

One way to transfer genes into plants is via the soil bacterium *Agrobacterium tumefaciens,* which causes tumors, also called galls, in many plants. Molecular biologists from the United States, Belgium, and the Netherlands were the first to show that *A. tumefaciens* attaches to a wound site and transfers its genes to plant cells. They showed that these genes carried information necessary to produce auxin and cytokinin. The abnormally high concentration of these hormones found in tissues as a result of this gene transfer led to uncontrolled growth. This resembled the situation in plant tissue culture where auxin and cytokinins are in continuous supply in the growth medium to promote cell proliferation and callus formation. When the genes are transferred from a single bacterium to a single plant cell and integrated into the chromosome of the plant cell, the cell is transformed. When the transformed cells containing the bacterial genes divide, all the cells produced contain the bacterial genes incorporated into their DNA. After molecular biologists realized that the bacterium could transfer its genes into plants, they quickly found that they could substitute other genes for the genes that the bacterium transferred to the plant cell. This enabled them to introduce any gene into plants. In addition, they could regenerate a whole new plant from a transformed cell and have a transformed plant in which every cell carried the gene. The steps involved in *Agrobacterium* transformation of a plant are shown in Figure 13-10.

Transferring via Particle Bombardment

Although *A. tumefaciens* is very useful in the transformation of a large number of plants, many plant species, especially grasses and cereal grains, are not susceptible to infection by *Agrobacterium*. To overcome the problems associated with *Agrobacterium* transformation, J. C. Stanford and his colleagues at Cornell University developed a microprojectile gun to deliver DNA directly into plant cells by shooting it through the cell wall and the cell membrane. They coated small tungsten particles with DNA and made an apparatus that could accelerate the particles with enough speed to penetrate the first cell layer of the plant tissue. After insertion into the plant cells, the DNA was transcribed into RNA, and the RNA was translated into protein, showing that the introduced DNA was genetically active. When the transformed cells divide and pass on their DNA, the introduced DNA is also present in all progeny of the cell that

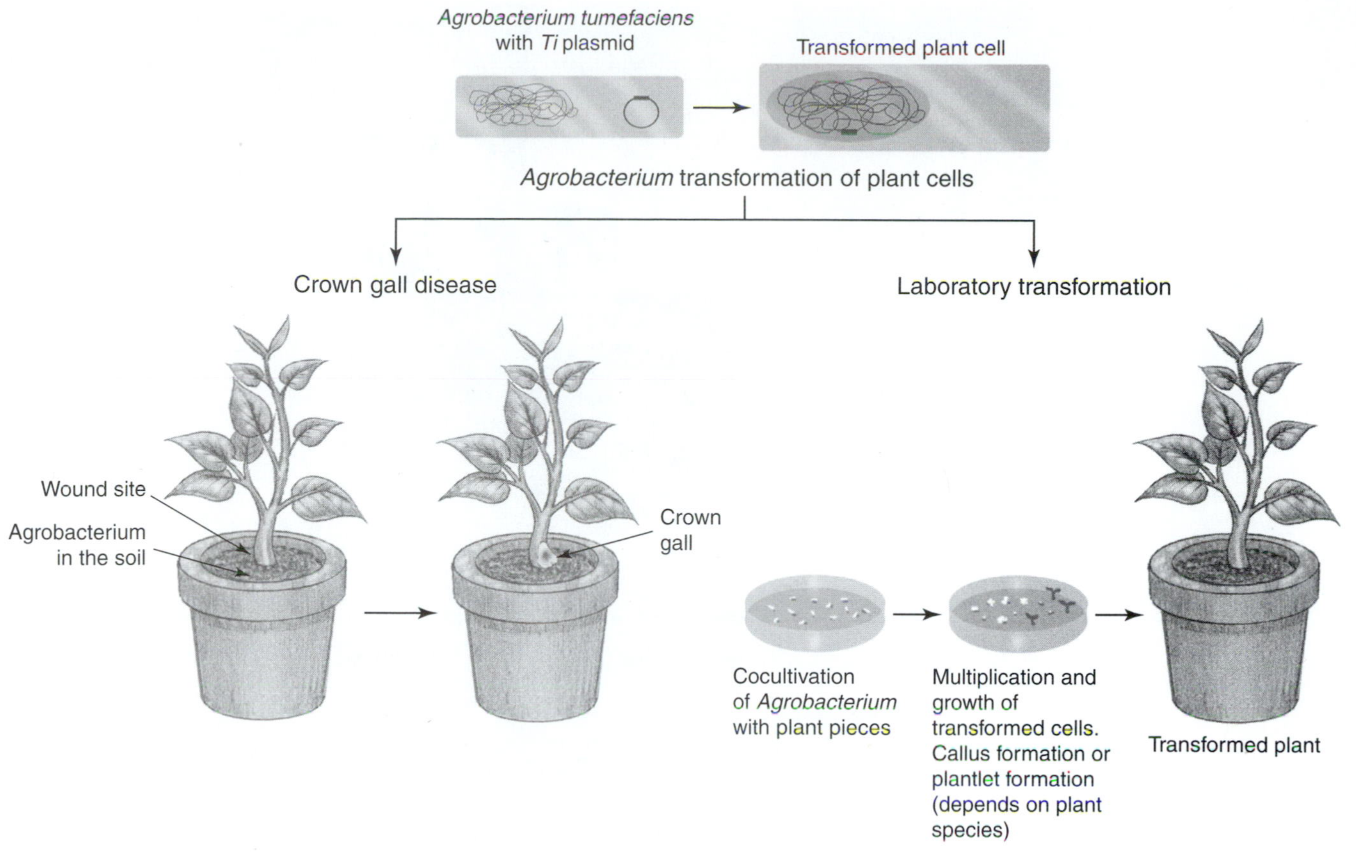

Figure 13–10 Steps involved in a *Agrobacterium* transformation of a plant

was originally transformed. A modified device used to accelerate the microprojectiles into a plant's tissue is shown in Figure 13-11.

Transferring via Electroporation

Electroporation is a totally different way of introducing DNA into plant cells than *Agrobacterium* or microprojectile bombardment. In electroporation, plant protoplasts are exposed to a sudden electrical discharge that opens up pores in the plant cell membrane and enables DNA to enter. After the electroporation process has taken place, the next step is to culture the protoplasts and try to regenerate them into plants. This is the most difficult step because many plant species cannot be regenerated easily from a single protoplast. Recently, scientists have used electroporation to insert genes into organized meristems. Using normal selection procedures, they successfully obtained transformed shoots that could be regenerated into plants containing the inserted gene. This method holds great promise for transforming important crop species that cannot be regenerated readily using single protoplasts.

Crop Improvements Through Genetic Engineering

The main goal of plant genetic engineering is to improve crop plants by introducing foreign genes into them. Current technology enables many desirable characteristics to be added by inserting a single gene. In theory, inserting a gene into a plant is not very different from what plant breeders have been doing for many years. When plant breeders want to make a wheat plant resistant to a

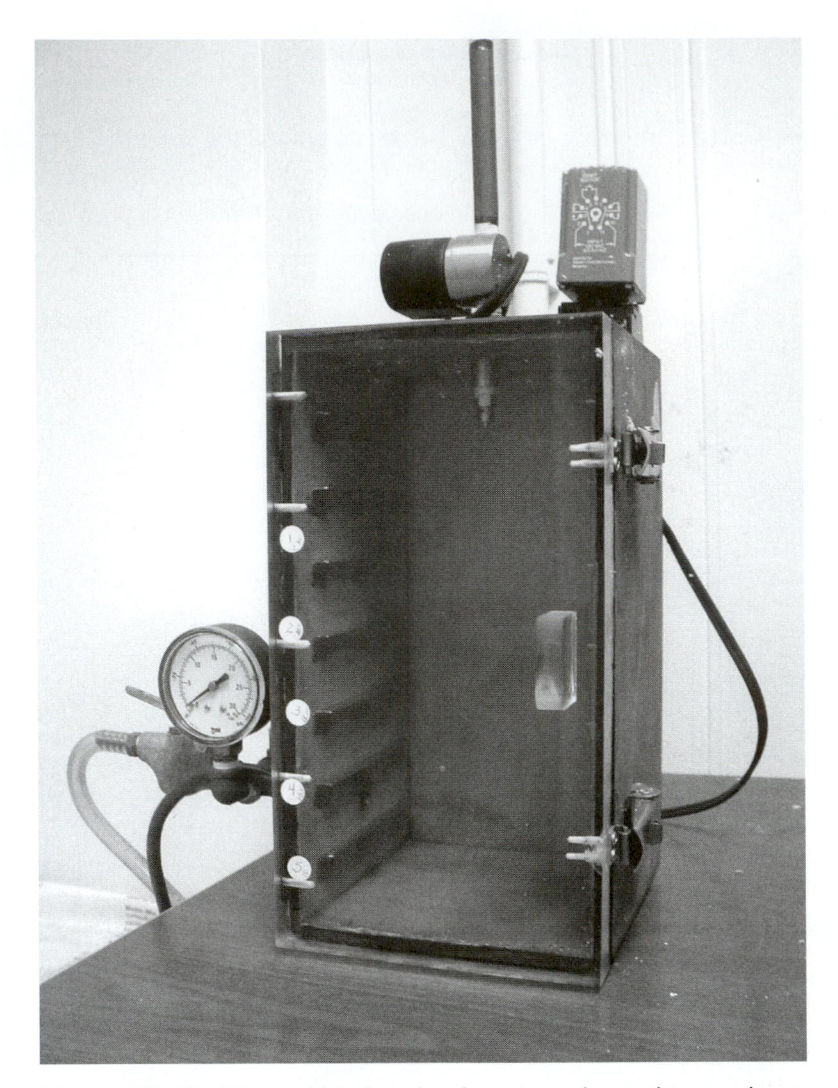

Figure 13–11 The particle bombardment gun is used to accelerate microprojectiles into plant tissue.

fungus that infects it, they search for a wild wheat variety that is resistant to the fungus. By making the necessary crosses, they can successfully transfer the resistance gene from the wild wheat to the cultivated wheat. When this is done, as opposed to using genetic engineering technology, hundreds of genes are transferred at the same time as the resistance gene. The extra genes usually do no harm to the plant, but they are not needed by the new strain.

What are some of the problems associated with improving plants via gene transfer? One of the main problems is that many useful genes have not been identified precisely. The power of genetic engineering is that there are no restrictions on which genes can be transferred between very different organisms. Therefore, useful genes can be found anywhere, not just in plants. For example, the bacterium *Bacillus thuringiensis* (Bt) carries a gene that makes an insecticidal protein. When the Bt gene is transferred into plants, the plant's cells make the insecticidal protein and any insect larvae that eat the leaves or roots die. The Bt gene is an example of a gene that has been identified and can be used effectively for crop improvement. However, in many cases, scientists have no idea which gene(s) would have to be isolated to promote important crop improvements, such as disease resistance (fungal, bacterial, and viral), stress tolerance (cold, heat, drought, and salinity), or improved nutrient, flavor, and postharvest qualities.

Another major technical problem for producing genetically engineered plants is the ability to regenerate plants from transformed cells. Growing a whole plant from a single transformed cell is very easy for tobacco or tomatoes. Transforming rice, cotton, or canola plants is not difficult; however, it is very difficult to transform corn, wheat, or soybeans.

Production of Biopesticides

As you know, fungal diseases, insects, nematodes, and weeds adversely affect the quality and quantity of crops produced. Currently, solid Integrated Pest Management (IPM) programs using cultural, mechanical, biological, chemical, and genetic methods are used to control pest problems. However, in recent years, larger doses of chemicals have been used in many countries. The use of pesticides in the United States has declined, but this overall statistic may be deceiving because there has been a dramatic drop in the use of pesticides in cotton, a crop that in the past has required very large amounts. On a worldwide scale, the use of pesticides is increasing continually, which is detrimental to the environment. Biotechnology and gene transfer can be used to select strains of natural enemies or genetically modified strains that are effective in controlling pests. The following are examples of biopesticides:

- **Mycoinsecticides.** More than 400 known species of fungi produce diseases in insect pests such as caterpillars, aphids, grasshoppers, and a variety of insect larvae. Several mycoinsecticides are used effectively in third-world countries; however, they act very slowly.
- **Nematode-bacteria complexes.** There are 24 known species of nematodes that contain symbiotic bacteria and can be used as biopesticides. Nematodes used as biopesticides can penetrate openings in the insect body where the bacteria is released. The bacterial infection kills the insect, and the nematodes multiply in the insect carcasses. These nematodes are highly specific to insects and do not harm plants or mammals.
- **Bacteria.** *Bacillus thuringiensis* (Bt) is the success story in this area, as few species of bacteria have shown potential as microbial insecticides. *B. thuringiensis* produces toxic proteins that act as poisons in the gut after the bacteria is eaten by insect larvae that feed on plants with this bacteria growing on them. The gene that encodes for the toxin produced by *B. thuringiensis* has been cloned and is incorporated into plants such as corn.
- **Viruses.** Seven major groups of viruses infect insects and have the potential to be developed as biological control agents. The known insect viruses typically have high specificity and are harmless to plants, mammals, and other animals. This specificity is very different from the broad-spectrum insecticides that are used to control insects. Some estimate that use of baculoviruses, which is the best studied of the insect viruses, could reduce the use of chemical insecticides in California by 60 percent and in Central America by 80 percent. Cost-effective control has been obtained with 30 different baculoviruses against lepidopteran pests. Although baculoviruses are safe and effective, they are not widely used due to their narrow host range.

CONCERNS RESULTING FROM GENETICALLY ENGINEERING PLANTS

Agricultural biotechnology using genetic engineering technology is focused on crop improvement to produce a large amount of a low-value product for human consumption. Plants that have been genetically engineered are called transgenic plants or genetically modified organisms (GMOs), which are organisms containing a foreign gene or genes. Some examples of traits that have been genetically modified in plants are insect resistance, disease resistance, herbicide resistance, modified fruit quality, and enhanced nutritional content. Because genetically engineered plants are in many cases to be used for human consumption, many questions arise.

Will Genetically Engineered Plants Be Safe to Eat?

Currently, no solid scientific evidence suggests that genetically engineered plants are not safe to eat. Genetically modified plants will have an additional gene or several genes present in all cells. On a daily basis, people now eat about 100,000 genes that are efficiently broken down by the human intestinal tract. Scientific evidence shows that genes added to plants by gene transfer are also efficiently digested. In addition to the foreign gene, the plant will also contain a new protein encoded by that particular gene. In many cases, the gene product may not be present in the part of the plant that people eat. For example, if the Bt toxin gene is expressed with a root-specific control region, then the protein is only expressed in the roots. This controls insects that feed on the roots of tomato or corn plants. In other cases, the protein is intended to be in the part of the plant that is eaten to increase nutritional value, such as increased protein or amino acid content. The detrimental effect of proteins produced by genetically engineered plants on humans must be evaluated on a case-to-case basis. Proteins produced in genetically engineered plants may also be produced naturally and already be consumed in large quantities by humans. For example, the gene for alpha amylase inhibitor found in beans is transferred to other legumes to inhibit the development of bruchid beetles. This inhibitor blocks human alpha amylase as well as insect alpha amylase. However, this protein is already eaten in large quantities by millions of people all over the world and its transfer into other crops is probably safe as long as the foods are cooked prior to being eaten. In fact, beans need to be cooked well before eating precisely because they contain alpha amylase inhibitor, as well as phytohemagglutinin, which is another plant-defense protein.

Although there is no reason to believe that GMOs are harmful if eaten, this does not mean that they will be accepted readily by consumers. Currently, there is debate as to whether or not genetically modified plants should be labeled. The FDA made its policy public in May 1992. This policy states that food obtained from genetically modified plants does not need to be labeled as such. The FDA has specified that what is important to the consumer is the actual content of the food, such as the nutritional content, allergenic responses, pesticides used, and others, not the process used to generate the plants. For example, wheat flour is not labelled based on breeding processes used to generate common wheat lines or the sources of genes in improved lines. Rather, consumers need to know the flour's protein content, presence of gluten and/or allergens, bran content, and added ingredients such as vitamins.

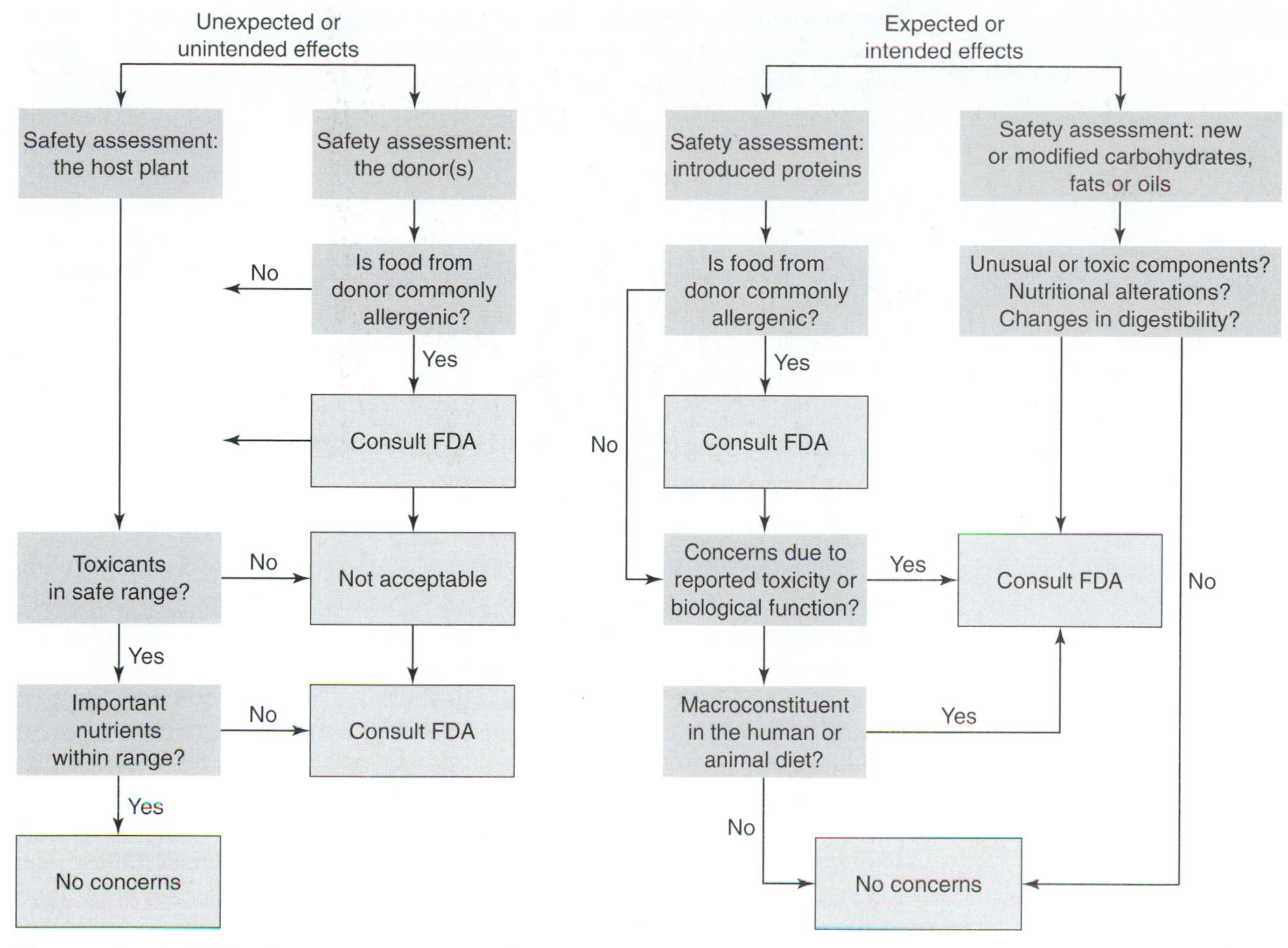

Figure 13–12 FDA safety assessment procedures

One of the main concerns of the FDA and consumers are foods that produce allergic reactions in humans. Genetically engineered food must be tested, and the company that wants to market the new food must demonstrate that it is not allergenic. The FDA has a specific set of safety assessment procedures that must be followed, as shown in Figure 13-12. This FDA assessment chart applies to foods from new plants produced by traditional or molecular techniques.

Can Genetically Engineered Plants Become Weeds or Transfer Genes to Other Plants to Make Them Weeds?

Weeds cause problems by competing with crop plants and spreading diseases and insects, thereby reducing yields. Weed biologists have identified 13 characteristics that make a plant a weed, and the most serious weeds have 11 or 12 of the 13 characteristics. Crop plants typically have only between 5 and 6 of these characteristics. The addition of a single gene that is totally unrelated to weediness is unlikely to turn a crop plant into a problematic weed; however, there are cases where crop plants may become weeds. This becomes a problem when several crop species are made resistant to the same herbicide, and these crop species are used sequentially in a crop rotation. There is considerable danger that "volunteers" from the first crop will become weeds in the subsequent crop.

This problem already occurs on a small scale; however, it would be greatly aggravated by herbicide-tolerant crops. One possible solution to this problem is to use crops tolerant to different herbicides in crop-rotation programs.

Another potential problem is the transfer of herbicide resistance genes to wild-type relatives, which creates problematic weeds. At present, it is difficult to say whether gene transfer to wild relatives will be a real problem. Even for herbicide resistance, it can be argued that past practices have already led to herbicide-resistant weeds. The only real way to control herbicide-resistant plants is to use alternative cultural practices and/or alternative herbicides. Therefore, herbicide-resistant weeds arising from gene flow to wild relatives may not be any more trouble than the weeds that we already have; however, only time will tell.

Genetically Engineered Plants' Environmental Impact on Agricultural Practices

Agricultural practices can negatively impact the environment due to the use of chemicals. Genetically engineered plants may overcome or reduce the need for chemicals. For example, producing plants that are resistant to insects, fungi, bacteria, viruses, or other pathogens would reduce the need for the use of chemicals. In addition, plants could be made to use nitrate from the soil more efficiently or to dissolve rock phosphate without needing to convert it to superphosphate in a chemical process. This reduces nitrate contamination in the groundwater and also lessens the need to use large amounts of energy to produce and transport nitrate and phosphate fertilizers. Another example of a potential area that would benefit from genetic engineering is the transfer of symbiotic nitrogen fixation capability from legumes to cereal grains. This reduced need for nitrogen fertilizers would reduce the production of nitrogen fertilizers, which is an energy-intensive process. Transferring symbiotic nitrogen fixing capacity form a legume plant to a cereal plant, such as corn, would require transferring a large number of genes and is unlikely to occur in the near future; however, the potential still exists.

Although the production of GMOs has the potential to reduce the need for chemicals, the potential for increased use of chemicals in agriculture also exists. Companies produce and market herbicide-resistant crop plants that tolerate specific herbicides, such as Monsanto's Roundup®-resistant crop plants, Hoechst's Basta®-resistant crops, and Dupont's Glean®-resistant plants. These crops make it attractive for the growers to use those specific herbicides as the most convenient method of weed control. This approach to weed control could lead to an increased need for chemicals in agriculture. Scientists at large companies have argued that plants made resistant to biodegradable herbicides such as Monsanto's Roundup® will lessen the impact of the chemicals that are needed. This is not always the case, however, because several companies are making plants resistant to herbicides that are not biodegradable and are very toxic. The government also gives substantial incentives that push agricultural practices in one direction or another, making it important to elect responsible individuals who will not promote excessive use of chemicals.

Opposition to Genetically Engineered Foods

A number of countries, groups, and individuals oppose the use of GMOs. This opposition is directed toward microorganisms, plants, and animals that either have a direct or indirect role in food production. The reasons for this opposition can be broken down into five general categories, which are summarized in the following sections (Brandt, 2003).

Ethical Considerations

Those opposing the use of GMOs state that gene transfer between organisms not from the same species is unethical because human beings should not alter an organism in this way. The counterpart to this argument is the theory that humanity has had a profound effect on the evolution of many species, including gene transfer between species through the use of plant breeding. In addition, gene transfer between unrelated organisms occurs in nature all the time.

Safety Considerations

One safety concern is that releasing GMOs into the environment could have many unforeseen ecological consequences. Others say that moving organisms between continents could create more problems than GMOs because there is no known biological control to bring them back into equilibrium when these organisms are relocated. Another concern is that GMOs may not be safe to eat, which is possible. However, others say that when a GMO is produced, we know exactly which gene is being introduced so it is highly unlikely that a problem will arise, as compared to traditional plant breeding. When new plants are produced via traditional plant breeding, large segments of DNA with many genes are transferred between plants, which can cause many unknown problems. The possibility exists that the GMOs used for food will contain novel allergens; however, this can be readily tested for prior to making GMOs available to the general public.

Anticorporate Arguments

Opponents of GMOs state that the purpose of corporations is to make money, not to protect the welfare of the general public. They also say that the general public should not rely on corporations to give them correct information about GMOs and that any information presented will be twisted to fit the corporations' needs. Although true in some regard, this does not present the full story because more often than not corporations promote human welfare by making new products. In fact, they change their products in response to demand, which may be for better nutrition or health. Therefore, to say all biotechnology companies are socially irresponsible is an incorrect statement.

Sustainability Considerations

Those who are against biotechnology say that it is driving toward high-input agriculture and will never contribute to making agriculture sustainable. Opponents of GMOs say that GMOs will probably lead to an increase in chemicals by introducing herbicide-resistant plants. Supporters of GMOs say that a combination of technology, government regulations, and tax laws determines the direction of agriculture, not the technology alone.

Philosophical Considerations

Opponents of GMOs say that we must return to an ecologically based stewardship of Earth instead of exploiting it as is the current trend. Although this sounds reasonable, remember that 5 billion people must be fed currently and 10 billion more must be fed in the future; therefore, we must work toward a world that uses all resources wisely and efficiently.

GENETICALLY MODIFIED CROPS PRODUCED IN THE UNITED STATES AND THE WORLD

The use of GMOs for food and in agriculture has generated a considerable amount of interest and controversy in the United States and around the world. Many people are strong supporters of gene-transfer technology, although others raise questions about environmental and safety issues. Crop varieties developed via genetic engineering were first introduced into commercial production in 1996. Today, these crops are planted on more than 109.2 million acres worldwide. As shown in Figure 13-13, the United States accounts for over two-thirds of all GMOs produced globally. The principal states growing the major genetically modified crops of corn, soybean, and cotton are shown in Figure 13-14. Genetically modified food crops grown by farmers in the United States include corn,

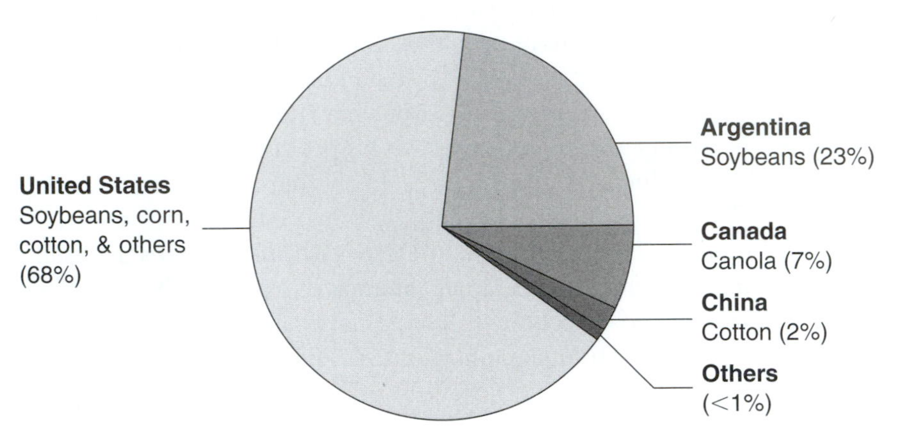

Figure 13–13 Breakdown of genetically modified crops in the United States and the world

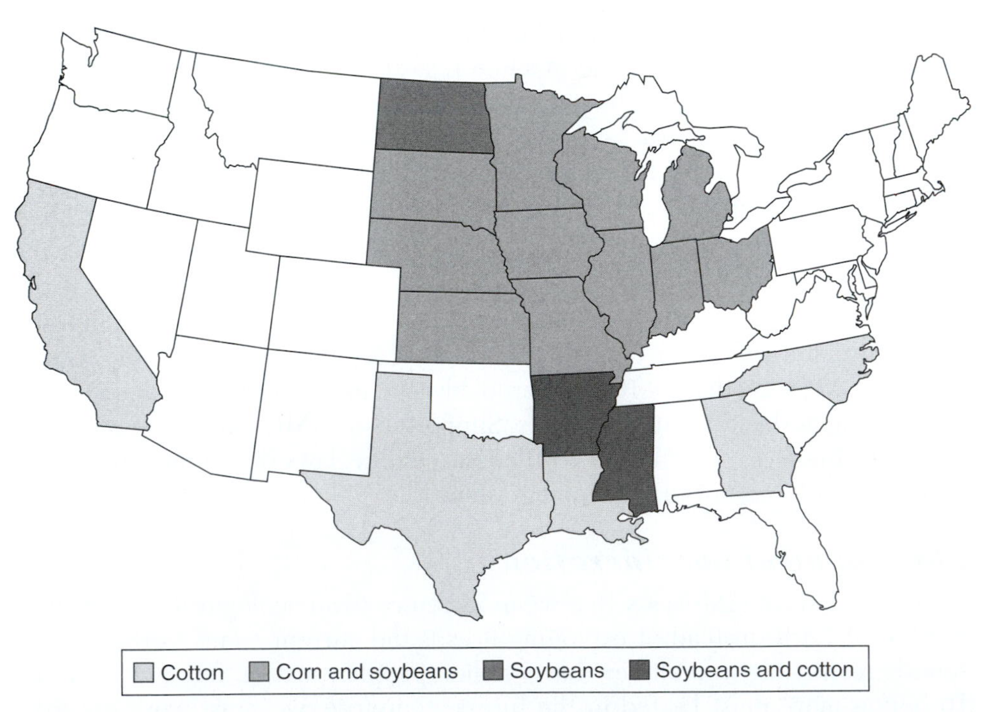

Figure 13–14 Principal states growing genetically modified crops and the crops grown

cotton, soybeans, canola, squash, and papaya. Other major producers of GMOs are Argentina (primarily GMO soybeans) and Canada (canola). Worldwide, about 670 million acres of land are under cultivation, of which 16 percent was used for GMOs as of the year 2000. The four countries that grow 99 percent of the global GMO crop as of 2000 are the United States with 74.9 million acres, Argentina with 24.7 million acres, Canada with 7.4 million acres, and China with 1.2 million acres. Other countries with significant acreage in GMOs, although much less than the four major countries, which are South Africa, Australia, Mexico, Romania, Bulgaria, Spain, Germany, France, Uruguay, and Indonesia (Emrich, 2003).

U.S. REGULATION OF GENETICALLY MODIFIED FOOD AND AGRICULTURAL BIOTECHNOLOGY PRODUCTS

The regulation of biotechnology products in the United States is under the same U.S. laws that govern the health, safety, efficacy, and environmental impacts of similar products derived by more traditional methods such as plant breeding. The FDA, the Department of Agriculture, and the EPA have the primary responsibility for regulating biotechnology products.

The FDA is responsible for the safety of food and animal feed and for the safety and efficacy of human and animal drugs and biological materials. Within the FDA, the four centers with responsibilities for biotechnology products include the Center for Food Safety and Applied Nutrition (CFSAN), the Center for Veterinary Medicine (CVM), the Center for Drug Evaluation and Research (CDER), and the Center for Biologics Evaluation and Research (CBER).

The EPA is responsible for regulating the use of pesticides, setting allowable levels (tolerances) of pesticide residues in food, and regulating nonpesticidal toxic substances, including microorganisms.

The USDA is responsible for the safety of meat, poultry, and egg products. It also regulates potential agricultural plant pests, noxious weeds, and the safety and efficacy of animal biologics. Within the USDA, the Animal and Plant Health Inspection Service (APHIS) has the major responsibility for regulation of biotechnology products, with additional potential responsibilities for the Food Safety and Inspection Services (FSIS). The major statutes under which the FDA, EPA, and USDA have been given regulatory or review authority include the following:

- The Federal Insecticide, Fungicide, and Rodenticide Act (FIFRA) (EPA).
- The Toxic Substances Control Act (TSCA) (EPA).
- The Food, Drug, and Cosmetics Act (FFDCA) (FDA and EPA).
- The Plant Protection Act (PPA) (USDA).
- The Virus Serum Toxin Act (VSTA) (USDA).
- The Public Health Service Act (PHSA) (FDA).
- The Dietary Supplement Health and Education Act (DSHEA) (FDA).
- The Meat Inspection Act (MIA) (USDA).
- The Poultry Products Inspection Act (PPIA) (USDA).
- The Egg Products Inspection Act (EPIA) (USDA).
- The National Environmental Policy Act (NEPA).

SUMMARY

You now have a general appreciation of plant biotechnology and the use of GMOs in agriculture. You learned about different types of biotechnology used today, including hydroponics, tissue culture used for micropropagation, the production of specialty chemicals, and the use of single cells as a source of genetic variability for plant improvement. In addition, you learned about genetically engineering plants—including methods for transferring foreign genes into plants such as *Agrobacterium tumefaciens*, particle bombardment, and electroporation—crop improvements through genetic engineering, and the production of biopesticides. You realize the pros and cons of genetic engineering by addressing commonly asked questions such as, Will genetically engineered plants be safe to eat? Is it possible that genetically engineered plants will become weeds or will transfer their genes to other plants thereby making them weeds? What impact do genetically engineered plants have on agricultural practices? Opposition to genetically engineered foods comes in the form of ethical considerations, safety considerations, anticorporate arguments, sustainability considerations, and philosophical considerations. In addition, you learned about genetically modified crops produced in the United States and the world, and you learned that now the United States regulates genetically modified food and agricultural biotechnology products.

Review Questions for Chapter 13

Short Answer

1. What are five commonly used methods of tissue culture?
2. What are four uses for tissue culture?
3. What are three methods for transferring foreign genes into plants?
4. What are five general categories for the opposition to genetically modified foods?
5. What are the three main agencies that are responsible for the regulation of biotechnology products in the United States?

Define

Define the following terms:

biotechnology
hydroponics
explants
meristem
clone
somatic embryogenesis
callus
totipotency
protoplasts
somaclonal variation
transformed
electroporation
transgenic plants
genetically modified organism (GMO)

True or False

1. A commonly used commercial method for growing plants hydroponically is the nutrient film technique.
2. The growth of single cells in tissue culture can be used as a source of genetic variability for plant improvement.

3. Somaclonal variation occurs when somatic embryos are derived from single cells and are grown into mature plants.
4. *Agrobacterium thuringiensis* is an example of a bacterium that is used to produce an insecticidal protein that kills any insect larvae that eat the leaves or root of that plant.
5. The use of genetically engineered plants has the potential to increase the need for more chemicals in agriculture.

Multiple Choice

1. The use of hydroponics requires a large amount of capital and energy; therefore, only high-value crops such as ________ and ___________ are grown using this method.
   A. tomatoes and peppers
   B. potatoes and corn
   C. wheat and barley
   D. None of the above
2. After callus is formed, the manipulation of the ratio of the plant hormones auxin and cytokinin can be used to promote
   A. callus.
   B. roots.
   C. roots and shoots.
   D. All of the above
3. Which of the following countries accounts for two-thirds of all genetically modified crops globally?
   A. China
   B. Europe
   C. United States
   D. England

Fill in the Blanks

1. The main goal of all agricultural research is to increase ________________.
2. ________________ is second in the production of GMOs on a worldwide basis.

Activities

Now that you understand plant biotechnology and GMOs, you will have the opportunity to explore this exciting area in more detail. In this activity, you will search Web sites that contain information about plant biotechnology and GMOs, including pros and cons. Select three sites and write a description for each site that includes the following information:

- the Web site address.
- the purpose of the site.
- an explanation of the area of plant biotechnology and/or GMOs.
- the pros and cons of plant biotechnology and/or GMOs, plus any other interesting information that you want to share in this area.

References

Brandt, P. (2003). Overview of the current status of genetically modified plants in Europe as compared to the United States. *Journal of Plant Physiology, 160;* 735–742.

Chrispeels, M. J., & Sadava, D. E. (1994). *Plants, genes, and agriculture.* Boston: Jones and Bartlett Publishers International.

Emrich, R. (2003). Discussion of current status of commercialization of plant biotechnology in the global marketplace. *Journal of Plant Physiology, 160;* 727–734.

Sonnewald U. (2003). Plant biotechnology: From basic science to industrial applications. *Journal of Plant Physiology, 160;* 723–725.

CHAPTER

14

Greenhouse Structures

Objectives

After reading this chapter, you will be able to

- provide background information on when greenhouses were first used and the variety of purposes they are used for today.
- discuss different basic greenhouse structure types and the advantages and disadvantages of each.
- discuss steps prior to laying the foundation, the type of foundation to use, and the basic components/functions of a simple greenhouse.
- provide the four major classes of greenhouse coverings used today and give the advantages and disadvantages of each.
- discuss the factors involved in locating a greenhouse range.
- recognize the proper ways to orient a greenhouse range and the different types of bench orientations to use in the greenhouse.
- discuss the four major methods for heating greenhouses and provide the advantages and disadvantages of each.
- list the different types of ventilation and cooling systems commonly used in greenhouses today.

Key Terms

attached greenhouse
conduction
connected greenhouse
freestanding greenhouse
greenhouse
greenhouse range
headhouse
infiltration
radiation

ABSTRACT

This chapter discusses the origins of greenhouse usage and the wide variety of purposes for which greenhouses are used today. Different basic greenhouse structures are used today; each structure has its own advantages and disadvantages. The structural components of a greenhouse have specific functions. Four major classes of greenhouse coverings are commonly used; each has advantages and disadvantages. The factors involved in selecting a location for a greenhouse range include market, accessibility, climactic conditions, topography, drainage, water, utilities, zoning regulations, labor supply, and land for expansion. Inside the greenhouse, issues to consider include how to orient the range properly, the different type of bench arrangements within the greenhouse, and the types of benches. Greenhouse temperature is controlled using heating, ventilation, and cooling systems.

INTRODUCTION

During the winter months, the early Egyptian and Roman civilizations used **greenhouse** structures to protect tender crops such as fruits and vegetables. Today, greenhouses are not limited to just food production but are also used in the ornamentals industry. Greenhouses can be

Figure 14–1 Connected greenhouses with cooling fans

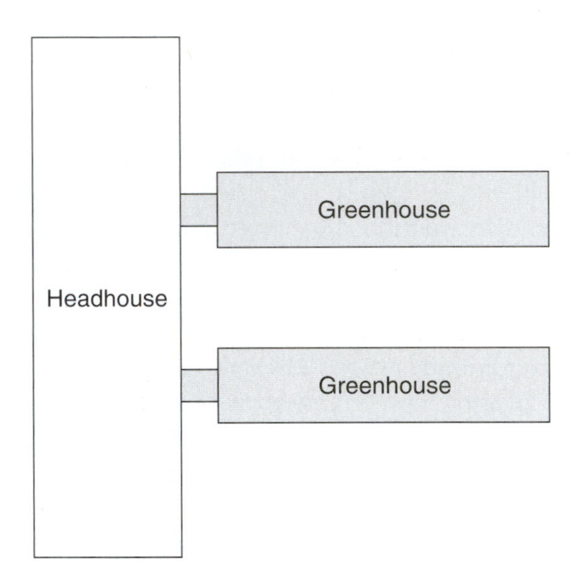

Figure 14–2 The location of a headhouse in relationship to the greenhouse

found in homes, offices, hospitals (horticulture therapy), public gardens, research institutions, and commercial organizations (wholesale or retail). Greenhouse structures are covered with a transparent material that allows sufficient sunlight to enter for the purpose of growing and maintaining plants (Figure 14-1). A key part of a greenhouse is a **headhouse,** which is a central building used for offices, storage, and workspace with attached greenhouses (Figure 14-2). The three basic types of greenhouses are attached, detached (freestanding), and connected. Each of these three basic types can be constructed in several different styles, including even span, uneven span, lean-to, Quonset arch, or Gothic arch. **Attached greenhouses** are attached to an existing structure. They should be located on the south side of the building to take advantage of sunlight. Detached greenhouses (or **freestanding greenhouses**) are separate from other buildings or greenhouses. The main advantage of detached greenhouses is that environmental control can be regulated easily and programmed to suit different needs; however, the main disadvantage is that more land is needed, thereby increasing cost. **Connected greenhouses** have several greenhouses joined together.

Various styles of connected greenhouses include ridge-and-furrow, barrel-vault, and sawtooth greenhouses. Connected greenhouses cost less overall, are more efficient because workers and equipment can move throughout the greenhouse with ease, and require less heat and less land. The main disadvantage is that accommodating different crops is difficult when such crops require different environmental conditions.

Prior to laying the foundation for a greenhouse, the water and electrical lines should be located and placed. After this is done, the proper type of foundation must be chosen that provides adequate structural support and is level. Construction materials for greenhouses must be strong, durable, easy to maintain, inexpensive, and cast as little shadow as possible (the frame). Two basic frame designs are used for greenhouses: the A-frame and arched frame. The structural components of a simple greenhouse include the ridge (top of the greenhouse), anchor support posts (provide the main structural support), trusses (structural support), purlins (structural support), ventilators (cooling), and cooling fans (cooling).

A variety of greenhouse coverings are available today. The most important function of a greenhouse covering is to allow the maximum amount of light into the greenhouse for the growth and development of plants. The five major types of greenhouse coverings are glass, plastic films, fiberglass-reinforced plastic, acrylic, and polycarbonate; each has advantages and disadvantages. Acrylic and polycarbonate are the most popular rigid plastic covering materials used today, whereas rigid plastics and glass are the most expensive coverings used today. Acrylic can last up to 25 years; polycarbonate can last for between 10 and 15 years. Both acrylic and polycarbonate have become very popular because they are lightweight, easy to install, and require less support, thereby reducing shading caused by sash bars. In addition, acrylic and polycarbonate both have very good light transmission and excellent heat insulating ability. The only major disadvantage is that these materials are flammable.

Success in the greenhouse industry begins with good planning. When selecting the proper site for a greenhouse, important factors to consider are market, accessibility, climactic conditions, topography, drainage, water, other utilities, zoning regulations, labor supply, and adequate room for future expansion.

After a location has been selected, the **greenhouse range** and benches within them must be properly laid out. The greenhouse range must be oriented to maximize the light entering the greenhouse and to minimize heat loss. The benches must also be oriented correctly for peak efficiency. Three bench arrangements commonly used in greenhouses are longitudinal, cross benching, and peninsular.

After properly orienting the greenhouse range and the benches contained within, selecting the proper system for controlling greenhouse temperatures is the next step. Good temperature control is important to maintain high-quality crops. Heat loss from greenhouses through **conduction, infiltration,** and **radiation** must be minimized. The three types of fuel used in greenhouses are natural gas, oil, and coal. The heating system must be selected based on the particular operation. Heating systems commonly used in greenhouses are steam heating, hot water heating, forced-air heating, and radiant heating (such as infrared radiant heaters and solar radiation systems). To control greenhouse temperatures, good ventilation and cooling systems are also necessary. Some examples include natural ventilation, fan-tube ventilation, fan and pad cooling systems, and shading. When establishing a greenhouse range there are many factors that should be considered carefully to ensure success.

BASIC GREENHOUSE STRUCTURE TYPES

Greenhouses are built or manufactured in many design types, sizes, and environmental control systems depending upon the amount of space available, types of plants to be grown, geographical location (differences between Florida and Alaska, for instance), and cost of construction materials. Some very simple designs can only control temperature and light, whereas some state-of-the-art computer-based facilities can control humidity, light, temperature, nutrients, and soil moisture and also have alarms to notify the grower of a problem. The problem may be taken care of remotely by computer; however, this depends on the problem.

Specific Design Types

The three basic types of greenhouses are attached, detached (freestanding), and connected. Each of these three basic types can be constructed in the following common styles: even span, uneven span, lean-to, Quonset arch, and Gothic arch (Boodley, 1981).

An attached greenhouse is connected to a building. The three basic styles include lean-to, attached even span, and window mounted (Figure 14-3). The attached greenhouse style should be included within the initial design of the house or careful thought should be given to greenhouse placement when adding to a house. Typically, these greenhouses are located on the south side of the building to take advantage of sunlight. Attached greenhouses typically have a simple design and are easy to build because they are connected to existing structures. The size and use of an attached greenhouse depends on the characteristics of the building it is being attached to. Attached greenhouses are typically found in homes, garden centers, and other commercial or private buildings where plants are grown.

A detached (freestanding) greenhouse is separate from other buildings or greenhouses. Detached greenhouses can come in four different styles: even

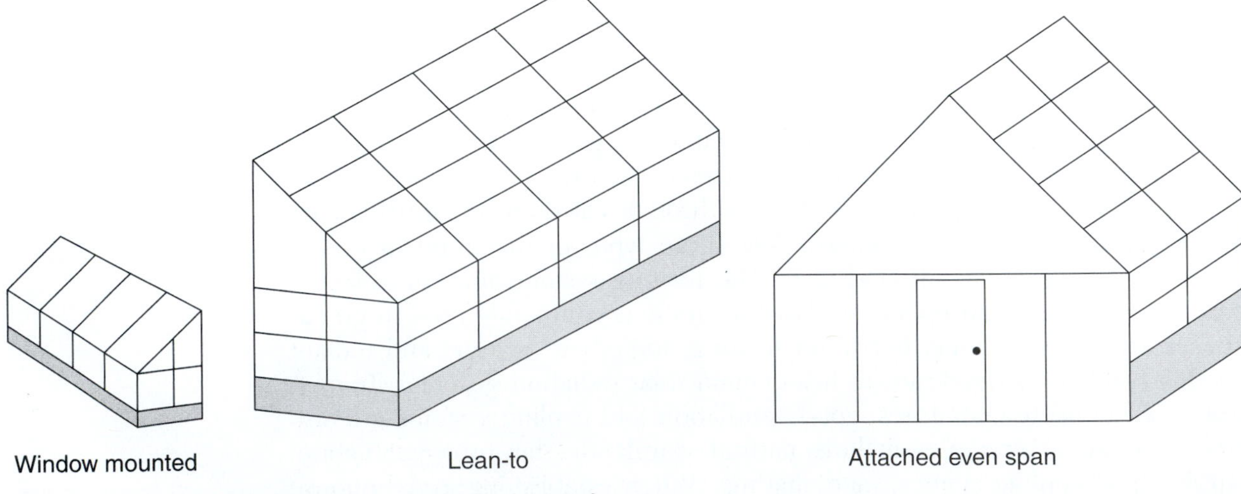

Figure 14–3 Three basic styles of attached greenhouses

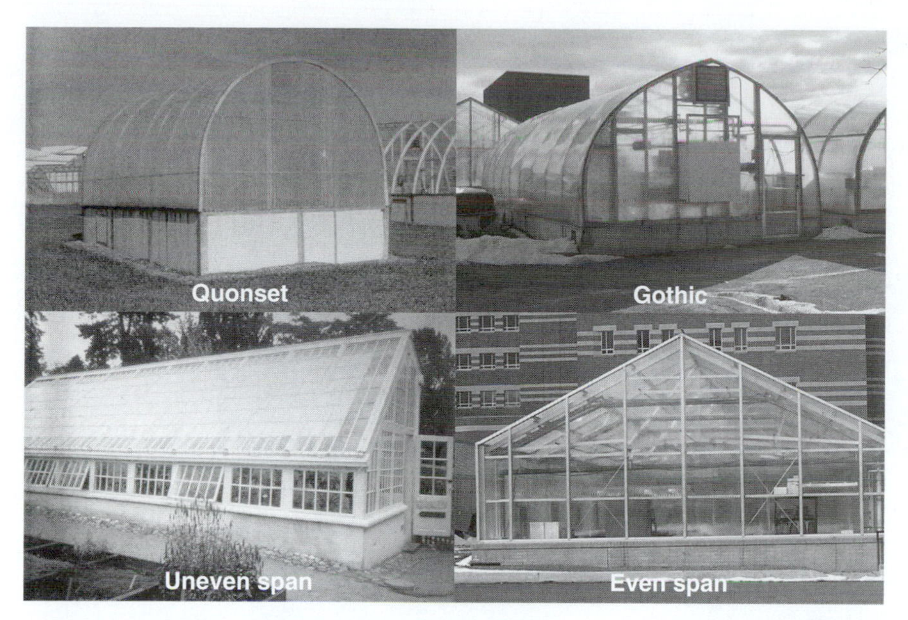

Figure 14–4 Four different styles of detached greenhouses. *Courtesy of Dr. E. Jay Holcomb, Department of Horticulture, The Pennsylvania State University.*

span, uneven span, Gothic, and Quonset (Figure 14-4). The most common style of detached greenhouse is the detached even span, which has a symmetrical roof with slopes having an equal pitch and width. Another type is the detached uneven span, which has an asymmetrical roof with slopes having unequal pitch and width. This type of design is used when the land is located on a hillside or where the land slopes; it does not lend itself readily to large-scale use and is only used in specific cases. Greenhouse designs also can have an arched roof, such as with the Quonset and Gothic arch designs. The Quonset design is more commonly used than the Gothic design because it is easier to install. The main advantage of detached greenhouses is that environmental control (temperature, light, relative humidity, and so on) can be regulated easily and can be programmed to meet specific needs. The main disadvantage is that more land is needed, thereby increasing costs.

Connected greenhouses are actually several greenhouses joined together. Various styles of connected greenhouses include ridge-and-furrow, barrel-vault, and saw-tooth greenhouse ranges (Figure 14-5). Greenhouse ranges are two or more greenhouses located together. The ridge-and-furrow design has a potential problem from snow and ice accumulating in the gutters connecting the houses. This problem is typically taken care of by installing heating pipes to melt away any accumulation in these areas. The ridge-and-furrow design is best suited for large greenhouse production enterprises that require similar environments. When smaller projects require specific environmental conditions, this type of greenhouse must be partitioned. When Quonset greenhouses are connected, they are called barrel-vault greenhouses; when lean-to greenhouses are connected, they are called saw-tooth greenhouses.

Connected greenhouses have many advantages. The overall cost is less, workers and equipment can move throughout the greenhouses easily without going outside, less heat is required because there are more inside walls, and less land is needed. The main disadvantage to connected greenhouses is the difficulty with accommodating different crops when the crops require different environmental conditions, such as temperature, humidity, and light.

Figure 14–5 Three different styles of connected greenhouses. *Courtesy of Dr. E. Jay Holcomb, Department of Horticulture, The Pennsylvania State University.*

Small greenhouse construction information can be obtained from books and cooperative extension service agencies. However, large commercial greenhouse structures are typically contracted out to specific companies that produce prefabricated and preassembled greenhouses.

STRUCTURAL PARTS OF A GREENHOUSE

Prior to laying the foundation, the water and electrical lines should be located and placed. The foundation should be level and provide adequate structural support for the greenhouse. The material used for greenhouse construction must be strong, durable, easy to maintain, and inexpensive. The frame should also cast as little shadow as possible. In early designs, wood was used for framing the greenhouse because it was inexpensive; however, it was also found to be less durable so it has been almost completely replaced by metal frames. Iron frames are heavy, and they have the potential to rust, which requires continual painting to prevent rust formation. Aluminum frames, on the other hand, are

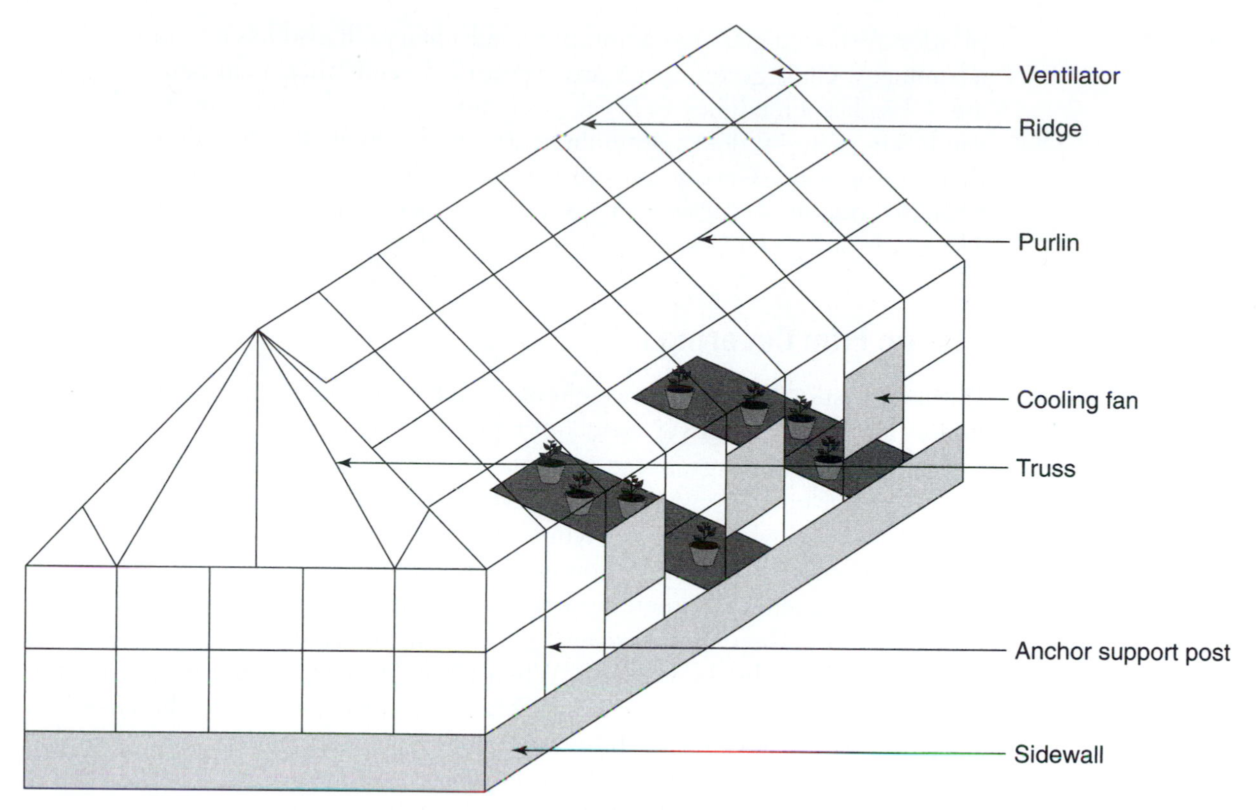

Figure 14–6 The structural components for a simple A-frame greenhouse

lightweight and rust resistant. The two basic frame designs are the A frame and the arched frame. The structural components for a simple A-frame greenhouse are shown in Figure 14-6. These components include the ridge (top of the greenhouse), anchor support posts (provide the main structural support), trusses (structural strength), purlins (structural strength), sidewalls (structural strength), ventilators (cooling), and cooling fans (cooling).

GREENHOUSE COVERINGS

The most important function of a greenhouse covering is to allow the maximum amount of light into the greenhouse for the growth and development of plants. No material is available that can transmit 100 percent of the light that strikes its surface. Light that strikes the surface of a greenhouse is either transmitted through or reflected back into the atmosphere. This section describes several types of greenhouse coverings.

Glass Coverings

Glass provides the best light transmission of any greenhouse covering by transmitting about 90 percent of the light that strikes its surface. Glass contains about 70 percent silica dioxide plus other oxides such as iron trioxide. The amount of light that glass transmits depends on its iron content; the higher the iron content, the lower the light transmission. Glass does not transmit light in the UV light range of the light spectrum. UV light does not reduce the transmission of light through glass over time as it commonly does with other materials, such as plastics. Although glass has a number of advantages, it also has a number of disadvantages. Glass greenhouses are expensive to construct, maintain, and operate. Glass has a tendency to break easily under normal conditions, is subject to vandalism, is heavy, does not retain heat well, and limits the greenhouse designs that can be used. Glass comes in several grades and weights, such as single-strength, double-strength, heavy-sheet, polished-plate, and heavy-plate glass. The grade and weight used will be determined by the grower's needs.

Plastic Film Coverings

Flexible plastic films are lightweight and can be used on a wider range of greenhouse designs than glass. A variety of types of flexible plastic films can be used as greenhouse coverings. Some examples of flexible greenhouse coverings used today are polyethylene, woven polyethylene, polyvinyl fluoride, polyvinyl chloride, ethylene-vinyl acetate copolymers, and polymethyl methacrylate. This flexible plastic film material is becoming increasingly more popular because it is inexpensive to put up a greenhouse with this type of material, and fuel costs are typically lower. The main disadvantage is the material's short life span, which is reduced considerably by UV light. When UV light block inhibitors are incorporated into the plastic film, the degradation is reduced; however, the coverings must still be replaced every three years. Another potential disadvantage is that condensation may build up on the inside surface of the greenhouse, which can lead to diseases and reduced sunlight entering the greenhouse.

Fiberglass-Reinforced Plastic Coverings

Fiberglass-reinforced plastic is a semirigid greenhouse covering that can be bent. This type of material is used on Quonset and even-span greenhouses. One of the biggest advantages of fiberglass-reinforced plastic is that it allows the same amount of sunlight through as glass. The fiberglass content disperses the light, which makes the intensity of the light transmitted into the greenhouse more uniform. Greenhouses with this covering material are also easier to cool than greenhouses covered with glass. Although a number of advantages are associated with fiberglass-reinforced plastic, its popularity has decreased in recent years because it is very susceptible to UV light, dust, and pollution degradation. In addition, it is flammable, so insurance rates may be higher.

Acrylic and Polycarbonate Coverings

Acrylic and polycarbonate are the most popular rigid plastic materials used today. Rigid plastic resembles glass panes when cut into pieces for covering greenhouses. These materials are available in single- and double-layer rolls. Rigid plastic and glass are the most expensive greenhouse coverings used today. Acrylic can last up to 25 years whereas polycarbonate has an expected life of between 10 and 15 years. Both acrylic and polycarbonate greenhouse coverings have almost completely replaced the use of fiberglass-reinforced plastic and glass because they are lightweight and easy to install. They require less support and are more adaptable to wider sash-bar spacing, which reduces the amount of shading. In addition, they provide very good light transmission and excellent heat insulation. The main disadvantage of these materials is that they are flammable.

FACTORS INVOLVED IN LOCATING A GREENHOUSE RANGE

Success in the production of plants in greenhouses starts with good planning. When selecting a site for a greenhouse range, it is important to consider as many factors as possible. The following are important factors to consider when deciding on the proper location for a greenhouse range (Biondo, 2004; Nelson, 1985):

- **Market.** Identify the size of the potential market and the distance from your production site.
- **Accessibility.** The greenhouse range should be located as close as possible to the potential market. Retail outlets should be close to the primary customer base. If possible, the greenhouse range should be highly visible to the general public, for example, located on or near a major highway. This also places the greenhouse closer to pollution and potential vandalism, however. Wholesale outlets need reliable and convenient transportation between the greenhouse site and markets. Greenhouses should be highly accessible to receive large trucks delivering soil mixtures, fertilizers, and other materials. The greenhouse operation should also be accessible to customers by having good roads and parking facilities.
- **Climactic conditions.** Consider the environmental conditions, such as the amount of sunlight, temperature (highs and lows), amount of snowfall, areas of high winds, and other conditions that will affect the greenhouse operation.
- **Topography and drainage.** The topography profoundly affects drainage. Greenhouses use high amounts of water and must be at a location with adequate soil drainage. The land selected must have a minimum slope, which makes greenhouse operations much easier.
- **Water and other utilities.** Because large quantities of water are used in greenhouses, a reliable source of good-quality water must be available year round. Consider whether the water at this location is suitable to the crops that will be grown. Higher levels of fluoride, chlorine, and salts may adversely affect certain crops. A nearby source of dependable fuel for heating and electricity for providing artificial light is also vital. As part of an active greenhouse operation, a large amount of waste is generated so the area needs access to a dependable waste disposal service. Dependable telephone service is also a must to remain accessible to the customer at all times.
- **Zoning regulations.** Zoning regulations are important to know for a particular location regarding agricultural production and building codes.
- **Labor supply.** Both skilled and unskilled labor should be available.
- **Expansion.** Adequate room should be available for future expansion.

LAYOUT OF THE GREENHOUSE RANGE

After the location of the greenhouse is determined, planning the layout of the greenhouse range is critical for maximizing efficiency.

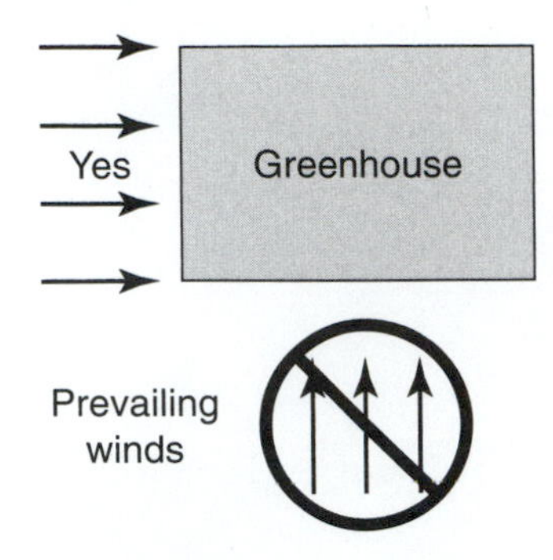

Figure 14–7 Orientation of the greenhouse to minimize heat loss

Greenhouse Range Orientation

The range must be oriented to allow for maximum light intensity to enter the greenhouse. Knowing the direction of the prevailing winds also helps to orient the greenhouse to minimize heat loss (Figure 14-7). After the greenhouse is oriented to maximize light and minimize heat loss due to wind, the areas around the greenhouse should have entrances and drives that are accessible to large trucks, customers, and employees.

Bench Orientation in the Greenhouse

Plants are grown in greenhouses either in ground beds or on benches. Ground beds are in the ground, and production takes place directly on the floor of the greenhouse. With benches, the plants are on raised platforms. The design and layout for these methods should accommodate greenhouse operations and make efficient use of greenhouse space. Three common types of bench arrangements (Figure 14-8) are as follows:

- **Longitudinal bench arrangement.** For this type of arrangement, beds or benches are constructed to run the full length of the greenhouse in several rows. Typically this type of arrangement is used for fresh cut flower production. This arrangement allows for mechanization, but it is difficult for employees to move across the greenhouse.
- **Cross benching.** This type of arrangement is similar to longitudinal except that the benches are arranged crosswise. With this type of arrangement, the benches are shorter creating more aisles. This arrangement

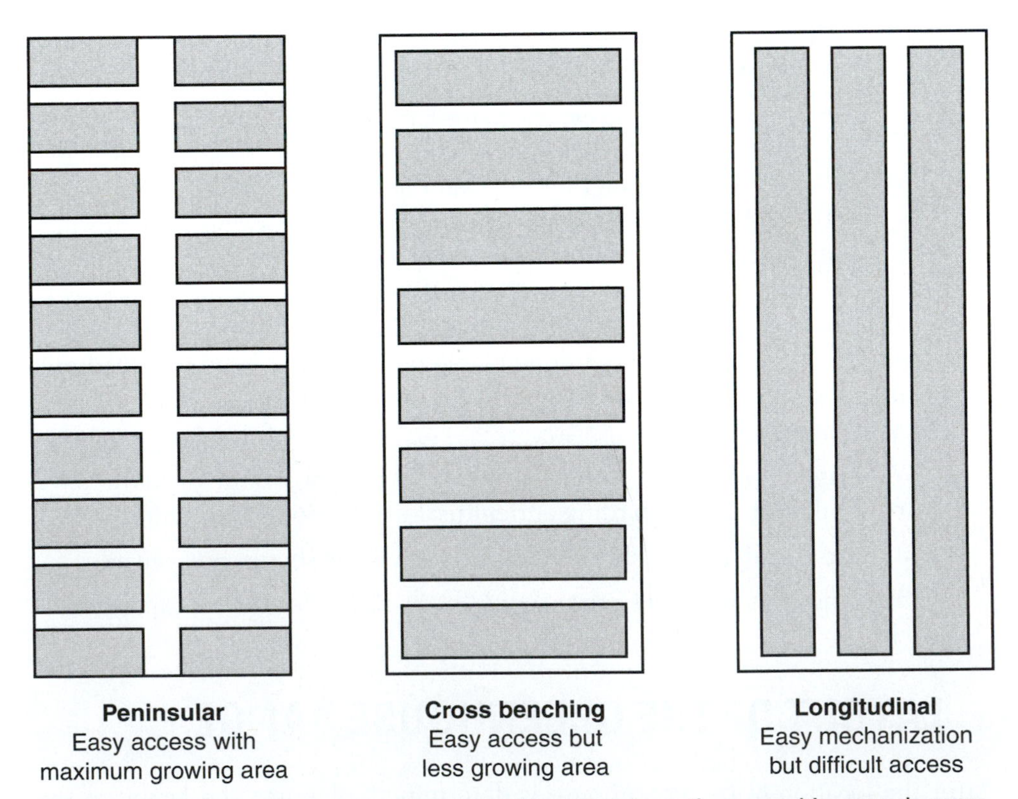

Figure 14–8 The three common types of bench orientations used in greenhouses

provides for easy accessibility to all benches but does not maximize growing space as the extra aisle space reduces efficiency.

- **Peninsular.** The peninsular bench arrangement has one central aisle that runs the entire length of the greenhouse and provides accessibility to employees while maximizing the growing area.

Types of Benches

Most greenhouses use raised benches in a wide range of designs and materials. The two main categories of raised benches are fixed and moveable benches (Figure 14-9). Fixed benches can be produced from many materials and can be temporary, consisting of cinder blocks for legs and moveable bench tops, or they can be permanent (Figure 14-10). The top of the bench may be wood, concrete, wire mesh, or molded plastic, which is used for ebb and flow watering systems (Figure 14-11). The bench must have a system that enables it to drain properly. If wood is used, it must be treated to prevent rot, and any construction materials, such as nails, must be treated to prevent rust. The height of the bench should permit cultural practices to be done with ease. In addition, the bench should be narrow enough (approx. 3 to 6 feet) to enable workers to reach pots

Figure 14–9 Moveable benches

Figure 14–10 Fixed benches

Figure 14–11 Ebb and flow bench. *Courtesy of Dr. E. Jay Holcomb, Department of Horticulture, The Pennsylvania State University.*

located in the middle of the row. The bench should also allow for good air circulation. Moveable benches allow the grower to change the bench arrangement as needed and are popular in container production enterprises. To maximize space, moveable benches are used with one central aisle; the benches are close together except when working on the bench. When work is completed, the bench is moved to permit enough space for the worker to complete the operation on the next bench (Figure 14-12).

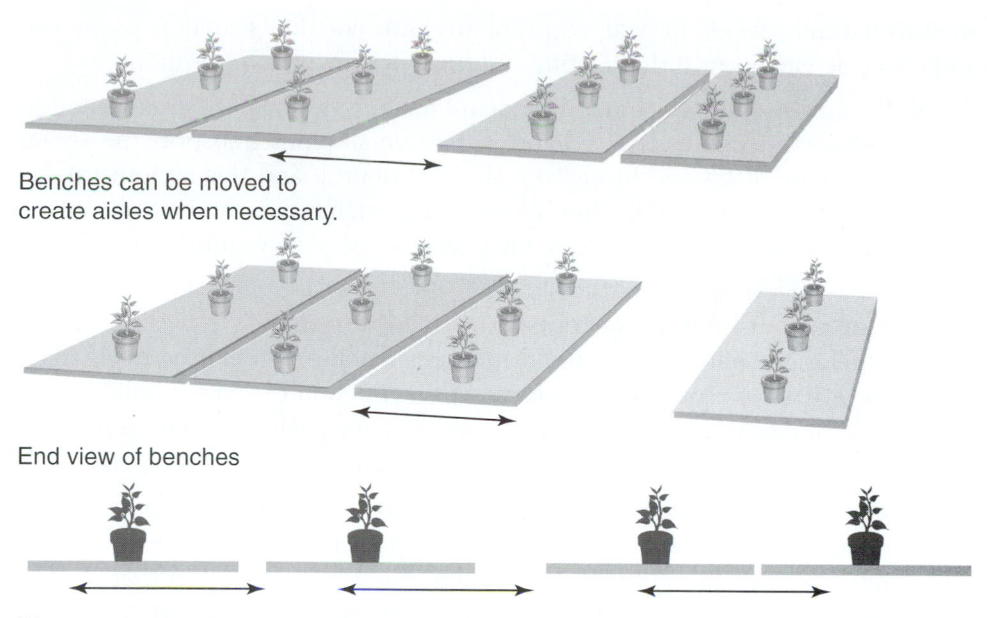

Figure 14–12 Moveable bench designs

CONTROLLING GREENHOUSE TEMPERATURE

The main reason for using greenhouses is for controlling temperatures, and good temperature control is vital to maintaining a high-quality crop. Mimicking the outdoor environment maximizes the growth of plants in greenhouses. The day-time temperature in greenhouses is typically 10 to 15 degrees higher than night-time temperatures, which are generally between 55 and 65°F. In a common greenhouse operation, labor is the most expensive operational expense; however, this is closely followed by the cost of maintaining temperature. Understanding how heat can be lost from greenhouses helps the grower control temperature in a cost-effective and efficient manner. Greenhouses experience heat loss through conduction, infiltration, and radiation. Conduction is heat loss by transmission through a greenhouse covering, infiltration is heat loss through cracks or holes that occur in a structure, and radiation is the loss of heat from a warm surface, such as a plant leaf, to a cooler surface, such as the greenhouse covering.

The three types of fuels used to heat greenhouses are natural gas, oil, and coal. Natural gas has the advantage of burning clean, and can be delivered to a greenhouse via pipes, which eliminates the need for a storage facility and delivery costs. Fuel oil is also commonly used, but has the disadvantage of requiring storage tanks and must be delivered. An additional problem with the use of oil is that the viscosity of oil is affected by temperature and oil does not flow properly at low temperatures when heat is required the most. The least commonly used form of fuel is coal because of the pollution produced when coal is burned. When choosing the fuel to be used at a greenhouse range, important considerations include availability, cost for delivery and transportation, storage, and any other special needs required by the fuel.

Heating Systems Commonly Used in Greenhouses

Heating systems used in greenhouses are steam heating, hot water heating, forced-air heating, and radiant heating (such as infrared radiant heaters and solar radiation systems). Each of the systems, except radiant heating systems, uses fuel such as natural gas, oil, or coal, so problems with pollutants such as propylene and ethylene can occur if the heating system is not maintained properly.

- **Steam heating systems.** For steam heating systems, water is heated to temperatures between 212 and 215°F. Small pipes transport the steam over long distances throughout the greenhouse and this system can be efficiently used in large greenhouses. Although steam can be transported long distances, it condenses on pipes, so provisions must be made to drain and recirculate water for reheating.
- **Hot water heating systems.** This type of system is typically used in smaller greenhouses. The disadvantage is that the system used to circulate the hot water requires large pipes to bring the water from the boiler to the greenhouse and back. The system is adapted for small greenhouses because it is difficult to transfer water over long distances without losing heat.
- **Forced-air heaters.** This type of system, commonly used in commercial greenhouse ranges, uses localized heater units that force air directly into the greenhouse or through long perforated plastic tubes that run the length of the greenhouse. A problem associated with this type of system is the dryness of the heat produced.
- **Infrared radiant heaters.** The major advantage of this system is that it conserves energy and has no problems with pollutants. A disadvantage is the initial higher cost in comparison with other systems. In addition, monitoring the actual temperature the plant is receiving may be difficult because air temperature is not a good indicator. This type of heater warms plants and other objects in the greenhouse, but it does not warm the air temperature to the same extent. Another potential disadvantage is that the sources of radiation must be located directly above the plant, or pockets of low temperature may occur.
- **Solar radiation systems.** The main advantages of this system are that it conserves energy and creates no problems with pollutants. The initial cost of solar collectors is high; however, once installed, they are cost effective. The major disadvantage of solar radiation systems is their dependency on weather, as clouds limit their effectiveness.

Ventilation and Cooling Systems

Greenhouses need to be ventilated and cooled throughout the year. During the summer and even the winter months, the air temperature in the greenhouse may become too high and inhibit plant growth. In addition to cooling, ventilating the greenhouse renews the supply of carbon dioxide needed for photosynthesis, circulates the air to reduce the chance of diseases, and prevents the buildup of pollutants such as ethylene. The following are commonly used ventilation systems in greenhouses.

- **Natural ventilation system.** Air is exchanged through open ridge and side vents (Figure 14-13). This is one of the oldest methods of cooling. Chilling injury to plants located near the side vents might occur. When using this system, screens are necessary to prevent unwanted pests from entering the greenhouse.
- **Fan-tube ventilation system.** The air is distributed through a plastic tube with holes that run the length of the greenhouse. This system is used in conjunction with the heating system but can also be used for circulating cool air (Figure 14-14).

Figure 14–13 Natural ventilation system using side and ridge vents

Figure 14–14 Fan-tube ventilation

Figure 14–15 Fan and pad ventilation cooling system

- **Fan and pad cooling system.** Creates summer cooling by drawing air through pads soaked with water (Figure 14-15). This system cools based on the principle of evaporative cooling.
- **Fog evaporative cooling system.** This system uses a fog-generation system inside the greenhouse; this is similar to the system used to cool football players on the sideline of a game (Figure 14-16).
- **Shading.** Shading can be accomplished by spraying on whitewash to reflect light and reduce light intensity. Liquid shading compounds are often applied to glass as well as polyethylene-covered greenhouses to reduce light intensity and temperature in the spring and summer (Figure 14-17). Other times of the year when heat buildup is not a problem, this material is washed off to allow more light into the greenhouse. A screen made from fabric is another way to reduce heat if only a small section of the greenhouse requires shading (Figure 14-18). Fabrics can be used to shade from 20 to 90 percent; however, 50 percent is the most common. Some greenhouses also have retractable shades that are used as needed.

Figure 14–16 Evaporative cooling system. *Courtesy of Dr. E. Jay Holcomb, Department of Horticulture, The Pennsylvania State University.*

Figure 14–17 Greenhouse with shading material on it to reduce heat

Figure 14–18 Shading material is used above plants to shade and cool them.

SUMMARY

You now have a basic knowledge of greenhouse structures and how to establish a successful greenhouse range for growing plants. You learned about the advantages and disadvantages of the different basic greenhouse structures used today and how the structural components of a greenhouse serve specific functions. You also learned the four major classes of greenhouse coverings commonly used together with their advantages and disadvantages. Important factors involved in selecting a location for a greenhouse range include market, accessibility, climactic conditions, topography, drainage, water, utilities, zoning regulations, labor supply, and land for expansion. Properly orienting the range and understanding the different types of bench arrangements and types of benches is also important. Accurately controlling greenhouse temperatures using heating, ventilation, and cooling systems is vital to a successful greenhouse.

Review Questions for Chapter 14

Short Answer

1. What civilizations were the first to use greenhouse structures?
2. Where can greenhouses commonly be found today?
3. What are four factors that affect the grower's choice when selecting the best size and type of greenhouse?
4. What are four key factors that should be considered when locating a greenhouse range?
5. What are three basic greenhouse design types and three examples of each?
6. What are the advantages and disadvantages of freestanding and connected greenhouses?
7. Prior to laying the foundation for a greenhouse, what should be done?
8. List the six major structural components of the greenhouse and provide a function for each.
9. What is the most important function of a greenhouse covering?
10. List the four major types of greenhouse coverings and provide advantages and disadvantages of each.
11. What eight factors should be considered when selecting a site for a greenhouse range?
12. What are three factors to consider when orienting a greenhouse range on a piece of land?
13. List three commonly used types of bench arrangements found in greenhouses, and then provide the benefits of each.
14. What are three ways that heat can be lost from greenhouses?
15. What are three types of fuels used in greenhouses?
16. List four types of heating systems used in greenhouses and provide the advantages and disadvantages of each.
17. What are four types of cooling systems used in greenhouses?
18. In addition to cooling, what are three advantages provided when ventilating a greenhouse?

Define

Define the following terms:

greenhouse range
headhouse
attached greenhouse
freestanding greenhouse
connected greenhouse
conduction
infiltration
radiation

True or False

1. Early Egyptian and Roman civilizations used greenhouse structures to protect tender crops such as fruits and vegetables during the winter months.
2. A headhouse is a central building that is used for offices, storage, and workspace without attached greenhouses.
3. The greenhouse design types and sizes used depend upon space available, geographical location, types of plants to be grown, and cost of construction materials.
4. Attached greenhouses should be included in the initial design of the home or workplace, or problems can occur.
5. Freestanding greenhouses use less land.
6. Freestanding greenhouses have better environmental control than connected greenhouses.
7. An advantage of connected greenhouses is that they are more efficient because workers and equipment can move throughout the greenhouse with ease.
8. A major disadvantage of connected greenhouses becomes clear when growing crops that require different environmental conditions.
9. Water and electrical lines should be located and placed before the foundation for a greenhouse is constructed.
10. Trusses and purlins are used for structural strength in greenhouses.
11. Glass is the best greenhouse cover because it is cheap and allows maximum light transmission.
12. Flexible plastic films are lightweight and have a number of other advantages, but they cannot be used on a wide range of greenhouse types.
13. Greenhouses covered with fiberglass-reinforced plastic are easier to cool than greenhouses covered with glass.
14. Longitudinal bench arrangements in the greenhouse are typically used because this design gives employees greater accessibility and maximizes the growing area.
15. Cross-benching arrangements in the greenhouse are typically used for cut flowers because such arrangements allow for mechanization.
16. The peninsular bench arrangement provides accessibility to employees while maximizing the growth area.
17. Most greenhouses use raised benches in a wide range of designs and materials.
18. When greenhouses are cooled using a natural ventilation system, chilling injury to plants can occur.
19. Greenhouse temperatures can be reduced by using shading materials on the greenhouse.
20. Shading materials are often applied to glass as well as polyethylene greenhouses in the spring and summer to reduce the temperature within the greenhouse.
21. One of the disadvantages of infrared radiant heaters is that it may be difficult to monitor air temperatures accurately because this type of heater warms plants and other objects in the greenhouse but not the air temperature to the same extent.
22. The major advantage of infrared heaters is that they conserve energy; however, there are problems with pollutants.

Multiple Choice

1. Greenhouse structures are built or manufactured in many design types and sizes. The type used depends on:
   A. space available.
   B. type of plants to be grown.
   C. cost.
   D. All of the above
2. Which of the following are common styles of connected greenhouses?
   A. Dutch houses
   B. Sawtooth greenhouses
   C. Barrel-vault greenhouses
   D. All of the above
3. Which of the following are basic greenhouse designs?
   A. Attached
   B. Freestanding
   C. Connected
   D. All of the above
4. Which of the following is an example of a freestanding greenhouse?
   A. Saw tooth
   B. Barrel vault
   C. Gothic arch
   D. All of the above
5. Which of the following is not an example of a connected greenhouse?
   A. Saw tooth
   B. Dutch house
   C. Quonset
   D. Barrel vault
6. The most important function of a greenhouse covering is to:
   A. be strong enough to withstand vandalism.
   B. provide good support while being inexpensive.
   C. allow maximum light into the greenhouse.
   D. None of the above
7. Which of the following greenhouse covers has the disadvantage of condensation building up on the inside surface?
   A. Fiberglass-reinforced plastic
   B. Polyethylene
   C. Acrylic
   D. Polycarbonate
8. Which of the following greenhouse covers allows in the same amount of light as glass?
   A. Acrylic
   B. Polyethylene

C. Fiberglass-reinforced plastic
D. Polycarbonate

9. Which of the following are common types of bench arrangements used in greenhouses?
   A. Longitudinal
   B. Cross benching
   C. Peninsular
   D. All of the above
10. In addition to cooling, ventilation of a greenhouse provides which of the following?
   A. A reduction in diseases
   B. Ethylene buildup prevention
   C. A renewed supply of CO_2
   D. All of the above

Fill in the Blanks

1. Properly orienting the greenhouse range is important to allow maximum ______________ to enter the greenhouse.
2. It is important to know the direction of the ______________ to orient the greenhouse to minimize heat loss.
3. Most of the heating systems used in greenhouses use oil or propane to fuel the boiler, which may cause problems with pollutants such as______________.
4. Label the structural components found in the greenhouse shown below.

Terms: Ridge, ventilator, purlin, side vent or cooling fans, sidewall, anchor support post, truss

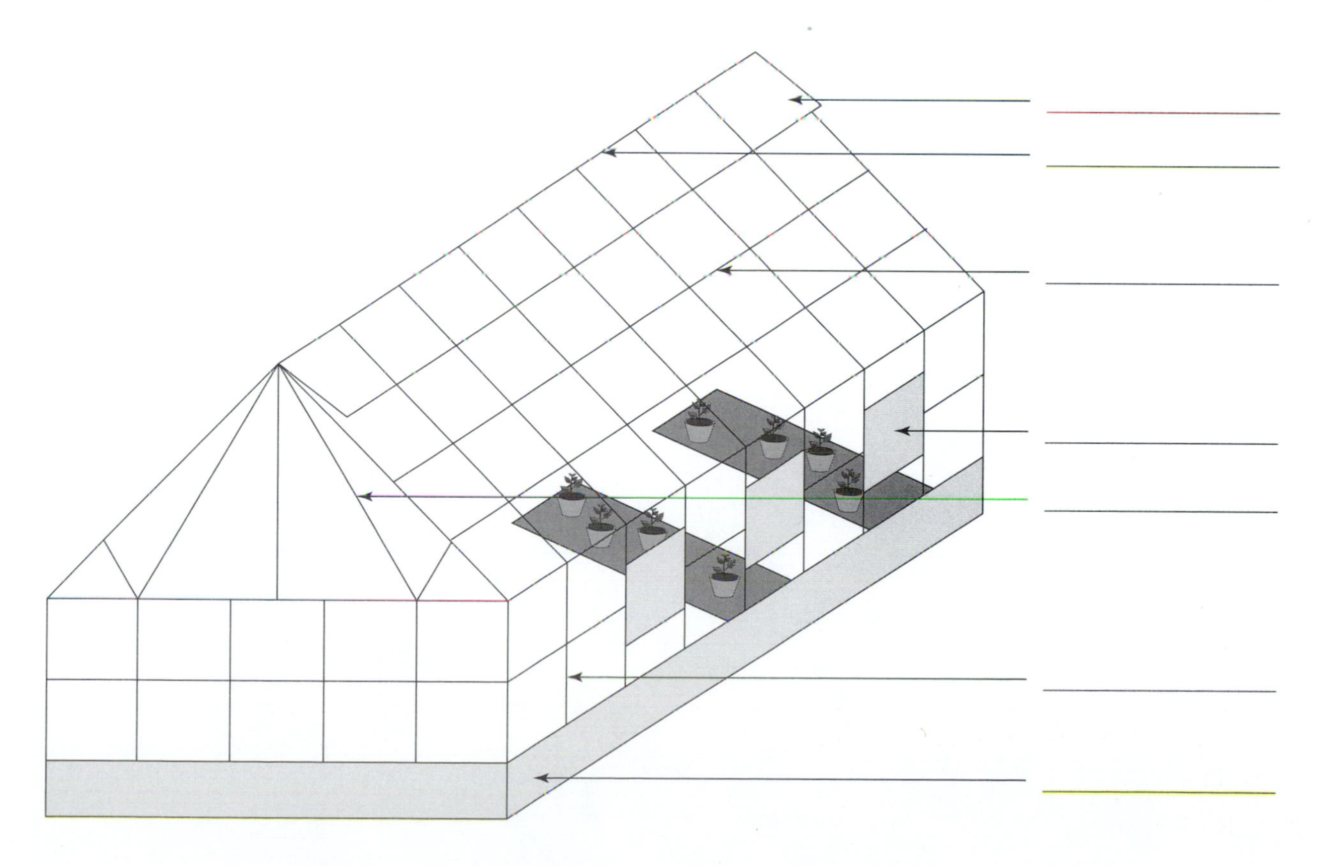

Activities

Now that you understand greenhouse structures, you will have an opportunity to explore them in more detail. In this activity, you will visit a local greenhouse and gather the following information:

- the type of greenhouse design.
- the type of greenhouse covering.
- the heating and cooling systems.
- the irrigation system(s).
- the type of bench arrangements.
- the type of plants grown.

In addition to the information you were asked to gather, include your thoughts on the rationale behind how the greenhouse is set up. If you do not have access to a greenhouse, surf the Internet for two sites that present information on greenhouse structures and summarize these; be sure to provide the address for the Web site where you found your information.

References

Biondo, R. J. (2004). *Greenhouse production*. Upper Saddle River, NJ: Prentice Hall.

Boodley, J. W. (1981). *The commercial greenhouse handbook.* New York: Van Nostrand Reinhold.

Nelson, P. V. (1985). *Greenhouse operation and management*. Reston, VA: Reston Publishing.

CHAPTER

15

Growing Crops in the Greenhouse

ABSTRACT

Growing ornamental crops in the greenhouse is becoming more technical and specialized as consumer demands for quantity and quality increase. Five environmental factors should be taken into consideration when growing plants in the greenhouse. The different types of containers for growing greenhouse crops include rooting, bedding plant, and green and flowering plant containers. The growing media also play an important function. Three important issues to consider when watering plants include timing, amount to apply, and careful watering. Seven types of irrigation systems are commonly used in greenhouses. Plants grown in the greenhouse are typically grown in soilless media contained in pots and thereby require fertilization, especially micronutrients. Fertilization of greenhouse crops is typically accomplished using inorganic fertilizers. Organic fertilizers are not typically used because applying exact concentrations of nutrient elements is difficult and bad odors are produced. Greenhouse integrated crop management includes different ways to prevent pests from entering the greenhouse together with the most effective way to get rid of pests after they have entered the greenhouse.

Objectives

After reading this chapter, you should be able to

- discuss environmental factors that should be considered when growing plants in a greenhouse environment.
- discuss what factors should be considered when selecting containers for crops.
- provide the three types of containers used for growing plants in greenhouses.
- discuss the important functions and components of growing media.
- discuss important considerations when watering plants.
- recognize the different types of irrigation systems used in greenhouses.
- understand how plants are fertilized in a greenhouse environment.
- list five different ways to prevent pests from entering greenhouses and describe how to get rid of them after they have entered the greenhouse.

Key Terms

azalea pot
bulb pan
day-neutral plants
DIF
hanging pots
inorganic fertilizer
light duration
light intensity
light quality
long-day plants
negative DIF
organic fertilizer
peat pellets
peat plugs
peat pots
photoperiodism
plastic foam
positive DIF
short-day plants
standard size pot
tensiometers
thermoperiodism
thermotropism

INTRODUCTION

This chapter surveys general practices used in producing crops in the greenhouse (Figure 15-1). To promote maximum crop growth and development, an optimal greenhouse environment must be maintained. Environmental factors that must be carefully controlled are temperature, light (including **light quality, intensity,** and **duration**), air, humidity,

Figure 15–1 A greenhouse environment was used to promote maximum crop growth and development of these regal geraniums.

and water. Environmental controls for heating and cooling can be very simple, such as using a basic thermostat, or they can be as complex as using a computer. In general, the night temperatures in the greenhouse should be kept between 55 and 65°F with the day temperature being 10 to 15°F warmer. In addition to optimizing shoot temperatures in the greenhouse, it is also important to optimize the root zone temperature. Supplying supplemental bottom heat reduces the need for higher above-ground temperatures in the greenhouse, thereby conserving energy. When deciding what form of supplemental lighting to use, the light quality, intensity, and duration required by the crops that will be grown must be considered.

In addition to maintaining the proper temperature and light levels, the atmospheric composition of the greenhouse must also be controlled. In a well-ventilated greenhouse, the addition of supplemental carbon dioxide is typically unnecessary. However, during the heating season when greenhouse vents are typically kept closed, carbon dioxide levels may become limiting, which makes it important to add supplemental carbon dioxide. Sufficient light must also be available when supplemental carbon dioxide is being added. Another problem that can occur in a greenhouse during the winter months is the buildup of unwanted gasses. This problem can be overcome by properly servicing the heating system to make sure it runs at peak efficiency and, in some instances, using air purification units. Maintaining the proper relative humidity is also very important in a greenhouse environment. If relative humidity is too low, the plants become stressed; if it is too high, plants are susceptible to leaf and flower diseases. Successful greenhouses also require good-quality water with low levels of contaminants—such as salt, chlorine, fluoride, or others—to minimize problems when growing plants.

In addition to optimal environmental conditions in the greenhouse, it is important to select a container that will promote maximal growth. Both advantages and disadvantages are associated with the different types of available containers. The three main materials used to produce pots are clay, plastic, and styrofoam. In the past, clay was the most popular; however, plastic pots are the most commonly used today. Clay pots are porous, which enables water to evaporate from the pot to avoid waterlogging. Clay pots are also heavier, stronger, and can support large plants without falling over. The main disadvantage is that clay pots are more expensive than plastic pots. Plastic pots are the most commonly used today because they are inexpensive and come in a variety of sizes and shapes. They are nonporous, so waterlogging can be a problem. Styrofoam containers have become popular because they insulate well to keep the soil warm, and they are lightweight and come in a variety of sizes and shapes. Two disadvantages are that they break easily and are difficult to clean. Today, the grower has many choices in containers and must carefully weigh the advantages and disadvantages to decide what will work best in a particular situation.

A variety of containers are available for different uses, which can be broken down into the following three groups: rooting containers, bedding plant containers, and green and flowering plant containers. Rooting containers come in organic and inorganic forms. Organic forms include **peat pellets, peat plugs,** and **peat pots.** Inorganic forms are made of **plastic foam** and are compressed plastic or inert inorganic materials in the form of a cube or block similar to peat plugs. Bedding plants are generally grown in bedding plant containers or cell packs, which can vary in size but typically come in sets of two, four, six, or more cells. The main advantage of bedding plant containers is that they lend themselves to mechanization. However, the main disadvantage is that the cells are typically small and plants may become root bound if left too long, which may lead to transplant shock. Most green and flowering plants are grown in plastic containers because they are less expensive. The four different types of pots commonly used are **standard-size pots, azalea pots, bulb pans,** and **hanging pots.** The type of pot used depends on the growth habit of the plant.

Growing media used for potting plants is commonly used to accommodate specific needs such as mixes used solely for promoting germination. Although differences exist in the types of growing media on the market, all mixtures should be reproducible, available, and have basic physical, chemical, and biological properties. Some important functions that all growing media should include are good moisture- and nutrient-holding capacity, physical properties that permit rapid water infiltration, aeration, and drainage. A good growing medium should contain materials that do not rapidly decompose, flow easily to facilitate rapid pot filling, and be free of toxins. In addition, growing media should provide support for the plant and have good cation exchange capacity (CEC) and a pH in the range between 5.5 and 6.0 or according to the needs of the specific plant being grown. A good growing media has both organic and inorganic components. Some key components in growing media are peat moss, wood byproducts, bark, and vermiculite, which are used because of their high moisture- and nutrient-holding capacity, while sand and perlite are used to promote good aeration and drainage. Selecting the proper materials for the preparation of growing media maximizes plant growth and development. Factors that should be taken into consideration are the quality of the ingredients, reproducibility and availability, cost, use, and ease of preparation.

Watering plants is a major responsibility because most horticultural crops are between 90 and 98 percent water. Both underwatering and overwatering can adversely affect the quality of a crop. Overwatering is the most common problem, which leads to oxygen deprivation in the root zone and ends in root rot. Important considerations for watering plants include timing, amount to apply, and watering carefully. Plants can be watered in many ways, so it is important to choose one that works best for the particular purpose of the greenhouse. Seven types of irrigation systems are commonly used in greenhouses; the oldest type is manually watering with a hose. Automated systems have been developed that differ in cost, efficiency, and flexibility and include tube irrigation, capillary mat, ebb and flow, overhead irrigation, perimeter irrigation, and soaker hose irrigation.

Plants grown in a greenhouse are generally grown in restrictive pots, with less growing media than when plants are grown in the field. Plants grown in the greenhouse typically use soilless growing medium, which requires fertilization, especially with micronutrients. The two types of fertilizers that can be used are organic and inorganic. **Organic fertilizers** are typically used by specialty growers and are not used in greenhouses for a variety of reasons. For instance, organic fertilizers cannot be easily adapted to automated watering systems, and applying the exact amounts of nutrient elements is difficult. Organic fertilizers also produce volatiles that have bad odors, making their use in enclosed conditions unpleasant, and these volatiles can also have adverse effects on plant growth. **Inorganic fertilizers** are synthetic nutrient compounds that are derived from mineral salts. They are convenient to apply using a variety of application methods. Another benefit is that they can be applied in exact amounts as needed. Both dry and liquid fertilizers can be used in greenhouses; however, liquid forms are most commonly used because they can be applied through the irrigation system and are immediately available. Dry formulations come as fast release and slow release. Fast release, as the name implies, is available immediately and can be used to take care of a deficiency problem. Slow-release fertilizers are commonly used in greenhouses because they are released over a long period of time. Media and foliar testing should be done to avoid deficiency problems in greenhouse crops. This is important because as soon as visual symptoms are observed, the plant is being stressed, which reduces quality and yields.

All the necessary precautions must be taken to prevent pests from entering the greenhouse. Many practices are often used in greenhouses that would be very difficult under field conditions. One way to prevent pests from entering greenhouses is by putting screens on vents and taking precautions to minimize pests from entering on the shoes or clothing of workers. Another way is to practice weed control outside the greenhouse, thereby minimizing their entry. Sanitation coupled with crop inspection is also effective for minimizing disease incidence in greenhouses. If possible, growers can manipulate the environment to adjust conditions so that they are not favorable for pests. If pests enter the greenhouse, biological control works well unless the population of pests gets too high, in which case a combination of chemical, genetic, cultural, and mechanical control should be used.

GREENHOUSE ENVIRONMENT

The greenhouse environment must be optimal to promote maximum crop growth and development. Environmental factors that must be carefully controlled to mimic the outdoor environment are explained next (Biondo, 2004; Nelson, 1985).

Temperature

Maintaining the proper temperature in greenhouses is critical to the success of the crop being grown. Two important terms to be familiar with when exploring the literature on temperature effects in plants are **thermotropism,** which is the plant's response to temperature, and **thermoperiodism,** which is the plant's response to changes in day and night temperature. Environmental controls for cooling or heating greenhouses can be simple or complex. The more complex systems can control heating and cooling plus a variety of other environmental conditions, such as fertilization, irrigation, light, humidity, and air circulation. In general, a night temperature of 55 to 65°F and a day temperature 10 to 15°F higher than the night temperature must be maintained for most greenhouse crops. However, this is just a rule of thumb because different plants have different requirements. Green foliage plants and some orchards require night temperatures as high as 70°F, whereas roses, mums, and azaleas do best at a night temperature of 63°F and carnations, snapdragons, cinerarias, and calceolarias do best at a night temperatures of 50°F. Therefore, growers must be familiar with their particular crops. Traditionally, greenhouse growers grew plants at lower night temperatures to reduce respiration, thereby conserving carbohydrates produced during the day. Interestingly, scientists have found that growing plants at a lower temperature during the day is an effective strategy for reducing growth and elongation of plants. The concept of using the differences in day and night temperatures to modify plant height is called **DIF** (day temperature minus night temperature). A **negative DIF** occurs when the day temperature is lower than the night temperature; it can be used in the same way as growth retardants are used to reduce internode elongation. A **positive DIF** occurs when the day temperature is higher than the night temperature promoting internode elongation. Although many species do not show any adverse effects of negative DIF, the potential still exists for problems so the DIF range should be less than 8°F to prevent any problems.

Both the root and shoot temperature are equally important for optimal plant growth. Researchers have shown that supplying bottom heat to the plant reduces the need for higher air temperature in the greenhouse and conserves energy.

Light

The quality, quantity (intensity), and duration of sunlight affects plant growth, so all three factors must be taken into consideration when deciding what form of supplemental light to provide.

Light quality refers to the wavelength or color of light. Light quality has a dramatic effect on photosynthesis and can also affect plant shape. For photosynthesis, plants efficiently use the light spectrum in the red and blue range. Both red and far-red light dramatically affects plant growth. Plants grown under high levels of far-red light have long internodes, are less branched, and are taller than plants grown under lower light levels; whereas plants grown under high levels of red light have less internode elongation, are shorter, and are typically darker green.

Light intensity is the actual quantity of light being supplied. Maximum light intensity in greenhouses can be achieved in several ways. First use greenhouse coverings that allow maximum light transmission. Properly designing and planning the greenhouse or range layout and routinely cleaning and maintaining greenhouse coverings also maximizes light intensity. The higher the light intensity, the greater the rate of photosynthesis up to a point where light saturation occurs. Supplemental lighting in greenhouses is commonly used to maximize growth. The most common form of supplemental lighting used in greenhouses is high-intensity discharge (HID) lamps because they provide maximum light intensity and light quality (Figure 15-2). High light intensity promotes shorter plants, whereas a lower light intensity promotes taller plants.

Figure 15–2 A high-intensity discharge lamp provides supplemental light for sunflower plants.

In some cases, light intensity must be reduced to release heat buildup in the greenhouse. This is accomplished by applying shading compounds or installing shading fabric. Shading material is used only during spring and summer months when light levels are high, and it must be washed off for the fall and winter months when light levels are typically low.

Light duration involves the length of exposure to light in a 24-hour period. In order to simulate the outdoor environment, plants are exposed to different day/night regimes.

Photoperiodism is the total light energy that affects growth and development of plants. The three photoperiodic classes of plants are day-neutral, short-day, and long-day plants. **Day-neutral plants** (geranium) flower in response to the genotype, which basically means that the plant flowers based on its genetics. **Short-day plants** (chrysanthemums, azaleas, and poinsettias) flower when the dark period is longer than a specified length. Short-day crops grow vegetatively when grown under long days and flower under short days. The length of the dark period is critical because even brief flashes of light during this period will modify the photoperiod. **Long-day plants** (shasta daisies and *Gypsophila*) flower when the dark period is shorter than a specified length.

Air

Photosynthesis is a series of chemical reactions in which carbon dioxide and water are converted in the presence of light to carbohydrates and oxygen. Carbon dioxide is found in the atmosphere at low concentrations—typically around 350 ppm. In greenhouses that are well ventilated, carbon dioxide is typically not limiting. However, during the winter months when greenhouse vents are closed for more efficient heating, carbon dioxide levels can become limiting, which causes a reduction in photosynthesis and thus a reduction in crop yields. When carbon dioxide is limiting, supplemental carbon dioxide can be added, but it is important to have adequate levels of light and heat. Another problem that occurs during the winter months is the potential for a buildup of unwanted gases. This problem can be avoided by properly servicing the heating system to make sure that it runs at peak efficiency and, in specific cases, to utilize air purification units.

Humidity

Humidity is the water vapor in the air and is linked to temperature. A hygrometer is used to measure humidity. In general, a 60 percent relative humidity is satisfactory for most house plants. Low humidity places the plant under stress, whereas high humidity leads to an increase in leaf and flower diseases. Optimizing the humidity helps overcome these problems. Humidifying a greenhouse can be accomplished using commercial humidifier units; however, when this is not possible, good watering procedures can lessen problems associated with low humidity.

Water

Water quality is key to an efficient greenhouse operation. Water used to irrigate plants should be analyzed before use by a local extension agent, a commercial analysis enterprise, or by the grower using a handheld electrical conductivity meter. The quality of water used is important for growing high-quality plants, and the grower should be aware of the detrimental effects that salt, chlorine, fluoride, and other contaminants commonly found in water can have on plants.

SELECTING CONTAINERS FOR CROPS

When selecting containers, the grower should be aware of the advantages and disadvantages associated with the different types available. The three main materials used to produce pots are clay, plastic, and styrofoam. Many years ago, most greenhouse crops were grown in clay pots; however, today many crops are grown in plastic pots. Clay is a natural, porous material that enables water to evaporate from its surface, thereby reducing the danger from waterlogging caused by overwatering. Clay pots are heavier, stronger, and can support large plants without falling over (Figure 15-3). Although clay pots have a number of advantages, the number one disadvantages associated with them is that they are more expensive. Other disadvantages include that they are bulky to handle and are breakable, they absorb mineral salts that produces unsightly white marks on the containers, and potbound growth occurs more rapidly. When reusing clay

Figure 15-3 Decorative clay pots are used for growing plants.

Figure 15–4 Styrofoam trays have insulating capabilities.

pots, they should be cleaned properly to remove any excess salt and steam sterilized to remove any potential diseases.

Plastic pots are the most popular today because they are less breakable, plant roots grow more evenly, they are lightweight and easy to handle, and they come in a variety of sizes and shapes. However, they are not porous, and plants grown in them are prone to waterlogging because of their inability to absorb moisture and lose it by evaporation. Even though overwatering plants in plastic pots may be a problem in the winter when drying is slow, plastic pots have the advantage in the summer when water stress is a problem because they retain the water. Another disadvantage is that plastic pots are light and fall over easily when plants get taller or when the medium begins to dry. When plastic pots are reused, they must be cleaned, disinfected using a commercial disinfectant solution, rinsed, and air-dried. Note that chemical sterilization is not as effective as steam sterilization. Styrofoam containers are now becoming very popular in the horticulture industry (Figure 15-4) because styrofoam is a good insulator, keeps the soil warm, is lightweight, and comes in a variety of sizes. Some disadvantages include that styrofoam containers break easily and are difficult to clean.

Today, the grower has many choices in containers. These choices are based on a variety of factors; the number one consideration is the size of the container, closely followed by the cost. Other important factors to consider are drainage, durability, ease of handling for mechanization, color, and shape. In addition, growers must closely consider the advantages and disadvantages associated with each of the different types of containers available and decide what works best for their particular operation.

TYPES OF CONTAINERS

A variety of containers are available for different uses, which can be broken down into three groups: rooting containers, bedding plant containers, and green and flowering plant containers.

Rooting Containers

Some rooting containers are made from organic compounds such as peat moss. Three examples of this are peat pellets, peat plugs, and peat pots.

Peat pellets are made of compressed peat moss that expands when moistened to become small peat filled pots. As shown in Figure 15-5A, the dry, compressed peat pellet is flat, saving space during storage; upon hydration, it expands and is ready for seeding or sticking a cutting. After several weeks, the grower has a rooted cutting or seedling ready for transplanting (Figure 15-6). The advantage of using peat pellets is that the seedling or rooted cutting can be transplanted directly without removing it from the moss medium, thereby reducing transplanting shock and increasing transplant success.

Peat plugs consist of compressed peat moss that expands when moistened to become small peat-filled plugs. As shown in Figure 15-5C, compressed plugs can be in styrofoam trays that need to be hydrated prior to use and can be used for planting seedlings or sticking cuttings. The smaller cells are designed for growing plants for a short period of time because they contain small amounts of media.

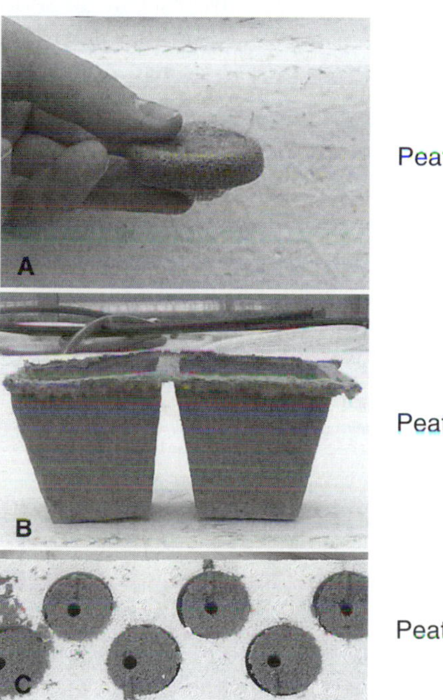

Figure 15–5 Three forms of peat rooting containers

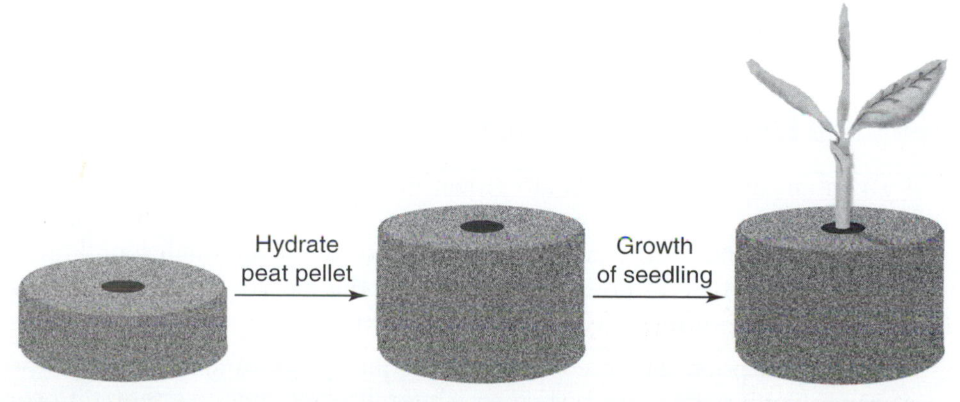

Figure 15–6 A peat pellet is hydrated, allowing for subsequent growth of the seedling.

Figure 15–7 Foam plugs in styrofoam trays

The advantage of using peat plugs is the same as for peat pellets. However, an additional benefit is that plugs come in different cell sizes, and the number of cells per tray can be from several to several hundred. Seedlings or cuttings can be planted mechanically for subsequent transplanting, which minimizes time and labor.

Peat pots are biodegradable, compressed peat moss pots (Figure 15-5B). They are generally treated with a fungicide and nitrogen fertilizer to aid in seed germination. The advantage of using peat pots is that the seedling can be transplanted directly, thereby reducing transplanting shock and increasing the success of the seedling. When planting seedlings in peat pots, no part of the peat pot should be above the soil surface or wicking action will cause the soil in the pot to dry out and the plant will die.

Rooting containers are also made from inorganic materials, such as plastic foam. Plastic foam is compressed plastic or inert inorganic materials in the form of a cube or block, which looks similar to peat plugs and has the same associated benefits (Figure 15-7).

Bedding Plant Containers

The majority of bedding plants are grown in plastic cell packs. Plastic cells may be individual cells, which vary in size, but generally they come in sets of two, four, six, and more. Each cell has a drainage hole in the bottom. This type of system facilitates transplanting because the roots of a plant are confined to a single cell. In large-scale mechanized systems, the appropriate soil mix, peat plugs, or plastic foam cubes or blocks can be used to save time and labor. Another advantage of using cell packs is that the seedling or rooted cutting can be transplanted directly when the plant reaches a salable size, thereby reducing transplanting shock and increasing transplant success. One disadvantage of these small cell packs is that the plants may become root bound if left too long leading to transplanting shock. Bedding plant seedlings are typically sold in cell pack units.

Green and Flowering Plant Containers

Most green and flowering plants are grown in plastic containers because they are less expensive. The shape of the pot is generally based on personal preference. Plastic containers are available in square and round, although round is the most popular. The type of pot used depends on the growth habit of the plant (Figure 15-8). Four different types of pots are commonly used:

- **Standard pots.** These containers are equal in width and height. These pots should not be used when growing tall plants because they tip over easily.

Figure 15-8 From left to right, standard pot, azalea pot, and bulb pan

Figure 15-9 Plastic hanging pots are used to grow flowering plants.

- **Azalea pots.** These containers are slightly shorter than the standard pots; they are three-fourths as high as they are wide. Many flowering plants are grown in azalea pots. This type of pot is more stable, so it is more commonly used when stability is a requirement.
- **Bulb pans.** These containers are half as high as they are wide; they are also called half pots. Bulb pans are typically used for propagating plants. They may also be used to grow plants with shallow roots to maturity or for plants grown from bulbs such as hyacinth, daffodil, and tulip.
- **Hanging baskets.** These containers can be made of wood, wire, ceramic, or plastic; however, the majority of containers are plastic because they are lighter. Ceramic, wire, and wood pots are also used but to a lesser extent. Like potted plants, the proper size basket, proper drainage, and saucers are important for improved customer satisfaction (Figure 15-9).

GROWING MEDIA

Growing media used for potting plants are produced to accommodate specific needs. For example, some mixes are used solely for promoting germination, whereas others are for vegetative growth. Although there are differences in growing media, all mixtures should be reproducible and available and have basic physical, chemical, and biological properties. All growing media should:

- have good moisture- and nutrient-holding capacity.
- contain physical properties that permit rapid water infiltration (easily wetted), aeration, and drainage.
- decompose slowly.
- flow easily to facilitate pot filling.
- be free of toxins.
- have good CEC (between 50 and 100 mEg/100 g of soil) and buffer capacity. The pH should be between 5.5 and 6.0 or according to the needs of the specific plant being grown.
- provide good support for the plant.

The materials used in formulating a good growing medium include organic and inorganic components. Some key components used in growing media today are peat moss, wood byproducts, bark, sand, perlite, and vermiculite. Each of these components has a specific function. Peat moss, wood byproducts, bark, and vermiculite (heat-treated mica) are used for their high moisture- and nutrient-holding capacities. Bark and wood byproducts may be used as a substitute for peat moss because they are less expensive. Sand and perlite, which is heat-treated lava rock, are used to promote aeration and drainage.

Other components that are added to growing media are fertilizer, limestone, and polystyrene pieces. In general, most components in soilless media do not supply plants with nutrition. Some fertilizer is necessary to provide enough nutrition to get plants started: slow-release fertilizers are typically used for this purpose. Limestone is added to growing media to maintain the correct pH in the media, thereby enabling plants to readily take up nutrients. To reduce the cost of transport of soilless media, manufacturers use dry ingredients because they are lightweight; however, materials such as dry peat moss repel water. To overcome this problem, manufacturers include wetting agents to facilitate the infiltration of water into the growing media. Polystyrene pieces are often added to the growing media to reduce the weight of the mixture; however, when the mixture is watered, the polystyrene floats to the surface and pools above the surface of the soil.

The proper materials must be selected for the preparation of a growing mixture to maximize plant growth. What factors should be considered in choosing materials for a good growing mixture? Many ready-to-use mixtures can be purchased from reputable suppliers. However, at times growers may want to produce their own specific growing mixtures. When this is the case, growers must consider many factors. These factors include the quality of the ingredients, reproducibility and availability of materials, cost, use, and ease of preparation.

IRRIGATION PRACTICES

Watering plants is an important responsibility because most horticultural crops are between 90 and 98 percent water. Water is critical for the production of high-quality crops grown under greenhouse conditions. Both underwatering and overwatering can adversely affect plant growth and development. Overwatering is one of the most common causes of injury to plants, because it creates an oxygen deficit in the root zone resulting in root rot. Plants that are underwatered typically grow at a slower rate and eventually wilt, thereby reducing their productivity. Plant leaves under continual water stress are generally small, and the overall plant stature is stunted with shortened internodes. At one time, underwatering was a method for controlling crop heights and to induce flowering. Today, growth retardants and touch (brushing) are commonly used methods to reduce growth.

Important considerations for watering plants include the following:

- **Timing.** Water should be applied before the plant shows signs of wilting because wilting indicates stress resulting in premature senescence and reduced crop yields. The proper time to water can be determined by simply feeling the weight of the pot, feeling the soil, using indicator plants, or using **tensiometers,** which are instruments used to measure the amount of water in a given pot.
- **Amount.** Thoroughly water each time plants are irrigated, but don't overwater. About 10 percent of the water applied to a pot should drain out the bottom, which ensures that the soil is thoroughly wet and helps flush out excessive salts that may have accumulated in the pot.
- **Water carefully.** To prevent disease, apply water to the growing medium and not to the foliage or flowers.

TYPES OF IRRIGATION SYSTEMS

Plants can be watered in a variety of ways. To choose the one that works best for a particular purpose, growers should consider the pros and cons for each. The types of irrigation systems commonly used in greenhouses are summarized next.

Manual Watering

The hose watering method has been used for many years and is still commonly used for small-scale watering. Prior to automated watering systems, hand watering was the only method used for watering plants. However, today automated systems are commonly used for large-scale commercial operations.

Automated Watering Systems

A variety of automated watering systems are used that differ in cost, efficiency, and flexibility. The different types of automated watering systems are described next.

Tube Irrigation

The tube irrigation system is the most commonly used method for watering potted plants. Two commonly used tube irrigation systems are the Chapin system and the Stuppy system. The only major difference between the two systems is the weight at the end of each tube; for the Chapin system, a lead weight is at the end of the tube (Figure 15-10), whereas for the Stuppy system, the weight is plastic (Figure 15-11).

Figure 15–10 Chapin tube irrigation system uses a lead weight. *Courtesy of Dr. E. Jay Holcomb, Department of Horticulture, The Pennsylvania State University.*

Figure 15–11 Stuppy tube irrigation system uses a plastic weight.

One of the advantages of this method is that water is applied directly to the soil without splashing plants with a soil/water mixture every time the plants are watered, thereby reducing the spread of diseases. Another advantage of this system is that individual pots can be watered automatically at the correct amounts and at the right frequency, thereby maximizing plant growth and development. Tube irrigation also enables the grower to apply water-soluble chemicals such as fertilizers and pesticides. Although a number of advantages are associated with tube irrigation, it is expensive to install and is not flexible. In addition, pots on the bench have to be uniform in size, and the tube must be inspected on a periodic basis to make sure no blockage exists due to salt buildup.

Capillary Mat System

The capillary mat system is a type of subirrigation where pots are placed on a mat saturated with water that is taken up by the growing medium via wick

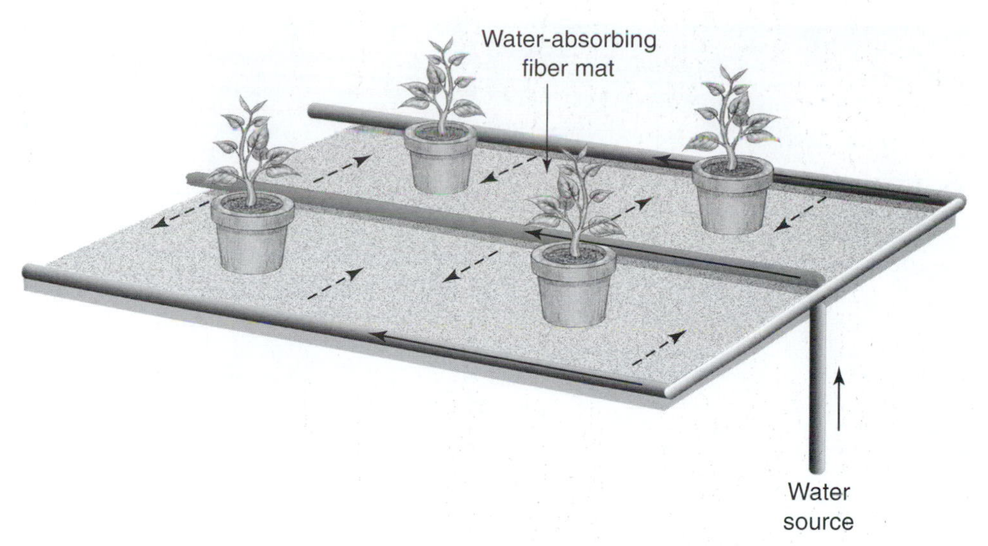

Figure 15–12 Capillary mat system must be on a level surface.

action through holes in the bottom of the pot (Figure 15-12). The mat is kept moist by tubes installed on the bench. For the system to work properly, the mat must be placed on a level surface. The capillary mat system is flexible and easy to install, can readily accommodate different pot sizes, and plants can be rearranged as required throughout the plant's life cycle. One disadvantage of this system is the potential for algae growth, which causes a number of other problems.

Ebb and Flow Method

The ebb and flow method is a form of subirrigation that pumps water into a watertight bench at regular intervals, filling the benches with water and enabling the growing medium to take up water via wick action through holes in the bottom of the pot. After the pots have taken up a sufficient amount of water, the benches are drained (Figure 15-13). Advantages of the ebb and flow system are similar to the capillary mat system. The ebb and flow system is flexible and easy to install, accommodates different pot sizes, and allows plants to be rearranged easily as required throughout the plant's life cycle. One disadvantage of this system is that it is more expensive than the capillary mat system.

Figure 15–13 Ebb and flow systems for watering plants are flexible and easy to install.

Overhead Irrigation

In overhead irrigation systems, water is applied over the canopy of plants using spray nozzles (Figure 15-14). This method is commonly used on bedding plants. The problem with this method is that wet foliage increases the chances of disease; therefore, watering should be done in the morning to give the plant foliage enough time during the day to dry.

Perimeter Irrigation

Perimeter irrigation is a surface form of watering (Figure 15-15). Pipes are placed around the perimeter of the flowerbed with nozzles located below the foliage, and they deliver a flat spray of water over the growing medium, ensuring that cut flowers and foliage do not get wet. The water pattern used must cover the entire surface of the bench, as shown in Figure 15-16. The

Figure 15–14 Overhead irrigation system is commonly used for watering bedding plants.

Figure 15–15 Perimeter irrigation system for watering plants. *Courtesy of Dr. E. Jay Holcomb, Department of Horticulture, The Pennsylvania State University.*

perimeter irrigation system is commonly used in the cut flower industry (roses and carnations) where blemishes on the flowers or foliage are not acceptable.

Soaker Hose Method

The soaker hose method for watering plants is a surface form of irrigation commonly used on flowerbeds. This method uses a series of hoses that are placed on the surface of the flowerbed and water oozes or drips out of the line.

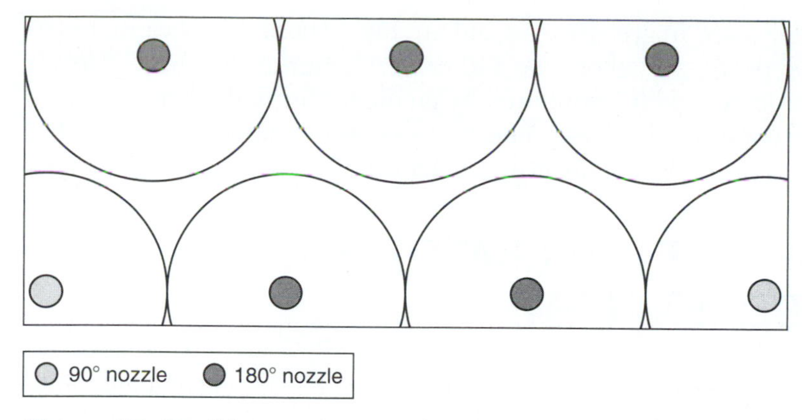

Figure 15–16 Water pattern used to cover entire surface of the bench with the perimeter irrigation system for watering plants

FERTILIZATION

Plants grown in a greenhouse environment are typically grown in pots, which is much more restrictive. The volume of growing medium in pots is also much smaller than when plants are grown in the field. In addition, plants grown in greenhouses generally use growing media that are soilless, thereby requiring fertilization, especially micronutrients. Plants grown in a greenhouse environment typically receive 100 times more fertilizer than those grown in the field.

Organic fertilizers are naturally occurring nutrient materials derived from plants or animals. Examples are bloodmeal, bonemeal, and manures. Organic fertilizers are typically used by specialty growers or hobbyists and are not generally used in greenhouses for a variety of reasons. Organic fertilizers are not easily adapted to automated watering systems so applying the exact concentrations of nutrient elements is difficult. Organic fertilizers also produce volatiles, which are unpleasant to use in enclosed conditions due to bad odors and which can have adverse effects on plant growth.

Inorganic fertilizers are synthetic nutrient compounds that are derived from mineral salts. They are convenient to apply and can be applied easily using a variety of application methods. Another benefit of inorganic fertilizers is that they can be applied in exact amounts as needed. Both dry and liquid fertilizers are used in greenhouses, although liquid fertilizers are more commonly used because they can be applied readily with the irrigation water. Liquid fertilizers can be applied as a constant feed, which involves giving low concentrations of fertilizer each time the plant is irrigated. This method mimics slow-release fertilizer action and is the most popular in greenhouse operations because it provides a fairly constant supply of nutrients to the plant, resulting in sustained growth and development.

Another method for applying liquid fertilizer is intermittent feed, which provides fertilizer on a periodic schedule that can be weekly, bimonthly, or monthly depending upon plant needs. The main disadvantage of this method is that shortly following application, gradual decreases begin and continue until the next application, thereby not providing a constant source of nutrition, which leads to fluctuations in plant growth.

Dry formulations of inorganic fertilizers come as fast release or slow release, which can be incorporated or applied by top dressing. Fast-release fertilizers are available immediately, as the name implies; therefore, if there is a deficiency problem, the problem can be taken care of right away. Slow-release fertilizers are also commonly used in greenhouses and are highly desirable because nutrients are released over a period of time and can be better used. Nutrient levels in greenhouse crops should be monitored by media testing and foliar analysis to avoid deficiency symptoms because when visual symptoms appear, the plant has already been stressed, reducing quality and yields.

GREENHOUSE INTEGRATED PEST MANAGEMENT (IPM)

Controlling pests in greenhouses starts with prevention. Preventing pest entry into greenhouses is an easier task than in the field. Pest control is discussed more completely in Chapter 12. To help prevent pests from getting into greenhouses, a number of practices are often used that can be difficult under field conditions:

- **Prevent pest entry.** Putting screens on greenhouse vents prevents insects, pathogens, and weed seeds from entering the greenhouse depending on the mesh of the screen used. Another way to minimize pest entry into the greenhouse is to disinfect shoes and to have workers change into clean overalls prior to walking into the greenhouse. This minimizes tracking in pests on the shoes or clothing.
- **Weed control.** Remove weeds from inside as well as outside the greenhouse. If required, use only chemicals approved for greenhouse use; for example, Roundup® is commonly used.
- **Sanitation practices.** Common greenhouse diseases are root rot, damping-off, botrytis, blight, powdery mildew, and root-knot nematodes. Sanitation effectively minimizes disease incidences in greenhouses. Dead leaves, flowers, and stems should be removed from the greenhouse, making sure the greenhouse is always kept clean. After each crop cycle, the greenhouse benches should be cleaned and sterilized. Sterilized media and pots should always be used in the greenhouse.
- **Crop inspection.** Daily inspection of crops is a very important way to prevent the spread of insects and diseases in the greenhouse. If a diseased or insect-infested plant is found in the greenhouse, make sure to discard it and implement some form of control. In some cases, the grower has no choice but to use chemicals. When this is the case, the safest products recommended for greenhouse use must be selected.
- **Environmental manipulation.** If possible, adjust conditions to make them unfavorable for pests and thus prevent pest problems.
- **Biological control.** If prevention of pests from entering the greenhouse is unsuccessful, biological control works well unless the population of pests gets too high, in which case a combination of chemical, genetic, cultural, and mechanical control should be used (as described in Chapter 12).

SUMMARY

You now understand how growing ornamental crops in the greenhouse has become more technical and specialized due to increased consumer demands for quantity and quality crops. Controlling all environmental factors in the greenhouse, including temperature, light (light quality, intensity, and duration), atmospheric composition, humidity, and water, is required to produce high-quality crops in the greenhouse. To grow high-quality crops, the proper container is also important. When selecting containers for growing greenhouse crops, the proper size consideration is closely followed by cost. Other factors that must be considered are drainage, durability, and ease of handling for mechanization, color, and shape. The different types of containers commonly used in greenhouses include rooting, bedding plant, and green and flowering plant containers. Along with containers, growing media also have important functions. In addition, three important factors should be taken into consideration when watering plants, including timing, amount to apply, and careful watering. There are seven commonly used types of irrigation systems used in greenhouses. Greenhouse plants are typically grown in soilless media contained in pots and thereby require fertilization, especially micronutrients. Fertilization of greenhouse crops is typically accomplished using inorganic fertilizers. Organic fertilizers are not typically used because exact concentrations of nutrient elements are difficult to apply and bad odors are produced. The chapter concluded with a discussion of greenhouse integrated crop management, including different ways to prevent pests from entering the greenhouse together with the most effective way to get rid of pests if prevention is unsuccessful.

Review Questions for Chapter 15

Short Answer

1. What are five environmental factors that must be considered when growing plants in the greenhouse?
2. What is a general rule of thumb used when selecting temperatures, which must be maintained for most greenhouse crops?
3. When supplying supplemental light, what are three factors that must be considered?
4. What are two ways to maximize light intensity in greenhouses?
5. What are two advantages of using clay pots and one major disadvantage?
6. What are the first and second considerations when selecting a container? After these decisions have been made, what are four other factors that must be taken into consideration?
7. What are four different types of containers used to grow plants in greenhouses?
8. What are four examples of rooting containers?
9. What are seven important functions of growing media?
10. List the six components commonly used in growing media and provide the function of each.
11. What are three important considerations that must be made when watering plants?
12. What are seven types of irrigation systems commonly used in greenhouses?
13. What are two types of fertilizers available and which of these two is most commonly used?
14. What are five different ways to prevent pests from entering greenhouses? After pests have entered the greenhouse, what are several ways to get rid of them?

Define

Define the following terms:

thermotropism	negative DIF	light intensity	tensiometers
thermoperiodism	positive DIF	light duration	organic fertilizer
DIF	light quality	photoperiodism	inorganic fertilizer

True or False

1. Supplying supplemental carbon dioxide to greenhouses is only effective when adequate light is available.
2. Some plants are highly sensitive to low concentrations of fluoride commonly put in water systems to prevent tooth decay.
3. Photoperiodism is the length of the light period that influences plant growth.
4. Light intensity is the wavelength or color of the light.
5. Thermotropism is the plant's response to changes in day and night temperatures.
6. Water containing chlorine and fluoride is beneficial to plants because it decreases the occurrence of disease.
7. In greenhouses, the night temperature should be cooler than the day temperature to mimic natural conditions.
8. Light intensity is the actual quantity of light being supplied.
9. Light quality has little effect on the overall size of the plant.
10. Clay containers are porous, which allows more aeration to the roots, and heavier, which provides more stability for top-heavy plants.
11. Some rooting containers are manufactured from organic compounds such as peat moss.
12. Some rooting containers are manufactured from inorganic compounds such as plastic.
13. Peat pellets and peat pots are commonly used as rooting containers.
14. Perlite is a good substitute for peat moss.
15. One of the important functions of a growing medium is that it must be reproducible and available.
16. An important consideration for watering plants is to be sure to water shortly after the plant shows signs of wilting.
17. Perimeter irrigation is a commonly used irrigation system in greenhouses.
18. Ebb and flow irrigation systems are not a commonly used irrigation system in greenhouses.
19. The best way to ensure proper nutrient levels of greenhouse crops is to only fertilize at the beginning of each month.
20. Inorganic fertilizers are synthetic nutrient compounds that are derived from mineral salts.
21. Organic fertilizers are nutrient materials derived from mineral salts.
22. IPM practices are only important to use for field crops, not greenhouse crops.
23. Techniques such as weed control, crop inspection, and environmental manipulation are effective means of preventing pests from entering greenhouses.

Multiple Choice

1. Which of the following is one of the first considerations when selecting containers for crops?
   A. Cost
   B. Durability
   C. Size
   D. Ease of handling
2. Many years ago, most greenhouse crops were grown in clay pots. Today, a majority of the greenhouse crops are produced in plastic containers. Which of the following is a reason for this change?
   A. Clay pots provide less aeration for the roots.
   B. Clay pots are more expensive then plastic containers.
   C. Clay pots come in a limited number of sizes.
   D. All of the above
3. Azalea pots are
   A. half as high as their widths.
   B. slightly shorter than standards.
   C. equal in width and height.
   D. None of the above
4. Bulb pans are
   A. equal in width and height.
   B. half as high as their widths.
   C. slightly shorter than standards.
   D. None of the above
5. Proper watering of container-grown plants is one of the hardest tasks for employees to learn. Irrigation should be based on
   A. demand of the crop.
   B. when the plant wilts.
   C. when plants are turgid.
   D. None of the above
6. Which of the following type(s) of irrigation system(s) are commonly used today?
   A. Hose watering
   B. Perimeter irrigation
   C. Capillary mat system
   D. All of the above

Fill in the Blanks

1. Thermotropism is a plant's response to ____________________.
2. Thermoperiodism is the plant's response to changes in ____________________ and ____________________ temperatures.

3. Clay pots are not commonly used anymore because they are more ________________ than plastic pots.
4. It is important to apply water to the growing medium and not to the foliage or flowers to prevent the spread of ________________.

Matching

Note: Some letters may be used more than once.

1. _____ Peat moss
2. _____ Vermiculite
3. _____ Sand
4. _____ Perlite
5. _____ Bark

A. A substitute for peat moss
B. Aeration and drainage
C. Moisture- and nutrient-holding capacity

Activities

Now that we have completed our discussion on growing plants in greenhouses, you will have an opportunity to explore this area in more detail. In this activity, you will visit a local greenhouse and gather the following information about the greenhouse operation:

- how the grower controls temperature, light, air, humidity, and water for the different crops grown.
- the type of containers commonly used for growing plants.
- the types of growing media used for the different greenhouse operations.
- the irrigation practices and system(s) used.
- the type of fertilizers used and how they are applied.
- the methods used to control pests.

In addition to the information you were asked to gather, include your thoughts on the rationale for how plants were grown. If you do not have access to a greenhouse, surf the Internet for two sites that present information on growing plants in the greenhouse and summarize them. Make sure to provide the Web site address where you found your information.

References

Biondo, R. J. (2004). *Greenhouse production*. Upper Saddle River, NJ: Prentice Hall.

Nelson, P. V. (1985). *Greenhouse operation and management* (3rd ed.). Reston, VA: Reston Publishing.

CHAPTER

16

Nursery Site Selection, Development, and Facilities

ABSTRACT

The environmental factors that should be considered when selecting a suitable nursery site include temperature, light, rainfall, wind, soil, topography, water, air pollution, and plant pests. The economic factors that should be considered when selecting a suitable nursery site include land cost and availability, labor, transportation, utilities and services, and competition. Developing a nursery for container- and field-grown nursery crops involves land leveling, road construction and design, irrigation, and land drainage, along with potting areas, greenhouses, coldframes, hotbeds, shadehouses, and overwintering houses. Storage and other facilities are also important to a successful nursery, including cold storage facilities, shipping areas, offices, pesticide storage and mixing areas, and storage buildings.

Objectives

After reading this chapter, you will be able to

- discuss the environmental and economic factors that should be considered when selecting a nursery site.
- discuss the proper way to develop a nursery for container and field-grown nursery crops.
- provide information on structures used for propagating and growing nursery crops.
- provide information on storage and other facilities that are important to a successful nursery.

Key Terms

air pollution
container nursery
field nursery
topography

INTRODUCTION

This chapter provides information on nursery site selection, development, and facilities required for a successful nursery operation. To select the proper nursery location, both environmental and economic factors must be considered. Environmental conditions such as temperature, light, rainfall, wind, soil texture, soil drainage, topography, soil fertility, water, **air pollution,** and plant pests at the potential location must maximize plant growth and development. After a site has been selected with suitable environmental conditions for high-quality plant growth, economic factors that affect a nursery operation must be considered. Economic factors that must be considered are land cost and availability, labor, transportation, utilities and services, and competition. The most suitable type of land to select depends on what works best for the specific nursery and the price range. For example, typically the best nursery sites are high-priced farmland, although this type of land may not be the most economical. Undeveloped land may initially be less expensive, but the cost of development is too high; therefore, it is important to

Figure 16–1 The pot-in-pot method is one way to grow container nursery crops

consider what works best for the grower. A dependable supply of skilled and nonskilled labor is also essential to nursery operations. Good transportation facilities should be available to facilitate the transport of products to and from the nursery. Dependable telephone, waste-removal facilities, electrical power, and other utilities and services are also essential to the everyday functioning of the nursery. Competition from other nurseries located in the vicinity of the potential site being evaluated should be taken into consideration to minimize problems down the road.

After careful consideration of environmental and economic factors, strategically organizing the nursery layout is important to maximize the production of crops. The two main types of nursery layouts are **container** and **field nurseries.** Container nurseries require two to three times less land per plant than field-grown plants (Figure 16-1). After the nursery layout has been established, the land must be prepared properly. Major objects such as trees, stumps, and other debris should be removed, and the land leveled. After the land is leveled, irrigation and drainage systems need to be installed as necessary. The road system should be installed to provide ample space for movement of nursery stock, equipment, and personnel. A variety of structures are used for propagation and subsequent growth of nursery crops, including potting areas, greenhouses, coldframes, hotbeds, shadehouses, and overwintering structures. In addition to having structures for propagating and growing plants, a nursery needs a cold storage facility, a centralized building for shipping, office space, pesticide storage and mixing areas, and storage buildings.

NURSERY SITE SELECTION

The proper location is one of the most important factors when selecting a suitable site for a nursery. As is the case with selecting the proper location for a greenhouse range, it is important to consider carefully the environmental and economic factors. Therefore, one of the first steps in selecting the proper site for a nursery is to evaluate the environmental issues of a given location (Landis et al., 1995).

Environmental Factors

A nursery site's environment can affect high-quality plant growth. Growers cannot afford to neglect any of the following environmental factors when selecting a nursery site.

Temperature

The type of nursery crop being grown determines the importance of temperature in site selection. All nursery growers should be familiar with the USDA Plant Hardiness Map of the United States (Figure 16-2). The Plant Hardiness Map identifies 11 zones by average annual minimum temperatures for each zone. It also identifies 12 different zones based on the average number of days above 86°F, because some plants that can tolerate low temperatures cannot tolerate extended summers and high temperatures. After an initial site selection has been made using a Plant Hardiness Map, the grower should talk to local extension people and weather agencies to further confirm the selection.

Light

The location must have adequate sunlight to maximize plant growth. Supplemental light cannot be supplied to nursery crops grown under field conditions, so the grower must rely solely on sunlight. Plants vary in their need for light, so the location should be best for the particular crop that will be grown. Most trees require full sunlight to establish and grow at peak performance. Interestingly, an area that is continually subjected to dim lights from highways, streetlights, or other sources will stimulate vegetative growth if the proper temperature is available. The increased growth delays the onset of fall dormancy, which puts the new growth at risk for being injured by cold temperatures resulting during the winter months.

Rainfall

Knowing the rainfall pattern for a potential nursery location is also important. The amount of rainfall for a given location affects the type of nursery crops that can be grown. In locations where summer rainfall is limiting, the grower must select species that can tolerate moisture stress and be prepared to irrigate when supplemental water is needed. High rainfall during critical times for nursery operations is another important issue to be considered when selecting a potential nursery site. When rain comes at an inopportune time, the soil becomes saturated, which leads to the inability to get machinery into the field, delays the digging of trees and seedlings, and delays production schedules. Machinery on saturated soil causes damage to the soil structure and promotes soil compaction. Another problem that can occur when excessive rainfall comes at the wrong time is soil erosion.

Wind

Areas with high winds can cause nursery plants to fall over, resulting in damage. Other problems that result from excessive winds are soil erosion, especially with sandy soils; interference with important nursery operations, such as spraying; and others. To overcome problems with high winds, a nursery should be

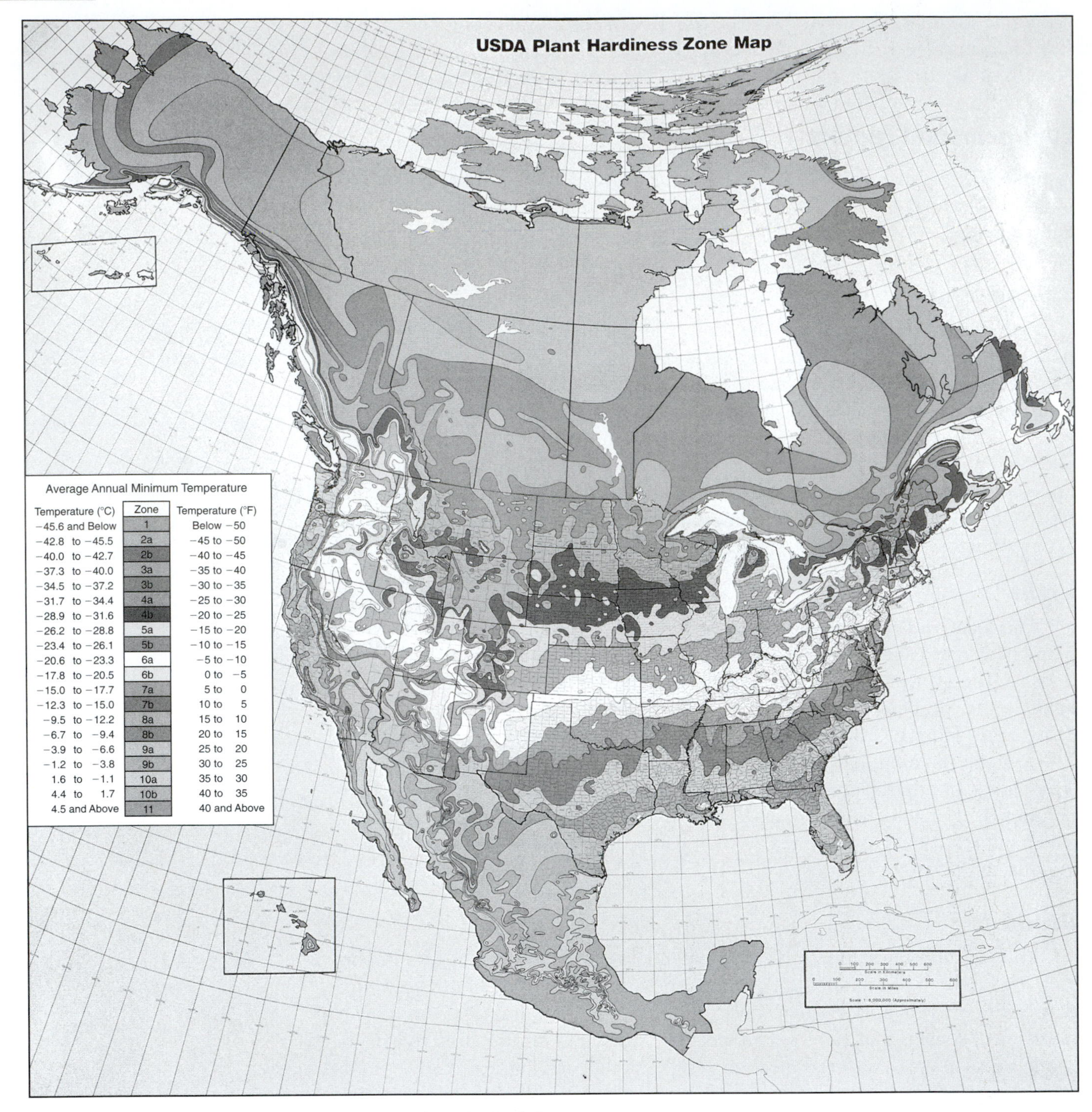

Figure 16–2 USDA Plant Hardiness Map of the United States

located in an area where natural windbreaks are present to protect the nursery or where artificial windbreaks can be installed.

Soil Texture, Drainage, Topography, and Fertility

The type of soil present at the site will determine the method used for digging plants in the nursery. The balled and burlapped (B&B) method (Figure 16-3) requires soil with a higher clay content that holds together when the plant is being dug, whereas the bare root production method requires sandy or loam soils. A soil with the proper texture and structure has a good moisture- and nutrient-holding capacity while providing good aeration and drainage. The potential nursery site should be naturally well drained, or, if necessary, through artificial drainage, which will add to production costs. Poor drainage leads to problems with soil aeration that may cause soilborne diseases. Another problem with poor drainage is that the soil takes longer to warm up in the spring and delays nursery operations.

Topography refers to the surface features of an area. Land should be relatively level with a slope of 1 to 2 percent for drainage. Steep or irregular land causes problems with soil erosion, land preparation, efficient use of nursery machinery, and the installation of roads and irrigation systems. Soil fertility is affected by the soil quality, which includes organic matter; proportion of sand, silt, and clay; nutrients; and other factors. The higher the quality of soil, the less input is necessary.

Figure 16–3 A balled and burlapped tree

Water

The nursery should have a dependable supply of good-quality irrigation water, which is low in salts, fluoride, chlorine, and other contaminants. Most nurseries cannot be completely dependent on rainfall, and water must be supplied to maximize plant growth and, in some cases, to prevent plants from dying. When selecting a site, it is important to know what options are available from groundwater, lakes, rivers, and others. These sources vary in quality, cost, accessibility, and reliability of supply, all of which should be considered when selecting a site.

Air Pollution

Harmful or degrading material in the air causes problems for nurseries. Some common air pollutants are ozone (O_3), sulfur dioxide (SO_2), and ethylene. In highly populated areas and locations that are highly industrialized, the grower should pay close attention to the nursery crops that will be grown to avoid problems with pollution found at that specific location.

Plant Pests

The grower should not select a site where there are numerous pests that will cause major problems; the grower's crop is their most desirable host.

Economic Factors

After a site has been selected with suitable environmental conditions for high-quality plant growth, the economic factors must be considered.

Land Cost and Availability

The best nursery sites will also be high-priced farmland, but in most cases, this type land is not economical. Undeveloped land may initially be less expensive, but the cost of developing it may be very high. Another factor to consider is that land near areas of high population may be very expensive. Although land in rural areas may be less expensive, transporting materials and products both to and from the nursery may be very costly. Therefore, the piece of land selected should be best for the grower's particular needs and budget.

Labor

A dependable source of skilled and nonskilled labor close to the nursery is essential. A small number of permanent staff is necessary for nursery operations; however, a seasonal labor pool should also be readily available to meet needs on specific occasions.

Transportation

Good transportation facilities—such as truck, rail, bus, or airport—are an essential part of any nursery. Therefore, the nursery should be close to major highways and airports but not too close because pollution may become a problem.

Utilities and Services

Dependable telephone, waste-removal facilities, electrical power services, and others are essential to everyday functioning of a nursery. If a telephone line goes down frequently, telephone orders cannot be made, leading to a loss of revenue. If power goes off during the winter months, the loss of heating can result in the loss of tender crops.

Competition

Before placing a nursery, the grower should be aware of other nurseries located in the vicinity. Competition from other nurseries should be minimized.

NURSERY DEVELOPMENT

After careful consideration has been given to the environmental and economic factors, an informed decision can be made on the proper site for a nursery location. After the proper site has been selected, the nursery layout must be determined to maximize crop production. The two types of nursery layouts are container nurseries and field nurseries.

A container nursery grows nursery crops to a marketable size in containers that differ in size and shape according to the species and the marketable size desired (Figure 16-4). Container-grown plants require two to three times less land per plant as compared to field-grown plants.

A field nursery grows nursery crops to a marketable size in the field (Figure 16-5). The type of layout should be chosen prior to establishing the nursery because both have different requirements.

After the nursery layout has been detailed and properly planned out on paper, the next important step is land preparation. Major objects such as trees, stumps, rocks, and other debris should be removed and the land leveled. After land leveling has been properly taken care of, irrigation line and drainage systems should be installed. Being able to provide water when needed is important, so a reliable source of water, sufficient pumping and pressurizing capacity, and a system that provides a uniform distribution of water is necessary. Good drainage is also a must so there is no standing water after a heavy rainfall, which would interfere with nursery operations.

After irrigation and drainage systems are in place, roads should be installed. The road system should provide ample space for movement of nursery stock, equipment, and personnel. Primary roads should be made out of concrete or asphalt, although secondary roads can be out of gravel.

Figure 16–4 Nursery crop in a container

Figure 16–5 A nursery crop in the field

NURSERY FACILITIES AND EQUIPMENT FOR CONTAINER AND FIELD NURSERIES

A variety of structures are available for propagating plants and supporting subsequent growth, so the method of propagation to be used must be known prior to selecting a structure. Structures for propagating and growing plants include the following:

- Potting areas should be centrally located and be free of disease and weeds.

Figure 16–6 A greenhouse provides the best environmental control.

- Greenhouses have the best environmental control for growing plants; however, they are also the most expensive (Figure 16-6).
- Coldframes consist of a wooden or concrete block frame with heat supplied by solar radiation through glass or another transparent covering. This is the simplest and most economical outside propagation structure (Figure 16-7).
- Hotbeds are similar to coldframes except that electric or hot water is used for heating.
- Shadehouses protect plants from environmental factors such as wind, temperature, hail, heavy rains, and solar radiation.
- Overwintering structures have a permanent frame, can be produced in a variety of sizes, and are covered annually with polyethylene to prevent nursery crops from being damaged during the winter months (Figure 16-8). The coverings are typically removed during the summer months to prevent heat buildup. This type of house can be constructed at lower investment costs than greenhouses, and this is the reason for their popularity.

Figure 16–7 A coldframe is the simplest outside propagation structure. *Courtesy of Dr. E. Jay Holcomb, Department of Horticulture, The Pennsylvania State University.*

Figure 16–8 A greenhouse is covered with polyethylene to prevent damage during the winter.

Structures for storage and other facilities include the following:

- **Cold storage facility.** Temperatures between 34 and 40°F and a high relative humidity are typically used for storing bare root material. These plants are generally defoliated to reduce moisture loss during storage and lessen the spread of disease. A simple, inexpensive way to defoliate these plants on a small scale is to place a few buckets of apples in the storage area that will give off ethylene and cause defoliation. On a larger scale, sprays with the ethylene-releasing compound Ethrel® have been shown to be effective. After plants are defoliated, the leaves must be discard to avoid disease problems.
- **Shipping.** A centralized building is used to store material until ready for shipment. A lot of space is necessary to permit large trucks to maneuver and load plants.
- **Offices.** The main building with offices is an important sales tool and should be well landscaped because it is the first impression people have

of the business. Sales staff, managers, accounting personnel, and others use these spaces. Parking areas outside this building should provide plenty of room for customer parking separate from employee parking.

- **Pesticide storage and mixing areas.** This area should closely follow EPA standards to avoid legal problems. All chemicals should be stored in approved containers and labeled appropriately. Pesticide storage buildings should have concrete floors that will contain spills if they occur. These buildings should be well ventilated and have temperature control to avoid problems with chemicals freezing or overheating. Materials that can absorb spills should be on hand in chemical mixing areas, together with facilities for employees to wash hands and clean up after mixing chemicals. All protective clothing and respirators should be stored in a room separate from the chemicals.
- **Storage buildings.** Equipment can be kept in these buildings during poor weather or when being repaired. Larger nurseries have workshops associated with these buildings for repairing and maintaining equipment. These buildings can also be used for dry storage of a variety of supplies, including fertilizers, pots, and tools.

SUMMARY

You now understand what is necessary to select a nursery site properly, develop the site, and create the necessary facilities for a successful nursery. Careful evaluation of environmental and economic factors is necessary when selecting a potential nursery site. Developing a nursery for container- and field-grown nursery crops includes leveling land, constructing roads, and implementing irrigation and land drainage systems. Structures used for propagating and growing nursery crops include potting areas, greenhouses, coldframes, hotbeds, shadehouses, and overwintering houses. Storage and other facilities important to a successful nursery include cold storage facilities, shipping areas, offices, pesticide storage and mixing areas, and storage buildings.

Review Questions for Chapter 16

Short Answer

1. Describe what a USDA Plant Hardiness Map is used for and explain the information it provides.
2. List four of the nine environmental factors that should be considered when selecting a suitable nursery site.
3. List three of the five economic factors discussed in the text that should be considered when selecting a suitable nursery site.
4. What are four common problems encountered in areas with high rainfall during critical times for nursery operations?
5. What are two types of nursery layouts?
6. What are four factors that should be considered in preparation of land for nursery development?

7. What are five structures for propagating and growing nursery plants?
8. What is a cold storage facility used for in a nursery?

Define

Define the following terms:

topography | air pollution | container nursery | field nursery

True or False

1. A Plant Hardiness Map of the United States defines areas where certain plants will or will not survive.
2. Areas of high rainfall during critical times of nursery operation should be avoided.
3. Common problems encountered when soil becomes saturated during critical nursery operations are delays in digging trees and seedlings.
4. Undeveloped land is inexpensive and always the first choice when selecting land for a nursery site.
5. Farmland is always the best choice when selecting a nursery site.
6. Plants grown in a container nursery require two to three times less land per plant as compared to field-grown plants.
7. A coldframe is a wooden or concrete block frame with heat supplied by solar radiation through glass or other transparent coverings.
8. Hotbeds are similar to coldframes except that solar radiation is used for heating hotbeds.
9. Greenhouses provide the best environmental control for propagating and growing plants; however, they are the most expensive.
10. Coldframes are the simplest and most economical outside propagation structure whereas greenhouses are the most expensive.
11. Shadehouses protect plants from environmental factors such as wind, temperature, hail, heavy rain, and solar radiation.
12. A cold storage facility is typically used to store bare root materials in nurseries. These plants are typically defoliated to reduce moisture loss during storage by treating with abscisic acid.

Multiple Choice

1. Areas with high rainfall during critical times of nursery operations should be avoided. Common problems encountered when soils become saturated include
   A. waterlogging.
   B. damage to soil.
   C. inability to get machinery into the field.
   D. All of the above
2. A container nursery should be
   A. compact for total ease of operation.
   B. spread out to maximize work conditions.
   C. located in major cities.
   D. All of the above

3. A field nursery requires
   A. two to three times more land then a container nursery.
   B. two to three times less land then a container nursery.
   C. the same amount of land as a container nursery.
   D. None of the above

Fill in the Blanks

1. The proper ____________________ is one of the most important factors when selecting a suitable site for a nursery.
2. ____________________ are commonly used to protect sites from excessive wind damage.

Activities

Now that we have completed our discussion on nursery site selection, development, and facilities, you will have an opportunity to explore this area in more detail. In this activity, you will visit a local nursery and gather the following information about the nursery site and its facilities:

- the mean temperatures for that region, annual rainfall, and other environmental factors that you feel are important.
- the type of facilities at the nursery, such as greenhouses, coldframes, and so on.
- the type of nursery (container or field).

In addition to the information you were asked to gather, include your thoughts on the rationale for the selection of this particular nursery site. If you do not have access to a nursery, surf the Internet for two sites that present information on nursery sites, development, and facilities, and then summarize them. Provide the Web site address where you found your information.

Reference

Landis, T. D., Tinus, R. W., McDonald, S. E., & Barnett, J. P. (1995). Nursery planning, development, and management. In *The Container Tree Nursery Manual, Vol. 1. Agric. Handbook 674.* Washington, DC: Forest Service, USDA.

CHAPTER

17

Producing Nursery Crops

ABSTRACT

This chapter begins with a brief background on nursery production in the United States, followed by a discussion on the production of container- and field-grown nursery crops. Eight important factors to consider when growing nursery crops in containers are the product mix, propagation media, propagation container, bed design, overwintering plants, fertilization, watering, and pest control. The five key factors to consider when growing nursery crops in the field are the product mix, site preparation, plant spacing, nursery crop production systems, and crops maintenance.

Objectives

After reading this chapter, you should be able to

- provide background information on nursery production in the United States.
- discuss container nursery production.
- discuss field nursery crop production.

Key Terms

agricultural limestone
balled and burlapped (B&B)
bare root method
dolomitic limestone
green manure
indicator plants
pot-in-pot method
product mix

INTRODUCTION

This chapter provides information on producing both container- and field-grown nursery crops. Nursery crop production in the United States is primarily concerned with wholesale and retail sales of woody ornamental plants such as trees, shrubs and vines, ground covers, and sod for the establishment of turf (Figure 17-1). In addition to ornamentals, nursery growers have expanded to sell fruit and nut trees plus small fruits and perennial vegetables. The top-selling nursery crops are given in Table 17-1. The nursery industry is one of the fastest growing segments of agriculture, with California being the leading state for production. In the nursery industry, the trend is toward producing container-grown nursery stock because it is less labor intensive, the grower has more control over the media used, and higher plant densities are possible. Another reason why container-grown crops are becoming more popular to consumers is that plants recover more quickly than field-grown crops after transplanting, and they can be marketed year round.

After selecting a nursery site, the first step is to run a marketing survey to decide what **product mix** (the types of plants to be grown) should be used. Typically, around 70 percent of the plants grown should be industry standards such as shade trees and others as shown in Table 17-1.

Figure 17–1 Christmas trees being grown in a field nursery

TABLE 17-1 NURSERY PLANT STANDARDS SOLD IN THE UNITED STATES

Shrubs	Deciduous Shade Trees	Flowering Trees
Roses	Oak	Crabapple
Sporeas	Red maple	Crape myrtle
Hibiscus	Ash	Dogwood
Hydrangeas	Japanese maple	Flowering cherry
Vibrunum	Birch	Magnolia
Lilacs	Honey locust	Flowering plum
Buddleias	Norway maple	Redbud
Weigelas	Sugar maple	
EVERGREEN		
Broad Leaved	**Needled**	
Holly	Juniper	
Azaleas	Pine	
Rhododendron	Spruce	
Euonymus	Arborvitae	
Boxwood	Yew	
Privet	Cypress	
Magnolia	Fir	
Cotoneaster	Cedar	

The remaining 30 percent should be new plants that have potential, but not high-risk plants.

The proper selection of media and containers is important to the successful growth of plants. Nursery bed designs vary in design and size depending on a variety of factors that include the size of the pots, container spacing, production practices, irrigation design, and coverage. Nurseries in many parts of the United States need to protect plants during the winter months. The roots of container-grown plants are not insulated against the cold like field-grown crops, so container-grown crops must be overwintered properly. Fertilization and watering of container-grown crops are very important to make sure that the proper amounts of nutrients and water are provided to maximize plant growth and development. Pest control in container nurseries is the same type that has been stressed throughout this text; the first line of defense is prevention. When all efforts to prevent have failed, the pest should be identified and a good Integrated Pest Management program (IPM)—including biological, mechanical, cultural, genetic, and, when necessary, chemical control—should be implemented.

The same general rules apply to field nurseries when deciding on the product mix—70 percent of the crops grown should be standards and 30 percent should be those that have potential. The main difference between container- and field nurseries is that field nursery plants are grown to a desired size in the ground instead of in containers. Good site preparation is vital for high-quality nursery stock grown in the field. The soil should first be tilled to loosen it, and any large objects that surface must be removed. Soil samples should be taken to analyze for nutrients as well as pathogenic soil fungi, insects, and weeds so the proper control measures can be taken prior to planting. Depending on the location, it may be necessary to install drainage and irrigation systems. Plant spacing is another important consideration when deciding on the proper placement of nursery crops. The proper distance between the rows should be determined based on the final size of the plant because, unlike plants in container nurseries, after plants are in the ground, it is time consuming and labor intensive to move them.

Another important factor to consider is the type of equipment required for harvesting and other nursery operations. Two major types of production systems are commonly used to remove crops grown in the field. The **balled and burlapped** (B&B) production system is when trees are grown in the field, dug up with a ball of soil surrounding the root system, and covered with burlap material. This method minimizes transplant shock but is expensive. The second type of production system is the **bare root** method, which describes trees grown in the field and dug without taking soil. Smaller trees and shrubs are harvested using this method. In the bare root method, plants are lightweight and thus shipping is more economical and the initial cost per plant to the grower is lower. However, the main disadvantage is a problem with transplanting shock which is a period of no growth following planting.

Properly maintaining field-grown crops is important for maximizing their growth and development. This is accomplished by routinely running soil tests to determine when to fertilize and how to adjust the pH to insure that nutrients are readily available to the plant. The correct moisture levels must be supplied after the nursery crop is planted because too much or too little water will have adverse effects. Supplemental irrigation following establishment of nursery crops is necessary when moisture levels in the soil become limited. Pest control should be taken care of using a good IPM program.

NURSERY PRODUCTION IN THE UNITED STATES

Nursery crop production is primarily concerned with wholesale and retail sales of woody ornamental plants. Woody materials grown in nurseries can be deciduous or evergreen. In addition to the ornamentals, nursery growers are producing fruit and nut trees plus small fruits such as strawberries, grapes, and perennial vegetables such as asparagus. The top nursery crops include deciduous shade trees, deciduous flowering trees, broadleaf evergreens, coniferous evergreens, deciduous shrubs, ground covers, vines, palms, fruit and nut plants, and miscellaneous crops with potential that varies among nurseries.

The horticulture industry is one of the fastest growing segments of agriculture. Ornamental crop retail sales in 1998 were almost $9.3 billion; $3.1 billion was accounted for by nursery crops, including both woody and herbaceous crops. California ranked number one with respect to sales, which were nearly 28 percent of all nursery production in the year 2000. California, Oregon, and Florida together accounted for 57 percent in the year 2002. The top 10 states for nursery production in terms of sales dollars in the United States in 1998 were California, Oregon, Florida, Ohio, Texas, Michigan, North Carolina, Illinois, Pennsylvania, New Jersey, and New York.

Nurseries can be broken down into wholesale, retail, landscape, and mail-order types. Wholesale nurseries specialize in growing a few crops on a large scale, and they then sell their material to other nurseries, landscapers, and lawn and garden centers. In the past, retail nurseries grew their own plant material; however, today they purchase their plant material from wholesale growers for distribution to the consumer. Landscape nurseries typically sell their retail plants, plus they offer a wide range of landscape services such as design, installation, and maintenance. Mail-order nurseries sell on the national level for people to purchase via catalogs and the Internet. Nurseries can be broken down further into container nurseries and field nurseries (Landis et al., 1995).

Container Nursery Production

The trend in the nursery industry is toward container-grown nursery stock because it is less labor intensive, the grower has more control over the media used, and higher plant densities are possible. Container-grown stock is better accepted by consumers, the plants recover more quickly after transplanting, and they can be marketed year round (Figure 17-2). Container-grown plants are not without problems, however. Water and fertilizer must be carefully tended to, winter protection is required because the root systems are not protected as they would be in the field, containers and potting labor are costly, and root stress and rootbound plants can become a problem. The main reason container crops are so popular is that container-grown nursery crops have 8 to 10 times as many sales per acre as do field-grown plants.

Determining the Plants to Be Grown

The proper product mix is important for a successful nursery operation. The product mix is basically what crops will be grown, which is determined by an extensive marketing analysis. Typically, around 70 percent of the plants grown are industry standards and 30 percent are new plants that have potential. Examples of standards are shrubs (roses), shade trees (red maple), flowering trees (dogwood), broadleaf (holly), and needled evergreens (juniper); other examples were shown previously in Table 17-1.

Selecting the Proper Media

Because nursery crops are in containers for a long time, growing media are very important. A good propagation medium should firmly hold cuttings in place, be free of insects and diseases, have good water- and nutrient-holding capacity,

Figure 17–2 Nursery crops in containers

and provide excellent aeration and drainage. If the medium does not have all these attributes, it will be difficult to grow high-quality crops, as is the case for greenhouse crops.

Selecting the Proper Container

Because container-grown nursery crops remain in their containers for a long time, the type of container is important. The factors to consider include size, cost, adequate drainage, proper color, weight, and ease in handling while being durable to facilitate shipping and handling. Containers should be free of toxic substances and able to prevent root circling (Figure 17-3).

Factors to Consider when Designing Container Beds

In container nurseries, beds are produced to enable plants that are similar in size and cultural requirements to be grouped together to maximize production. Beds vary in design and size depending on the size of the pots, container spacing, production practices (such as fertilization and pruning), irrigation design, and coverage. After the site is leveled, there are two potential designs. One design has the center of the bed raised (Figure 17-4A) to allow any excess water during rain or watering to drain to the outside. A second version has the slope going toward the center of the bed to allow the water to drain out of the way of workers (Figure 17-4B).

Overwintering Container-Grown Plants

Nurseries in most parts of the country need to protect plants from low temperatures during the winter months. The roots of plants are not as cold hardy as the shoots, making container-grown plants susceptible to winter damage because the root system is not insulated by the soil like the roots of field grown crops are. For sensitive crops or in areas where winter protection is required,

Figure 17–3 Root circling

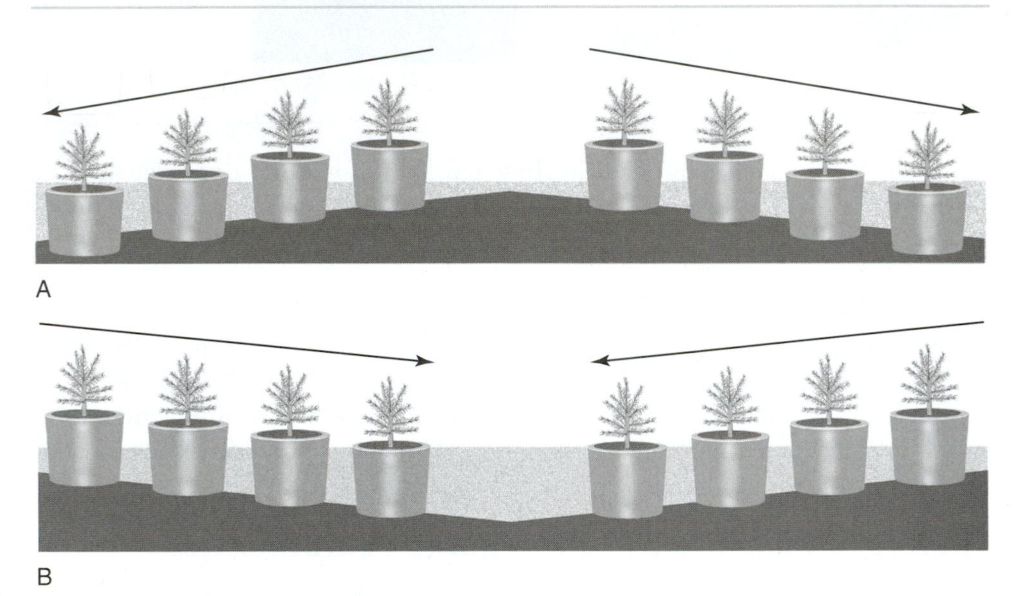

Figure 17–4 Two main bed designs are used to grow nursery crops in containers. A. Center of the bed is raised allowing water to drain to the outside. B. The slope is higher on either end allowing water to drain toward the center of the bed.

there are many ways plants can be protected. One way to provide winter protection is to place pots close together and wrap them with plastic around the outside edge (Figure 17-5). Another way is to put a border row of containers filled with soil but no plants to protect the pots containing plants (Figure 17-6). A combination of the two methods—using plastic and a border row of containers with soil only—can also be used to protect plants during the winter months. Overwintering houses covered with white polyethylene are frequently used for winter protection; based on location, these overwintering structures can either have supplemental heat or only solar heating.

A production practice that can be used to protect roots of plants against extremes in temperature is the **pot-in-pot method.** With this method, larger

Figure 17–5 Winter protection is provided to container-grown nursery crops by putting them close together and wrapping them with plastic around the outside edge.

Figure 17–6 Winter protection is provided to container-grown nursery crops by putting a border row of containers filled with soil but not plants to protect the pots containing plants.

pots (holder pots) are placed in holes in the ground and buried up to their lips. Containers with plants are then placed into the holder pots (Figure 17-7). By using this method, there is typically no need to protect the roots from winter injury. During the summer months, temperatures in the ground may get too high; therefore, precautions must be taken to overcome this condition at locations where high temperatures are a potential problem.

Maintaining Container-Grown Nursery Crops

Proper maintenance of container-grown nursery crops includes fertilizing, watering, and controlling pests.

Fertilization of container-grown plants. Most growing media used in nurseries contain none or only a few nutrients, so it is important to use preplant

Figure 17–7 Pot-in-pot method for growing nursery crops. *Courtesy of Dr. James Sellmer, Department of Horticulture, The Pennsylvania State University.*

amendments. Preplant amendments include **dolomitic limestone** (calcium and magnesium) for pH control and a controlled-release fertilizer. The nutritional status of the growing media and the plant must be regularly monitored through foliar and soil analysis. Three types of media tests are available: electrical conductivity (EC), pH, and nitrate analysis. Fertilizers and other dissolved salts change the capability of a solution to conduct electricity. Pure water is a poor conductor; however, as the salt content increases, the conductivity of the solution also increases. Today portable conductivity meters can be used by the grower to measure the EC of solutions. The growing media should be tested every one to two weeks for salts, nitrates, and pH. Portable EC, pH, and nitrate meters are excellent for this purpose. A chart of the pH, EC, and nitrate levels should be kept to determine when fertilization or pH control is necessary. Visual symptoms are not good indicators of when to apply nutrients, because when symptoms appear, plant growth and development have already been adversely affected. If deficiency symptoms occur, quick-release fertilizers should be applied to get rapid results. Quick-release fertilizers must be applied in the proper amount. If too much fertilizer is added, it will burn the plants' root systems, and if not enough is applied, the deficiency symptoms will persist.

Watering container-grown plants. Proper watering of container-grown plants is one of the hardest tasks for new employees to learn. Irrigation should be based on the demand of the crop, which can be determined in several ways. One way an experienced grower can determine when water is needed is by feeling the weight of the pot or actually weighing the pot. Another method is simply to feel the growing medium, although it takes experience to know the correct feel. **Indicator plants** show signs of wilting prior to crop plants, thereby signaling when crop plants should be watered. Moisture sensors strategically placed in pots are also a very good method for determining when to water container-grown plants. Irrigation should take place during the early morning hours to minimize evaporative losses. On hot, sunny days it may be necessary to apply water more than one time per day.

Pest control. The type of pest control should be the same as stressed throughout this text. The first line of defense, if possible, is prevention. When all efforts to prevent have failed, the pest must be properly identified and then a good IPM program must be implemented that includes biological, mechanical, cultural, genetic, and chemical control (when necessary).

Field Nursery Production

Factors to consider when setting up a field nursery include determining the plants to be grown, site preparation, plant spacing, and the production system to be used to remove the nursery crop from the field.

Determining the Plants to Be Grown

The same rules that are used to determine the product mix for a container nursery apply to field nurseries. Based on an extensive marketing analysis, typically 70 percent of the nursery crops produced are standards such as shade trees, flowering trees, and other examples (shown previously in Table 17-1) and 30 percent are plants that have potential. The 30 percent that rounds out the operation is not high risk; rather it is truly plants that have potential as determined based on a market survey.

Site Preparation

The main difference between container and field nurseries is that field nursery plants are grown to a desired size in the ground, not in containers. Good site preparation is essential for high-quality nursery stock grown in the field. When preparing the site for nursery crops, many tasks must be completed. The soil should be plowed to loosen it, and any large objects uncovered during the plowing operation should be removed. To determine what nutrients need to be applied, soil samples should be collected and analyzed. In addition to soil analysis for nutrients, the soil should be tested for pathogenic soil fungi, insects, and weeds so that the proper control measures can be taken prior to planting. Organic matter and/or sand may need to be added to improve the tilth of the soil. Cover crops, also called **green manure,** can provide a good source of organic matter while preventing erosion. In some cases, installing drainage and irrigation systems may be necessary depending on the location.

Plant Spacing

When deciding the distance between the rows of plants, growers must consider the final size of the plant being grown. Once plants are in the ground, it is time consuming and labor intensive to move them. Another important factor to consider is the type of equipment required for harvesting and other nursery operations. For example, if plants are to be harvested by hand, less space is needed than if mechanical units are used.

Nursery Crop Production Systems

The two commonly used systems for removing nursery crops from the field are the balled and burlapped (B&B) and bare root. In the B&B production system, trees are grown in a field, dug out with a ball of soil surrounding the root system, and covered with burlap material (Figure 17-8). This can be done by hand or mechanically. The main advantage of this method is that transplanting shock is minimized, whereas the main disadvantage is the cost of

Figure 17–8 The balled and burlapped method for planting nursery crops

plant removal and shipping. Soils with a higher clay content typically work best for this method. The second type of production system is the bare root method, in which trees grown in a nursery or field are dug out without taking soil. Smaller trees and shrubs are harvested using this method. The advantages associated with this method are that the plants are lightweight, making shipping more economical, and the initial cost per plant to the grower is lower; however, the main disadvantage is that there is a problem with transplanting shock. Soils with a higher sand content typically work best for this method.

Maintaining Field-Grown Nursery Crops

Proper maintenance of field-grown nursery crops includes pH adjustment and fertilization, irrigation, and pest control.

pH Adjustment and fertilization. In some areas of the United States, the soils are acidic so a soil test is necessary to determine the pH and what adjustments need to be made. Lime materials are applied to raise the soil pH; **agricultural limestone** contains calcium, whereas dolomitic limestone contains calcium and magnesium. When liming agents are added, first they supply calcium and increase the pH, which reduces the effects of calcium leaching. Second, when dolomitic limestone is applied, magnesium is supplied and phosphorous, molybdenum, and magnesium are increased. Third, the availability of aluminum, manganese, and iron is reduced, which decreases their harmful effects. Fourth, the pH change enhances favorable microbial activity, making essential elements more available for plant growth and development. Fifth, the soil structure is also improved. In most areas of the United States, the soils are not acidic, so limestone is not required. However, in areas where the soil pH is high, sulfur compounds are used to lower the pH.

Species differ in their capability to take up and accumulate elements, so it is difficult to make a generalized statement on nutrient requirements. Prior to planting nursery crops in the field, a soil test must be run to determine what nutrient elements are required. The best way to supplement these elements is to grow cover crops one or two years prior to planting. The combination of fertilizer and cover crops will improve both the nutrient levels and the structure of the soil. Fertilizer applications should be based on the soil test and the kind of cover crops grown. Fertilizers high in phosphorus should be applied prior to planting nursery stock to provide a source of phosphorus for several years. Applying the proper cover crop together with the correct amounts of fertilizer and the proper pH maximizes the growth and development of nursery crops.

Irrigation. The correct moisture levels must be supplied after the nursery crop is planted, because too much or too little water will have an adverse effect. Supplemental irrigation is important during transplanting and establishment phases because in both of these phases it has been shown to affect the survival and growth rate of field-grown plants. The proper way to establish an irrigation schedule for nursery stock is through the use of tensiometers and close evaluation of environmental conditions. In general, 1 inch of water is typically required to irrigate nursery stock after it has been determined that water is required.

Pest control. Nursery crops must be protected from pests that will minimize their growth and development. Although difficult, prevention is the first line of defense for your nursery; however, if this is not possible, the pest should be properly identified and IPM practices should be implemented, including biological, mechanical, cultural, genetic, and chemical control (when necessary).

SUMMARY

You now better understand nursery production in the United States in both container- and field-grown forms. Eight important factors must be considered when growing nursery crops in containers: product mix, propagation media, propagation container, bed design, overwintering plants, fertilization, watering, and pest control. The five key factors to consider when growing nursery crops in the field are the product mix, site preparation, plant spacing, nursery crop production systems, and crop maintenance.

Review Questions for Chapter 17

Short Answer

1. What are eight important factors that should be considered when growing nursery crops in containers?
2. What is the proper product mix for a successful nursery and how is it determined?
3. What are four factors that a good propagation medium should have?
4. What are seven factors that should be considered when selecting the proper propagation container?
5. How can root circling be prevented?
6. What are five factors that should be considered when designing a bed for container-grown crops?
7. Why is it important to overwinter container-grown crops, and what are three ways to protect container-grown crops from winter damage?
8. What are five important factors that should be considered when growing nursery crops in the field?
9. What are four factors that should be considered in site preparation for field-grown nursery crops?
10. What are two major factors that should be considered when planning the proper plant spacing for a field nursery?
11. What are the pros and cons of balled and burlapped and bare root methods of nursery crop production?

Define

Define the following terms:

product mix
indicator plants
balled and burlapped
bare root
pot-in-pot method

True or False

1. Container-grown plants are susceptible to winter damage because their root systems are not insulated by the soil like the root systems of field-grown plants are.
2. One method to determine when a container-grown nursery crop should be watered is to use indicator plants.
3. A method commonly used to determine when a container-grown nursery crop should be watered is moisture sensors.
4. A method used to determine when a container-grown nursery crop should be watered is simply to feel the weight of the pot.
5. A method to determine when a container-grown nursery crop should be watered is simply to feel the soil.
6. Irrigation should be based on the demand of the crop.
7. To conserve water, irrigation applications should take place during the early morning hours; doing so minimizes evaporative losses.
8. Copper compounds are used to prevent root circling in container-grown plants.
9. The main advantage of the balled and burlapped method is the initial lower cost per plant to the grower.
10. The same general rules used to determine the product mix for a container nursery also apply to field nurseries.
11. Good site preparation is essential for high-quality nursery stock plants; therefore, it is important to control pathogenic fungi, insects, and weeds prior to planting.
12. The main advantage of the balled and burlapped method of production is that it is cheaper than other methods.
13. The balled and burlapped method of production typically requires that the soil in which plants are grown has a high clay content.
14. The main advantage of the bare root system of production is that it is cheaper and there is little problem with transplanting shock.

Multiple Choice

1. In 1998, the number one state for nursery production in terms of sales dollars in the United States was
   A. Oregon.
   B. New York.
   C. California.
   D. Florida.
2. For a nursery to be successful, it must have the proper product mix. Which of the following describes the proper product mix?
   A. Grow 50 percent of standard plants with 50 percent of new plants that have potential.
   B. Grow a mixture of all standard plants.
   C. Grow 70 percent of standard plants with 30 percent of new plants that have potential.
   D. None of the above

Fill in the Blank

1. Dolomitic limestone contains ____________________ and ____________________.

Activities

Now that we have completed our discussion on growing nursery crops, you will have the opportunity to explore this area in more detail. In this activity, you will visit a local nursery and gather the following information:

- determine whether it is a container or field nursery.
- determine the product mix (types of plants being sold).
- list the different plants being grown and include information about how they are being grown.
- if it is a container nursery, describe the containers.
- if it is a field nursery, discuss plant spacing and the nursery production system.

In addition to information you were asked to gather, include your thoughts on the production practices used. If you do not have access to a nursery, surf the Internet for two sites that present information on growing nursery crops. Summarize these sites and provide the Web site address where you found your information.

Reference

Landis, T. D., Tinus, R. W., McDonald, S. E., & Barnett, J. P. (1995). *Nursery planning, development, and management.* In *The Container Tree Nursery Manual, Vol.1. Agric. Handbook 674*. Washington, DC: Forest Service, USDA.

CHAPTER

18 Floral Design

ABSTRACT

Flowers have played a major role in all civilizations since the beginning of recorded history. Background information on the uses of cut flowers, the different types of permanent flowers, and commonly used flowering plants for cut flowers is provided in this chapter. Flowers must be harvested at the proper time, and after the flowers are harvested, they must be handled properly. The five principles of floral design include proportion, scale, balance, rhythm, and dominance. The seven design elements are color, line, form, shape, space, texture, and pattern. The flower designer must follow six key design rules to create a successful design. The tools and materials commonly used for designing flowers include containers, cutting tools, floral foam, straight wire, ribbon, water tubes, floral tape, and floral clay.

Objectives

After reading this chapter, you should be able to

- discuss ways in which flowers were used in past civilizations.
- provide background on the uses of cut flowers.
- provide background information on the different types of permanent flowers.
- recognize flowering plants that are commonly used for cut flowers.
- discuss the proper time for harvesting certain types of flowers.
- outline the steps involved in conditioning flowers and foliage after receiving them in the retail or wholesale outlet.
- list the principles of floral design.
- discuss design elements that are physical characteristics of plant materials.
- discuss the six key design rules that are required in order to have a successful floral design.
- understand the commonly used tools and materials for designing flowers.

Key Terms

balance
cut flowers
dominance
floral form
floral garland
florists
form
hand pruner or floral shears
Ikenobo Design
line
proportion
rhythm
scale
shape
space
texture
vase life

INTRODUCTION

This chapter provides some of the basics of floral design. Flowers have played a major role in all civilizations from the start of recorded history. The first written record of the use of flowers was by the Egyptian civilization in 2800 B.C.; Egyptians are known for their large showy arrangements. The Greeks followed the Egyptians in around 600 B.C.

Figure 18–1 Display of assorted arrangements at a flower shop

and are known for scattering flower petals on special occasions and for wearing or carrying wreaths. The Romans then conquered the Greeks and are known for the use of **floral garland.** During the same period of time as the Greek and Roman civilizations, Oriental floral design emerged in India and then spread to China and Japan. Around the fall of the Roman Empire, Europe was formed, which led to designs that were distinct for the different regions: Renaissance, Baroque, Flemish, and Victorian. Early settlers from Europe who came to the United States brought with them their European cultures and floral designs.

Cut flowers are detached from the parent plant for use in floral arrangements, corsages, bouquets, and more (Figure 18-1). Fresh cut flowers have a very short life span, so permanent flowers such as artificial, silk, and dried flowers have increased in popularity; however, fresh cut flowers are still the most popular and are used for a variety of occasions. Both woody and herbaceous plants are used for cut flower production, and different species can retain their desirable qualities for different periods of time. Plants grown for cut flowers should have vigorous growth and be disease free. At what stage of floral development the flower should be removed from the parent plant depends on the species. After flowers or foliage are detached from the parent plant, they must be properly cared for and handled. Steps for conditioning flowers and foliage include unpacking and inspecting; prioritizing the order of processing flowers; removing sleeves, ties, and any foliage below the water line; recutting stems and putting them in warm water containing floral preservatives; placing in the light at room temperature; and then placing in the cooler.

Flower arranging is an art; however, there are five principles of floral design: **proportion, scale, balance, rhythm,** and **dominance.** In addition, the seven key design elements are physical characteristics, including color, **line, form, shape, space, texture,** and pattern. A successful floral designer pays careful attention to six key design rules that affect the design from top to bottom. The floral designer must also be aware of the correct tools and materials now available that make designing flowers easier and make designs look their best. Some key tools and material include containers, cutting tools (such as pocketknives, floral shears, ribbon shears, and wire cutters), floral foam, straight wire, ribbon, water tubes, floral tape, and floral clay.

HISTORY OF FLORAL DESIGN

Written records trace the first use of flowers to the Egyptian civilization in the year 2800 B.C. The Egyptians used flower arrangements to beautify their homes and in ceremonies as offerings to their gods and to their dead. They are known for their showy flower arrangements in large vases (Figure 18-2). A typical Egyptian arrangement included roses, water lilies, violets, and narcissus flowers arranged together with a variety of types of foliage such as ivy and palm leaves. Around the year 600 B.C., the Greeks began to use flowers. Unlike the Egyptians, however, the Greeks were the first to scatter flower petals on the ground during weddings and other festive occasions. In addition, Greeks were the first to carry

Figure 18–2 Ancient Egyptians would have liked this showy flower arrangement in a large vase.

Figure 18–3 The Greeks were first to wear floral wreaths and scatter petals. *Courtesy Getty Images Inc.*

and wear floral wreaths (Figure 18-3). The Romans conquered the Greeks and took over many of the Greek customs. In fact, the Romans modified the floral wreath to make the floral garland, which is a noncircular wreath made of flowers, foliage, or a combination of the two (Figure 18-4).

During the same time period as the Greek and Roman civilizations developed their design skills, Oriental floral design emerged. Buddhists priests in India began placing flowers on altars in religious ceremonies. Their religious beliefs prevented them from killing live plants, but did allow them to use flowers that had broken off the plant naturally. Around the beginning of the first century, the Chinese adopted the Buddhist religion from India. They thought it was not proper just to lay flowers on their altars, so they developed a practice of placing the flowers in large containers. Around 600 A.D., Buddhism was introduced to Japan. The Japanese culture created a style of its own called the linear asymmetrical balance. At the beginning of the eleventh century, a Buddhist priest named Ikenobo developed a specific method for Japanese floral arrangements called **Ikenobo Design,** in which each flower has a specific meaning and an exact location in the arrangement.

Figure 18–4 The Romans created floral garland, shown here as decoration.

After the fall of the Roman Empire around 300 A.D., the emerging countries of Europe including France, Italy, Germany, and England all developed the use of flowers distinct to their region, such as Renaissance, Baroque, Flemish, and Victorian. Early settlers from Europe who came to America brought with them their European cultures. The Victorian design became the pattern for early American floral arrangements. The term *Williamsburg design* describes the American version of the Victorian style. The Williamsburg style is a large, round-shaped floral arrangement commonly used in the United States today. In the 1920s, a small bouquet for women to wear, known as a corsage, became very popular and is still very popular in the United States today.

BACKGROUND ON CUT FLOWERS

Cut flowers are grown for the sole purpose of removing them from their parent plant to display them in containers, use them in corsages, and more. **Florists** are individuals who use cut flowers in their trade. Cut flowers can be sold individually or in bunches; however, to add to their value, florists make floral designs for a variety of special occasions. Fresh cut flowers have a very short life span, so dried flowers and permanent flowers, such as silk or polyethylene, are also used in certain cases. Cut flowers can be used for a variety of occasions; for example, weddings, funerals, Valentine's Day, Mother's Day, and Easter. Today, working in a florist shop is a lot of fun; however, it is also a tough profession. For example, there are long hours on holidays, and work comes in flushes for weddings and other special occasions.

PERMANENT FLOWERS

During the early 1950s, technology became available for the inexpensive production of flowers made of polyethylene plastic (at this time they were called plastic flowers). To improve the image of this type of flower, florists later called them artificial flowers, and these are still used today in specific cases. In the 1970s, silk flowers became very popular. Although called silk flowers, they are actually made of polyester fabric.

In the 1980s through today, dried flowers and weeds have become very popular. Flowers may be dried either by natural or artificial means, depending on the species being dried. To dry flowers naturally, they are cut and then hung upside down in bunches or singly in a well-ventilated dry room until they become brittle. Examples of species that can be dried naturally are *Gypsophila* (baby's breath), *Eryngium* (holly), and *Limonium* (statice). Plants with fruits or pods such as poppy, foxglove, and Chinese lantern are also excellent examples of flowers that can be dried naturally. Artificial or chemical drying uses special drying agents to dry flowers rapidly. Some examples of drying agents include silica gel or a mixture of silver, sand, and borax. Flowers are dried by placing the chemical used for drying in a container, placing the flower on top of the chemical, and then adding more drying agent on top of the flower. The container is then placed in a warm place and allowed to dry for three to four days depending upon the flower. Examples of flowers that can be dried by this method are *Delphinium, Pelargonium,* fuchsia, and rose. A commonly used method for drying leaves from trees, such as horse chestnut, oak, and mountain laurel, is to put the end of the petiole in a 1:2 mixture of glycerine and water.

COMMONLY USED FLOWERING PLANTS FOR CUT FLOWERS

Both woody and herbaceous plants are used for cut flower production. The most commonly used cut flowers are roses, carnations, and chrysanthemums. In the cut flower industry, annual herbaceous flowers are most commonly used, and these require planting every season and high maintenance to produce a high-quality product. Woody plants used for cut flowers have the disadvantage of requiring a long time to establish the crop before flowers can be harvested; however, as perennials, they require less maintenance. Each species and cultivar has a different **vase life,** which is the amount of time a cut flower retains its desirable qualities prior to deteriorating. Some examples of commonly used herbaceous cut flowers are baby's breath (*Gypsophila elegans*), carnation (*Dianthus caryophyllus*), chrysanthemum (*Chrysanthemum* spp.), freesia (*Freesia refracta*), gerbera daisy (*Gerbera jamesonii*), snapdragon (*Antirrhinum majus*), zinnia (*Zinnia elegans*), nasturtium (*Tropaeolum majus*), and globe amaranth (*Gomphrena globosa*). Examples of commonly used woody cut flowers are rose (*Rosa* spp.), hydrangea (*Hydrangea arborescens*), pussy willow (*Salix matsudana*), forsythia (*Forsythia x intermedia*), and holly (*Illex* spp.).

HARVESTING FLOWERS

Plants grown for cut flowers should have vigorous growth and be disease free. Cut flower plants must be provided with optimal growing conditions throughout their life cycles to maximize their postharvest life. In addition to growing plants for cut flowers under the proper conditions, harvesting at the optimal stage of development is also critical. For example, chrysanthemums are harvested when their flowers are fully open, whereas roses are harvested when they are partially open (first two petals starting to unfold) and peonies and German irises are harvested when the buds begin to show color.

CARE AND HANDLING OF FLOWERS AND FOLIAGE AFTER HARVEST

To increase the vase life of cut flowers, both growers and consumers must take care of them properly. Immediately after harvest, the grower should remove the field heat because wilting greatly reduces the vase life of cut flowers. By rapidly reducing the storage temperature to between 32 and 35°C, respiration and water loss is minimized. After removal of the field heat, maintaining a high relative humidity of approximately 90 percent is critical together with proper air circulation to minimize water droplets forming on flowers and leaves. When water droplets form, flowers are subject to fungal attacks such as the gray mold caused by *Botrytis*. Following harvest, cut flowers must have good-quality water. Hard water can be harmful to many cut flowers. Lastly, during shipment and handling, it is important to minimize the exposure of cut flowers to ethylene, which is a very powerful plant hormone that leads to the deterioration of most cut flowers.

Retailers should use the following steps to conditioning flowers and foliage:

1. **Unpack and inspect.** Immediately after receiving flowers, unpack and inspect them as quickly and efficiently as possible.
2. **Prioritize the order of processing flowers.** Prioritize based on the expense of the flowers and then on condition.
3. **Remove sleeves, ties, and any foliage that will be below the water line.** Leaves are removed to prevent them from deteriorating in the water and ties are removed to overcome potential restriction in water uptake. The sleeves should be removed to provide the flowers with space and to release any unwanted gasses, such as ethylene, that may have become trapped during shipping.
4. **Recut all stems and put in a plastic or glass container (do not use metal) with warm water between 100 and 105°F.** This facilitates water uptake and fully hydrates the cut flower.

5. **Add floral preservatives.** Floral preservatives such as Floralife (silver thiosulfate) should be dissolved in the warm water mentioned in the previous step. The main function of silver thiosulfate is to block ethylene action and prevent bacterial growth.
6. **Place cut flowers at room temperature in the light.** This should be done to allow for the uniform uptake of water containing preservatives.
7. **Place cut flowers in a cooler.** The cooler should be set at between 34 and 38°F with high humidity, good air circulation, and constant lighting.

PRINCIPLES OF FLORAL DESIGN

Flower arranging is an art; however, there are guidelines to help a floral designer create a beautiful arrangement of flowers. Principles of design include the following:

- **Proportion and scale.** This is the relationship between size and shape, including specific characteristics of the flowers, the container used, the table or shelf, and the room they are in. Scale is the part of proportion dealing with the relative size among things only, not shapes. The arrangement should be one and one-half times the container height or width. The flower arrangement should not overpower the room or be too small for a given location.
- **Balance.** This is the physical and visual stability of a floral design. A symmetrical design is equal in size and shape on both sides of the central axis of the arrangement, whereas asymmetrical designs have two sides that are not equal (Figure 18-5). Typically centerpieces are symmetrical,

Figure 18–5 Symmetrical arrangement on left and asymmetrical flower arrangement on right

which enables people sitting at the table to see the same beauty no matter where they sit.

- **Rhythm.** The arrangement should have an ordered flow. In other words, all aspects of the arrangement should be tied together to create a pleasing effect.
- **Dominance.** One design element or characteristic should be more noticeable than others, thereby creating a focal point.

Design Elements

Design elements are the physical characteristics of the plant materials that a designer uses. The following are design elements:

- **Color.** Color is the most important element in the visual arts. All colors can be made from the three primary colors—red, yellow, and blue. When two primary colors are combined, they create secondary colors—green (made by combining yellow and blue), violet (combining blue and red), and orange (combining red and yellow). In an arrangement, you can create a feeling of warmth by using red and yellow colors or a cooling effect by using blue and green colors. When color is used wisely, the floral design turns out beautiful and can also help mask problems associated with a given design (Figure 18-6).
- **Line.** This is the movement between two points within a design, which can be created with linear or round plant material (Figure 18-7). Changing the heights of linear flowers can produce lines in a floral design. A line within an arrangement can also be made with round flowers by using the same-size flowers and alternating their position in the line, or round flowers can create a line by starting with small flowers at the top and becoming larger as they progress toward the bottom of the arrangement. Round flowers can also be used to create a line by starting with smaller, light-colored flowers at the top of the arrangement and then using larger and darker flowers as they get closer to the container.

Figure 18–6 Floral arrangement with the proper use of color

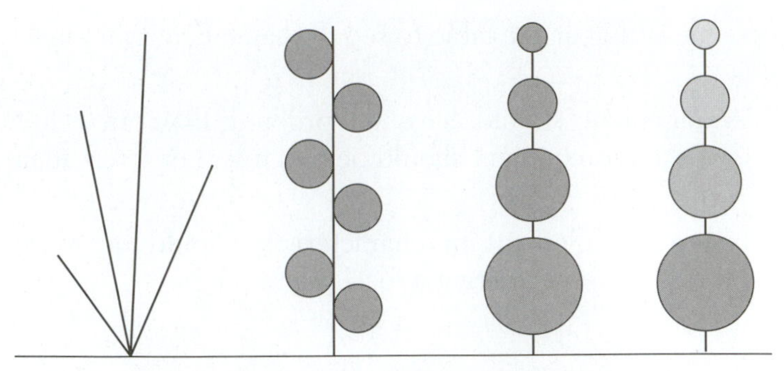

Figure 18–7 Both linear and round plant materials can be used to create a line.

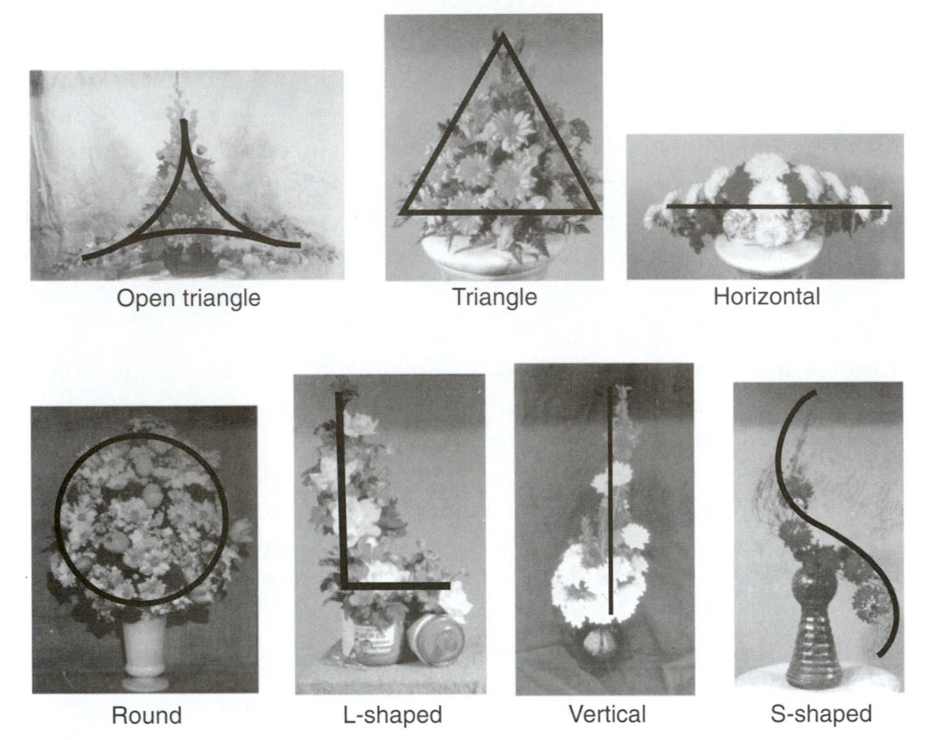

Figure 18–8 Different shapes of floral designs. *Courtesy of Dr. E. Jay Holcomb, Department of Horticulture, The Pennsylvania State University.*

- **Form.** Form is the three-dimensional shape of the outline of the floral design.
- **Shape.** Shape is the two-dimensional term for form. Examples of basic floral design shapes are triangle, symmetrical triangle, open triangle, L-shaped, vertical, Hogarth or S-curve, horizontal, and round shapes (Figure 18-8).
- **Space.** This is the distance between plant materials found in the arrangement. In a floral design, some parts should be very dense and other parts should be open spaces to feature uniquely shaped flowers such as irises and daisies. Generally, more space is left at the top of the arrangement with the flowers getting progressively more dense as they near the top of the container.
- **Texture.** This is determined by the surface quality and placement of plant parts in the design. Surface texture qualities include smooth, rough, shiny, velvety, and others.

Figure 18–9 Four main shapes or forms used in floral arrangements

- **Pattern.** This is determined by the physical characteristics of the plant material, such as the arrangement of leaves and petals. Plant material used in an arrangement can be classified into four main shapes or forms (Figure 18-9): line, mass, form, and filler flowers.

Six Key Design Rules

The following design rules should be followed strictly to create a successful flower design:

1. The arrangement height begins with the proper placement of the first flower, as it is important to work from the top of the arrangement to the bottom.
2. Smaller flowers and/or buds should be placed at the top of the arrangement, whereas large fully opened flowers should be located near the bottom.
3. Light-colored flowers should be placed near the top of the arrangement and dark-colored flowers should be placed near the bottom.
4. Flowers should be placed farther apart at the top of the arrangement and get progressively closer at the bottom.
5. Placement of flowers should be from the back of the arrangement toward the front.
6. A focal point should be established at the bottom of the arrangement.

TOOLS AND MATERIALS USED FOR DESIGNING FLOWERS

The floral designer should be aware of the correct tools and materials, especially because new materials and supplies are coming on the market that make designing flowers easier and make designs look their best. The most commonly used tools and materials include the following:

- **Containers.** A variety of basic container design styles are commonly used. Each design requires a specific type of container with different

Figure 18–10 Different types of containers used for arranging flowers

geometrical shapes, colors, textures, and materials, such as glass, plastic, and styrofoam (Figure 18-10).

- **Cutting tools.** The pocketknife is one of the most important tools available to the floral designer for processing flowers for storage, cutting flowers to a desired height for an arrangement, cutting floral foam to be put into a container, and much more. **Hand pruners or floral shears** are used to cut thick or woody stems that cannot be cut readily with a pocketknife. Ribbon shears have long, sharp blades for cutting ribbon, fabrics, or other similar materials. Utility shears are used for cutting paper and plastic, trimming leaves or small branches off stems, and more. Wire cutters are also used for cutting wire used in making corsages or bouquets. A variety of types, sizes, and shapes of floral shears, ribbons shears, wire cutters, and utility shears are used by florists depending on the task and the designer's preference. All cutting tools must be sharp to maximize their efficiency.
- **Floral foam.** Water alone does not provide enough support when arranging flowers, so flowers are arranged in a block of absorbent material called floral foam. Floral foam comes in two basic formulations: one for fresh flowers and the other for dried plant materials. The standard size of a block of floral foam is about the size of brick (Figure 18-11). Floral foam can be cut to fit any size container with a pocketknife or a piece of wire. The foam should be soaked with water prior to placing it into the container. The foam should be placed into the container so that it extends above the top edge of the container. In some containers, wire mesh is put over the top of the floral foam to provide additional support.
- **Straight wire.** Wires are used for a variety of purposes, including making corsages and bouquets. Commonly used wire sizes are 18 to 30 gauge; the smaller the gauge, the thicker the wire. Different gauge

Figure 18–11 Block of floral foam used for arranging flowers

wires are used for different purposes; for example, #26 is used for bows, #22 for medium-sized flowers, #20 for larger flowers, #28 for filler flowers, and between #24 and #30 for corsages and bouquets.

- **Ribbon.** Ribbons come in different widths, including #1, #1.5, #3, #9, and #40; they also come in a variety of types such as satin, cotton, silk, sheer, burlap, or paper. Typically, smaller sizes (such as #1 to #3) are used for corsages and bouquets, larger types (such as #5 and #9) are used for potted plants, and extra large (such as #40) is used for funeral arrangements or larger arrangements.
- **Water tubes.** Water tubes are small rubber-capped plastic tubes that are used for holding water for a single flower or cluster of flowers. The rubber cap has a small hole for insertion of the flower stem. These tubes are used to extend the life of flowers used in funeral sprays, wreaths, and others. This method is a little more expensive than just wiring the flowers to a short stick or piece of metal, but the flowers last much longer because the water tube keeps the flowers hydrated.
- **Floral tape.** Floral tape is parafilm-coated paper, used to cover wires and stems and thus to enhance the beauty of the arrangement. Prior to being stretched, parafilm will not adhere to a container; however, once stretched, it adheres to the container and itself. Floral tape comes in half-inch and 1-inch widths in a wide variety of colors, such as green, yellow, black, red, white, and gray.
- **Floral clay.** Floral clay is used to hold materials in place, such as a piece of styrofoam in the bottom of a container used for a dried flower arrangement, or to hold a piece of chicken wire in place to support flowers.

SUMMARY

You are now familiar with the world of floral design. Flowers have played a major role in all civilizations since the beginning of recorded history. Cut flowers and the different types of permanent flowers are both used in floral design. Flowers must be harvested at the proper time and handled properly. The five principles of floral design are proportion, scale, balance, rhythm, and dominance. The seven design elements are color, line, form, shape, space, texture, and pattern. The chapter concluded with a discussion of the six key design rules that are required for a successful design and the tools and materials commonly used for designing flowers.

Review Questions for Chapter 18

Short Answer

1. Based on written records, which culture first used flowers?
2. What type of floral designs are Egyptians, Greeks, and Romans known for?
3. Oriental floral design began in three countries. List the three countries, designating which was first, second, and third to use flowers. Then, provide their contribution to floral design.
4. After the fall of the Roman Empire, many countries emerged in the region now called Europe. List two of the four design types created in those countries.
5. What design became the pattern for early American floral arrangements? What term describes the American version of this type of design?
6. What are three types of permanent flowers?
7. What are silk flowers made of?
8. What are the most commonly used cut flowers?
9. List the six steps that should be followed to condition flowers and foliage properly.
10. What are the five principles of floral design?
11. What are the seven design elements?
12. What are the six key design rules?

Define

Define the following terms:

floral garland	vase life	rhythm	shape
Ikenobo Design	proportion	dominance	space
cut flowers	scale	line	texture
florists	balance	form	pattern

True or False

1. According to written records, the Egyptian civilization was the first to use flowers.
2. Corsages were very popular in the Baroque era.

3. In Ikenobo Design, flowers are distributed randomly throughout the arrangement.
4. The Renaissance design became the pattern for early American floral arrangements.
5. Egyptians are known for their showy flower arrangements in large containers.
6. Greeks did not arrange flowers in vases or jars; they are known for scattering flower petals at weddings and wearing and carrying wreaths.
7. Romans were the first to use the floral garland.
8. Each of the European floral design styles has its roots in the traditional Greek and Roman cultures.
9. Silk flowers are made of polyester fabric.
10. Silver thiosulfate prevents bacterial growth and blocks ethylene action in cut flowers.
11. Prior to conditioning flowers and foliage that have been received, the order of processing should be prioritized based on the condition and expense of the flowers.
12. One of the six rules of floral design is to work from the top of the arrangement toward the bottom.
13. One of the six rules of floral design is to use smaller flowers or buds at the top of the arrangement.
14. One of the principles of floral design is proportion, which is the relationship between size and shape.
15. One of the principles of floral design is scale, which is a part of proportion dealing with size only, not shape.
16. Design elements are the physical characteristics of plant materials, such as color and line.
17. The pocketknife is one of the most important tools available to the floral designer.
18. Commonly used wire sizes are 5 to 10 gauge.

Multiple Choice

1. Egyptians, Greeks, and Romans incorporated the use of flowers into their cultures. Which of the following were the Greeks known for?
   A. Showy arrangements
   B. Floral garland-type arrangements
   C. Scattering flower petals on the ground during festivals
   D. All of the above
2. Buddhist priests in which of the following countries were the first to use flowers in religious ceremonies?
   A. India
   B. Japan
   C. China
   D. None of the above
3. Steps for conditioning flowers and foliage include
   A. prioritizing the order of processing the flowers based on the amount of each type.
   B. recutting all stems and placing them in cold water.
   C. adding floral preservatives such as Floralife.
   D. placing them in a cooler with low humidity.
4. Basic floral design shapes include
   A. asymmetrical triangle.
   B. symmetrical triangle.

C. open triangle.

D. All of the above

5. Which of the following is *not* a key rule of floral design?

A. Use light-colored flowers near the top.

B. Place small flowers or buds at the top of the arrangement.

C. Start flowers at the front of the arrangement and work toward the back.

D. Establish the focal point at the top edge of the container.

Activities

Now that we have completed our discussion of floral design, you will have the opportunity to explore this fascinating area in more detail. In this activity, you will visit a local florist and gather the following information:

- the types of flowers that are sold, including fresh, artificial, and dried.
- care and handling procedures used for cut flowers.
- types of arrangements made and whether the arrangements followed the six rules of floral design.

In addition to the information you were asked to gather, include your thoughts on how the flower shop was run in general. If you do not have access to a flower shop, surf the Internet for two sites that present information on floral design and summarize them. Provide the Web site address where you found your information.

CHAPTER 19

Interiorscaping

ABSTRACT

This chapter provides background information on interiorscaping and the environmental factors that should be considered when growing plants indoors, including temperature, humidity, water, light, growing media, fertilization, acclimation, and pest control. Choosing the proper plants to be used in the interiorscape requires evaluating plant characteristics and considering personal preferences. The chapter concludes with a discussion on terrariums, including putting a terrarium together and subsequent care and maintenance.

Objectives

After reading this chapter, you should be able to

- provide background information on interiorscaping.
- discuss the eight factors that should be considered when growing plants indoors.
- describe terrariums and how to create and care for them.

Key Terms

foot candles	interiorscaping	tensiometer
humidity	lux	terrarium
hygrometer	soluble fertilizers	

INTRODUCTION

Interiorscaping is the use of ornamental plants for functional and aesthetic purposes (Figure 19-1). Interiorscaping uses foliage plants inside buildings to create a feeling of outdoors. Interiorscaping began in the early 1970s to create an outdoor-like environment for office workers and mall shoppers. Plants in the indoor landscape filter the air, reduce noise, look pretty, and more. Today, the interiorscaping industry has become a big business because plants in the landscape have been proven to make workers more productive. When deciding what plants to grow indoors, the plants' environment must be evaluated, including temperature, **humidity,** water and light, growing media, fertilization, acclimation, and pest control. The plant characteristics and personal preferences must also be considered.

Terrariums are miniature landscapes growing in a covered glass or plastic container that has a high capacity to retain moisture and is used for displaying or storing plants. Terrariums were initially invented around 1850 for use in transporting living plants from the far parts of the world, when sea voyages took months or years to complete. Today

Figure 19–1 A cactus garden in a container is an example of use of plants for aesthetic purposes.

with modern transportation, terrariums are used solely for decorative purposes. Terrariums can be maintained for long periods of time with minimal attention; however, they still require care. Establishing a terrarium requires a number of steps. The first consideration is the container; any glass, plastic, or clear container can be used for a terrarium. After a container is selected, the correct plants must be selected. The plants in the terrarium will be subjected to tropical conditions, such as high humidity and low light conditions, so the plants should be selected with this in mind. After the container and plants are selected, the proper tools and the optimal plant medium are also required. The first step is to make sure the inside of the container is clean. Then 1 to 2 inches of gravel are added to the container followed by 1 to 2 inches of planting medium. A hole is made in the planting medium, and the plant minus half of its root ball is inserted into the hole. This is done to make room for the roots in the pot and to slow the plant's growth. Then soil is firmed in place around the plant roots. Mulch is put on the soil surface and then misted lightly. The inside of the container should be cleaned if necessary and the container is closed. Care and maintenance of the terrarium includes watering, providing adequate light, maintaining proper temperature, applying fertilizer, and preventing pest control. When the terrarium is doing well, plants will grow and soon become crowded, so excess foliage should be pruned routinely.

BACKGROUND ON INTERIORSCAPING

Interiorscaping began in the early 1970s by using plants to create an outdoor-like environment for office workers and mall shoppers. Today plants are commonly used in buildings to make an environment more appealing. Interiorscaping is becoming a major industry because plants in the interior landscape have been proven to make workers more productive.

When growing house plants and deciding what plants to use in the interiorscape, factors to consider include the plants' environment, growing media, fertilization, acclimation, pest control, plant characteristics, and personal preferences.

Plant Environment

Growing plants indoors or in an unnatural environment requires knowledge of the environmental conditions necessary for optimal plant growth. This section describes the factors to consider to optimize the environment.

Temperature

Most foliage plants belong to the tropical plant group; they generally prefer temperatures between 64 and 75°F for vigorous growth and development. Temperatures that fluctuate dramatically can have adverse effects on houseplants. For example, plants grown near windows or on windowsills are at risk of large shifts in temperature that can injure the plants. In addition, chilling injury can damage tropical plants at temperatures between 35 and 45°F, which is well above freezing. Chilling injury can be caused by cold drafts, temperature drops when transporting plants, and many other factors. High temperatures can also cause problems that are not a direct cause of the temperature. For example, woodstove and other heating systems cause damage as a result of moisture loss that leads to desiccation.

Humidity

Humidity is the amount of moisture in the air. Humidity and temperature work together. For example, when the temperature is high, relative humidity is low. When the relative humidity is high and the temperature is reduced, water will bead on the leaf surface and provide a place in which disease organisms can grow. The amount of moisture in the air can be determined by using a **hygrometer.** In rain forests, where tropical plants grow naturally, the humidity is almost always above 80 percent. In an air-conditioned building, the humidity is usually between 40 and 65 percent. During the heating season, humidity is typically between 15 and 20 percent; if a woodstove is used as a source of heat, humidity levels can be close to 0 percent.

Humidity levels should be maintained at 60 percent or higher to maximize plant growth and development. Many houseplants are stressed at a relative humidity below 40 percent. Plants with thinner leaves, such as palms, exhibit leaf tip burning at a lower relative humidity (Figure 19-2), whereas plants with thicker leaves, such as aloe (Figure 19-3), can tolerate a lower relative humidity. Good watering and misting practices will reduce plant damage caused by low humidity.

Water

Both the homeowner and professional have problems with overwatering indoor plants, which leads to poor plant growth. Underwatering plants is

Figure 19–2 Palm plant exhibiting leaf tip burning as a result of low relative humidity

Figure 19–3 Aloe plants can tolerate lower humidity.

typically less likely to cause problems with indoor plants. A variety of factors determine the water needs of houseplants. First, consider the temperature of the room the plants are grown in; if the room is warm, plants will lose water more rapidly due to transpiration and evaporation and so need more water, whereas if the room is cool, less water is lost and less water is needed. Second, plants that naturally transpire more because they have a large surface area tend to lose more water due to transpiration than plants with a smaller surface area. Third, actively growing plants require more water than dormant plants or plants that naturally grow at a slow growth rate. Fourth, the size of the container and the size of the plant also have a dramatic effect on how often plants need to be watered. Fifth, if the growing media has a high water-holding capacity, it will require less water, whereas if it has a poor water-holding capacity, it will require more frequent watering. Sixth, the material used to make the container profoundly affects when the plant should be watered; for example, plants grown in plastic pots need to be watered less than plants grown in clay pots, which breathe. To determine when to water, the grower can feel the weight of the pot, touch the soil, use an indicator plant, or use a **tensiometer,** which is a moisture gauge. Plants can be watered from above to the soil surface or from below to a saucer where the plant is sitting. When watering plants from above, the grower should take care not to splash water on the leaves, which can lead to disease.

Light

Light determines the rate of plant growth. On a bright, sunny summer day, there is 130,000 **lux** or around 12,000 **foot candles** of light. Lux is the metric unit for expressing the illumination falling on all points of a surface measuring one meter square, each point being one meter away from a standard light source of one candle; 1 lux equals 0.093 foot candles, which are units for measuring illumination. In the wild, plants grown in the rain forest receive around 1,000 foot candles of light, whereas in a typical shopping mall, they receive between 50 and 100 foot candles. Light levels of between 50 and 100 foot candles are adequate for plant maintenance, which means that they do not grow, but they do not die. Above 100 foot candles, plants grow, meaning new leaves and branches are produced although at a slow rate. Plants that can grow under poor light include the snake plant (*Sansevieria trifasciata*), pothos (*Scindapsus aureum*), heart philodendron (*Philodendron scandens*), and dumbcane (*Dieffenbachia* spp.). Plants that require a well-lit location include the rubber plant (*Ficus elastica*) and gardenia (*Gardenia jasminoides*). Plants typically need 12 to 16 hours of light per day to grow normally. However, in a normal house or business, lights may be on for long periods of time on some days while on other days not at all; therefore, houseplants may need supplemental lighting.

Artificial light sources. Artificial light sources can be used for decorative purposes and for proper growth of the plant. Lights can be used to showcase plants with unique features or to enhance the decor of a given room. Plants need lights to grow and develop normally, so in dimly lit rooms plants need supplemental light. Three general sources of light can be used: incandescent, fluorescent, and mercury vapor lights.

Light bulbs, also called incandescent lamps, emit light in the orange-red spectrum. Plants grown with their only light source coming from light bulbs are small and spindly because incandescent lamps do not provide the proper light quality and give off large quantities of heat. Incandescent light sources are typically used to supplement fluorescent lights or for decorative purposes.

Fluorescent lamps provide the quality of light necessary for plant growth but work best when used in conjunction with incandescent bulbs to provide red light. Fluorescent lamps are commonly used because they are energy efficient, cost less than the other types to operate, and produce little heat so they can be placed close to plants without scorching them. Fluorescent lights come in a variety of colors, which provides decorative flair. In addition, fluorescent lights come in a variety of different light qualities; therefore, it is important to read carefully the label designating the spectrum produced. For example, daylight fluorescent tubes provide mainly blue light and very little red light, which works well for foliage plants but is less acceptable for flowering plants. The best fluorescent lights produce a reddish hue; however, these are typically more expensive than daylight fluorescent tubes, which provide the same benefit when supplemented with incandescent bulbs. Higher light is required with orchids, and the best fluorescent bulbs for this purpose are very high output (VHO) fluorescent tubes. VHO lamps produce more heat than cool white fluorescent bulbs, so this should be taken into consideration. Fluorescent bulbs age and lose intensity slowly, so light levels should be monitored carefully and the bulbs should be replaced when light levels drop below a certain point.

If very high light intensity is necessary, mercury vapor lights are commonly used because they emit less heat than incandescent lights. The disadvantages associated with mercury vapor lamps are that they operate on high power (a minimum of 250 watts is needed), and they are expensive.

Natural Lighting

One of the ways to enhance light levels indoors is by installing a skylight in the roof and thus enabling more light to enter the room. The material covering the skylight should be translucent, not transparent, because translucent material allows the incoming light to be distributed better over a large area without hot spots. It is better to have too little than to have too much light, which can damage the foliage.

Growing Media

Potting mixtures differ dramatically in their makeup and what they are used for; some are designed for specific purposes and others are used for general purposes. The ideal growing medium should have good drainage and aeration; have good moisture- and nutrient-holding capacity; be disease, insect, and weed free; and provide support for the plant.

Fertilization

All plants should be properly fertilized to grow and develop normally. When fertilizing houseplants, the grower should be aware that soilless mixtures are typically deficient in nutrients, especially micronutrients; therefore, fertilizing with both macro- and micronutrients is important. Fertilizers come in several forms: solids, liquids, powders, crystals, and granules. **Soluble fertilizers** contain powders or crystals that have been dissolved in water first and then applied to the soil as a liquid. Solid fertilizers are applied as fertilizer spikes, sticks, or pills, which are inserted into the soil and are released over time. Granules placed on the soil surface are rapidly released. Other granules are coated with a special material applied to the soil surface or incorporated into the soil mixture and released slowly over a long period of time. Although most nutrients are applied to the growing medium, some are applied as a foliar spray. Indoor plants must be grown under the proper growth conditions, such as temperature, light, and water, for fertilizers to be taken up and used effectively. One of the most common mistakes when taking care of houseplants is overwatering followed by inadequate light and improper temperature, all of which can be misinterpreted to be a nutrient deficiency. Fertilization is necessary when a plant is growing actively; however, fertilization should be reduced or eliminated when a plant is growing more slowly. Indoor plants should be fertilized in the spring and summer; however, during the winter or fall because light levels are inadequate for plant growth, fertilization can cause more harm than good.

Acclimation

In Florida, growers aggressively produce foliage plants under high-light conditions and constant fertilization. Plants must be acclimated properly to the home or office environment to overcome problems with low light and humidity typically found in these locations. Problems frequently occur when plants are not acclimated properly prior to reaching the home or office environment. Plants should be acclimated gradually from high to lower light and humidity before introducing them into the home or office environment.

Integrated Pest Management (IPM)

The first step in IPM is prevention. Plants and humans live together in small and enclosed places for mutual benefit. However, if pests invade this environment, the first line of attack should be to identify the pest and implement IPM, which is a combination of cultural, biological, chemical, mechanical, and genetic control. Pesticides should be used only as a last resort because they are toxic to humans as well as pests. Typically, diseases and pests on houseplants are due to improper growing conditions, such as inadequate light levels, improper temperatures, poor nutrition, and overwatering or underwatering. Therefore, prior to controlling the problem, it is important to make sure that the conditions for plant growth are adequate. Cleanliness should be the number one priority when growing houseplants; leaves should be cleaned regularly to remove the dust that will reduce photosynthesis and harbor diseases and insects. Many houseplants can be cleaned with soapy water, but the leaves should be rinsed thoroughly. Some common disease problems on popular foliage plants are leaf spot, crown and root rot, virus, and powdery mildew; some common insect problems are mealybug, scale, spidermite, and whitefly.

Common problems with houseplants are caused by improper watering, light, nutrition, and temperature, together with pathogenic and pest problems. Water-related problems include the following:

- Plant wilting can be caused by excessive watering and poor drainage conditions. This problem can be overcome by reducing the amount of water applied and, if necessary, repotting the plant.
- Stunted plant growth with small leaves can be caused by inadequate amounts of water. This problem can be overcome by proper watering practices.

Light-related problems include the following:

- Insufficient light levels typically cause spindly, elongated growth. This problem can be overcome easily by moving the plant to an area with more light or by supplying supplemental light.

- Discoloration or bleaching of leaves can be caused by too much sunlight and can be overcome by moving the plant to an area with indirect sunlight.

Problems related to plant nutrition include the following:

- When the whole leaf exhibits uniform chlorosis, the plant should be fertilized with nitrogen because this is a sign of nitrogen deficiency.
- Stunted plant growth is a sign of inadequate fertilization and can be overcome by applying the proper amount of fertilizer.
- When the edges of the leaf and/or the leaf tip are burned, this is a caused by the application of too much fertilizer. This can be overcome by flushing the soil with large amounts of water and adjusting the fertilizer rate.

Temperature-related problems include the following:

- Slow growth can be caused by low temperatures and can be overcome by moving the plant to a warmer location.
- Leaf abscission can be caused by a rapid drop in temperature. This can be prevented by protecting the plant from drafts or rapid shifts in temperature.
- Chlorosis followed by abscission can be caused by temperatures that are too high. This can be overcome by reducing the temperature or moving the plant to another location.

Pathogenesis-related problems include the following:

- Rotting at the base of the stem and blackening is a sign of gray mold, which is a common problem in houseplants that have been overwatered and have poor drainage. If caught in time, this problem can be overcome by repotting in well-drained soil and watering less frequently; otherwise, the plant should be discarded.
- Mottled and distorted leaves are probably caused by a viral infection caused by an insect, such as an aphid; this plant should be destroyed to prevent other plants from being infected.

Common pest problems include the following:

- A fine web is observed on the underside of the leaf and yellowing is caused by spider mites, which like hot, dry air conditions. Cutting off the part of the plant that has been infected and misting the plant to avoid further dry conditions can overcome this problem.
- Cotton-like material on the underside of the leaf and yellowing is likely to be caused by mealy bug. The way to overcome this problem is to remove the cotton-like material and swab with alcohol.

Plant Characteristics

The plants' characteristics should be considered when selecting plants to be put into the interiorscape. The plants should have aesthetically pleasing foliage and/or flowers, including when the plant matures. For example, a plant in its juvenile stage may look very appealing; however, after it matures, the plant may not have the same aesthetic properties. The growth cycle of the plant is important to consider because some plants are only pleasing to the eye when they flower but have very unattractive foliage. Another important plant characteristic to consider is the growth rate of the plant; some plants grow very rapidly to achieve a desired effect, whereas others take longer.

Personal Preference

Plant selection should be based on appeal and not necessarily because they are on sale. Plant buyers should visit several stores with a variety of houseplants and make an informed decision based on personal preferences, whether they are purchasing a flowering plant or a foliage plant. Another important issue is how much care certain plants require. Some examples of easy-to-grow plants are the rubber plant (*Ficus elastica*), spider plant (*Chlorophytum comosum*), snake plant (*Sansevieria trifascata*), and philodendron (*Philodendron* spp.).

TERRARIUM OR BOTTLE GARDEN

A terrarium is a miniature landscape growing in a covered glass or plastic container that has a high capacity to retain moisture and is used for displaying or storing plants (Figure 19-4). The terrarium was invented around 1850 and initially used as a way of transporting living plants from the far parts of the world, when sea voyages took months or years to complete. Today, terrariums are used for decorative purposes. Terrariums can be maintained for long periods of time with a minimum amount of attention, but they still require care.

This section explains the necessary steps for putting together a terrarium.

Figure 19-4 A simple terrarium

Figure 19–5 Different types of containers can be used for terrariums.

Container Selection

Any clear glass or plastic container can be used for a terrarium. Large-neck bottles or fish tanks make good terrariums, but real experts use challenging small-necked bottles. Containers come in all shapes and sizes, so the best choice is based on personal preference (Figure 19-5).

Plant Selection

For a terrarium to be successful, the correct plants must be chosen. For example, cactus and succulent-type plants do not do well under terrarium conditions. Plants should be chosen that thrive under tropical conditions, which include high humidity and low light levels found in a glass bottle. When planting several types of plants together in a bottle garden, it is important to select plants in a variety of colors, sizes, and shapes. Mosses, ferns, lichens, and ivy work well in terrariums.

Necessary Tools

When using a container with a wide opening, such as fish tank, common garden tools can be used; however, small-mouth containers require special tools for putting medium into the container, putting plants into the container, firming soil around the plants, and cleaning the inside of the container. Commonly used materials include a funnel, long forceps, a long-handled fork or spoon with a narrow head, a blunt-ended tool with a long handle for firming the soil around plants, cotton balls with flexible wire to clean the inside of the container, and a scalpel or narrow scissors for trimming.

Planting Medium

Selecting the proper medium for growing plants in a terrarium is the same as selecting plants in a greenhouse or in the field. Good drainage and aeration is important to prevent waterlogging and ensure the proper growth and development of the roots and thus better overall plant growth. Plant media should have good nutrient- and water-holding capacity; be free from insects, diseases, and other pests; and provide good support for the plant.

The basic steps for putting together a terrarium include the following:

1. Make sure the inside of the container is clean.
2. Add about 1 to 2 inches of gravel to the bottom of the container.
3. On top of the gravel, add 1 to 2 inches of a planting medium composed of 3 parts commercial potting soil and 1 part crushed charcoal, which removes unwanted chemicals that may otherwise accumulate in a closed environment
4. Dig a hole in the planting medium. Prior to inserting the plant into the hole, remove about half of the soil from the plant root ball, insert it into the hole, and then firm the soil in place around the plant roots.
5. Apply a layer of ground bark or other mulching material to cover the soil surface and mist lightly.
6. Prior to covering, make sure that any parts of the inside of the container that were soiled during the insertion of plants and other materials are cleaned.

Care and Maintenance of the Terrarium

After the terrarium is set up, proper care and maintenance is needed to keep the plants healthy. Important factors include watering, lighting, pruning, controlling temperature, fertilizing, and controlling pests.

Watering. The terrarium should only be watered when necessary; when in doubt, water should not be added. When water beads on the inside of the container occasionally, this shows that the system is working. If the terrarium has been overwatered, it should be unsealed and water should be allowed to evaporate. The frequency of watering increases with the increasing size of the opening of the container.

Light. Plants should not be placed on a windowsill receiving direct sunlight most of the day or they will be damaged. Terrariums receiving light predominately from one direction should be rotated to prevent phototropic curving toward the light. Additional light can be provided by cool white fluorescent lamps.

Pruning. When the terrarium is doing well, plants will grow and soon become crowded and require periodical pruning to remove excess foliage. All clippings should be removed from the container to avoid decay, disease, and odors.

Temperature. Because most terrariums contain tropical plants, they require warm temperatures from 65 to 75°F. Windowsills should be avoided in order to prevent exposure to abrupt changes in temperature.

Fertilization. Most terrariums will not need additional fertilization because slow growth minimizes the need for pruning and general upkeep. However, when necessary, organic fertilizers should be used; chemical fertilizers should not be used in closed terrariums because they cause problems with volatiles which will affect the plant's growth.

Disease and pest control. If the terrarium was set up properly, there should be no problem with diseases and pests. However, if pests such as mites, mealy bugs, and whiteflies become a problem, spraying with appropriate chemicals may be necessary.

SUMMARY

You now have basic knowledge of interiorscaping, including environmental factors that should be considered when growing plants indoors. The choice of plants to be used in the interiorscape is affected by different plant characteristics and personal preferences. The chapter concluded with a discussion of terrariums and the major steps involved in putting together a terrarium and subsequent care and maintenance. With proper care and maintenance, a terrarium will provide many years of enjoyment.

Review Questions for Chapter 19

Short Answer

1. What are eight factors that should be considered when optimizing a plant's environment for growth indoors?
2. What are two factors that should be considered when deciding what plants to use in the interiorscape?
3. Why were terrariums originally invented?
4. What criteria should be used when selecting plants for a terrarium?
5. What criteria should be used when selecting a container for a terrarium?
6. Provide the basic steps involved in putting together a terrarium and what factors should be considered after the terrarium is complete.

Define

Define the following terms:

interiorscaping	tensiometer	foot candles	terrarium
hygrometer	lux	soluble fertilizers	

True or False

1. Overwatering of indoor plants (resulting in root rot diseases) is common, both by the homeowner and the professional.
2. Most foliage plants belong to the tropical plant group and are sensitive to chilling.

3. Indoor plants should generally be fertilized in the spring and summer, not in the winter or fall.
4. Chilling injury occurs when plants are exposed to temperatures at or below freezing.
5. High temperatures are a direct cause of injury to houseplants.
6. Good watering practices will reduce plant damage caused by low humidity.
7. Foot candles are units for measuring illumination.
8. Plants grown with light bulbs as their only source of light are typically shorter and compact.
9. Incandescent bulbs provide red light.
10. Fluorescent lights are rich in the red light portion of the light spectrum.
11. It is important to fertilize plants during the winter and fall months.
12. A common disease in foliage plants is *Puccinia posophylli*.
13. Two common insects found on foliage plants are whiteflies and viromites.
14. Cactus and succulent-type plants are commonly used in terrariums.
15. Any glass or clear plastic container can be used for a terrarium.
16. The planting medium used in a terrarium should contain some crushed charcoal.
17. It is important to remove about half of the soil from the plant root ball prior to putting plants into a terrarium.

Multiple Choice

1. Which of the following are disease problems on popular foliage plants?
   A. *Erwinia amylovora*
   B. *Puccinia podophylli*
   C. *Venturia inaequalis*
   D. None of the above
2. Which of the following are insect problems on popular foliage plants?
   A. Mealy bug
   B. Scale
   C. Spidermite
   D. All of the above
3. Light bulbs are called incandescent lamps; they emit light in which of the following spectrums?
   A. Far red
   B. Red
   C. Blue
   D. All of the above
4. Terrariums were invented around 1850 for which of the following purposes?
   A. As a way of transporting living plants from the far parts of the world
   B. For decorative purposes
   C. For horticulture therapy
   D. All of the above

Fill in the Blanks

1. Foliage plants produced in Florida are grown under high light and proper humidity, temperature, and fertilization; therefore, prior to transfer to the home environment, they must be ____________________ to low light and humidity typically found in these locations.
2. A ____________________ is a miniature landscape growing in a covered glass or plastic container.

Activities

Now that we have completed our discussion on interiorscaping, you will have the opportunity to explore this fascinating area in more detail. For this activity, you will need a camera (digital or regular). Go out to a shopping mall or any other local businesses in your area and find examples of poor interiorscaping and what you consider to be good interiorscaping; take a picture of each. Be sure to note the plants and other materials used. If you cannot find interiorcaping anywhere, you will need to do a Web search to find an image of a poor and good interiorscape. Write a description for each landscape that discusses the overall design; the plants used; any other materials used, such as pools, rocks, and so on; the quality of the maintenance; and the reasons you selected the interiorscapes along with the pictures of the interiorscapes and written descriptions. Feel free to add any other details about the interiorscapes you are presenting.

CHAPTER

20

Designing Landscapes

Objectives

After reading this chapter, you should be able to

- provide background information on planning a landscape.
- list and explain some practical reasons for landscaping.
- discuss the different types of landscaping.
- list and explain responsibilities of individuals who prepare the landscape design, install the design, and maintain the design.
- discuss and explain the four elements of design.
- discuss and explain the five basic principles of design.
- provide and explain important factors that should be considered when preparing a landscape plan.

Key Terms

accent planting
asymmetrical balance
balance
bush planting
corner planting
focalization
form
foundation planting
hardscaping
landscape architect
landscape contractor
landscape design
landscape planning
landscaping
line
line planting
planting plan
private area
proportion
public area
rhythm and line
service area
simplicity
site analysis
site plan
sitescaping
symmetrical balance
xeriscaping

ABSTRACT

Practical reasons for landscaping include creating aesthetics, increasing property value, reducing noise, blending structures into the terrain, screening unsightly areas, providing privacy, preventing erosion, reducing pollution, controlling people/traffic, reducing damage/injury to people, creating places for outdoor activities, enjoying landscaping as a hobby, and modifying environmental factors. The different types of landscaping are home landscaping, public area landscaping, commercial landscaping, and specialty site landscaping. Several people are involved in preparing, installing, and maintaining the landscape design. The four elements of design include color, texture, form, and line. The five basic principles of design are simplicity, balance, proportion, focalization, and rhythm and line. Important factors to consider when preparing a landscape plan include ways to create the plan, client needs and purpose of the landscape, analysis of the site and its use, proper plant material selection, and plant arrangement in the landscape.

INTRODUCTION

Plants in the landscape fill important psychological needs; for example, the color green has a calming effect on people. In large cities, the few green spaces lead to stress in humans. Living plants promote human health and help to overcome the stress of everyday living, so the need for attractive landscapes is increasing continually (Figure 20-1). Good **landscaping** begins with a good design and includes installation and maintenance; when any of these three fail, the entire landscape also fails. The reasons to landscape include to increase property value,

Figure 20–1 A nicely landscaped home. *Courtesy of Dan T. Stearns, Department of Horticulture, The Pennsylvania State University.*

to provide privacy, to prevent erosion, and to reduce human injury. Landscaping can be broken down into several categories: home landscaping; **public area** landscaping, such as parks; commercial landscaping; and specialty site landscaping, such as zoos (Figure 20-1).

For a landscape to be successful, it must start with a well-conceived design followed by the proper installation and maintenance. The **landscape architect** establishes the design plan for plants and **hardscaping** features, such as pools and patios. The **landscape contractor** implements the **planting plan** and maintains the landscape. To establish a good **landscape design,** the landscape architect must use properly the four basic design elements: color, texture, **form,** and **line.** In addition to the four basic design elements, the landscape architect must use properly the principles of design **simplicity, balance, proportion, focalization,** and **rhythm and line.** Other important factors to consider when preparing a landscape plan are ways to create a landscape plan (by hand or computer), client need, purpose of the landscape, important features at the site, proper selection of plant materials, and arrangement of plants in the landscape.

BACKGROUND INFORMATION ON PLANNING A LANDSCAPE AND KEY DEFINITIONS

The use of plants for landscaping is not new; people have always liked to see attractive outside areas. Plants fill important psychological needs by promoting human health and helping people overcome the stress of living. In large cities, the lack of green spaces leads to a lot of stress and increases demand for attractive landscapes. Good landscaping begins with a good design and includes installation and maintenance. Failure to do any of these three can result in failure of the landscape. The focus of this chapter is design; the subsequent two chapters focus on installation and maintenance. Key definitions for this chapter include the following:

- **Landscaping.** The use of plants to make outdoor areas more attractive.
- **Landscape planning.** The preparation of details of how a site will be landscaped, including both the art and science involved.
- **Landscape design.** The preparation of a landscape plan for beautifying a site, including drawings and supporting information with known specifications.
- **Sitescaping.** Landscaping a small part of a larger area to meet the particular needs of people; for example, a small Chinese garden in a portion of the backyard.
- **Bush planting.** Landscaping by an untrained person with no knowledge of plant materials.

PURPOSES OF LANDSCAPING

Landscaping's most common use is to create an aesthetically pleasing outside environment for both homeowners and businesses. Landscaping definitely creates a more beautiful environment that can be enjoyed by homeowners and the public in parks and horticultural gardens, but those aren't the only purposes for landscaping:

- Landscaping can dramatically enhance the property value in a specific neighborhood. In fact, landscaping can transform simple boring structures into beautiful ones.
- Landscaping can be used to reduce noise from factories, highways, and other sources because plants buffer noice.
- Landscaping helps buildings and other structures blend into the terrain. At many universities, any new building that goes up is not allowed to affect how the campus looks. Alumni returning for a visit want to see the university the way it was 10, 20, or 50 years ago.
- Landscaping can be used to screen unsightly areas by providing a beautiful plant wall around dumpsters, storage areas, and other unsightly areas.
- Landscaping provides privacy in backyards of homes or shields private business locations from the general public.
- Landscaping prevents erosion to improve and conserve natural resources and reduces pollution to increase the quality of human life.
- Landscaping can be used to control people traffic and vehicle traffic. Barberry bushes have sharp thorns, which deter people from taking a shortcut through private property or through a location where they should not be walking. A big oak tree in front of a building can prevent an out-of-control vehicle from crashing into the building. In addition, trees and other plants strategically located on a median between roads prevent drivers from going over the median.

- Landscaping can reduce damage and injury. For example, shrubs along a roadway can absorb shock in an automobile crash.
- Landscaping can create places for recreation and outdoor activities.
- Landscaping can be used as a hobby or for horticulture therapy. Individuals can care for their gardens while getting exercise and enjoyment.
- Landscaping can be used to modify a variety of environmental factors. For example, trees can be installed as windbreaks to reduce wind. Trees can also be used to reduce light and reduce temperatures at a given location.

TYPES OF LANDSCAPING

Landscaping can be broken down into home landscaping, public area landscaping, commercial landscaping, and specialty site landscaping. Home landscaping is typically up to the individual homeowner; however, in some cases, developers establish a theme for a given development and landscape the area based on that particular theme. Landscaping public areas, such as parks, is done for a variety of reasons. For example, small parks can be put at strategic locations in the city to provide a place for workers to rest on their lunch hour. On a community level, parks are landscaped for specific purposes, such as parks with baseball fields, tennis courts, running trails, and other recreational facilities. On national or state levels, parks may be used to preserve wildlife while providing recreation areas. Commercial landscaping involves the public because most businesses are open to the public. Commercial landscaping is done at shopping malls, banks, churches, restaurants, and other places. Commercial landscaping should be functional and safe. Specialty landscaping includes gardens in zoos or botanical gardens, where special designs are used to establish a specific habitat, or at a golf course, where landscaping is used to provide obstacles for the game and aesthetic beauty.

INDIVIDUALS RESPONSIBLE FOR LANDSCAPE DESIGN, INSTALLATION, AND MAINTENANCE

A successful landscape begins with a well-conceived design, followed by the proper installation/implementation and maintenance. The landscape architect professional establishes design plans for the installation of plants and hardscaping (the permanent landscaping structures, including fences, patios, walks, driveways, retaining walls, and other inanimate features of the landscape) outdoors to fulfill aesthetic and functional purposes. The landscape architect discusses the client's desires; implements a **site plan** and planting plan for a given location, which provide detailed guidelines on how to install the landscape; and oversees the project until completion. A site plan is a drawing indicating the location of plants and hardscaping objects in the landscape. The planting plan is a drawing that uses symbols to specify the types and names of plants, their quantities, and their locations in the landscape. After the planting plan is complete and the client and architect are in agreement, the landscape architect contracts the nursery for plant materials to be used in the project and the appropriate supply stores for hardscaping materials. The landscape contractor then implements the design provided by the landscape architect and, in many cases, maintains the landscape. A landscape contractor uses the architectural plans to install plants and hardscaping features in the landscape. The landscape contractor and landscape architect work closely together to make sure that the plan is implemented correctly.

ELEMENTS OF DESIGN

A beautiful landscape is influenced by a variety of factors, including personal background, culture, and tradition. The landscape architect uses plants as the primary materials in the creation of a design. Different types of plants have specific characteristics that influence their use in the landscape. Plants individually are beautiful; however, when combined in a landscape, they must complement each other for a landscape design to be effective. The proper use of elements of design in a landscape can elicit a response by creating a specific feeling in the viewer.

For a landscape design to be successful, it must include considerations of the plants' features. Color is a design element that people respond to in different ways and should be used carefully to produce a successful design. Different colors elicit different responses; for example, reds, yellows, and oranges give the feeling of warmth and the appearance of coming toward the viewer, whereas blues and greens give a cool feeling and appear to the viewer as if they are in the background. Sources of color in the landscape are flowers; green, variegated, or seasonal leaves; and the bark on trees. These sources provide the landscape designer with a wide variety of options.

Texture is another design element and refers to the visual and physical characteristics of a plant. The physical texture characteristics can be felt by the sense of touch whereas the visual surface characteristics can project different images to the individual looking at the landscape because such characteristics can create a very smooth or a coarse feeling. When seasons change, so do a variety of textures; for example, when a deciduous tree has leaves, the tree provides a smooth texture, but when the leaves fall off, the tree provides a coarse texture. The bark of trees also varies in texture, with some being very smooth and others being very coarse (oak), or the bark may have a peeling appearance, such as birch. Different textures can be incorporated into the landscape by using inanimate objects such as mulch, stones, gravel, brick, and others. The distance of the viewer from the landscape can visually modify how coarse an object appears; the greater the distance, the less coarse an object will appear. There should be no abrupt changes from coarse to fine textures in a landscape.

Form is a three-dimensional element; for example, a tree shape can be round, weeping, horizontal, and a variety of other shapes (Figure 20-2). Form provides another element whereby the designer can produce an interesting landscape design. The last design element is line, which is a one-dimensional effect produced by arranging three-dimensional objects in a certain fashion. Line is a design tool used by the landscape architect to create or manipulate patterns in a given landscape. Lines can be straight, but they do not always have to be straight or continuous. Lines can also be contoured or curvilinear to create an interesting design.

Figure 20–2 Two examples of tree shapes

PRINCIPLES OF DESIGN

The principles of design help landscape architects develop good designs that are economical to install and easy to maintain. For a landscape design to be successful, the landscape architect must use the five basic principles of design: simplicity, balance, proportion, focalization, and rhythm and line. Simplicity is a landscape design principle that uses a number of strategies to reduce high levels of variation and general distractions in the landscape. The landscape architect should use simplicity, which basically means not being overly complex in a given design, thereby promoting overall unity. Most individuals enjoy a landscape that is easy to understand, so the elements used in a landscape should blend together. The use of a variety of plants in a landscape is important to a successful design. Too little variety is boring and too much variety is overwhelming. Seven to eight groups of plants should be used and each can be repeated in the landscape. For example, using a row of yew plants along a foundation is simplicity. Balance is a landscape design principle that uses equal weight of the elements of design to show uniformity. Balance can be **symmetrical balance,** where the same number and type of plants are on both sides of a landscape. Another type of balance is **asymmetrical balance,** which is when different numbers of plants are on both sides of a landscape. Proportion is the landscape design principle that refers to the relationship between the sizes of the different types of plants used in a landscape design. Proportion is not only used with respect to the plants in the landscape but also with respect to their relationship with inanimate objects in the landscape. For example, large trees should not overshadow a small house. Focalization is a landscape design principle that creates an accent in a particular arrangement. The landscape architect creates focal points in the landscape to direct where people look. Focalization is a very effective way to guide the viewer to areas in the landscape that the landscape architect does not want them to miss, such as fountains, sculptures and exotic plants. Another reason for the use of focalization is to redirect the viewer's attention from an unattractive location in the landscape. Rhythm and line is the landscape design principle that deals with flow throughout the landscape. Different sections in the landscape must be linked to create movement throughout the design. Planting bed shape, bed orientation, and heights and shapes of plant materials and inanimate objects in the landscape have a profound effect on rhythm and line. All parts of the landscape, including the landscape wall, ceiling, and floor, must fit together without being forced.

IMPORTANT FACTORS TO CONSIDER WHEN PREPARING A LANDSCAPE PLAN

Several factors must be considered when preparing a landscape plan. The first is how is the plan going to be created. The client's needs and purpose for landscaping are considered next along with an analysis of the site and its uses. The final factor is to determine the plant material to be used and the plants' placement in the landscape.

Creating a Landscape Plan

There are two ways to create a landscape plan: hand drawn or computer generated. Both methods produce excellent results.

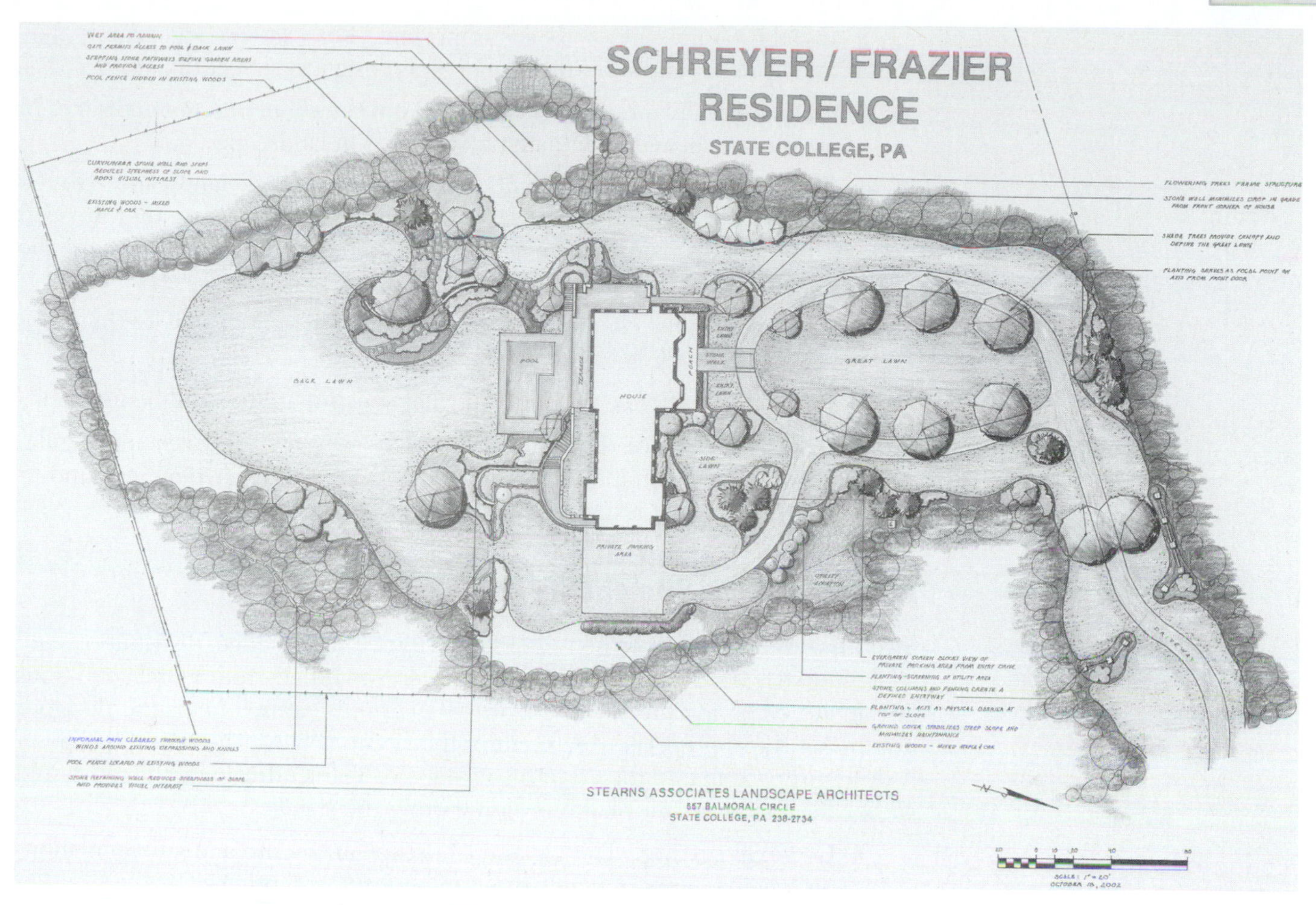

Figure 20–3 Planting plan. *Courtesy of Dan T. Stearns, Department of Horticulture, The Pennsylvania State University.*

- **Hand-drawn landscape plans.** Drawing instruments such as paper with a pencil or pen and basic drafting equipment (such as a ruler, drafting bench, and a few other pieces) are necessary to produce a planting plan (Figure 20-3).
- **Computer-generated landscape plans.** The computer-aided design (CAD) program is commonly used to produce computer-generated landscape plans. In recent years, the use of computers in the generation of landscape plans has become very popular; however, it is still important to be familiar with the preparation of both-hand drawn and computer-generated landscape plans to increase your job prospects.

Client Needs and the Purpose of the Landscape

The purpose of the landscape helps determine design; for example, the landscape may be used for aesthetics, to reduce noise, or to direct traffic. To satisfy the customer, the designer must learn about the customer's preferences and expectations for the completed project. At the initial session with the client, the following questions should be addressed:

- What is the purpose of the landscape?
- How many people will reside at the location and what are their ages? For a residential landscape, the family size, ages of individuals in the family,

and the family's life style are important. For example, the family may value privacy and want their landscape design to reflect this need.

- What are the plant preferences? Find out the client's likes and dislikes in plant material and inanimate objects used in landscapes.
- How much work are the clients willing to put into the landscape? Do the clients want the landscape to be high maintenance because they are avid gardeners or low maintenance because they do not like to work in the garden?
- How much is the client willing to spend on a project? Find out the range the client is willing to spend.
- Is a service entrance needed and what specific utilities will be needed?
- What rules govern the landscape design? Discuss the rules and regulations for development with the client so you both have a clear understanding of what is and what is not allowed in the development.

Analysis of the Site and Its Use

Site analysis is the survey of a site to determine the presence, distribution, and characteristics (both natural and synthetic) along with environmental conditions at the site. Studying the site to identify its natural features helps the designer determine the appropriate treatment of the site to achieve the client's purpose. A design should not be forced into a given site. The landscape architect should gather the following information when visiting the site:

- **General setting.** The general landscape at the site and any regulations on fencing, types of plants that can be used, and others.
- **Size of the property.** The size of the lot, dimensions, or space available for the project to determine the variety, number, and size of plants or inanimate objects, such as pools and fountains, that can be used.
- **Existing landscape.** The terrain, drainage, soil, climatic conditions, rocks, natural features, such as streams or ponds, and other features that will affect the landscape.
- **Architectural design of the structure.** The design and size of the structure to be landscaped for the proportionality of the design.

After the site characteristics have been determined, the landscape designer may begin to create the plan by breaking down the site into three general areas (Figure 20-4):

- **Public area.** The area that will be seen from the street. In general, the client is most concerned with the public area because it is the first impression visitors get. Mounding or terracing a flat lot can enhance the public area by creating interesting designs (Figure 20-5). Flower gardens can also be used to enhance the beauty of the public area. However, in both cases, the design should complement the architecture of the building (Figure 20-6). Note that some communities restrict what can be seen from the street in a given development, especially regarding walls, fences, and hedges.
- **Private area.** The area that is out of the view of the general public. The private area is typically where decks, hot tubs, and patios are located. Fencing or plant materials should be used to screen the private area to provide privacy. When the homeowner requires shading, the patio or

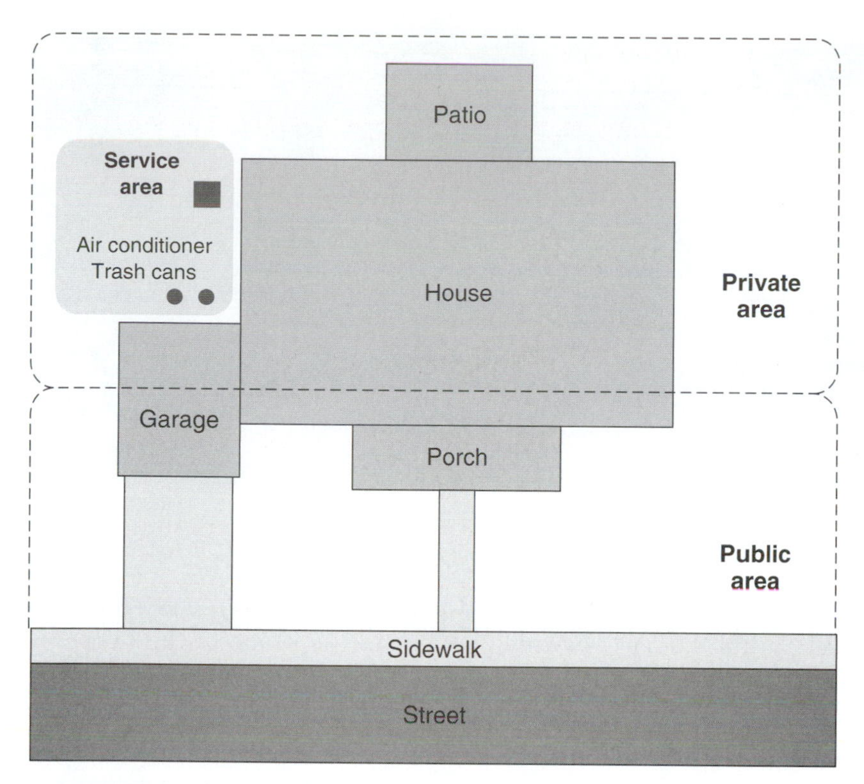

Figure 20–4 A planting site is broken down into three general areas: public, private, and service areas.

Figure 20–5 Use of terracing in the landscape

Figure 20–6 A wise use of flowers in a front yard

deck should be located on the east side of the house to receive the sun's warmth during the early morning hours and the shade of the house by noon. However, if full sun is required most of the time, the patio or deck should be located on the south side of the house. For the corporate client, this could include an area for employees.

- **Service area.** The area near the rear entrance that is relatively isolated from the public and private areas and is where utility accessories and unattractive items such as garbage cans are stored. This section is typically in the backyard and should be shielded from view by fencing or plant material. This area is not meant to be aesthetic; instead, it is functional for garbage cans, storage sheds, gardens, clotheslines, and other necessary areas that are not usually appealing. This area would also include loading docks, storage buildings, dumpsters, and other utility areas for the corporate client.

Plant Material Selection

Selection of plant materials when implementing a plan requires knowledge of plants. Five general categories of plant materials are used: trees, shrubs, groundcovers, vines, and flowers. Many sources of information must be considered carefully when selecting plants to be grown in the landscape:

- **Plant names.** Common and scientific names should be known. For example, a Yew (common name) *Taxus* (genus), *cuspidata* (species), L. (authority) plant is commonly used in a landscape, but there are a variety of types, so it is important to know the genus, species, and in some cases the cultivar.
- **Deciduous or evergreen.** Determine whether the plant being used is deciduous or evergreen. Deciduous trees provide privacy only during the summer and spring, and evergreens can be used for privacy all season long.
- **Growth rate and maturity.** Trees and shrubs are classified by height and spread, which is valuable information to the landscape architect. The growth rate of the plant determines how much training and trimming is required. Understanding when the plant will move from the juvenile to mature phase is also important for a variety of reasons, such as knowing when a given plant will flower.

- **Life span of the plant.** Whether the plant is perennial, biennial, or annual determines whether it needs to be replaced annually, biannually, or, in the case of a perennial, not for a number of years.
- **Characteristics of flowers and fruits.** It is important to know the special features of the flowers and fruits; for example, the fruits of the female Ginkgo have a foul odor that is not conducive to a residential landscape.
- **Color.** Color can come from leaves, flowers, bark, fruits, and inanimate objects in the landscape. Plants are selected for their year-round color, not just for their flowering colors. Color can be provided year round with shrubs, vines, and evergreen trees.
- **Soil, nutrient, and pH requirements.** Questions about the soil at the location include whether it is well or poorly drained, have good or bad nutrient-holding capacity, is typically acidic or basic, have many rocks that need to be removed, and so on. For example, if soils are typically acidic, then rhododendrons and azaleas will grow nicely, whereas carnations and dahlia will grow poorly.
- **Pest problems.** A knowledge of pests at a given location is extremely valuable information when selecting plant material. For example, if the designer knows that a specific pest is at the location, specific cultivars resistant to that pest will be a major benefit to the client.
- **Climatic requirements.** The hardiness zone and what plants can be grown in that zone is an important factor to consider. Light requirements for the plant should also be determined carefully: full sun, shade, or partial shade. Water requirements must be understood because different plants have different water requirements in order to live and grow. **Xeriscaping** is a form of landscaping that uses plants on the basis of their water requirements. Overall, in a xeriscape the emphasis is on the use of plants that have low moisture requirements. Plants are selected based on three water zones (amount of water needed by the plant). The very low water zone is made up of plants that require no water beyond what is naturally available; for example, cactus (Figure 20-7). The low water zone

Figure 20–7 Xeriscaping is based on water zones.

requires some irrigation; for example, shrubs and ground covers. The moderate water zone typically requires supplemental water; for example, annuals and other succulent materials.

- **Others.** Information such as poisonous foliage, thorns, and so on.

Plant Arrangement in the Landscape

Design of plantings begins with the outdoor room concept. An outdoor room has the same boundaries as an indoor room: outdoor floor, outdoor ceiling, and outdoor walls. The outdoor floor is the ground covering, which in most cases is turf, sand, gravel, brick, wood, water, and other materials, whereas the outdoor ceiling is the upper limit of the landscape, which consists of covered patios and trees. The outdoor walls of the room set the boundaries, which consist of flowerbeds, shrubs, and fences. Arrangement of plant materials in the outdoor room uses four kinds of plantings. **Corner plantings** create the frame of the outdoor room, **line plantings** create the walls of the outdoor room, **foundation plantings** are located along the walls or foundations of buildings, and **accent plantings** create an area of particular beauty or interest in a landscape (Figure 20-8). Today, computer graphics can show the final size of plants in a given landscape, which allows the landscape architect to see how the planting will look in 5, 10, or 20 years.

Line

Foundation

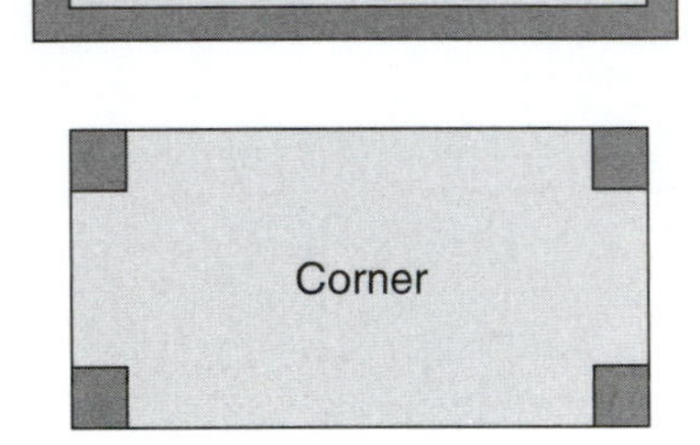

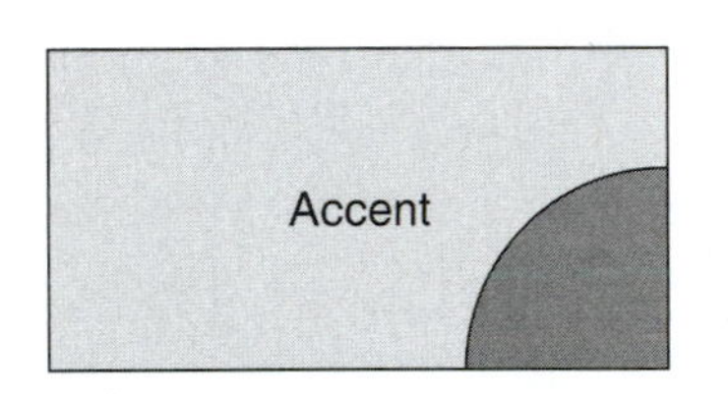

Figure 20–8 Arrangement of plant materials uses line, foundation, corner, and accent plantings.

SUMMARY

You now have a basic knowledge of what is involved in preparing a landscape design. There are many practical reasons for landscaping in both corporate and residential settings. The different types of landscaping include home landscaping, public area landscaping, commercial landscaping, and specialty site landscaping. The four elements of design include color, texture, form, and line. You learned the five basic principles of design used in landscaping: simplicity, balance, proportion, focalization, and rhythm and line. The important factors to consider when preparing a landscape plan include ways to create the plan, client needs and purpose of the landscape, analysis of the site and its use, proper plant material selection, and plant arrangement in the landscape.

Review Questions for Chapter 20

Short Answer

1. List nine purposes for landscaping.
2. List four categories of landscaping.
3. What are four elements of design?
4. What are the five principles of design?
5. What are six important factors that should be considered when preparing a landscape plan?
6. What are five general categories of plant materials commonly used in the landscape?
7. What are nine factors that should be considered when selecting plants for the landscape?

8. What are three general areas into which a site can be broken down based on its use?
9. What are three boundaries that are found in the outdoor room?
10. What are four kinds of planting commonly used for arranging plant materials?

Define

Define the following terms:

landscaping	site plan	symmetrical balance	private area
landscape planning	planting plan	asymmetrical balance	service area
landscape design	landscape contractor	proportion	xeriscaping
sitescaping	form	focalization	corner planting
bush planting	line	rhythm and line	line planting
landscape architect	simplicity	site analysis	foundation planting
hardscaping	balance	public area	accent planting

True or False

1. Successful landscaping begins with proper installation followed by routine maintenance.
2. Landscape planning is both an art and a science.
3. The principles of design help people develop good designs that are economical to install even though they may be difficult to maintain.
4. Skills in both computer-aided design and hard-drawn design maximize job prospects.
5. It is not very important to know the actual purpose of the landscape because there are many other more important factors that should be considered when preparing a landscape plan.
6. An important factor to analyze when preparing a landscape plan is the site; if this is done properly, any design can be used effectively.
7. Today the only acceptable landscape plans are created using a computer and CAD program.
8. In a xeriscape the emphasis is on the use of plants with low moisture requirements.
9. Designing plantings begins with an outdoor room concept where the room consists of a floor, ceiling, and walls.
10. When arranging plant materials, four kinds of plantings are commonly used. One of these is a line planting, which is planting along walls or foundations of buildings.
11. When arranging plant materials, four kinds of planting are commonly used. One of these is a corner planting, which creates the walls of the outdoor room.

Multiple Choice

1. Bush planting is another name for
   A. a person who plants bushes or shrubs for a living.
   B. an untrained person who tries to landscape without a knowledge of plant material.
   C. a landscape where bushes are used as the main portion of the design.
   D. All of the above

2. Important elements of design and effect in landscaping include
   A. balance.
   B. complexity.
   C. a lot of maintenance.
   D. All of the above

Fill in the Blanks

1. Successful landscaping begins with a good ____________ followed by the proper ____________ and routine ____________.
2. The scientific name of a specific plant is given here. For each numbered part of the name, write the taxon or, in the case of 4, the most appropriate term.

   Brassica oleraceae L. 'Union'

   Brassica 1. ____________ *oleracea* 2. ____________
   'Union' 3. ____________ L. 4. ____________
3. ____________ is a form of landscaping that uses plants on the basis of their water requirements.

Activities

Now that we have completed our discussion on designing landscapes, you will have the opportunity to explore this area in more detail. To find the necessary information for this activity, explore the Internet. In this activity, you will search university Web sites throughout the world for information about landscape design programs. (*Hint:* Pennsylvania State University has a landscape design program.) Choose two or three universities with landscape design programs that you find interesting or that you and your classmates should know more about. Write a summary about the programs, including why you selected them. Be sure to include the names of the university and the Web site address.

CHAPTER

21

Installing Landscapes

ABSTRACT

This chapter provides background information on the planting plan and some key definitions. Site preparation includes staking the site, hardscaping, and preparing the bed. Hardscaping features commonly used in the landscape include the different types of landscape surfacing materials: concrete and asphalt, brick and stone, and other surfacing materials. In addition, the different types of landscape fencing are chain-link type, rail type, and basket weave or board-on-board. Landscape lighting includes walk lighting, up lighting, and down lighting. Following the hardscaping installation, proper preparation of the planting bed is key to the successful installation of plants in the landscape. The chapter concludes by explaining how to install bedding plants, trees and shrubs, and ground cover and lawns in the landscape.

Objectives

After reading this chapter, you should be able to

- provide background information on the planting plan and some key definitions.
- discuss how to stake the site for plants.
- discuss hardscaping features commonly used in the landscape.
- provide background on proper preparation of the planting bed and how to install bedding plants, trees and shrubs, and ground cover and lawns in the landscape.

Key Terms

balled and burlapped
bare root
bedding plants
container-grown plants
continual bloom
down lighting
hardscaping
landscape construction
up lighting

INTRODUCTION

This chapter provides information on implementing a landscape plan (Figure 21-1). After a landscape design has been properly planned and completed as described in Chapter 20, proper installation is required. After the landscape designer and landscape contractor have discussed the planting plan, the first step in its installation is for the crew supervisor to call the local office for information on the location of underground utilities at the site. The location of all underground utilities is established and color-coded flags are placed where utilities such as electric, gas, oil, steam, telephone, cable, water, and sewer are located. The first step in site preparation is to stake the site for plants and **hardscaping** features such as walls, patios, and walks. Small wooden or plastic stakes with the plant names written on them are driven into the ground to show the exact center of the planting hole. Hardscaping materials should be installed prior to plants to prevent the plants from being damaged. The landscape contractor installs a variety of different surfacing materials. Material selection is based on cost. The most popular types of paving materials are concrete,

Figure 21–1 Beginning stages of installing plants in a landscape

asphalt, brick, and stone. The advantage of concrete and asphalt is that they are relatively inexpensive, although they do not create interesting or unusual designs. Brick and stone create interesting landscape designs; however, they are more expensive. Fencing is another hardscaping feature commonly used in the landscape. The chain-link fence gives little privacy and is not very pretty, but it provides security and is relatively inexpensive. Rail-type fences provide no security but is commonly used for decorative purposes. Another type of fencing is basket weave or board on board, which provides privacy and security but is expensive. To add beauty to the landscape and provide security at night, low-voltage lights are installed along sidewalks and among plants. Two functional features of night lighting are to provide security from vandalism or theft and to make light available so people can see their way at night. A variety of lighting is used for aesthetics; some examples include silhouette lighting, **up lighting,** and **down lighting.**

After the hardscaping features of the landscape are installed, the bed must be prepared properly prior to installing plants. The first steps in bed preparation are to remove large objects from the bed, till to a depth of 12 to 18 inches, and make any amendments necessary to provide adequate aeration, drainage, proper nutrition, and proper pH levels. After the bed is prepared properly, plants can be installed in the landscape. The three different categories of plants are **bedding plants,** trees and shrubs, and ground covers and lawns. Bedding plants refer to a wide range of plants, mostly annuals (herbs, vegetables, and flowering ornamentals); however, in many cases, for **continual bloom** (which means there are flowers year round), perennial plants and bulbs, rhizomes, corms, and tubers are used. Bedding plants can be direct seeded or they can be transplanted. Direct seeding is relatively inexpensive; however, the process is slower than using transplants. Although transplants are faster, the process is more expensive and transplanting shock may occur.

Trees and shrubs are sold in three ways: **bare root, balled and burlapped,** and **container grown.**

Spring and fall are the best time for planting trees and shrubs because they offer the best environmental conditions. Transplanting trees and shrubs in the summer is more challenging due to high temperatures and sunlight, which can dry out plant tissues. The steps for installing balled and burlapped or container-grown trees and shrubs are the same and should be done carefully to insure success. The steps involved in growing bare root trees and shrubs are similar to balled and burlapped plants and container-grown plants; however, there are several differences. The first difference is that bare root plants should be brought to the field in a container of water to prevent the roots from drying out. After bare root plants are in the field, they should be pruned carefully, placed in the hole, and held in place until the hole is filled with soil so the crown is above the soil level. After filling the hole, the soil must be packed firmly around the plant to make sure the tree or shrub is erect.

Ground covers are low-growing plants that spread and form a mat-like growth over an area. Low-spreading plants less than 2 feet tall are used as ground covers; they can be shrubs, vines, perennials, and grasses. Ground covers can be used to prevent soil erosion on steep grades, as an accent under trees, and throughout the landscape. Installation requires the proper bed preparation for good establishment. Ground covers can be planted using seeds, potted plants, or bare root plants. Hydro seeding is when a water slurry containing seeds is pumped under high pressure onto a slope or over a larger area to establish ground covers. Grass is a key component in almost any landscape design; the installation of a lawn is discussed in Chapter 23.

BACKGROUND INFORMATION ON THE PLANTING PLAN

To establish a planting plan, the landscape designer talks with the client about the client's specific needs and prepares a plan. After a landscape design has been planned properly and is completed as described in Chapter 20, proper installation is the next step. This plan is then turned over to landscape contractors, who implement the designer's plan. Materials identified in the planting plan must then be selected carefully and placed in the landscape as specified by the designer. The following are helpful definitions:

- **Landscape construction.** The execution of the planting plan and hardscaping features.
- **Hardscaping.** Landscape features such as fences, patios, walks, pools, and walls.

SITE PREPARATION

Before any plants are put in the landscape, the site must be prepared. The site needs to be staked, hardscaping must be installed, and planting beds must be prepared.

Staking the Site

Two days before starting the landscape job, the crew supervisor calls the local office to find out where all the underground utilities are located at the site. Most states require contractors to call for local information prior to digging.

The crew supervisor then marks where all the underground utilities are located at the site prior to staking the site for plant and hardscaping materials. Small wooden or plastic stakes with plant names written on them are driven into the ground to show the exact center of the planting hole and color-coded flags are used to indicate where utilities such as electric, gas, oil, steam, telephone, cable, water, and sewer are located. In addition to staking plant material locations, the supervisor also stakes the sites where patio, wall, fence, and other hard material will be located.

Hardscaping

Hardscaping features such as walls, patios, walks, and fences should be installed prior to plants to prevent the plants from being damaged. The landscape contractor installs different types of landscape surfacing materials. Material selection is based on cost. Concrete, asphalt, brick, and stone are the most popular types of paving material. Once installed, these provide the homeowner or corporate client with a durable, long-lasting, and low-maintenance surface for walks, driveways, and patios. The advantages and disadvantages of each are as follows:

- **Concrete and asphalt.** Both concrete and asphalt are relatively inexpensive to install, which is the main advantage. However, the main disadvantage of these materials is that they do not create interesting or unusual landscape designs.
- **Brick and stone.** These types of surfacing material create interesting landscape designs; however, they are more expensive than concrete and asphalt.
- **Other surfacing materials.** Crushed stone, marble chips, or wood chips are inexpensive alternatives to concrete, asphalt, brick, or stone; however, they are not permanent as they can be washed away in heavy rains or lost due to general maintenance such as raking.

Fencing is another hardscaping feature used in the landscape. The style of fence used in the landscape depends upon privacy and security considerations. Different styles of fencing are as follows:

- Chain-link-type fences provide little privacy and are not very pretty, but they are commonly used because they provide security and are fairly inexpensive (Figure 21-2).
- Rail-type fences give no security or privacy but are used to show where the property line is located or for decorative purposes (Figure 21-3).
- Basket weave or board-on-board type fences provide privacy and security; however, they are expensive (Figure 21-4).

Many other styles of fences are also available. All wooden fences should use treated fence posts to avoid problems with rot.

To add beauty to the landscape and provide security at night, low-voltage lights are usually installed along sidewalks and among plants. Electricity and damp soil is a hazardous combination, so a ground fault circuit interrupter (GFCI) is installed. The GFCI disconnects the electricity to a lamp when it

Figure 21–2 Chain-link fence

Figure 21–3 Rail-type fence

Figure 21–4 Basket weave fence (left) and board-on-board fence (right)

detects a short circuit. Lighting can be used for functionality and for aesthetics. One of the functional features of lighting is security; this type of lighting is typically used in conjunction with a motion sensor to deter vandalism and theft (Figure 21-5). Another form of functional lighting in the landscape is walk lighting, which provides light along walkways to help people see at night (Figure 21-6). A variety of types of lighting are used today for aesthetic purposes, including up lighting and down lighting (Figure 21-7). Up lighting is when light shines up at the base of the tree or interesting object to highlight it in the landscape. Down lighting is when light shines down on a tree or interesting object to create an interesting shadow pattern on the ground.

Bed Preparation

Proper preparation of the planting bed is key to establishing plants in the landscape. The first step is to remove large objects from the new landscape planting beds and make sure that the soil texture will support plant growth. For example, if the soil is very sandy, add peat moss to increase its nutrient- and water-holding capacity, or if there is too much clay, add sand to increase aeration and drainage. In general, the soil is tilled to a depth of 12 to 18 inches. During the tillage operations, the necessary peat moss or sand amendments can be made. A soil test should be run to determine necessary nutrient and pH adjustments; for example, most annuals prefer a pH around 6.5 whereas decorative evergreens enjoy a pH that is more acidic.

Figure 21–5 Floodlights provide light around the house for a variety of purposes.

Figure 21–6 Walkway lighting is helpful for safety concerns.

Figure 21–7 Up lighting and down lighting

INSTALLATION OF PLANTS IN THE LANDSCAPE

Plants to be installed in a landscape include bedding plants, trees and shrubs, and ground covers (Carpenter and Walker, 1990).

Bedding Plants

The term *bedding plants* refers to a wide range of plants (mostly annuals), including herbs, vegetables, and flowering ornamentals; however, some garden designs use perennials, bulbs, rhizomes, corms, and tubers for continual bloom. Homes built in the 1950s contained large flowerbeds in the front yard. Brightly colored flowers attract the public's eye but also detract from the home appearance. Today, flowers are used as accents. Annuals used as bedding plants include poppy, zinnia, marigold, petunia, pansy, ageratum, snapdragon, coleus, impatiens, tomato, and peppers. Bulbs, rhizomes, corms, and tubers are also used to provide early color in the fall and spring. Perennials are used to provide color in the spring and throughout the summer. Commonly used flowering perennials are carnation, Shasta daisy, Canterbury bells, columbine, delphinium, foxglove, day lily, iris, peony, and lupine. Commonly used bulbs, corms, tubers, and rhizomes are dahlia, caladium, crocus, amaryllis, hyacinth, iris, narcissus, gladiolus, tulip, and lily.

Bedding plants can be direct seeded or they can be transplanted. Direct seeding must be done in the early spring when there is no longer a danger of frost. Proper spacing and planting depth are important to insure proper plant growth and development. High-quality viable seeds are used to obtain a good stand. After the seeds are planted, they must be watered properly to prevent problems with rot caused by overwatering or drying out from underwatering. An alternative to seedings is the use of transplants, which requires proper scheduling. For example, seedlings must be started indoors, and they should reach the proper size when weather conditions permit their installation in the garden.

Trees and Shrubs

It is vital to obtain healthy plants for installation in the landscape. Trees and shrubs are sold in three ways: bare root, balled and burlapped, and container grown. Bare root is a seedling without a ball of soil around its roots. The bare root method has the advantage of being inexpensive, lightweight, and easy to transport. However, the disadvantage is problems with transplanting shock, which results in poor growth or death of the plant. Balled and burlapped plants are harvested for transplanting by digging a certain distance around the trunk and lifting the plant with a ball of soil around the roots. The soil is then wrapped tightly with a burlap material and held in place with twine (Figure 21-8). This process can be done by hand or mechanically. This method minimizes transplanting shock and the overall survival rate is very good; however, these types of plants are expensive. Trees and shrubs are also sold in containers.

Container-grown plants are grown in different-sized containers. Container-grown plants have the advantages of less transplanting shock, overall good

Figure 21-8 Balled and burlapped tree being planted

survival rate, and ease of transportation. Disadvantages associated with container-grown plants are that they may become overgrown and the plant may become rootbound, which can cause transplanting shock.

Planting Trees and Shrubs

After a high-quality plant has been selected, the success of the seedling depends on the timing of planting, preparation of the seedling, preparation of the soil, and planting technique. Spring is typically the best time for planting trees and shrubs because it offers the best environmental conditions that promote plant establishment and subsequent growth. These conditions include adequate moisture, warm soils, and relatively cool temperatures that minimize transpiration, resulting in less moisture loss and reduced transplanting shock. Fall is also a good time to plant trees and shrubs in locations where winters are mild. Although trees and shrubs can be planted during the summer, it is definitely more difficult to plant in summer than in the fall or spring because high temperatures and sunlight can cause plant tissues to dry out. When installing plants during the summer months, extreme care must be taken to ensure that the plants will survive and grow. Commonly used trees and shrubs are shown in Table 21-1.

TABLE 21-1 TREES AND SHRUBS COMMONLY USED IN LANDSCAPES

EVERGREENS

Common Name	Scientific Name	Hardiness Zone
Hicks yew	*Taxus media hicksii*	4
Sargent juniper	*Juniperus chinensis Sargentii*	4
English holly	*Illex aquifolium*	6–9
Privet	*Ligustrum japonicum*	6–7
Barberry	*Barberis darvinii*	4–9

DECIDUOUS TREES

Common Name	Scientific Name	Hardiness Zone
Aspen	*Populus tremuloides*	3–8
European white birch	*Betula pendula*	3–9
Flowering dogwood	*Cornus florida*	7–9
Western redbud	*Cercis occidentalis*	6–9
Weeping willow	*Salix alba*	3–10

DECIDUOUS SHRUBS

Common Name	Scientific Name	Hardiness Zone
Cotoneaster	*Cotoneaster horizontalis*	5
Flowering almond	*Prunus glandulosa*	6
Cranberry bush	*Viburnum opulus*	3
Winter berry	*Illex verticillata*	4

Installation Techniques for Trees and Shrubs

Planting plans show the plant by name and location on the plan. Designers recommend planting balled and burlapped (B&B), bare root, or container-grown plants based on the time of the year and the budget for each landscape job. The steps involved in planting trees and shrubs using either balled and burlapped or container grown-plants are as follows:

- Always dig the planting hole a minimum of 12 inches larger than the soil ball to be planted to allow enough space for new roots to grow from the ball. Fertilization is not necessary at the time of planting.
- The top of the ball should be at the top of the planting hole. To avoid problems with stem damage caused by rot organisms, never plant a tree or shrub deeper than the height of the soil ball.
- Place the plant into the hole without removing the plant from its container or any burlap or twine to make sure the hole is the proper size.
- Untie and remove all twine and burlap or carefully remove the plant from its container. Most burlap is made from plastic or nonbiodegradable material, so be sure it is removed completely unless labeled clearly as biodegradable. Then back-fill the hole by placing soil initially removed from the planting hole around the root ball. Make sure to break up all large clumps with a shovel and remove any stones prior to putting the soil back in the hole. In doing so, you remove any chances of air pockets around the root ball, which will cause problems with root growth. After the hole is filled, make a soil saucer slightly larger than the soil ball with extra soil from the hole to aid in trapping water when irrigating.
- Newly transplanted trees and shrubs require deep watering, so soil should be wet to a 12-inch depth.
- Trees 6 feet in height or taller require additional support or staking.
- Antitranspirants should be sprayed on the plant's foliage to reduce water loss following transplanting.
- Mulch or plastic should be placed around the base of newly planted trees to reduce soil water loss and to keep soil temperature warm around plant roots.
- The last and very important step is to clean up the site before leaving.

The steps involved in growing bare root trees and shrubs are similar to those described for balled and burlapped and container-grown plants. However, they are different in that bare root plants should be brought to the field in a container of water to prevent the roots from drying out. After the plants are brought to the field, they should be pruned carefully and placed in the hole and held in place until the hole is filled with soil so the crown is above the soil level. After filling, the soil should be packed firmly around the plant to keep the tree or shrub erect.

Ground Covers

Ground covers are low-growing plants that spread and form a mat-like growth over an area. Typically, ground covers are low-spreading plants that are less than 2 feet tall; they can be shrubs, vines, perennials, and grasses. Ground covers are used for a variety of purposes. For example, they can be used to

TABLE 21-2 GROUND COVERS COMMONLY USED IN LANDSCAPES

Common Name	Scientific Name	Hardiness Zone
English ivy	*Hedera helix*	5–9
Common periwinkle	*Vinca minor*	3–9
Japanese pachysandra	*Pachysandra terminalis*	4–8
Creeping juniper	*Juniperus horizontalis*	3–9
Crown vetch	*Coronila varia*	3–9

prevent soil erosion on steep grades, as an accent under trees, and for a variety of other purposes. Once properly installed, ground cover typically does not require regular maintenance. Examples of selected ground covers are shown in Table 21-2. Installation of ground cover requires properly preparing the soil for good establishment. If the soil is too sandy, organic matter should be added; if there is poor drainage, sand should be added. Soil tests should be run to correct for nutrient deficiencies and adjust the pH prior to installation. This is important with ground covers as with other plants used in the landscape because some plants, such as periwinkle (*Vinca minor*), will not grow properly in poorly drained soils with a pH below 6.5. A ground cover that can be established from seed is crown vetch (*Coronilla varia*). Ground covers grown in pots are expensive, but they are much more efficient than starting from seed in most cases. An alternative to potted ground covers is bare root stock ground covers because they are less expensive and are easier to establish than from seed. The disadvantage of using bare root ground covers is that they require more care and are subject to transplanting shock. Hydro seeding is a method used for seeding ground covers on steep slopes where manual planting is difficult or for a large area that needs to be planted. In hydro seeding, a water slurry containing seeds is pumped under high pressure onto the slope or over a large area.

Grass is a key component in almost any landscape design. The installation of a lawn will be discussed in detail in Chapter 23.

SUMMARY

You now have a basic knowledge of what is involved in the installation of plants in the landscape. Site preparation includes staking the site, hardscaping, and preparing the bed. Hardscaping features commonly used in the landscape include the different types of landscape surfacing materials, such as concrete and asphalt, brick, stone, and other surfacing materials. The different types of landscape fencing include chain-link type, rail-type, and basket weave or board-on-board. Accent lighting such as walk lighting, up lighting, and down lighting is commonly used in the landscape. Proper preparation of the planting bed is key to a successful installation of plants in the landscape. Installing bedding plants, trees and shrubs, and ground cover is the final step in the implementation phase of landscaping.

Review Questions for Chapter 21

Short Answer

1. What should be done prior to staking the site?
2. List the four major types of landscaping surfacing materials commonly used and the advantages and disadvantages of each.
3. List the three different styles of landscape fencing and what they are used for.
4. What is GFCI used for?
5. List the two types of lighting techniques and provide an explanation for each.

Define

Define the following terms:

landscape construction
hardscaping
up lighting
down lighting
bedding plants
continual bloom
bare root
balled and burlapped
container-grown plants

True or False

1. Hardscaping describes landscapes that only use rocks and cacti.
2. Hardscaping features in the landscape should be installed after the plants have been planted.
3. Today, low-voltage lamps provide security and beautify the landscape; therefore, GFCIs should be installed for protection.
4. The balled and burlapped method for harvesting nursery plants has the disadvantage of causing transplanting shock.
5. A tensiometer is one way to determine the location of electrical lines before digging.
6. When planting trees that have been balled and burlapped, it is important to make the planting hole the same size as the ball.
7. When using the balled and burlapped planting technique, the top of the ball should be approximately 6 inches below the soil surface.
8. Antitranspirants should be sprayed on the plant's foliage to reduce water loss following transplanting.
9. Homes in the 1950s contained only a few flowering plants in the front yard as an accent.
10. Rail-type fences not only provide security but also show where the property line is located.

Multiple Choice

1. Which of the following types of fencing provides security?
   A. Chain-link fences
   B. Rail type
   C. Board-on-board
   D. A and C

2. What is the main advantage of using asphalt as a paving material?
   A. Can be used to create interesting designs
   B. Relatively low cost of installation
   C. A wide range of available colors
   D. None of the above

Fill in the Blanks

1. The last and very important step when installing a landscape is to ________________ the site before leaving.
2. ________________ should be placed around the base of newly planted trees to reduce water loss and maintain soil temperature around the roots.
3. ________________ type fences give little privacy and are not very pretty but are commonly used because they provide security.
4. ________________ type fences give no security or privacy but are used to show where a property line is located.
5. ________________ type fences provide privacy.

Activities

Now that we have completed our discussion on installing landscapes, you will have the opportunity to explore this area in more detail. Go to a location near you and look for homes or other buildings being landscaped and take pictures of what you see at these sites. Write a description for what is being done at the landscaping site and include plants used, surfacing materials, methods of planting, and anything else you can think of regarding installation of a planting plan. If there is no landscaping going on in your particular area or you want to supplement your findings, search the Internet. Be sure to include the Web site address.

Reference

Carpenter, P. L., & Walker, T. D. (1990). *Plants in the landscape* (2nd ed.). New York: W. H. Freeman.

CHAPTER

22

Landscape Maintenance

Objectives

After reading this chapter, you should be able to

- provide background information on landscape maintenance.
- discuss the important aspects of pruning.
- list and discuss methods used for pruning different types of plants.
- discuss the proper care and maintenance of trees, shrubs, and flowering plants.
- discuss turfgrass care and maintenance.

Key Terms

arboriculture	head height	renewal pruning
central leader	heading back	scaffold branches
central leader method	modified central leader	thinning out
dead zone	open center tree form	topiary

ABSTRACT

This chapter begins with a brief background on landscape maintenance followed by the important aspects of pruning, including pruning goals, tree parts, pruning equipment, common questions about pruning, and general pointers in making pruning cuts. There are different methods for pruning different types of plants, including ornamental trees, fruit trees, grapes, flowering shrubs, evergreen shrubs, and deciduous trees. Proper care and maintenance of trees, shrubs, and flowering plants include proper watering; fertilization; and insect, disease, and weed control. The chapter concludes with a brief discussion of turfgrass care and maintenance.

INTRODUCTION

An attractive landscape begins with a good design followed by proper installation and routine maintenance (Figure 22-1). The landscape designer must understand the level of maintenance that the customer wants. Some landscapes require high maintenance, whereas others require very little maintenance. Low-maintenance landscapes feature plants that do not require extensive watering, pruning, pest control, and mowing. No matter how good the landscape design is, all landscapes require regular maintenance. The amount of annual maintenance that needs to be done depends on the season, kind and age of plant, and the desired effect of the landscape design.

Pruning is a common landscape maintenance activity for both low- and high-maintenance landscapes. The best pruning results are achieved when specific pruning goals are planned prior to pruning. Examples of pruning goals are to trim plants to maintain their natural beauty or to obtain a desired effect, to eliminate dangerous branches, to enhance the overall heath of the plant, and to enhance flower and fruit production.

Figure 22–1 Routine maintenance of a lawn, which is an important part of the landscape

Proper pruning requires knowledge of the major external parts of the tree, including the crown, scaffold branches, trunk, roots, terminal, and water sprouts. In addition to the external parts of the plant, knowledge of the internal portions of the tree—the xylem, phloem, cambium, and pith—is important. Sharp pruning equipment provides the best results, but it should always be used safely. Commonly used pruning equipment includes shears, manual saws, power saws, pole saws, and pruners.

Pruning must be done cautiously with sharp tools by starting with parts that pose a threat to the plant, selecting wide **scaffold branch** angles, not pruning too close to the trunk, leaving a small portion of stem above the bud, and pruning to direct branches outward.

The three major types of pruning are **thinning out, heading back,** and **renewal pruning.** The main goals for training ornamental trees are to make the tree trunk strong, produce an attractive plant form, and establish a set height. The **central leader** method is commonly used for ornamental trees. For fruit trees, the main pruning goals are to develop strong scaffolds to support the weight of fruits, properly space branches to facilitate harvesting, and facilitate pruning, spraying, and other necessary orchard operations. The three major training and pruning systems used in orchards are the **central leader, modified central leader,** and the **open center tree form** method. Grapes are generally grown on the many types of trellises used in the industry; two commonly used types are single-wire and double-wire trellises, which can be spur pruned or cane pruned. Flowering shrubs must be pruned every year to make sure they flower year after year. After a flowering shrub has stopped flowering, rejuvenation pruning should be done to promote flowering again. Both small and medium evergreens growing in a foundation planting or as hedgerows require annual pruning. The two types of pruning methods used for evergreens are formal and informal (the informal is more commonly used today). Deciduous trees typically require very little pruning; however, they do require pruning to remove damaged, diseased, or insect-infested branches.

Another important factor in caring for trees, shrubs, and flowering plants is proper watering. Typically, watering is not needed when there is adequate rainfall; however, in times of drought, supplemental watering is necessary. The frequency of watering depends on the type of plant grown, stage of plant growth and development, soil properties, time of year, and cultural practices used. Fertilization is also necessary to maintain good plant growth, which enables plants to resist low-level pest populations. Trees and shrubs can be damaged severely by insects and diseases, so a standard IPM system should be in place. A successful landscape must also control weeds because weeds grow faster than landscape plants. Turfgrass must also be kept attractive and healthy by managing nutrients properly, watering, mowing, removing clippings, providing aeration, and controlling pests; this will be discussed in more detail in Chapters 23.

BACKGROUND OF LANDSCAPE MAINTENANCE

A good design and proper installation reduces the annual maintenance required by a landscape. In the initial planning stages, the landscape designer must understand the level of maintenance required by the customer. Some landscapes require a high level of maintenance, require many hours of work,

Figure 22–2 Mulching around trees reduces the lawn size and the need for weeding.

and/or cost a lot of money. However, other landscapes look beautiful and can be maintained at a low cost. For a low-maintenance landscape, plants should be chosen that do not require extensive watering, pruning, pest control, or mowing. Irrigating the landscape is one of the major jobs in landscape maintenance; one way to reduce the workload for this task is to install an automatic watering system. If this type of system is too expensive, the lawn area can be reduced to decrease the need for irrigation. Reducing the lawn area also reduces the need for mowing. Ground covers are commonly used as a substitute when reducing the lawn size. Another way to reduce maintenance is to mulch regularly to reduce the need for weeding (Figure 22-2). In addition, many cultivars are resistant to a wide variety of diseases and insects that may be located in a particular location, thereby reducing maintenance. If raking leaves (Figure 22-3) is going to be a problem, evergreens should be substituted for deciduous trees. No matter how good the landscape design is, all landscapes require regular maintenance. The amount of annual maintenance required depends on the season, kind and age of the plant, and the desired effect of the landscape design. The following sections describe general maintenance activities required in all landscapes, including watering, controlling weeds, fertilizing, pruning, and controlling pests.

PRUNING

The study of trees, their growth, and culture is **arboriculture.** Pruning is a common landscape maintenance activity in both low- and high-maintenance landscapes according to a planned schedule or as needed. To obtain the best results it is important to have specific pruning goals in mind prior to pruning.

Figure 22–3 Beautiful fall foliage leads to a lot of raking.

Pruning Goals

Examples of good pruning goals are as follows:

- Trim plants to maintain their natural beauty. Prune plants to regulate their size without affecting their natural beauty.
- Trim plants to obtain a desired effect. Pruning to achieve and maintain a desired shape depends on the type of plant and the desired effect (Figure 22-4).
- Eliminate dangerous branches. Pruning is also done for safety purposes by removing dangerous branches hanging over houses and walkways, which may pose a threat to structures or people walking under the branches. To keep trees healthy, immediately remove any broken or damaged tree branches after windstorms.
- Enhance overall plant health. To maintain healthy, actively growing plants, remove dead or diseased branches to protect the rest of the plant and adjacent plants from becoming infected. Branches can also be thinned to increase light into the center of the tree canopy and thus promote even fruit ripening.
- Enhance flower and fruit production. Most flowering and fruit-producing plants require annual pruning to flower and produce high-quality fruit on an annual basis.

Figure 22–4 Yew plant being pruned to regulate height

Tree Parts

To prune trees properly, the pruner must know the major external portions of the tree (Figure 22-5), which are crown or canopy, terminal or leader, scaffold branches, trunk, and suckers or water sprouts. Knowledge of the key internal portions of the tree is equally important (Figure 22-5).

Pruning Equipment

Pruning equipment should always be sharp to get the best results. When pruning, eye and ear protection must be worn, especially when using power equipment. (Note that federal labor laws prohibit minors from operating power saws.) Pruning tools consist of a variety of saws and shears designed to cut different-size limbs at different heights and at different locations on the plant. Commonly used pruning tools are shears (hand, lopping, and hedge), manual saws (bow, tree surgery saw, two-edge saw, folding saw, ridge handle curved saw), power saws (electric power chain or gasoline powered), pole saws, and pruners.

Common Questions About Pruning

Is the use of tree topping a common practice today? In the past, tree topping—reducing the total tree height by one-third to one-half by removing upper scaffold branches—was common; however, it has since been found that this practice is

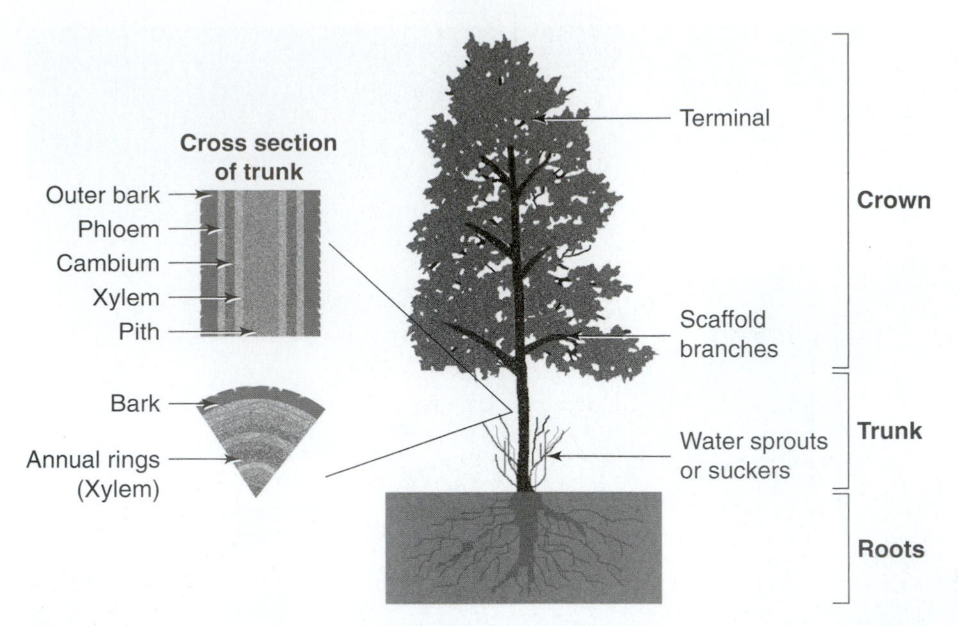

Figure 22–5 Key external and internal portions of a tree

not good for the tree. What is the best time of the year to prune? Late fall to early winter is the best time of the year to prune because few other farming operations are done at this time. Should tree paint be applied to cuts after pruning? Tree paint is no longer used commonly to protect wounds caused by pruning because it was not needed.

General Pointers for Making Pruning Cuts

A number of factors dictate the exact way a pruning method is implemented: type of plant, goal of pruning, the environment, and others. Regardless of the method, certain guidelines should be followed for successful pruning. Some general pointers are as follows:

- Define the pruning goals and the best time of the year to prune.
- Never take too much off; it is better to cut off too little and come back than to cut off too much.
- Use sharp tools to produce sharp cuts, which will heal faster than cuts that have been torn.
- Start pruning with parts that pose a threat to the plant, for example, any dead or dying materials.
- Select branches with a wide branch angle (45°) because branches with a narrow angle (20°) will split under windy conditions or from the weight of fruit.
- Do not prune too close to the trunk because doing so will make a large wound that will not heal properly; leave a small portion of the stem to create a smaller wound that will heal properly. Do not leave too much of a stub from the branch being pruned because disease can enter through the stub (Figure 22-6).
- Leave a small portion of the stem above the bud; too much stem left can result in disease entry and too little can result in desiccation of the bud.
- Prune branches that are growing inward toward the center of the tree, but leave outward growing branches. When pruning is done properly, bud growth can be directed by selecting buds.

Good branch angle ~45°

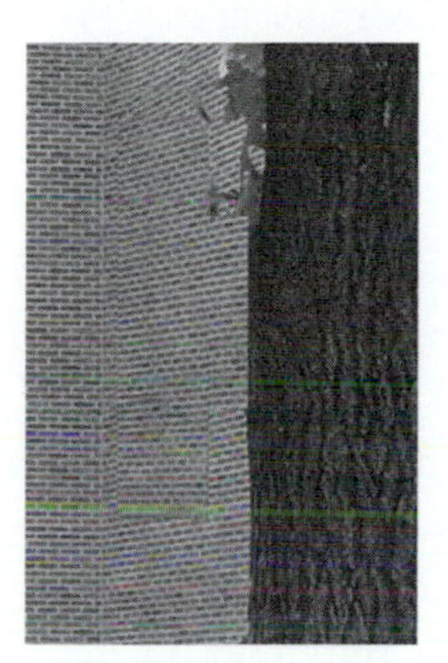
Too close
Large wound to heal

Too long
Disease can
enter stub

Good cut
Small wound

Figure 22–6 General pointers in making pruning cuts

PRUNING DIFFERENT TYPES OF PLANTS

The three basic strategies for pruning are thinning out, heading back, and renewal pruning. Thinning out is the removal of excessive vegetative growth to open the plant canopy and reduce the number of fruiting branches with the goal of promoting larger and overall better-quality fruits. By thinning out the plant, more light is able to penetrate to promote better fruit set, increased productivity, and overall quality. Heading back is the removal of the tips of terminal branches to promote secondary branching. This type of pruning makes the plant look fuller; its shorter size may even promote a new shape. Renewal pruning is used to rejuvenate old plants by removing old unproductive branches, thereby promoting vigorous growth. This is commonly used in flowering shrubs that have stopped flowering and fruit trees that have been neglected.

Ornamental Trees

The main goals of training ornamental trees are to make the tree trunk strong, produce an attractive plant form, and establish a set height. To develop a strong trunk, the central leader method is commonly used. The central leader method identifies and trains one strong upward-growing branch to grow as the central axis of the tree; the **central leader** is the main upright shoot of the tree. Ornamental trees take time to establish their form; however, they can be trained into

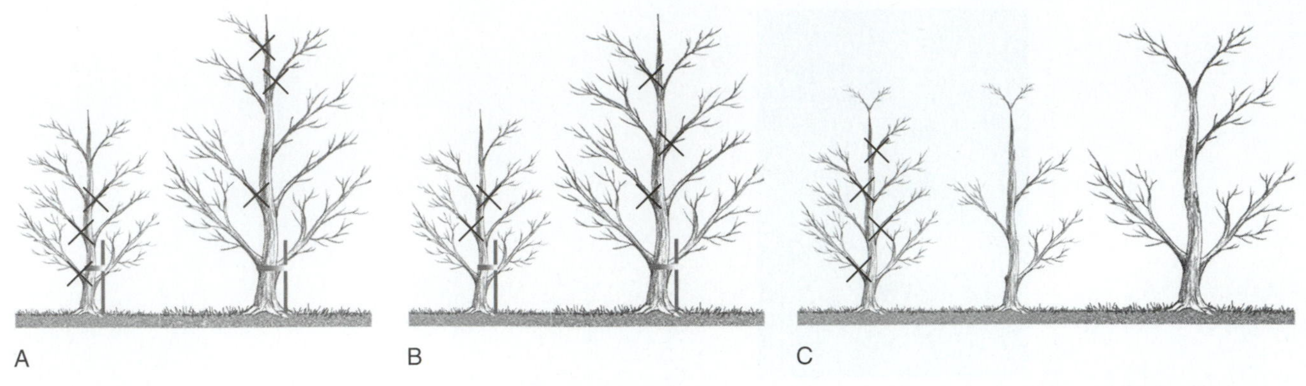

Figure 22–7 Feathered tree form (A), standard tree form (B), and branched-head standard tree form (C)

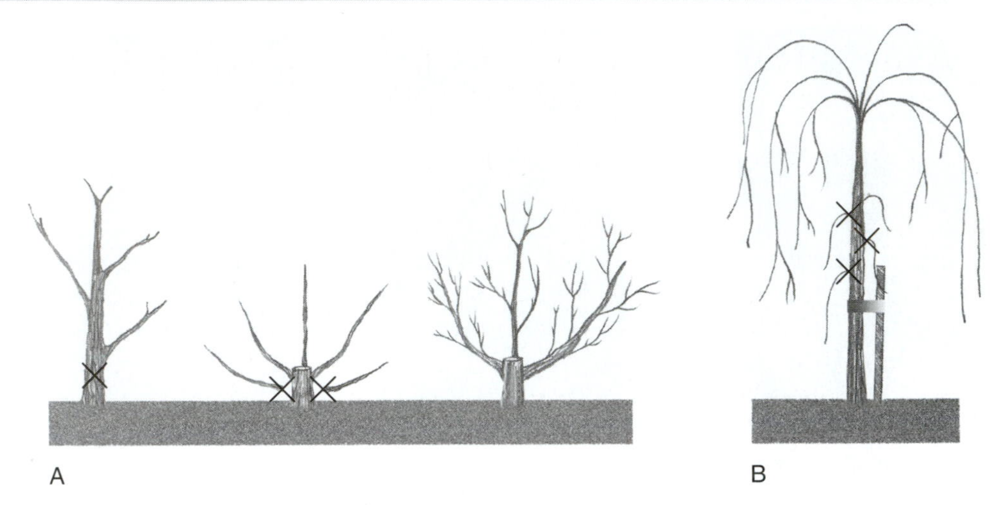

Figure 22–8 Multistem tree form (A) and weeping standard tree form (B)

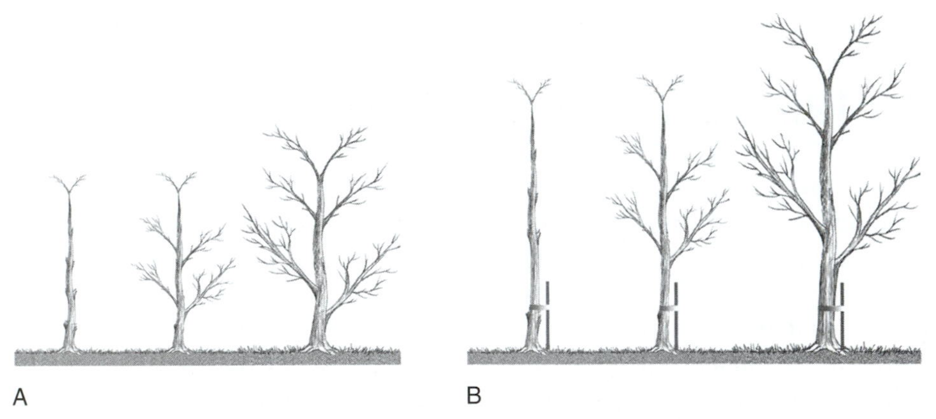

Figure 22–9 Development of heading height in trees; low heading height (A) and high heading height (B)

several forms: feathered, standard, branched-head, weeping standard, and multistem tree form (Figures 22-7 and 22-8). The **head height** is the height of the scaffold branches (the main branches growing from the trunk of the tree) above ground at the adult stage (Figure 22-9).

Fruit Trees

The pruning goals for fruit trees are to develop strong scaffolds to support the weight of fruits and properly space branches to facilitate harvesting, pruning,

Figure 22–10 Training for a central leader tree form

spraying, and other necessary orchard operations. Pruning can also control the time to first fruiting, promote fruiting on a yearly basis, produce attractive shapes especially when used in the landscape, and confine the tree to space available. One of the main reasons for training and pruning fruit trees is to increase productivity and to make harvesting and other operations in the orchard easier. The three major training and pruning systems used in orchards are the central leader, modified central leader, and the open center.

Central Leader Tree Form

The central leader method is commonly used for dwarf and semidwarf fruit trees, such as apples and sweet cherries, and produces narrow- and conical-shaped trees with several tiers of scaffold branches (Figure 22-10). The uppermost bud becomes the upright central leader of the tree. Then three or four lateral branches with wide-angle crotches that are spaced equidistant around the tree are selected for the frame of the tree. To prevent the central leader from bending and stopping terminal growth, fruiting on the upper third of the terminal should be discouraged. Water sprouts should be removed as they develop.

Figure 22–11 Training for a modified central leader. The "x" designates removal of the central leader.

Modified Central Leader Tree Form

The modified central leader system is the same as the central leader tree form system in the early stages to permit the formation of strong scaffold branches. After all the scaffold branches are established, the central leader is removed, creating an open center (Figure 22-11).

Open Center Tree Form

The open center tree form is commonly used for peaches, nectarines, plums, pears, and others. This method allows easy access to the tree for fruit thinning and allows for good sunlight penetration to help fruit ripen (Figure 22-12). The starting point for training a plant for the open center system is to head the plant. Heading promotes new growth required for the training period. During the first dormant season, three to four branches are selected to be scaffold branches, and the remaining branches are removed. The branches should be spaced equal distances apart, leaving enough room for growth.

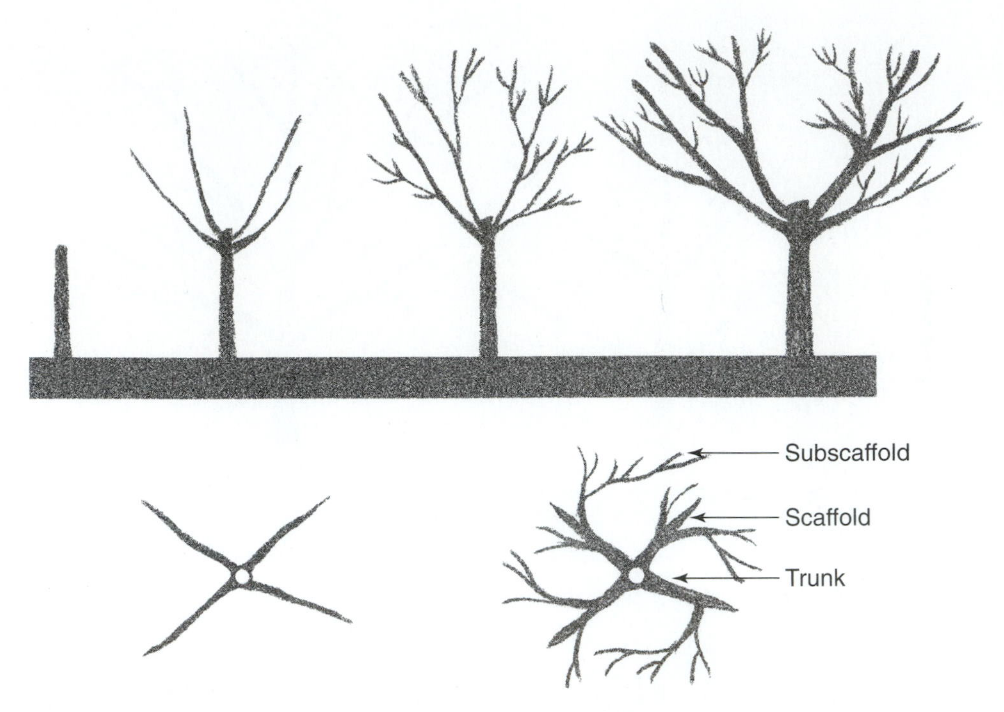

Figure 22–12 Training for the open center tree form

Grapes

Grapes are usually grown on the many types of trellises used in the grape industry. Two commonly used trellises are single-wire or double-wire trellis types, which can be spur pruned or cane pruned. Spur-pruned vines have a permanent trunk and arms. During the winter months, the previous season's growth should be cut back to two to four buds, thereby creating a short shoot resembling a fruit tree spur. In the spring, the buds will break dormancy, grow slightly, and then produce flowers and fruits on a lateral bud from the current season's growth (Figure 22-13). Cane-pruned vines differ from spur-pruned vines because they only have a permanent trunk. At the end of each season, the old canes are removed by pruning and three or four spurs are left at the top of the trunk to provide a place for new cane to form the next season. Annually, new canes are placed on the trellis as arms to support the fruit (Figure 22-14).

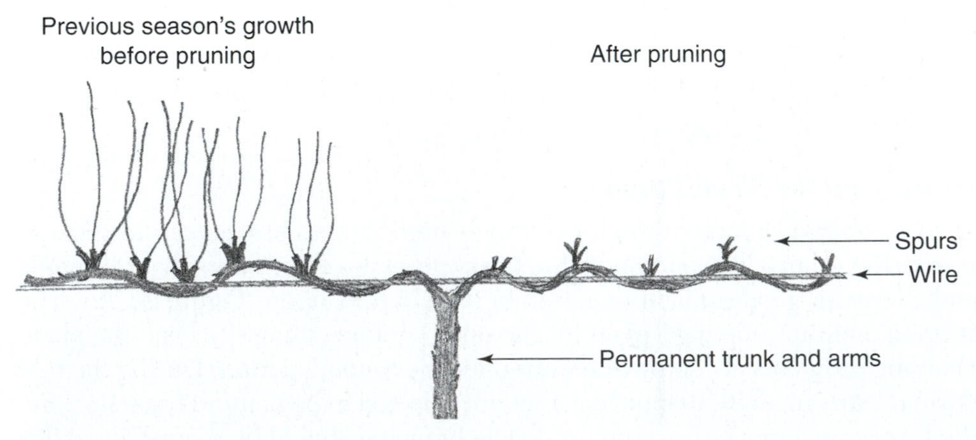

Figure 22–13 Spur-pruned grape vines

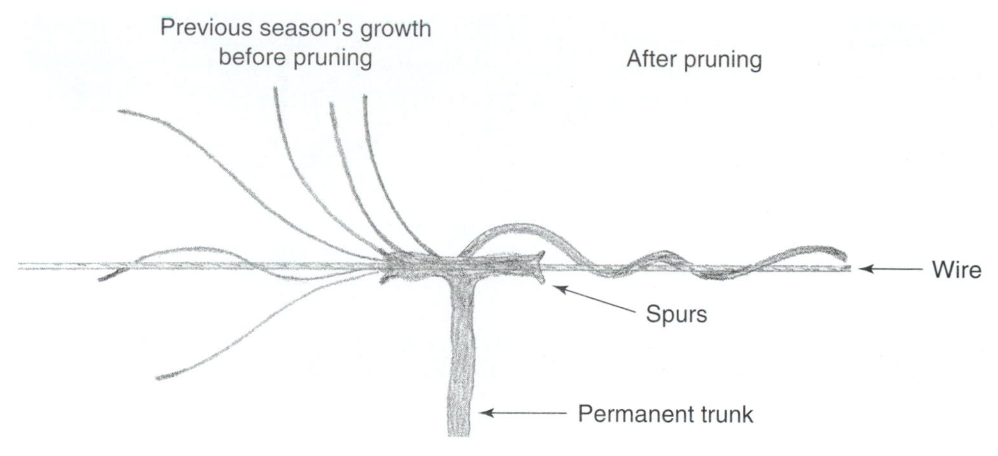

Figure 22–14 Cane-pruned grape vines

Flowering Shrubs

Proper pruning must be done every year to make sure that flowering shrubs flower year after year. Shrubs develop next year's flower buds immediately after flowering each year, so pruning must be done carefully to avoid removing the flowering buds. Rejuvenation pruning is used on shrubs that have not been pruned properly and have stopped flowering.

Evergreen Shrubs

Small and medium evergreens growing in a foundation planting or as hedgerows require yearly pruning. Landscape designers typically recommend yew (Figure 22-15), juniper, aborvitae, or boxwood evergreen varieties. There are two

Figure 22–15 Yew in the landscape

Figure 22–16 Topiary used as a design tool

Figure 22–17 Barberry plants used as a hedgerow

types of pruning methods for evergreens:

- **Formal.** Formal hedges are created by planting and pruning to form a specific geometric shape. A **topiary** is a plant that has been trained and pruned into formal geometric or abstract shapes. Yews and boxwood are commonly used for topiaries because they are easy to train, evergreen, and long-lived. Topiaries need to be maintained to keep their form and beauty (Figure 22-16). The degree of pruning depends on the plant type, plant vigor, form design, and demand by the owner.
- **Informal.** Today the formal look is not used widely; rather the informal look is used commonly because it emphasizes a natural look, which is more appealing. Informal hedges are allowed to grow freely in their natural shape (Figure 22-17) and are pruned only to restrict size and to remove damaged branches. Because evergreens are used commonly as hedges in the landscape, it is important to know that they have a **dead zone,** which is the area 6 to 12 inches below the green needles on the branch. Most evergreens cannot develop new leaf buds after pruning into the dead zone, so cuts should not be made in this region.

Deciduous Trees

Shade trees growing in the home landscape typically require little pruning. Most pruning of deciduous trees relates to removing storm- or disease/insect-damaged branches for safety reasons.

CARE OF TREES, SHRUBS, AND FLOWERING PLANTS

Proper care of trees, shrubs, and flowering plants includes watering, fertilizing, and controlling insects, diseases, and weeds.

Watering

Watering is typically not important when there is adequate rainfall; however, in times of drought, supplemental watering is necessary. Watering in a typical landscape is frequently performed improperly. Many gardeners feel that if the soil surface is wet, the plants have received adequate water; however, this is incorrect. The proper way to water is to wet the root systems of plants, so the grower needs to know where the roots of a given plant are located. Roots grow toward water, so if only the ground surface is watered, then the roots will stay near the surface making it necessary to water more frequently. However, if the plants are watered deeply, a deeper penetrating root system will be promoted. The frequency of watering depends on a variety of factors as described in the following paragraph.

The type of plant being grown is very important to consider when watering plants. For example, some plants are drought tolerant or have deep root systems that enable them to withstand fairly long periods of time without water. However, lawns and a variety of bedding plants need to be watered frequently in the absence of natural rainfall.

Another important consideration when watering plants is the stage of plant growth and development. When plants are being established in the landscape, more frequent watering is necessary; whereas after they are established, less frequent watering is necessary. Watering is also critical during flowering and fruiting because if water stress persists during this time, it will cause flower and fruit drop.

Soil properties have a profound effect on when plants need to be watered. For example, in sandy, well-drained soils, it is necessary to water more frequently than in soils with a high organic matter content.

The time of the year has a dramatic effect on when supplemental water needs to be added. During the early spring there is frequently more rainfall then during the summer months therefore more supplemental water will be needed during the summer months for plants that are sensitive to reduced water conditions.

Cultural practices also affect when plants need to be watered. For example, the use of mulch or plastic reduces evaporation, thereby improving water retention by the soil.

Several methods are commonly used for watering plants in the landscape hand watering, sprinklers, and drip irrigation. Hand watering is typically done on a small scale. Sprinklers are popular for watering lawns and flowerbeds because they can be used to water a larger area, however, water is wasted because the area cannot be confined and evaporation occurs. Drip irrigation in flowerbeds provides water to a specific location and makes efficient use of water because you are targeting specific plants and less water is lost due to evaporation. The downside of drip irrigation is that drip lines may become clogged (Figure 22-18).

Figure 22–18 Drip irrigation line

Fertilization

Fertilization is necessary to maintain good plant growth, which enables plants to resist many diseases and pests. Starter solutions are used to get plants established. After they are established, fertilizer is used to sustain active vegetative growth and flowering. Although fertilization can be used to increase flower production, it will also increase the shrub's total growth, which will necessitate additional pruning. Late summer or early fall fertilizer application should be avoided because it may delay the plants' preparation for the winter dormancy period and lead to plant damage. Fertilizer application methods include the following:

- **Top dressing.** Sprinkling granular fertilizer around the base of the plant.
- **Liquid.** Watering with fertilizer.
- **Fertilizer spikes.** Inserting solid spikes directly into the soil around the plant.

Insects and Diseases

Trees and shrubs can be damaged severely by insects and diseases. Insects with either chewing (caterpillars and bag worms) or sucking (aphids, scale insects, mites) mouthparts are problematic. Diseases caused by bacteria, fungi, and viruses are also problems in trees and shrubs. In most cases, pests do not cause significant problems in the landscape but they do cause yellowing of foliage or blemishes, which reduce the aesthetic value of ornamentals. A standard IPM system is important for controlling insects and diseases. When cultural, biological, mechanical, and genetic control procedures no longer control the pest, chemicals may be used. Table 22-1 lists problems caused by pests, control measures, and plants affected.

Weeds

Weeds grow faster than landscape plants and create numerous problems. Weeds must be controlled to prevent the landscape from becoming overrun. The use of herbicides should be a last resort because people and their pets play on the lawn and handle plants. An IPM program around homes should start with cultural methods for weed control. Weeds should not be allowed to grow to a point where they set seed. Weeds can be removed physically by hand (Figure 22-19) or using a hoe, which also improves the soil structure. The use of ground covers, mulch, and plastics is the most effective method of weed control in the landscape. When herbicides have to be used, the grower should read the safety sheet

TABLE 22-1 PROBLEMS CAUSED BY PESTS, CONTROL MEASURES USED, AND PLANTS AFFECTED IN THE LANDSCAPE

Pest	Plants Attacked	Control
Aphid	Cole crops	Diazinon or Malathion
Fruit or ear worm	Corn or tomato	Sevin or Diazinon
Damping-off	Tomato, pepper	Captan and Thiram
Downy mildew	Cole crops	Zineb or Maneb
Ants	Lawns	Diazinon or Sevin
Armyworm	Trees and shrubs	Diazinon or Carbaryl
Japanese beetle	Trees and shrubs	Diazinon
Caterpillar	Landscape plants	Diazinon or Sevin
Borer	Fruit trees and ornamentals	Lindane or Dimethoate
Mite	Annuals, perennials, trees, and shrubs	Malathion or Kelthane

Figure 22–19 Weeding by hand

and label the location properly so people know and understand that chemicals have been used.

TURFGRASS CARE AND MAINTENANCE

Turfgrass maintenance involves keeping a stand of turfgrass in an attractive and healthy condition by managing nutrients properly, watering, mowing, removing clippings, providing aeration, and controlling pests. For more details on turfgrass care and maintenance, refer to Chapter 23.

SUMMARY

You now have a basic understanding of what is necessary to maintain a landscape properly. The important aspects of pruning include pruning goals, tree parts, pruning equipment, common questions about pruning, and general pruning pointers. Different methods are used for pruning different types of plants, including ornamental trees, fruit trees, grapes, flowering shrubs, evergreen shrubs, and deciduous trees. The proper care and maintenance of trees, shrubs, and flowering plants includes watering, fertilization, and insect, disease, and weed control.

Review Questions for Chapter 22

Short Answer

1. What are five important pruning goals?
2. List the five important external portions of the tree and the four key internal portions of the tree.
3. What should be done to pruning equipment prior to use to obtain the best results and what are five commonly used pruning tools?
4. What are eight general pointers that should be followed when making pruning cuts?
5. What is the best time of the year to prune and why?
6. Should tree paint be used to protect wounds after pruning?
7. What is tree topping, and is it commonly used today?
8. What are two commonly used trellises and two pruning methods used for grapes?
9. When do flowering plants initiate their flower buds?
10. What are two pruning methods commonly used for evergreens, and which one of the two is more commonly used today?

Define

Define the following terms:

arboriculture	renewal pruning	head height	open center tree form
thinning out	central leader method	scaffold branches	topiary
heading back	central leader	modified central leader system	dead zone

True or False

1. A good design and proper installation does not mean there will be a reduction in the amount of annual maintenance.
2. It is not always important to have specific goals in mind prior to pruning to achieve the best results.
3. An important pruning goal is to maintain the plants' natural beauty.
4. An important pruning goal is to decrease flower and fruit production.
5. Tree paint is commonly used to protect wounds after pruning.
6. When pruning, it is important to select branches with narrow branch angles to achieve a more compact plant.
7. When pruning, it is very important to prune as close to the trunk as possible.
8. The best time of year to prune is during the late fall to early winter because few other farming operations are done at this time.
9. Tree topping, which reduces the total height of the tree one-third to one-half by removing upper scaffold branches, is a common practice today.
10. Grapes are usually grown on trellises.
11. Flowering shrubs develop next year's buds during the early spring.
12. Evergreens have a dead zone 6 to 12 inches below the green needles.

13. Shade trees growing in the home landscape require little pruning.
14. Fertilization can be used to increase flower production in trees, shrubs, and flowering plants.
15. In times of drought, trees and shrubs do not require watering.

Multiple Choice

1. Arboriculture is the term used to describe
   A. early civilizations or cultures that used plants in the landscape.
   B. the study of the culture of plants for their usefulness in medicine.
   C. the study of trees, their growth, and culture.
   D. None of the above

Fill in the Blank

1. ______________________ is the study of trees, their growth, and culture.

Matching

Place the proper term in the space provided for the internal portions of the stem.

Terms: phloem xylem cambium pith

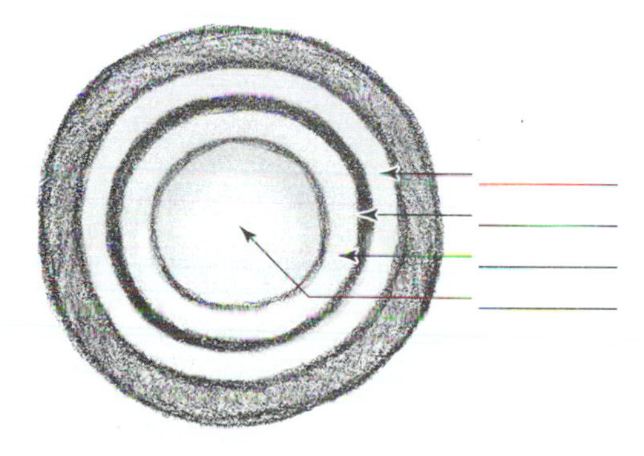

Activities

Now that we have completed our discussion on landscape maintenance, you will have the opportunity to explore this area in more detail. Go to a location near you and look for homes or other buildings being landscaped and take pictures of what you see at these sites. Write a description of what type of maintenance is needed and what is being done properly, plus anything else you can think of regarding maintenance of the landscape. If there is no landscape maintenance going on in your particular area or you want to supplement your findings, search the Internet. Be sure to include the Web site address for all the sites discussed.

CHAPTER

23

Warm- and Cool Season Turfgrass Selection, Establishment, Care, and Maintenance

ABSTRACT

The three major functions of turf include utility, ornamentation, and sports. The six factors commonly used to assess visual turf quality include density, texture, uniformity, color, growth habit, and smoothness. The functional turf quality is judged by the turf's rigidity, elasticity, resiliency, and recuperative potential. A good turf begins with the correct turfgrass, so it is important to consider climatic requirements, life cycle, usage, maintenance needs, visual and functional quality, and disease and insect resistance. The method used to establish turfgrass from seed depends on the kind of grass used, environmental conditions, and cost. Factors to consider when establishing turfgrass by seed include seed selection, source of seeds, seedbed preparation, planting, care, and first mowing. Turfgrass can also be established by vegetative propagation. Turfgrass maintenance is key to keeping a stand of turfgrass in an attractive and healthy condition. This is accomplished by properly managing nutrients, watering, mowing, aerating, and controlling pests. Landscapers should understand the different types of mowing equipment and the proper mowing frequency, mowing height, and method of removing clippings and thatch. The chapter then revisits scientific classification using a grass as an example. The chapter concludes by describing cool- and warm season turfgrasses.

Objectives

After reading this chapter, you should be able to

- list and discuss the three major functions of turf.
- provide the six factors used for assessing visual turf quality and the four factors used for judging functional turf quality.
- list and provide background information on the six factors that should be considered when selecting turfgrasses.
- describe how to establish turfgrass from seed and from vegetative propagation.
- explain how to maintain turfgrass.
- briefly discuss scientific classification by using a grass as an example.
- provide examples of warm- and cool season turfgrasses.

Key Terms

plug	sprig	turfgrass blend
seed purity	transition zone	turfgrass mixtures
seed viability	turf	winter overseeding
sod	turfgrass	

INTRODUCTION

The major functions of turf are utility, ornamentation, and sports. Turf quality—the actual excellence of turf—is assessed in two ways: visually and functionally. Density, texture, uniformity, color, growth habit, and smoothness are ways used to assess visual quality. The rigidity, elasticity, resiliency, and recuperative potential of turf are factors typically used to assess the functional quality of turf. A good turf begins with selecting the correct **turfgrass,** which can be accomplished by knowing the climatic requirements at the location, life cycle of the plant, usage of the turf (Figure 23-1), maintenance needs, visual and functional quality, and disease and insect resistance. Turfgrasses can be established either by sexual propagation or asexual propagation. Sexual propagation is most commonly used for cool season turfgrasses, whereas asexual propagation

Figure 23–1 Students enjoying a nice spring day on freshly mowed turf

is typically used for warm season turfgrasses. The seed used to establish turfgrass must be selected from a reputable supplier. Properly preparing the seedbed requires soil testing, controlling weeds, correcting problems with soil texture, removing large objects, installing drainage and irrigation systems, and leveling soil. After the seedbed is prepared properly, the seeds must be sown at the proper time using the proper seeding rate. Hand seeding is commonly used for small plots, whereas for larger plots mechanical seeders are used.

After the seeds are planted, they germinate and grow into turfgrass, which leads to mowing. The general rules of thumb when mowing turfgrass are never remove more than one-third of the turfgrass blade at one time and make sure to use a sharp lawnmower blade. Turfgrass establishment by vegetative propagation is similar to establishment by seed; however, the main difference is that **sod, plugs,** or **sprigs** are used. Turfgrass maintenance involves keeping a stand of turfgrass in an attractive healthy condition through proper management of nutrients, watering, type of mowing equipment, mowing frequency and height, removal of clippings and thatch, aeration, and pest control. Knowing the scientific classification of turfgrasses is important because the grower must be sure the correct turfgrass is being used. Warm season turfgrasses are commonly propagated vegetatively by sod, plugs, and sprigs. Although a large number of warm season turfgrasses are used, some common examples include bahiagrass, bermudagrass, buffalograss, centipedegrass, carpetgrass, Saint Augustinegrass, and zoysias. Although hundreds of different types of grasses exist, only a limited number will produce high-quality turf in the cooler portions of the United States. Examples of commonly used cool season turfgrasses are bluegrasses, bentgrass, fescues, and ryegrass.

MAJOR FUNCTIONS OF TURF

Before discussing the major functions of turf, it is important to distinguish between turf and turfgrass. *Turf* refers to a collection of plants in a ground cover and the soil in which roots grow. Turf may consist of one species of grass or a

Figure 23–2 Nicely landscaped home (top) and a poorly maintained landscape (bottom)

mixture and contains a high level of organization. *Turfgrass* refers to a collection of grass plants that forms a ground cover and must be regularly maintained. The three major functions of turf are utility, ornamentation, and sports.

- **Utility.** Establishing a turf at a given location can remove pollutants from the air or soil, cool the area during hot weather, stabilize the soil to reduce erosion, and cut down on dust.
- **Ornamentation.** The ornamental role of turf typically depends on the visual effect a landscape designer wants to achieve. One of the benefits of installing turf for ornamental purposes is to increase the property value (Figure 23-2).
- **Sports.** Used on playing fields for sports such as tennis, golf, soccer, football, and baseball.

TURF QUALITY

Turf quality is the excellence of the turf, which is based on visual and functional factors. Following are the six factors used to assess visual quality:

- **Density.** The number of aerial shoots per unit area; this is an obvious visual quality because a nice thick turf is very important in a landscape.
- **Texture.** The width of grass blades; wide blades are coarse, and narrow blades are fine. The desired effect determines the coarseness of the blade used.
- **Uniformity.** The evenness of turf distribution on a site. For example, uniform turf is very important on the putting greens of golf courses (Figure 23-3).
- **Color.** The measure of reflected light. A dark green turf is a general indicator that the turf is healthy.
- **Growth habit.** The type of shoot growth, which can be bunch, rhizomatous, or stoloniferous.
- **Smoothness.** The soil surface features, which are determined when the soil is prepared prior to planting. In most cases, it is important to have a smooth surface.

Following are the four factors used for judging functional quality:

- **Rigidity.** The capability to resist compaction and wear. An optimal turf should be rigid enough to resist compaction and wear, but it should not be so rigid that it makes walking uncomfortable.
- **Elasticity.** The capability to recover from compression. All turfgrasses compress when subjected to activity; however, the turfgrass must be elastic enough to recover from compression.
- **Resiliency.** The capability to absorb shock. This is particularly important when turf is used in sports; for example, the turfgrass should absorb some of the shock felt by a football player thrown to the turf.
- **Recuperative potential.** The capability to recover from damage. Turfgrasses must recuperate following wear from a sport such as football or golf.

Each of the factors is affected by the chemical and water content of the tissue, plant size, density, and the organic/inorganic constituents in the growth medium.

Figure 23–3 Putting greens require uniform turf.

TURFGRASS SELECTION

A good turf begins with selecting the right turfgrass, which can be accomplished by being knowledgeable about the issues discussed in the following sections.

Climatic Requirements

Turfgrass selection must be based on the capability of a species to grow and serve a useful purpose in a particular climate. Turfgrasses can be broken down into two groups:

- Warm season (Southern turfgrass) (Figure 23-4)
- Cool season (Northern turfgrass) (Figure 23-5)

Climatic conditions must be evaluated carefully when growing warm- and cool season turfgrasses. Warm season turfgrasses grow best at temperatures between 80 and 95°F and are limited by cold weather, which imposes winter dormancy. Cool season turfgrasses grow best at temperatures between 60 and 75°F and are limited by heat and drought, causing summer dormancy that turfgrasses may not be able to survive. A region called the **transition zone** is the area between definitive climate zones; this climate favors some warm- and cool season turfgrasses but is not ideal for either.

Figure 23–4 Warm season turfgrass that has coarse leaf blades

Figure 23–5 Cool season turfgrass

Life Cycle

The life cycle of the turfgrass must be known to maintain it properly for peak performance. During the life cycle of an annual grass, the grass goes through a vegetative, reproductive, and senescent stage, followed by the death of the plant. Perennial turfgrasses go through vegetative, reproductive, and dormant stages. Mowing keeps turfgrasses in their vegetative stage rather than following the complete life cycle through reproduction, which is preferred. This is preferred because following seedhead production, the plant is programmed to senesce and die.

Usage

The grower should know what the turfgrass will be used for to select the proper turfgrass for maximal performance. For example, a turfgrass used in a homeowner's lawn would not hold up on a sports field.

Maintenance Needs

Prior to selecting a turfgrass, its maintenance needs should be considered. For example, a homeowner may want to have a turfgrass with low maintenance needs, whereas at a golf course, playability is much more important than maintenance needs. Bentgrasses and bluegrasses are examples of turfgrasses that require high maintenance, whereas tall fescue and centipedegrass are relatively low-maintenance grasses.

Visual and Functional Quality

Each of the visual qualities—density, texture, uniformity, color, growth habit, and smoothness—should be considered along with the functional qualities of turfgrasses mentioned earlier.

Disease- and Insect Resistance

In addition to standard cultivars, seed companies use genetic engineering to develop endophyte-enhanced turfgrass cultivars. Endophytes are microscopic fungi that live within the host grass plant. These fungi produce chemicals that are toxic to certain turf insects and diseases. These genetically engineered cultivars require fewer pesticide applications, which is better for the environment and saves time and money.

TURFGRASS ESTABLISHMENT

The method used to establish turfgrasses depends upon the kind of grass used, the environmental conditions where it will be grown, and the amount of money in the budget. Two general propagation methods are used:

- **Sexual propagation.** Seeding is the most common method used with cool season grasses. Although seeding is inexpensive, more time is required to establish a high-quality turf (Figure 23-6).
- **Asexual propagation.** Vegetative propagation using sod, sprigs, and plugs is common with warm season turfgrasses. Although vegetative propagation establishes high-quality turf very rapidly, it is expensive (Figure 23-7).

Figure 23–6 Early stages of turf establishment from seed

Figure 23–7 Turf establishment from sod

Planting Turfgrass Using Sexual Propagation

Prior to discussing planting turfgrass using sexual propagation, you should be aware of the following definitions: **turfgrass blend, turfgrass mixtures, seed purity, seed viability,** and **winter overseeding.** A turfgrass blend is a combination of different cultivars (usually at least three) from the same species. Blends should have grasses with similar appearance and competitive capabilities; at least one cultivar in a blend should be well adapted to the planting site. Turfgrass mixtures are a combination of two or more different turfgrass species. Seed purity is the percentage of pure seed of an identified species or cultivar present in a particular lot of seed. Seed viability is the percentage of seed that is alive and will germinate under standard conditions. Note that storage reduces a seed's viability. Winter overseeding is planting one grass in another established grass without destroying the established grass to provide green color in the winter while the permanent lawn is dormant.

Seed Selection and Source of Seeds

Seed selection as discussed previously depends on climatic requirements, life cycle, usage, maintenance needs, visual and functional quality required, and resistance to diseases and insects. Seed should always be purchased from a reputable source; although these seeds may cost more, they are worth the price. The seed industry must provide the following information about the seed on the package: company name and address, cultivar name, percentage germination and date of testing, purity or usable grass seed, percentage nongrass seed, inert material, and seed lot number (although the information may vary). The seed should be free of all weeds, especially noxious ones, and the germination should be above 80 percent. The percentage seed germination should be verified prior to planting. Placing a known number of seeds on moistened paper and counting the number of seeds that germinate can easily determine the percentage seed germination. For example, if you have 10 seeds and 9 germinate, there is a 90 percent germination rate. If possible, freshly harvested seeds are better than seeds that have been stored because the germination rate decreases with time in storage.

Seedbed Preparation

Good seedbed preparation is essential for high-quality turf. Factors to be considered when preparing a seedbed are as follows:

- **Soil testing.** A representative soil sample should be taken and sent away for analysis. Fertilizer application and pH adjustment should be made based on the results of the soil analysis.
- **Weed control.** Prior to planting seed, all weeds must be controlled because they are more difficult to control after grass seeds are planted.
- **Soil texture.** It may be necessary to add organic matter and/or sand to improve the aeration, drainage, and/or nutrient- and water-holding capacity of the soil.
- **Removal of large objects.** The soil must be tilled and large objects such as stones should be removed.
- **Drainage and irrigation systems.** In some cases, such as golf courses, the installation of drainage and irrigation systems is necessary.
- **Soil leveling.** The last step prior to planting seed is to make sure the soil is level.

Time to Sow and Seeding Rate

The best time to plant seed is during the early spring, late summer, or early fall because these times have weather conditions that are ideal for optimal grass seed germination. Sowing seeds at the proper time allows grass seed to become established before weeds have a chance to become established. The seeding rate depends on the type of turfgrass. The proper seeding rate is important for quickly establishing a lush, dense turf. Overseeding leads to crowding, which causes stress and poor growth, whereas underseeding results in low plant populations, which delay the covering of the ground and allow weeds to take over. Reputable seed companies will include recommended seeding rates on the seed package.

Planting Seed

After the seedbed has been prepared properly and all planting requirements have been taken care of, it is time to plant the seed. For small plots, seeding can be accomplished easily by hand. This method is easy but typically results in uneven spreading. For larger areas, mechanical seeders should be used. Gravity feed and broadcast mechanical seeders are the most commonly used type of spreaders for seed dispersal (Figure 23-8). Both gravity and broadcast spreaders can be manually pushed or pulled by a tractor. Hydroseeding or hydromulching (Figure 23-9) is a method that mixes seeds in a water slurry with wood cellulose or peat moss and pumps this onto the soil surface under high pressure.

Figure 23–8 Broadcast spreader

Figure 23–9 Establishing turf using hydromulching

This method is typically used on steep slopes. To obtain uniform coverage of seeds, the recommended seeding rate should be divided by two so that one-half of the seed is distributed in one direction on the first pass and the other spread crosswise on the next pass. After the seed is spread, mulch should be applied to protect the seed from birds, to prevent the seed from washing away during watering, to conserve moisture during the germination process, and to control soil erosion. The mulch must be spread thinly to prevent grass seedlings from being impeded. Straw is commonly used as mulch (Figure 23-10); however, other materials can be used. After spreading the mulch, the area should be firmed with a lightweight roller; rolling establishes good soil/seed contact, which enables seeds to imbibe water efficiently and anchor to the soil. To complete the process, the area should be watered and continue to be watered based on demand; watering should be done less frequently but more thoroughly and deeply as the turfgrass plants mature.

First Mowing

The new lawn will be ready to mow when the turfgrass plants are higher than the height at which they will normally be maintained. For example, if the lawn is regularly cut at 2½ inches, then the grass is mowed for the first time when it is 3 inches tall. The general rule is never remove more than one-third of the turfgrass leaf blade at any one time. A sharp lawnmower blade is used to minimize damage to the turfgrass, which in turn reduces the grass's susceptibility to diseases and other problems.

Figure 23–10 Straw used as a mulch to protect newly seeded lawns

Planting Turfgrass Using Asexual Propagation

Vegetative (or asexual) propagation involves using parts of a growing plant (sod, plugs, or sprigs) to establish new turf.

Sod

Sod is the surface layer of the turf, including plants and a thin layer of soil used in propagating new turf. Sod roots in one to two weeks. Sod is grass that has been specially cultivated, mowed, and cut like pieces of carpet and is distributed in rolls, squares, or strips. After the site has been prepared, sod is laid out over the seedbed like rolling out a carpet, placing squares or strips together like a puzzle. Pieces of sod are pushed close together to make sure no space remains between the different pieces of sod. After putting the sod in place, it is rolled to place the root system in close contact with the soil and then watered frequently during the first several weeks to ensure proper establishment.

Plug

Plugs are small blocks or squares of turf. Plugs are placed in small holes equally distributed throughout the seedbed with space between each; in time, the space between plugs fills in to make a dense turf. After plugs are in place, the plot is rolled to provide good contact between the roots and the soil. The plot should then be kept well watered to ensure the plugs become well established and spread to fill in spaces between them. Establishing turf using plugs takes longer than sod but is quicker than sprigs or seed.

Sprig

A sprig is part of the grass plant without soil, such as rhizomes or stolons. Sprigs are placed in small holes equally distributed throughout the seedbed with space between each. The sprigs are covered leaving one-fourth of the material aboveground. The sprigs must be kept well watered to allow them to become established. Sprigs need about two months prior to the first frost to become established. With time, the space between sprigs fills in to make a dense turf. Establishment of turf using sprigs takes longer than by sod and plugs but is quicker than seed.

Turfgrass selection, seedbed preparation, and first mowing are essential for good-quality turf, as mentioned previously. Factors to be considered when selecting turfgrass, preparing the seedbed, and mowing are the same as described for sexual propagation.

MAINTAINING TURFGRASS

Turfgrass maintenance involves keeping a stand of turfgrass in an attractive and healthy condition (Figure 23-11) through proper management practices.

Proper Management of Nutrients

Soil testing should be used to determine the needed nutrients and the pH of the soil, and then fertilizer and lime should be added as required. For warm season turfgrasses, fertilizer and lime should be applied to established turf in the early spring and fall to encourage growth. Additional applications may be needed throughout the growing season and should be applied as needed; golf courses may require multiple treatments to make the turf more heat-resistant. For cool season turfgrasses, fertilizer applications should be done four times per year. The first application of fertilizer should be in early spring (March/April) to help break dormancy, heal injuries, and prevent the invasion of summer and annual weeds. The second application should be in late spring (May) to help the turf survive heat encountered later in the season. The third application should be in

Figure 23–11 Freshly mowed turf

late summer (August/September) to stimulate new tillers and rhizomes for the following year and to help maintain a vigorous and green turf until the fall. The fourth and final application should be in late fall (September/October) to promote root growth until the ground freezes and aid in early spring greening. Turfgrass maintenance fertilizers should be high in nitrogen. To avoid overstimulation of growth right after fertilization, a slow-release fertilizer should be used. The numbers on the fertilizer bag for N-P-K or nitrogen, phosphorous, and potassium designate the amount of each of these nutrients; for example, 16-4-8 is 16 percent nitrogen, 4 percent phosphoric acid, and 8 percent potassium. The fertilizer rate is the amount of fertilizer to be applied; this should be determined based on a soil test, the length of the growing season, the type of grass, and the amount of traffic. After the amount of fertilizer to be applied is determined, the fertilizer is spread uniformly over the turfgrass surface because these nutrients do not move laterally in the soil so missed spots will be apparent. To ensure that the fertilizer being applied is available to the plant, problems with soil acidity must be corrected. Acidic soils have a pH below 7.0; these soils should have their pH corrected by adding the proper amount of lime. However, if the soil is alkaline, it should be treated with sulfur-containing compounds.

Watering

After 7 to 10 days without rain, grasses will wilt and should be watered to a 4- to 6-inch depth to prevent permanent damage to the lawn. Watering should be done in the early morning and late afternoon if possible to minimize evaporation. If watering is done in the late afternoon, it should be done early enough to enable the grass leaves to dry by nightfall. The best time to water is when the grass needs it; this promotes deeper rooting and better drought tolerance.

Mowing

Most homeowners typically do not want to spend a lot of time taking care of their lawns. However, to have a nice-looking lawn, mowing with the proper equipment, at the proper frequency, and to the proper height is crucial.

Types of Mowing Equipment

For the homeowner, the two types of mowers are the reel and rotary; both come in a range of different sizes, power levels, and styles. Some mowers require pushing, whereas others are motorized and require only steering to operate. The rotary mower (one of the most common) has a horizontal blade fixed on a vertical axis that spins at high speeds (Figure 23-12). The reel mower has one stationary blade plus a set of blades arranged in a helical manner that gather up the grass and cut it off against the stationary blade. The reel-type mower produces a better-looking cut than the rotary type. The mulching mower is another common type of mower that cuts grass and reduces it to mulch. There are also dual-type mowers that can be switched from mulching to regular cutting. All mowers must be maintained properly; it is especially important to have sharp blades to avoid a rough cut that can lead to diseases.

Figure 23–12 Underside of a rotary mower

Two types of commercial mowers are the flail mower and the sickle bar mower. These are typically used to mow large areas when a fine finish is not required, such as mowing along roadsides.

Mowing Frequency and Height

Each grass species or cultivar has an optimal recommended height for cutting. The range of heights used to grow turfgrass varies from 1/2 inch to 4 inches. Based on the species or cultivar and season, it is very important to determine what height is necessary and develop a schedule to maintain it. Mowing should be done at regular intervals and as often as needed, depending on the rate of growth. The 1/3:2/3 rule should always be used; this rule is to cut only the top one-third and leave the remaining two-thirds intact after mowing. For example, the grass should be mowed at a height of 3 inches if a 2-inch height of grass is desired. When plants are cut too short, they do not have enough leaf area to photosynthesize efficiently, thereby turning pale green or yellow until new growth occurs. The mowing height depends on the type of turfgrass and the cultural conditions used to grow it.

The keys to successful mowing include the following:

- **Mowing pattern.** Mowing should not be done in the same direction every time; rather mow in one direction one time and at right angles to that direction the next time to reduce compaction.
- **Do not mow when the grass is wet.** For the mower to work at peak efficiency, the grass blades should not be wet. In addition, wet grass does not fall to the soil surface but instead forms clumps, which are detrimental to the turfgrass.
- **Always have sharp blades.** This prevents tearing of the turfgrass leaves, which gives a better appearance and prevents disease problems. Make sure that rocks and debris are removed from the lawn prior to mowing to keep the blade sharp for a longer period of time.
- **Overgrown lawn.** If too much time has gone between cuttings, mow the turfgrass in two directions, which is double the work but necessary to ensure an evenly cut lawn.
- **Overlap.** With each pass of the lawnmower, overlap to make sure that the entire area has been covered.

Removal of Clippings and Thatch

When the amount of clippings produced by mowing is not readily sifted down into the turf, it is best to remove clippings (Figure 23-13). Mulching mowers cut the grass into small pieces that do not lay on the turf but fall to the ground; these clippings eventually decompose and improve soil fertility. The plant organic matter that is not decomposed between the turfgrass and the soil surface is called thatch. Thatch is caused by an accumulation of excess grass stems, stolons, rhizomes, and roots in the turf. Grass clippings do not contribute to thatch; rather thatch is caused by overfertilization. Maintaining a moderate fertilization program will prevent thatch. Controlling thatch requires the physical removal (aeration, punching holes into the soil, or raking) of excess plant growth to permit air and water to penetrate into the thatch layer, which allows microbes in the soil to do their job.

Aeration

Soils should be aerified to maintain good levels of oxygen in the soil. This is accomplished by using an aerifying machine, which removes soil cores to depths of 2 to 3 inches.

Figure 23–13 Turf with clumps left after mowing

Pest Control

Good IPM practices should be used to control weeds, diseases, and insects in turfgrasses.

Weed Control

One of the keys to minimizing weeds is to care for the turf properly so that it becomes thick and able to compete with weeds. For example, if turfgrass is not mowed to the proper height and if there is poor nutrition and watering, this will weaken the turfgrass and favor weed growth. A well-planned IPM strategy must be used to reduce weeds. For successful control, weeds should be removed prior to setting seed. A popular chemical used in turf today is 2,4-D because it is selective for the control of broadleaf weeds such as dandelion, plantain, chickweed, clover, and others. Other examples of common weeds in turf are crabgrass, foxtail, annual bluegrass, barnyard grass, nutsedge, and quackgrass. The best way to choose the proper herbicide is to check with local authorities for recommendations at your particular location.

Disease and Insect Control

Diseases are typically fungal and are readily spread by wind, people, and pests that come in contact with the fungal spores in one location and transport to another location where the spores cause a problem. Both overfertilization and overwatering lead to problems with turfgrass diseases. Some common disease problems affecting turfgrasses are *Fusarium* blight, *Rhizoctonia* (brown patch), *Pythium* blight, *Sclerotinia* (dollar spot), *Helminthosporium* (crown rot), slime molds, and algae. The most devastating insect pest in turfgrass is the Japanese beetle during the larva stage, where it is also called a grub. The best way to choose the proper control measure for both diseases and insects is to check with local authorities for recommendations at your particular location.

SCIENTIFIC CLASSIFICATION (NOMENCLATURE)

The classification of plants was covered earlier in this text, but in this section a turfgrass example is presented to refresh your memory using *Cynodon dactylon* L. 'Cheyenne' as an example of a grass species. Knowing the scientific classification of all plants is important for the reasons presented earlier. The following is the proper classification of *Cynodon dactylon* L. 'Cheyenne':

Kingdom	Plantae
Division or phylum	Spermatophyta
Class	Angiospermae
Subclass	Monocotyledonae
Order	Poales
Family	Poaceae
Genus	*Cynodon*
Species	*dactylon*
Cultivar	Cheyenne

WARM SEASON TURFGRASSES

Sod, plugs, or sprigs are commonly used methods to propagate vegetatively most warm season turfgrasses. There are 600 genera and 7,500 species of grasses; of these, fewer than 50 are used as turfgrasses. Examples of commonly used warm season turfgrasses are as follows:

- **Bahiagrass.** Seed propagated, native to tropical Americas, tough, coarse texture, used more for hay than turf, and minimal irrigation required. Popular cultivars include Argentine, Pensacola.
- **Bermudagrass (*Cynodon* spp.).** Seed propagated, very popular turfgrass, wide range of cultivars, quick to establish itself, grows under a wide range of soil conditions, and not shade tolerant.
- **Common bermudagrass.** Medium-to-coarse texture. Popular cultivars are Arizona and Cheyenne.
- **African bermudagrass.** Used for many hybrid crosses, fine texture, and bright green.
- **Buffalograss (*Buchloe dactyloides*).** Fine texture, gray-green, and drought and temperature hardy.
- **Centipedegrass.** Native to China; utility turf and lawns, and sends out stolons.
- **Carpetgrass.** Native to Central America and West Indies; lawns used in utility spaces, and provides erosion control.

- **Saint Augustinegrass.** A coarse-textured aggressive turfgrass, adaptable to a wide range of soil conditions, not very cold tolerant, and grows well in shaded areas.
- **Zoysias.** Native to the Orient and the Phillipines; slow grower, makes beautiful turf, dense sod, is easily controlled, and has excellent cold tolerance.

COOL SEASON TURFGRASSES

Although there are hundreds of different species of grasses, only a limited number will produce high-quality turf in the cooler portions of the United States. Examples of commonly used cool season turfgrasses are as follows:

- **Bluegrasses (*Poa* spp.).** Drought and shade sensitive; common species are Kentucky bluegrass, which was introduced from Europe and Asia and is the most widely used turfgrass in the Northern regions. Kentucky bluegrasses are susceptible to damage with close mowing and are dormant during the summer months. Commercial varieties are Adelphi and Baron.
- **Bentgrass.** Creeping bentgrass (*Agrostic palustris*) was first introduced from Europe; it is commonly used on putting greens and other close-cut areas, but it is not used for lawns. Commercial varieties are Penncross and Penneagle.
- **Fescues (*Festuca* spp.).** Fescues were first introduced from Europe; they are drought and shade tolerant. Red fescue is not used for sod or sport turf. Besides Kentucky bluegrass, the fescues are the most common lawn turfgrass. Tall fescues are generally found in poorly maintained turf areas; repeated close cutting will kill tall fescue. Commercial varieties are Jamestown for red fescue and Jaguar for tall fescue.
- **Ryegrasses (*Lolium* spp.).** First introduced from Europe. Annual ryegrass grows very rapidly and is typically used as a temporary ground cover. Perennial ryegrass is a short-lived perennial generally used as a companion in mixtures with other more slowly growing species. Commercial varieties for perennial ryegrass are Blazer and Regal.

Most seed mixtures for cooler regions will contain a blend of two or three Kentucky bluegrass cultivars and a blend of two or three improved perennial ryegrass cultivars. For turf that is partially shaded, 25 percent by weight of an improved cultivar of red fescue should be added.

SUMMARY

You now have a basic background in warm- and cool season turfgrass selection, establishment, care, and maintenance. The three major functions of turf include utility, ornamentation, and sports. The six factors commonly used to assess visual turf quality include density, texture, uniformity, color, growth habit, and smoothness. Turf quality is judged per its rigidity, elasticity, resiliency, and recuperative potential. Selecting the proper turfgrass for your location requires understanding the climatic requirements, life cycle, usage, maintenance needs, visual and functional quality, and disease and insect resistance of the turfgrass. Growers must be aware of the factors to consider when establishing turfgrass from seed and how to determine the proper method used to establish turfgrass. Establishing turfgrass by seed involves considering seed selection, source of seeds, seedbed preparation, planting, care, and first mowing. Turfgrass can also be established by vegetative propagation. Turfgrass maintenance is key to keeping a stand of turfgrass in an attractive and healthy condition and is accomplished by properly managing nutrients, watering, mowing, aeration, and pest control.

Review Questions for Chapter 23

Short Answer

1. Provide three major functions of turf and give an example of each.
2. Provide six factors used to assess visual turf quality and briefly explain each.
3. Provide four factors for judging functional quality of turf and briefly explain each.
4. What are six factors that should be considered when making a turfgrass selection?
5. Classify *Cynodon dactylon* L. 'Cheyenne' using scientific classification; be sure to provide everything from Kingdom through Cultivar.

   Plantae ____________________
   Spermatophyta ____________________
   Angiospermae ____________________
   Monocotyledonae ____________________
   Poales ____________________
   Poaceae ____________________
   Cynodon ____________________
   Dactylon ____________________
   Cheyenne ____________________

6. What are three factors that should be considered when deciding which method of propagation will be used in establishing turf?
7. What are two major forms of propagation used in establishing turfgrasses and the advantages and disadvantages of each?
8. What are six factors that should be considered when preparing a seedbed for turfgrasses?
9. What are three times of the year that are ideal for optimal grass seed germination?
10. When should a new lawn be mowed for the first time? After the first mowing, what is the general rule of thumb that should be used when mowing turfgrasses?
11. What are three ways to establish new turf vegetatively?
12. How many fertilizer applications should be made each year for cool season turfgrasses and at what time during the growing season? Briefly explain the benefits associated with each fertilizer application.
13. When should grass plants be watered?
14. When should grass clippings be removed from the turf?
15. What is the cause of thatch? How can thatch buildup be prevented and controlled?
16. How are pests controlled in turf?

Define

Define the following terms:

turf
turfgrass
transition zone
turfgrass blend
turfgrass mixtures
seed purity
seed viability
winter overseeding
sod
plug sprig

True or False

1. The three purposes and functions of turf are utility, ornamentation, and sports.
2. Turf is a collection of grass plants that forms a ground cover and must be regularly maintained.
3. Turfgrass is a collection of plants in a ground cover and the soil in which roots grow.
4. Sod is a small block or square of turf.
5. A sprig is part of the grass plant without soil.
6. Sod is the surface layer of turf, including plants and a thin layer of soil used in propagating new turf.
7. Warm season turfgrasses are limited by cold weather, which imposes winter dormancy.
8. Cool season turfgrasses are limited by heat, which imposes summer dormancy.
9. Most warm season turfgrasses are propagated vegetatively by sod, plugs, or sprigs.
10. An example of a warm season turfgrass is ryegrass.
11. An example of a cool season turfgrass is zoysia grass.
12. Sexual propagation is the most common method used with cool season grasses.
13. Winter overseeding is planting one grass in another established grass without destroying the established grass.
14. The width of the grass blade determines the recuperative potential of turfgrass.
15. For mowing lawns, the best amount to cut off is two-thirds of the amount of grass growing.
16. Turfgrass blends are combinations of two or more different turfgrass species.
17. Turfgrass mixtures are a combination of different cultivars from the same species.
18. Grass clippings are the major contributor to thatch buildup.
19. Thatch is caused by an accumulation of excess grass stems, stolons, rhizomes, and roots in the turf.
20. Fertilization of cool season turfgrasses in the late spring will help them to survive heat encountered later in the season.
21. Fertilization of cool season turfgrasses should never be done in the late fall because it will damage plants when the first frost arrives.
22. After 7 to 10 days without rain, grasses will wilt and should be watered to prevent permanent damage.
23. The best time to water is when the grass needs it; this will promote deeper root growth and better drought tolerance.
24. It is always important to rake up grass clippings after mowing.

Multiple Choice

1. Which of the following is the definition of *turf*?
   A. A collection of grass plants that forms a ground cover and must be regularly maintained
   B. A collection of plants in a ground cover and soil in which roots grow

C. A collection of monocots and dicots, which form a ground cover

D. None of the above

2. Mowing is important because it keeps the grass in which of the following?

A. Senescent stage

B. Reproductive stage

C. Vegetative stage

D. Dormant stage

3. The best time for good grass seed germination is during the

A. early spring.

B. late summer.

C. early fall.

D. All of the above

4. A turfgrass blend is

A. a combination of different cultivars of the same species.

B. a combination of two or more different turfgrass species.

C. a mixture of four or more different cultivars of different species.

D. None of the above

5. Thatch buildup is caused by

A. overwatering.

B. overfertilizing.

C. aerification.

D. winter overseeding.

6. For cool season turfgrass, the recommended number of times to fertilize each year is

A. once.

B. twice.

C. three times.

D. four times.

7. For warm season turfgrasses, fertilizer and lime should be applied to established turf in the

A. early spring or fall.

B. spring and summer.

C. late spring and early fall.

D. None of the above

Fill in the Blanks

1. Mowing keeps grass in the ____________________ stage of the life cycle.
2. Seeding is the most common method used with ____________________ season grasses. The major advantage of seeding is that it is ____________________; however, the main disadvantage is that ____________________ time is required to establish a high-quality turf.
3. Vegetative propagation using ____________________, ____________________, and ____________________ is commonly used with ____________________ season turfgrasses.

Activities

Now that we have completed our discussion on warm- and cool season turfgrasses, you will have the opportunity to explore this area in more detail. In this activity, you will visit a local golf course and gather the following information:

- type of grass used.
- how the course is maintained, including watering, fertilizing, mowing, and controlling disease and pests.

In addition to the information requested, include your thoughts on how the golf course was set up and maintained. If you do not have access to a golf course, tour some local residential and commercial areas and look for differences in mowing, watering, quality of maintenance, and any other characteristics you feel are important.

CHAPTER

24

Olericulture

ABSTRACT

Olericulture is important to the homeowner and the commercial producer. Edible plant parts are used to classify vegetable crops by breaking them down into nine major categories. Vegetable crops are established starting with site selection and necessary facilities followed by cultivar selection based on climatic requirements (temperature, water, and light), life cycle, cultural requirements, usage, and quality. Many important factors must be considered when preparing the seedbed, establishing vegetable crops by seed and transplants, and maintaining vegetable crops. Commercial or traditional vegetable crop production is accompanied by nontraditional forms such as organic, sustainable horticulture, hydroponics, and biotechnology. Home gardens are also a part of the olericulture field.

Objectives

After reading this chapter, you should be able to

- describe the importance of olericulture to the homeowner and commercial producer.
- discuss how edible parts are used to classify vegetable crops by breaking them down into nine major categories.
- discuss how to select a site and necessary facilities and establish a vegetable crop.
- explain how to care for and maintain vegetable crops.
- briefly discuss commercial (traditional) and nontraditional vegetable crop production, including organic, sustainable horticulture, hydroponics, and biotechnology.
- briefly provide some important factors to consider when establishing a home vegetable garden.

Key Terms

base temperature
biotechnology
conservation tillage
degree-days
direct seeding
fluid drilling
fresh market vegetables
frost-free period
hybrid
hydroponics
indirect seeding (transplanting)
olericulture
organic vegetable production
processed vegetables
strip tillage
sustainable horticulture
transplanting shock

INTRODUCTION

The USDA food guide pyramid suggests that individuals consume three to five servings of vegetables per day. The American Cancer Society and the American Heart Association both endorse the use of vegetables in the human diet for a variety of reasons, mainly to reduce the incidence of cancer and heart disease. The branch of horticulture dealing with the production, storage, processing, and marketing of vegetables is called **olericulture.** Vegetables can be produced commercially or in the home garden. Commercial vegetable production is divided into two distinct

Figure 24–1 Vegetables being sold at a farmer's market

areas: fresh market (Figure 24-1) and processing. The total value of vegetable crops in the United States is estimated to be $14.3 billion per year and is on the rise. The top two vegetable-producing states are California followed by Florida. Although most of the vegetables are produced by traditional methods, a growing number of people prefer vegetables grown organically, which means without chemicals or inorganic fertilizers.

Vegetable crops can be classified in a variety of ways; the two most commonly used methods are edible parts and scientific classification. To produce a high-quality vegetable crop with high yields, the grower must select the best site, facilities, and the proper cultivar. After these selections are made, the seedbed must be prepared. The grower then establishes vegetable crops from seed or transplants, followed by proper care and maintainence. To select the proper site, the following environmental factors should be considered: temperature, light, rainfall, wind, soil, topography, water, air pollution, and plant pests. Economic factors to consider include land cost and availability, labor, transportation, utilities and services, and competition. Storage and other facilities needed include a cold storage facility, shipping areas, offices, pesticide storage and mixing area, and storage buildings.

Many **hybrid** cultivars are superior to standard cultivars. A hybrid is an improved plant developed by crossing parents of different genotypes for a beneficial trait. Cultivars are selected based on climatic requirements (temperature, water, and light), life cycle, cultural requirements, usage, and quality. The **frost-free period** is the average number of days from the last spring frost to the first frost in the fall. This is the main selection criterion when selecting a cultivar because without the proper temperature, there is no growth and the grower has limited means of affecting temperature under field conditions. Different cultivars also have different requirements for water, light, and sensitivity to air movement, so the grower must know the requirements by the specific cultivar chosen and the environmental conditions at the particular location. The life cycle of the cultivar selected may require one, two, or more years before crop harvest. Cultivar selection should also be based on cultural requirements of the specific crop, the usage of the crop, and quality desired to maximize performance.

Next, seedbed preparation is essential for the production of high-quality vegetables. A soil test should reveal pH and nutrient requirements. The grower must also control weeds and other pests at the planting site prior to planting, add organic matter and/or sand as necessary to improve soil structure, till the soil, and install irrigation systems as needed. After the seedbed is prepared properly, the final step is crop establishment, which can be done by **direct** or **indirect seeding.** The particular spacing depends on the crop and its intended use. When the vegetable crop has been established properly in the field, it must be maintained and cared for properly by managing pH levels and nutrients, watering, controlling pests, staking, and training as necessary. The grower needs to know how air movement affects the crop and when the crop is mature and ready to harvest. After the crop is harvested, the first step is to remove field heat rapidly and store at the proper temperature. A marketing plan will help ensure that the high-quality crop produced is sold to maximize profit.

In recent years, the complex commercial vegetable production industry has seen the number of commercial growers decline while the size of operations has increased dramatically.

In addition to traditional vegetable production, the nontraditional types of vegetable production include **organic production, sustainable horticulture, hydroponics,** and **biotechnology.** Organic production methods do not use or only use reduced amounts of chemical pesticides or inorganic fertilizers. Sustainable horticulture uses Integrated Pest Management (IPM) together with sound management practices to reduce inputs such as tillage, chemicals, inorganic fertilizer, and water. Hydroponic systems are used to grow plants in the absence of soil. This method is only used for high-dollar crops such as tomato and lettuce. Biotechnology is the manipulation of living organisms or substances obtained from living organisms for the benefit of humanity. The production of genetically modified organisms (GMOs) has a wealth of potential for vegetable crop improvement; for example, plants can be genetically modified to make them resistant to various pests and thereby reduce the need for pesticides.

Home vegetable gardens are very popular because large quantities of vegetables can be produced from a very small portion of land if the garden is planned properly and the grower is knowledgeable. Proper planning includes selecting the proper location, choosing what to plant, and deciding how the plants will be arranged. After the garden is planned, the soil must be prepared for high-quality/viability seed or transplants to be planted at the proper depth and distance apart. The next steps are controlling pests, watering, staking, training, and putting in windbreaks where necessary. To harvest at the proper time, the grower must know when the vegetables are mature and how to store or process them to maximize his or her time investment. At the end of the season, the area should be cleaned and all plant material chopped and then plowed into the soil or used for compost.

BACKGROUND INFORMATION FOR OLERICULTURE

Vegetables are very important to the human diet; in fact, nutrition experts recommend three to five servings per day for a healthy diet. Olericulture is the branch of horticulture dealing with producing, storing, processing, and marketing vegetables. Vegetables can be produced commercially or in home gardens. In commercial vegetable production, **fresh market vegetables** are packaged and sold as soon as possible after harvest (Figure 24-2); whereas **processed vegetables** are canned, frozen (Figure 24-3), or dried. The total value of vegetable crops in the United States is estimated to be $14.3 billion annually and is rising steadily. The leading vegetable production states are California, Florida,

Figure 24–2 Vegetables being sold fresh

Figure 24–3 String beans are available frozen, canned, and fresh.

Figure 24–4 USDA Recommended Food Guide

Arizona, Texas, and Georgia. California alone produces almost half of the vegetables grown in the United States. Although most are grown via traditional methods, many people prefer vegetables that are grown by organic vegetable production methods without chemicals or inorganic fertilizers.

People are becoming more interested in their health and increasing their quality of life, which has led to a rise in vegetable consumption in recent years. Vegetables are important sources of vitamins, minerals, and fiber and are low in fat and sodium. The American Cancer Society states that eating vegetables reduces cancer rates, and the American Heart Association states that eating vegetables reduces the incidence of heart disease. In addition, research has shown that eating vegetables reduces stress, insomnia, and problems associated with aging. The USDA includes vegetables in the food guide pyramid, which recommends between three and five servings of vegetables per day (Figure 24-4).

CLASSIFICATION OF VEGETABLE CROPS

Vegetable crops are classified either by edible parts or scientific classification. For information on scientific classification and other means of classifying plants, refer to Chapter 6. Classification of vegetable crops by edible parts can be broken down into nine major categories:

- potherbs and greens—spinach, collards.
- salad crops—celery.
- cole crops—cabbage, cauliflower.
- root/tuber crops—potatoes, beets, carrots, radishes.
- bulb crops—onions, leeks.
- legumes—beans, peas.
- cucurbits—melons, squash, cucumber.
- solanaceous—tomatoes, peppers.
- other—sweet corn.

ESTABLISHING A VEGETABLE CROP

To produce high-quality vegetable crops with high yields, it is important to select the proper location with the necessary facilities. After the location has been selected, the proper cultivar for this location should be chosen and then the seedbed should be prepared properly prior to planting the vegetable crop from either seed or transplants.

Selecting the Site and the Necessary Facilities

To select the proper site, the following environmental factors must be considered: temperature, light, rainfall, wind, soil, topography, water, air pollution, and plant pests. After the environmental factors have been considered carefully, the economic factors to evaluate when selecting a suitable site include land cost and availability, labor, transportation, utilities and services, and competition. Storage and other facilities also important to a successful vegetable-producing location include a cold storage facility, shipping areas, offices, pesticide storage and mixing areas, and storage buildings.

Selecting a Cultivar

The proper vegetable cultivar should be selected based on the proper climatic requirements (temperature, water, and light), life cycle, usage, cultural requirements, yield, quality, horticultural characteristics (color, size, flavor, shape, and appearance), and resistance to pests. Hybrid cultivars of most vegetable crops have now been developed and generally have superior yield, quality, and resistance to pests than standard cultivars. A hybrid is an improved plant developed by crossing parents of different genotypes for a beneficial trait. Cultivars should be selected based on climatic requirements, life cycle, cultural requirements, and usage and quality.

Climatic Requirements

Every living thing must have a suitable environment in which to sustain life. The climatic factors that influence vegetable growth and production are temperature, water, and light.

Temperature. All plants have an optimum temperature range for peak efficiency, which generally differentiates between warm- and cool season crops. Cool season crops are asparagus, broccoli, cabbage, celery, garlic, onion, pea, spinach, carrot, lettuce, and Irish potato; warm season crops are cucumber, eggplant, lima bean, okra, melons, squash, peppers, sweet corn, sweet potato, and tomato.

Vegetable crops differ in their susceptibility to freeze/frost injury; in warm season crops, damage can occur at 50°F or lower, whereas in cool season crops, injury occurs at around 45°F and lower. Freeze/frost injury causes a reduction in growth and production and, in extreme cases, death. The single most important factor in determining a crop's suitability for a particular climate is the length of the growing season required to maturity prior to frost damage. This period of time is referred to as the frost-free period, which is the average number of days from the last spring frost to the first frost in the fall.

Water. Vegetable crops demand a consistent supply of water as even minor water stress will lead to reductions in crop yields. Vegetable crops can receive water via precipitation or irrigation. Water is important to maintain the plant's turgidity, cooling effect via transpiration, absorption of nutrients, photosynthesis, and a variety of other factors required for survival. Excessive water can promote disease and drown the plant by preventing oxygen uptake by the root system, so it is important to select a cultivar that is adaptable to the location.

Light. Vegetable crops must have the proper light intensity, quality, and duration to maximize growth and development. The major role of light is to provide energy for photosynthesis, so the proper amount of light produces high-yielding vegetable crops with excellent quality. When selecting a cultivar, the grower should know its light requirements and whether they fit with the location.

Life Cycle

Most vegetable crops are grown as annuals, although many can survive more than one season. Examples of annuals, biennials, and perennials are corn, cauliflower, and asparagus, respectively. Understanding how long it will take for a specific cultivar to produce a vegetable crop that is salable helps the grower prepare to market the crop.

Cultural Requirements

Prior to selecting a vegetable crop, the grower should know the cultural requirements to maximize performance.

Usage and Quality

Knowing what the vegetable crop is to be used for (processing or fresh market) helps the grower select a cultivar for maximal performance. In addition, the yield; disease resistance; and horticultural characteristics, such as color, size, flavor, shape, and appearance, are all very important factors to consider when selecting vegetable cultivars.

Preparing the Seedbed

Good seedbed preparation is essential for high-quality vegetables. Important factors to consider when preparing a seedbed are as follows:

- Test the soil to determine the proper amounts of nutrients to add and whether any pH adjustments are necessary.
- Control weeds and other pests at the planting site prior to planting.
- Add organic matter and/or sand as necessary to improve the soil structure and maximize aeration and drainage as well as nutrient- and water-holding capacity.
- Till the soil. Tillage is generally one of the earliest operations performed in crop establishment (Figure 24-5). Tillage improves the soil environment by

Figure 24–5 Tractor tilling the field prior to crop establishment

increasing aeration, increasing moisture infiltration, controlling weeds, and incorporating crop residue. The initial tillage buries crop residue and loosens the soil to enhance root growth, which is typically followed by disking to smooth the seedbed and incorporating fertilizers and weed control chemicals. **Conservation tillage,** also called minimal tillage, involves tillage practices that leave 30 percent or more crop residues on top of the soil to prevent soil erosion. The most common form of conservation tillage used in vegetable production is **strip tillage,** which is the process of leaving strips of vegetation between tilled areas to prevent soil erosion. Tillage continues after a crop has been planted but it is then called cultivation. Cultivation is primarily used to control weeds, to incorporate fertilizer, to break soil crusts, and to improve the infiltration of water.

Figure 24–6 Drip irrigation, a form of surface irrigation

- Install irrigation systems. Adding and conserving soil moisture is very important for the production of high-quality vegetables. Several methods of irrigation are used to supply moisture to the soil:
  - subsurface irrigation, commonly used in association with a shallow water table.
  - surface irrigation, such as furrow or drip irrigation (Figure 24-6).
  - overhead irrigation, such as sprinklers.

Planting Vegetable Crops

After seedbed preparation is complete, the final step is planting the seed or transplanting. The particular spacing depends on the crop and its use. The ideal planting density should be low enough to prevent competition for resources between crop plants but high enough to produce a canopy that will cover the ground.

Direct seeding is the placement of seed directly into the field. The main objective of direct seeding is to achieve the best stand or optimal population of plants, which requires proper seedbed preparation, attention to environmental factors, high-purity seed, and high seed germination rates. Seeds can be planted manually by hand and or by using precision seeders. Manual seeding is done on small-scale plantings because it is quick, easy, and inexpensive; however, it is not cost effective to use this method on a large scale. Precision seeders are commonly used for large-scale commercial plantings. Vegetable seeds come in a variety of shapes and sizes, which makes them difficult to use with precision seeders. To overcome this problem, vegetable seeds are often coated to make them a more uniform size, thereby facilitating the use of precision seeders for a wide range of vegetable crops. **Fluid drilling** is a method that uses seeds that have been treated, pregerminated, suspended in a gel, and placed in precision seeders that deposit a designated number of seeds per hole in the soil by an automated process. Fluid drilling is usually used in conjunction with plastic culture (Figure 24-7).

Indirect seeding is the transfer of seedlings (transplants) from a seedbed or container (Figure 24-8) to the field. The main advantages are to assure a good stand of crop plants and extend the length of the growing season, especially in cooler climates. The main disadvantages are that transplants cost more and are subject to **transplanting shock,** which is a one- to two-week period following planting in which no growth occurs due to shock.

Figure 24–7 Different-color plastics used in the field

Figure 24–8 Seedlings ready for transplanting

CARE AND MAINTENANCE OF VEGETABLE CROPS

After a vegetable crop has been established properly in the field, the vegetable crop must be cared for and maintained to promote high quality and yield.

Proper Management of Nutrients

Vegetable crops require additional fertilization as they develop throughout the growing season. Overfertilization can be equally damaging as underfertilizing.

Foliar analysis of nutrients should be done throughout the growing season to determine the needed nutrients, and then fertilizer should be applied as needed. Fertilizer applied by side dressing should be a couple of inches from the plant, and dry fertilizer should never be applied to the foliage of plants because it will cause burning.

Watering

Watering should be based on demand by the crop because more water is typically required than is available from the environment. Water can be added in a variety of ways, including by drip irrigation, by sprinklers, or by soaking the ground between the rows. Irrigation should be done with three-quarters of an inch of water two times a week; this way is better than daily applications, which do not promote a deep root system. Watering in the early afternoon cools the plant during the hotter part of the day, prevents interference with pollination, and enables the foliage to dry prior to nightfall to prevent the spread of diseases.

Pest Control

Common pests in vegetable crops are insects, weeds, diseases, and animals. As stressed throughout this text, the grower should first try to prevent the pest from attacking the crop; if this is not possible, the pest should be identified properly. After the pest is identified, an IPM strategy, including cultural, biological, chemical, genetic, and mechanical methods should be used to control pests.

Staking and Training

Some vegetables must be staked or trained in some manner, for example, pole beans and tomatoes (Figure 24-9). If staking is not done, the fruits will lay on the ground and lead to problems with rot. Staking also maximizes available space.

Protection from Damaging Temperatures

Modifying temperatures for the successful growth of crops can be achieved in several ways. The first method is to grow plants in a greenhouse; however, this is not a cost-effective method for many vegetable crops. Other methods used for frost protection are row covers, sprinkler irrigation, and fogging, whereas in extreme cases, heaters, smudge pots, and fans are used; neither of these steps is effective with deep frosts. Plasticulture or plastic mulch method to modify soil temperatures is typically used in conjunction with drip irrigation. Plastic mulch has become very popular because of the many associated advantages, such as reduced weeds and increased moisture control. Plastics now come in a variety of colors to suit specific needs in addition to temperature control.

Vegetable crops are also susceptible to heat injury. In warm season crops, temperatures between 80 and 95°F cause injury, whereas in cool season crops, damage occurs between 75 and 85°F. The type of injury observed in both warm season and cool season crops is premature flowering and reduced crop yields. Vegetable crops can be protected against heat injury in the same way they are protected from low temperatures (described previously).

Figure 24–9 Tomato plants being trained

Air Movement

Adequate air movement is required for good plant growth because it decreases humidity, reduces the incidence of disease, helps cool the plant, makes the plant develop a thicker cuticle to protect itself against pests, facilitates pollination, replenishes beneficial gases such as carbon dioxide, and provides a variety of other benefits. However, excessive air movement has a negative effect on photosynthesis and crop yields, can cause physical damage, and spreads diseases, weed seeds, pollutants, and airborne pathogens. Wind protection can be achieved by planting trees, hedgerows, or tall grasses as windbreaks.

CROP MATURATION, HARVESTING, STORING, AND MARKETING

In addition to growing the vegetable crops, the grower must also consider when and how to harvest, conditions during storage such as cooling and relative humidity, and ultimately marketing of the crop.

Crop Maturation

Growers should know when crops are at peak color, flavor, and overall quality. Vegetables are harvested at different stages of maturity for different uses and markets. For example, in the fresh market, tomatoes to be used in distant markets are picked in the green stage, whereas tomatoes that are to be sold that day are typically picked when they are red ripe. Vegetables to be used for processing are often picked based on acid, sugar, and color levels, which can be measured using instruments. The time it takes for a vegetable crop to mature can be measured in two ways. The first way is to count the number of days from planting to maturity; however, the rate of maturity for each crop depends on temperature and rainfall so this is not the most accurate way to determine the maturation time. The second method measures the time to maturity more accurately by using **degree-days:** add the daily high and low temperature together, divide by two, and then subtract the **base temperature** for the specific crop, which is the lowest temperature at which growth can occur.

Harvesting

Many vegetable crops are now picked mechanically (such as tomatoes and peppers); however, most large vegetables (such as watermelons) are still picked by hand. After vegetables are harvested, the area should be cleaned and any stalks or vines should be chopped and put into a compost pile or plowed back into the soil.

Vegetable Cooling and Storage

Vegetable crops are still living when detached from the plant. The first step following removal from the plant is to remove field heat rapidly and to place the vegetable under the appropriate storage conditions. For more information, refer to the postharvest physiology information in Chapter 11.

Marketing

Marketing vegetables is very important because even an excellent product won't sell without marketing. The consumer may be unfamiliar with certain vegetables or may not be aware of new varieties as they become available.

COMMERCIAL VEGETABLE PRODUCTION

The commercial vegetable industry is a complex system involving growers, buyers, shippers, wholesalers, retailers, brokers, chemical and seed industry personnel, researchers, consultants, and consumers. The number of commercial growers has declined in recent years, and the size of operations has increased. Due to the complexity of commercial vegetable production, it is not covered in this text; however, to learn more about commercial vegetable production, check out several books on commercial vegetable production from the local library or contact your local extension agent (Decoteau, 2000; Thompson & Kelly, 1959; Wave & McCollum, 1980).

NONTRADITIONAL VEGETABLE PRODUCTION

Nontraditional vegetable production includes organic vegetable production, sustainable horticulture, hydroponics, and biotechnology. Each method has its advantages and disadvantages.

Organic Production

Vegetable crops produced in the United States are monitored closely for harmful chemicals and have been found to be very safe. However, many people prefer vegetables that have been grown using organic vegetable production, which does not use chemical pesticides or inorganic fertilizers. Organic production is more costly and generally results in lower yields and quality of vegetables because of disease and insect problems; however, in recent years, considerable strides have been made to increase yields and quality. A variety of organic production systems is used ranging from those that do not use any synthetically produced inputs to those that merely reduce synthetically produced inputs. Instead of inorganic fertilizers, organic production uses cover crops, manures, compost, and sludge. Pest control used for organic production is an IPM program without the use of chemicals. Sanitation, crop rotation, biological control, crop resistance, and natural chemical compounds are generally important measures used for organic production. The benefit of organic vegetable production is that no chemical inputs are leached or run off into the water supply and soil; the main disadvantage is the higher cost to the consumer.

Sustainable Horticulture

Sustainable horticulture production systems use IPM and sound management practices to reduce inputs such as tillage, chemicals, inorganic fertilizer, and water. Using drip irrigation and plastic culture reduces water inputs. In addition, sustainable horticulture production systems use plants that are resistant to pests, use windbreaks, avoid less productive sites, and provide a variety of other management tools to reduce inputs.

Hydroponics

Hydroponic systems are used to grow plants in the absence of soil. Although many advantages are associated with this method, it is an expensive means of producing vegetables. Therefore, hydroponics is only used for crops that can generate a higher dollar amount (such as tomatoes, lettuce, and basil) (Figure 24-10).

Figure 24–10 Hydroponically grown basil

Biotechnology

Biotechnology is the manipulation of living organisms or substances obtained from living organisms for the benefit of humanity. The production of genetically modified organisms (GMOs) has a wealth of potential for vegetable crop improvement. Genetic engineers alter plant DNA by adding genes, removing genes, or changing genes that are already in the plant. Advantages of genetic engineering include the production of plants that are disease and insect resistant, have better yields, have improved quality and shelf life, and are herbicide resistant. In the United States, one of the first transgenic plants to be marketed was Flavr-Savr tomato, which was introduced by the Calgene biotechnology company. The Flavr-Savr tomato was reported to have the same horticultural characteristics as standard tomatoes but had a longer shelf life; however, it never became very popular due to issues with flavor. The latest transgenic vegetable plants approved by the U.S. FDA are squash varieties that were made resistant to zucchini yellows mosaic virus, cucumber mosaic virus, and watermelon mosaic virus II. Many genetically engineered plants are used today, and this number will increase. Although the science and technology is available now, the major limitations are the ethical concerns and the legal battle ahead concerning who owns the specific technologies for genetically engineering plants. Other examples and the ethics involved are discussed in Chapter 13.

HOME VEGETABLE GARDEN

Large quantities of vegetables can be produced from a very small portion of land if the garden is planned properly and the grower is knowledgeable about planting and growing vegetables (Figure 24-11). The steps to a successful home vegetable garden are described next.

Figure 24–11 Home vegetable garden

Planning the Garden

The garden location should be near the home but in the open where it gets full sun, near a water source, have fertile soil, and have good aeration, drainage, and nutrient- and water-holding capacity. In general, a sandy loam soil works best for most vegetable crops.

Besides choosing plants due to preference, the grower must choose cultivars that work well in the growing location. The garden should be arranged in rows to maximize the amount of sunlight that the plants receive; make sure large plants do not shade the smaller ones.

Planting and Growing Vegetables

The principles used for a home vegetable garden are similar to commercial vegetable production practices with the major difference being the size of the operation. When preparing a home garden the following steps should be taken:

- **Soil preparation.** Run soil tests and add the necessary fertilizer, lime, organic matter, or sand during tillage to loosen the seedbed.
- **Planting.** Plant high-quality/viability seed or high-quality transplants the right depth or distance apart.
- **Pest control.** Use an IPM strategy, read the label when using pesticides, and use all the safety factors outlined in the pest control information in Chapter 12.
- **Watering.** Avoid conditions where the soil is too wet or too dry.
- **Staking and training.** Make sure vegetables, such as tomatoes and beans, are properly staked.
- **Harvesting.** Harvest at the proper time; harvesting too early or too late can lead to problems.
- **End of season.** When the season has ended, clean the area and chop up all plant material for use as compost or plow the material into the soil.

A good garden requires continuous attention, so the grower should inspect the garden on a regular basis.

SUMMARY

You now understand the importance of olericulture to the homeowner and commercial producer. Edible plant parts are used for classifying vegetable crops by breaking them down into nine major categories. Vegetable crop production begins with selecting cultivars based on climatic requirements (temperature, water, and light), life cycle, cultural requirements, usage, and quality. Many important factors must be considered when preparing the seedbed and during establishment of vegetable crops by seed and transplants. Beyond commercial or traditional vegetable crop production, there are nontraditional vegetable crop production methods such as organic, sustainable horticulture, hydroponics, and biotechnology. Some key steps are necessary when establishing a home vegetable garden as well.

Review Questions for Chapter 24

Short Answer

1. What two major categories can commercial vegetable production be divided into?
2. What state produces almost half of the vegetables grown in the United States?
3. What are two major benefits of eating vegetables?
4. List the nine major categories used to classify vegetable crops by edible parts and provide an example of each.
5. What are five major criteria used in the selection of a cultivar?
6. Provide an example of a cool season and warm season vegetable crop.
7. What is the number one factor determining a crop's suitability for a particular climate?
8. Provide two ways to determine how long it takes for a vegetable crop to mature.
9. What are four ways to provide frost protection in the field?
10. What are five factors that should be considered when preparing a seedbed for vegetable crops?
11. Explain the difference between tillage and cultivation.
12. What are two disadvantages of using transplants?
13. What are six important factors that are required for the proper maintenance and care of vegetable crops?
14. What are four types of nontraditional vegetable production?
15. What are three factors that should be considered when planning a vegetable garden?
16. When preparing a home garden for growing vegetable crops, what are seven factors that should be considered?
17. Provide three of the five horticultural characteristics that are important when selecting vegetable crop cultivars.

Define

Define the following terms:

olericulture
fresh market vegetables
processed vegetables
organic vegetable production
hybrid
frost-free period
base temperature
degree-days
conservation tillage
strip tillage
direct seeding
fluid drilling
indirect seeding (transplanting)
transplanting shock
sustainable horticulture
hydroponic
biotechnology

True or False

1. Measuring maturation time according to heat units is a more exact manner than counting the days to maturity.
2. Row covers are used in cooler climates to increase temperature in the crop environment and to prevent infestation with insects earlier in the season.
3. Strip tillage involves tillage practices that leave more than 30 percent or more crop residues on the top of the soil to prevent soil erosion.

4. An example of subsurface irrigation is drip irrigation.
5. Fluid drilling is a method for precision planting of vegetable seeds.
6. When establishing vegetable crops by transplanting, there are problems with transplanting shock.
7. Vegetables used for processing are often picked based on acid, sugar, and color levels, which can be measured by instruments.
8. Vegetables are harvested at different stages of maturity for different uses and markets.
9. The first step following removal of a vegetable from the parent plant is to remove field heat rapidly.
10. There are a variety of organic production systems for vegetables ranging from those that do not use synthetically produced inputs to those that merely reduce synthetically produced inputs.
11. Tomato, lettuce and basil are the primary crops produced via hydroponics.
12. The science and technology are now available for the production of GMOs; however, the major limitation is the legal battle ahead.
13. In the home garden, it is important to select vegetable crops that the gardener prefers.
14. The principles used for the home vegetable garden are similar to commercial vegetable production with the major difference being the size of the operation.

Multiple Choice

1. The state that produces almost half of the entire vegetable crop in the United States is
   A. California.
   B. Pennsylvania.
   C. Florida.
   D. Georgia.
2. Tillage is used for
   A. soil improvement.
   B. seedbed establishment.
   C. weed control.
   D. All of the above
3. After a crop is in the field, "tillage" is then called
   A. irrigation.
   B. conservation tillage.
   C. cultivation.
   D. aeration.

Fill in the Blanks

1. The leading U.S. vegetable-producing state is ________________.
2. A ________________ is an improved plant developed by crossing parents of different genotypes for a trait.
3. Recently ________________ has been used to increase soil temperature in the field.
4. ________________ horticulture is the use of IPM and sound management practices to reduce inputs.

Activities

Now that we have completed our discussion on growing vegetable crops, you will have the opportunity to explore this area in more detail. In this activity, you will visit a local vegetable farm, farmer's market, or a neighbor's vegetable garden and prepare a summary of what you learned there. Following are some hints:

- Find out what vegetables are grown and the method used to grow them, for example, traditional, organic, hydroponic, and so on.
- If it is a commercial operation, find out what type of inputs there are and whether the vegetables are to be used for fresh market or processing.
- If it is a farmer's market, inquire how the vegetables were grown and compare the prices with a local supermarket. If you feel adventurous, buy vegetables from the farmer's market and the local supermarket and then run a taste test to see which one tastes better to you.
- Go to your local cooperative extension office and find out what vegetables are recommended for your area.

In addition to information you were asked to gather, include your thoughts on the production practices used. If you do not have access to any of these locations, surf the Internet for two sites that present information on growing vegetable crops and then summarize these sites. Provide the addresses for the Web sites where you found your information.

References

Decoteau, D. (2000). *Vegetable production.* Upper Saddle River, NJ: Prentice Hall.

Thompson, H. C., & Kelly, W. C. (1959). *Vegetable crops* (5th ed.). New York: McGraw-Hill.

Ware, G. W., & McCollum, J. P. (1980). *Producing vegetable crops* (3rd ed.). Danville, IL: The Interstate Printers and Publishers.

CHAPTER

25

Pomology: Fruit and Nut Production

ABSTRACT

The three major types of fruits are simple, aggregate, and multiple fruits. The important aspects of fruit crop establishment are selecting the site and necessary facilities, choosing cultivars, selecting rootstocks, preparing the land, and planting fruit crops. The proper care and maintenance of fruit crops includes the proper management of nutrients, watering, pest control, pruning, fruit thinning, protection from damaging temperatures, and the effects of air movement on fruit crops. The chapter concludes with a discussion of crop maturation, harvesting, storage, and marketing of fruits and nuts.

Objectives

After reading this chapter, you should be able to

- provide background information and discuss the importance of pomology.
- know the three major types of fruits and key aspects about each.
- recognize the important aspects of fruit crop establishment.
- understand how to care for and maintain fruit crops by learning proper management skills.
- understand the important aspects of crop maturation, harvesting, storing, and marketing of fruits and nuts.

Key Terms

advective freezes
aggregate fruits
ecodormancy (quiescence)
endodormancy (rest)
fertigation
fresh market fruits
fruit
heat unit requirement
horticultural characteristics
multiple fruits
nuts
pomology
processed fruits
radiation freezes
rootstock
simple fruits
tensiometer
winter chilling requirement

INTRODUCTION

The USDA food guide pyramid suggests that individuals eat **fruits** and **nuts** two to four times a day. The American Cancer Society and The American Heart Association both endorse eating fruits and nuts for a variety of reasons, including the reduction of cancer rates and heart disease. The branch of horticulture dealing with producing, storing, processing, and marketing fruits and nuts is called **pomology** (Figure 25-1). Fruits and nuts can be produced commercially or by the homeowner. Commercial fruit and nut production can be broken down into two distinct areas: fresh marketing and processing. The total annual value of the fruit crop in the United States was approximately $12 billion in 2000, and the nut crop was almost $1.5 billion; both of these numbers are steadily on the

Figure 25–1 Apple trees grown on trellises

rise. The United States is one the world leaders in the production of fruits and nuts. Some of the leading states for the production of fruits and nuts are California, Florida, Washington, Pennsylvania, Texas, Georgia, Arizona, and Hawaii.

The three major types of fruits are **simple fruits, aggregate fruits,** and **multiple fruits.** The ideal fruit-growing area depends on the type of fruit being grown. Therefore, the proper location with necessary facilities should be selected first, followed by carefully considering the proper cultivar, **rootstock,** land preparation, and way to plant fruit crops. Site selection depends on environmental and economic factors. Necessary facilities include cold storage facilities, shipping areas, offices, pesticide storage and mixing areas, and storage buildings. After the proper location is selected, cultivar selection is based on climatic requirements (temperature, water, light), usage, cultural requirements, yield and quality, and resistance to pests. In many cases, the cultivar must be grafted onto the proper rootstock to overcome certain problems. A rootstock is a root system and base of a tree on which the fruiting top or scion is budded or grafted. Rootstocks are used to make plants tolerate poorly drained soil conditions, to protect from nematodes and soil diseases, and to control size (e.g., apples and pear trees).

Next, the land must be prepared properly prior to installing the fruit crop. Soil tests are run to determine the amount of nutrients to add and whether pH adjustments are necessary. The soil should have good water-holding capacity, and if not, it should be amended to correct any problems. An adequate supply of water is very important in fruit production, so installing irrigation and drainage systems may be necessary at certain locations depending on the **tensiometer** readings (used to measure soil moisture). Large rocks and other objects should be removed and the soil plowed to remove any hardpans, to level the land, to increase soil aeration and moisture infiltration, to control weeds, and to incorporate crop residue and fertilizer or lime as necessary. Roads within the orchard should be carefully planned and designed to maximize efficiency prior to installing fruit crops. Finally, good installation practices must be used that are appropriate for the fruit crop being grown.

After the crop has been installed, proper care and maintenance helps ensure high quality and yield. Proper care and maintenance includes managing nutrients, good watering practices, timely pest control, proper pruning practices, fruit thinning, and protecting from damaging temperatures and wind. Recognizing when the crop is mature is important in order to harvest at peak color, flavor, and overall quality. Fruits are harvested at different stages of maturity for different uses and markets. During the winter months, fruit trees experience winter dormancy, so knowing when a fruit tree's buds will break is important for adequately caring for and maintaining the orchard. This is accomplished by first determining the **winter chilling requirement** and the **heat units requirement;** with this information, the grower will know when the buds will break and be prepared for problems with freezing temperatures and problems when the crop reaches maturity. After fruits are mature, they must be harvested properly. Certain types of fruits used for processing are often picked mechanically, such as raspberries, blackberries, cherries, and others. Most large fruits are picked by hand. After the fruit is harvested, it is still living, so the first step is to remove the field heat rapidly and place the fruit in the appropriate storage conditions. Prior to harvesting, be prepared to market the fruits properly and thus give the products a chance to sell.

BACKGROUND INFORMATION ABOUT POMOLOGY

Just like vegetables, fruits and nuts are key components in the human diet (the USDA recommends two to four savings per day). Pomology is the branch of horticulture dealing with the producing, storing, processing, and marketing fruits and nuts. Fruits are defined botanically as a mature ovary of a flowering plant or tree that includes the seed, its covering, and any parts closely connected to it, whereas nuts are seeds, such as the walnut. Whether a tomato is a fruit or a vegetable is a common question. The tomato is botanically defined as a fruit; however, because pomology deals with fruit and nut trees and olericulture deals with annual plants, the tomato is also classified as a vegetable. Fruits and nuts are produced commercially and by the homeowner. Commercial fruit and nut production can be divided into two categories: fresh market or processed. **Fresh market fruits** are used as soon as possible after harvest (Figure 25-2), whereas **processed fruits** are canned, dried (Figure 25-3), or frozen.

Figure 25–2 Fresh fruit on display at a supermarket

Fruits are important to the human diet for a variety of reasons. First, they have a high nutrient content, which varies between plant species: In general, fruits are very important sources of vitamins A, C, B_6, and folacin, plus minerals such as potassium, magnesium, copper, and iron. Second, they are rich in certain phytochemicals that are known to have a variety of health benefits. For example, elagic acid, abundant in strawberries and muscadine grapes, has been shown to reduce the risks of certain forms of cancer. Third, fruit pectins play an important role in reducing dietary cholesterol. Fruit pectins trap dietary cholesterol, which ultimately prevents it from depositing in the linings of blood vessels. Fourth, fiber from fruit can prevent colon cancer and reduce constipation. Fifth, many fruits are high in antioxidants that reduce free radicals, which are responsible for aging and different types of cancer.

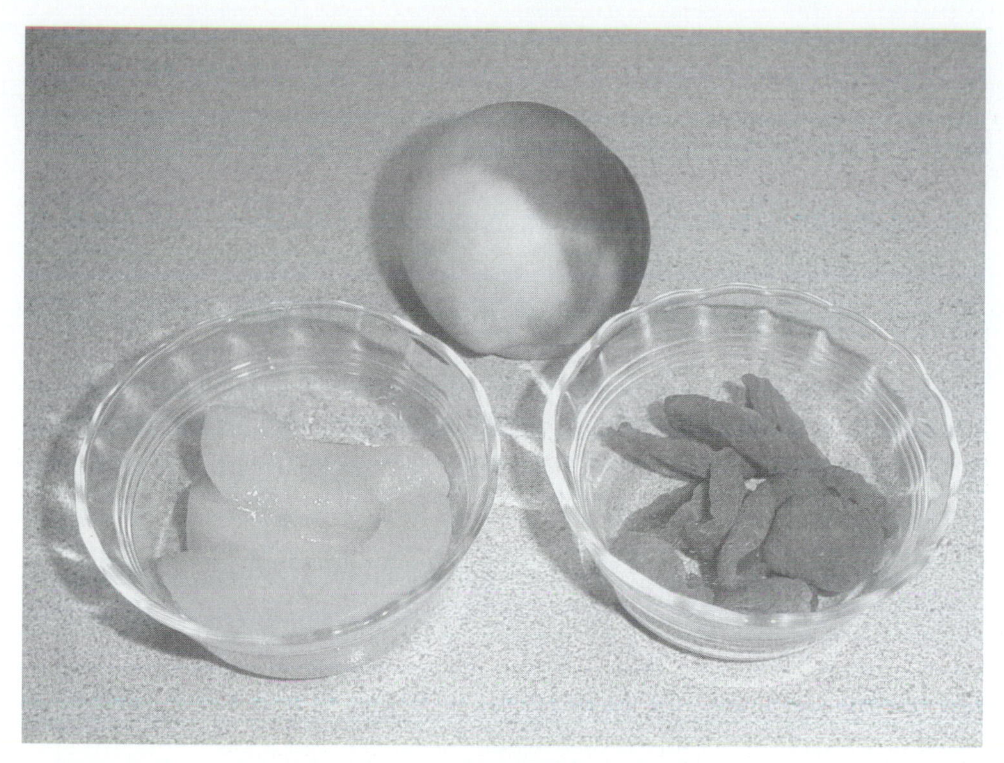

Figure 25–3 Examples of fresh, canned, and dried fruit

TYPES OF FRUITS

Fruits have many different forms and shapes. Simple fruits are comprised of a single ovary with or without some other flower parts, which have developed as part of the fruit. Simple fruits can be broken down into two groups: fruits with a fleshy pericarp (for example, berries) and fruits with a dry pericarp (for example, walnuts). Another type of fruit is an aggregate fruit, which is comprised of a single receptacle (base of the flower) with masses of similar fruitlets (for example, blackberries and strawberries). The last type of fruit is a multiple fruit, which is comprised of ovaries of many separate but closely clustered flowers (for example, pineapple) (Figure 25-4).

Figure 25–4 An apple is a simple fruit (top), raspberry is an aggregate fruit (middle), and pineapple is a multiple fruit (bottom).

FRUIT CROP ESTABLISHMENT

The ideal fruit-growing area varies with the type of fruit being grown. To produce high-quality fruits, the grower must select the proper location with the necessary facilities, select the proper cultivar, select the proper rootstock (for fruit trees), prepare the soil, and be knowledgeable about the proper way to plant fruit-producing crops.

Selecting the Site and the Necessary Facilities

To select the proper site, these environmental factors must be considered: light, rainfall, wind, soil, topography, water, air pollution, and plant pests. There are economic factors to consider as well: land cost and availability, labor, transportation, utilities and services, and competition. Finally, storage and other facilities are important to a successful fruit-producing location, including cold storage facilities, shipping areas, offices, pesticide storage and mixing areas, and storage buildings.

Selecting the Cultivars

Establishing plants that produce fruit is a long-term commitment, so selecting the proper cultivar is crucial. Selecting the cultivar should be based on the proper climatic requirements (temperature, water, and light), usage, cultural requirements, yields and quality (including **horticultural characteristics** such as color, size, flavor, shape, and appearance), and resistance to pests.

Climatic Requirements

All living things must have a suitable environment to survive. Several climatic factors influence fruit production and growth.

Temperature. The fruit crop must be readily adapted to the climate at the location selected because to grow at peak efficiency, all plants must be grown within their optimum temperature range. Fruit crops can be grown in any climate, but only certain species of fruit will grow and produce well. It is very important to consider winter hardiness or how low a temperature the flower buds and plant can withstand without damage. In North America, winter freezes present a significant hazard in fruit production. The USDA Hardiness Zone Map can be used as a guide to determine what species will do best in a particular location.

Water. Fruit crops demand a consistent supply of water, and any stress leads to a reduction in quality and yields. The selected cultivar should be adaptable to the location selected.

Light. Sunlight is needed for photosynthesis and color formation in fruits, so high light intensity is important for high yields and quality. Day length determines when some fruit crops flower; for example, June-bearing strawberries form flower buds in response to long days. Therefore, the cultivar selected should grow well at the light intensity, quality, and duration of light at the chosen location.

Usage

When selecting the cultivar, knowing what the fruit crop is to be used for—processing or fresh market—helps the grower select for maximal performance.

Cultural Requirements

Prior to selecting a fruit cultivar, knowing the cultural requirements helps the grower select for maximal performance. Cultural requirements include watering, fertilization, light, and other factors required for vigorous plant growth.

Yields and Quality

A cultivar should be selected that will produce good yields and have the desired horticultural characteristics, such as color, size, flavor, shape, and appearance. The cultivar should also ripen in the correct market window.

Resistance to Pests

The cultivar selected should have good resistance to pests and diseases at the location in which it will be grown.

Selecting the Rootstocks

After a cultivar is selected, often it must be grafted on a rootstock to overcome certain problems associated with a specific cultivar. A rootstock is defined as the root system and base of the tree on which the fruiting top or scion cultivar is budded or grafted. Rootstocks are typically used to enable plants to tolerate poorly drained soil conditions, to protect from nematodes (small parasitic soil worms) and soil diseases, and to control size.

Preparing the Land

When preparing the land prior to the installation of a fruit crop, the following factors should be considered:

- Soil tests should be run to determine the amount of nutrients to add and if pH adjustments are necessary. Fruit crops typically perform best with a soil pH in the range of 6 to 6.5, with 6.5 considered near optimum. Below a pH of 6.0, the availability of the major nutrients decreases and problems with calcium and magnesium deficiency increase. Dolomitic limestone, which supplies both calcium and magnesium, is typically used to increase soil pH in orchards. At a pH of 7.0 or above, problems with iron or zinc deficiencies can occur. To reduce the pH of the soil, sulfur is typically added.
- For most fruit crops, the soil should be well drained with a good nutrient- and water-holding capacity. Sandy loams or loamy soils are best for growing fruit crops. Fruit trees can be grown on shallow rocky or calcareous soils using **fertigation,** which is the application of fertilizer through an irrigation system. An adequate supply of water is very important in fruit production so irrigation and drainage systems should be installed as necessary at a given location.
- Large rocks or other objects should be removed and the soil should be plowed to remove hardpans, to level the land, to increase soil aeration, to increase moisture infiltration, to control weeds, and to incorporate crop residue and fertilizer or lime as necessary.
- Roads within the orchard should be carefully planned and designed to maximize the efficiency of the orchard prior to the installation of fruit crops.

Planting Fruit Crops

The rows should be laid out according to the type of fruit crop being grown. Fruit trees can be laid out in a checkerboard or diamond pattern to maximize light penetration. The best orientation is north-south, which allows both sides of the plant to receive sun. The east side receives sun during the morning hours, whereas the west side receives sun during the afternoon. Fruit crops should be installed by standard methods such as soil preparation, fertilization, and others (Childers, Moore, & Sibbett, 1995; Krawczyk, 2005).

MAINTENANCE AND CARE OF FRUIT CROPS

The proper care and maintenance of fruit crops is very important to promote high quality and yield. After the proper site is selected and the fruit crop installed properly, the following tasks should be done to maximize the plants' productivity.

Manage Nutrients Properly

Commercial fruit crops should be fertilized in a scientific manner starting with extensive soil sampling and testing. Soil sampling should be followed with leaf analysis to determine the nutrient content of the leaves and to decide how many nutrients to apply to promote good growth but not overstimulate the plant.

Water

Plants should be watered based on demand; however, it is important to know that crop demands will vary with the species, time of year, and plant size.

Control Pests

Good pest control is also necessary for high yields, so an IPM approach should be used to control pests.

Prune

Pruning is one of the most expensive and labor-intensive aspects of fruit production. Pruning increases fruit quality, such as color, by allowing more sunlight and air movement through to the plant; reduces disease problems and allows pesticide applications to better penetrate the tree; and increases fruit and overall plant size. For more information on pruning methods, refer to Chapter 22.

Thin Fruits

Fruit thinning can be done either by hand, which is a very labor-intensive practice, or by using chemical thinning agents, which is a commonly used practice in the United States. In fruit trees, thinning eliminates problems with biennial bearing, which is when trees fruit on alternate years; prevents physical damage to the tree; and enhances fruit size, shape, color, and overall quality.

Protect from Frosts and High Temperatures

Unanticipated frosts damage fruit crops annually, whereas fruit crops are more tolerant to higher temperatures. In some cases, nothing can be done to prevent damage; however, in many cases, severe damage can be prevented by being knowledgeable about the type of freeze and how to prevent damage:

- **Spring freezes.** Many fruiting plants have flowers that emerge early in the spring. After flowers have emerged, damage typically occurs at temperatures at or below 28°F or −2.2°C. For this reason site selection in relation to spring freezes is very important in fruit production. The two types of freezes that commonly occur are advective and radiation:
  - **Advective freezes.** These type of freezes are accompanied by wind and can be overcome by the proper site selection or through the use of windbreaks.
  - **Radiation freezes.** These type of freezes occur on still, clear nights. Protection against this type of freeze is achieved by using overhead sprinklers, covering the crop, or using heaters.

- **Heat damage.** Heat tolerance is the capability of a plant to tolerate high temperatures and still produce well. High temperatures can also cause severe damage to fruit crops; however, this not typically a problem in most fruit-growing areas.
- **Air movement.** Adequate air movement is required for good plant growth and development because it decreases humidity, which reduces the incidence of disease; helps cool the plant; makes the plant develop a thicker cuticle to protect itself against pests; facilitates pollination; replenishes beneficial gasses such as carbon dioxide; and provides a variety of other benefits. However, excessive air movement has a negative effect on photosynthesis and crop yields, causes physical damage, and spreads diseases, weed seeds, pollutants, and airborne pathogens. Wind protection can be achieved by planting trees or hedgerows as windbreaks.

CROP MATURATION

When crops are mature, they are at peak color, flavor, and overall quality at the time of harvest. Fruits are harvested at different stages of maturity for different uses and markets. For example, in the fresh market, peaches to be used in distant markets are picked in their firm mature stage, whereas peaches to be sold that day are typically picked when they are slightly soft. Fruits to be used for processing are often picked based on acid, sugar, and color levels, which can be measured using instruments.

During the winter months, fruit trees experience winter dormancy. The two types of dormancy are **endodormancy** and **ecodormancy.** Endodormancy (rest) is the dormancy of seeds and buds imposed by internal blocks, which are removed by winter chilling, and is regulated by endogenous levels of plant hormones. Ecodormancy (quiescence) is a state in which the plant is not growing and will not grow until external conditions are satisfied, usually with warmer temperatures. Both endodormancy and ecodormancy work together to prevent plants from breaking bud dormancy during the wrong time of the year. This is not a perfect system, however, because under unusual weather conditions, fruit trees can be fooled; for example, the citrus crop in Florida has been lost on a number of occasions, which drives up prices.

Knowing when fruit tree buds will break helps the grower adequately care for and maintain an orchard. This is accomplished by first determining the winter chilling requirement, which is calculated in most areas of the United States by how many hours at and below 45°F (7.2°C) are required during the winter for the plant to break its winter resting period and develop normally when temperatures rise in the spring. There is a range of chilling requirements for each species that differs between cultivars within the species; for example, the chilling requirement for apples is between 50 and 1,700 hours to break dormancy. After a tree has received its winter chilling requirement, it requires a specific heat unit requirement for bud break to occur. The heat unit requirement is the number of hours of warmth needed for a certain growth phase to occur. Heat units are calculated by taking the average temperature for the day minus 50°F (which is the minimum temperature for growth of fruit trees). For example, if the average temperature for the day is 68°F, the grower subtracts 50 from this number to get 18 heat units, whereas if the average temperature for the day is 40°F, the grower subtracts 50 to get 0 heat units for the day.

HARVESTING, STORING, AND MARKETING

Certain types of fruits used for processing are often picked mechanically, for example, raspberries, blackberries, and cherries. Most large fruits are still picked by hand.

Fruit is still living when detached from the plant. The first step following removal from the plant is to remove field heat rapidly and to place the fruit under the appropriate storage conditions.

As with any crop, marketing is also important for fruit crops. The consumer may not be aware of new varieties of fruits and nuts introduced in the marketplace.

SUMMARY

You now understand the importance of pomology. The three major types of fruits are simple, aggregate, and multiple fruits. The important aspects of fruit crop establishment are site selection and necessary facilities, cultivar selection, rootstocks, land preparation, and proper planting. Fruit crops also require proper care and maintenance. Growers must understand the intricacies of crop maturation, harvesting, storage, and marketing to create a successful product.

Review Questions for Chapter 25

Short Answer

1. What two major categories can commercial fruit production be divided into?
2. Give two main reasons why fruits are important.
3. List the three major types of fruits.
4. What are four factors that should be considered when selecting the proper fruit cultivar?
5. Give three reasons why rootstocks are commonly used.
6. What is the best soil pH for growing fruit crops?
7. What are two types of spring freezes that commonly occur and what is one way to overcome each?
8. What are three important reasons for pruning?
9. What are three benefits that can be obtained from fruit thinning?

Define

Define the following terms:

pomology
fruits
nuts
fresh market fruits
processed fruits
simple fruits
aggregate fruits
multiple fruits
horticultural characteristics
rootstock
fertigation
tensiometer
advective freezes
radiation freezes
endodormancy (rest)
ecodormancy (quiescence)
winter chilling requirement
heat unit requirement

True or False

1. Fruit pectins trap dietary cholesterol, which ultimately prevents it from depositing in the linings of blood vessels.
2. Endodormancy is the same as rest, which is dormancy imposed by external blocks.
3. Ecodormancy is the same as quiescence, which is dormancy imposed by external blocks.
4. The winter chilling requirement is calculated by determining the number of hours at and below 45°F (7.2°C).
5. Heat units are calculated by taking the average temperature for the month minus 50°F.
6. Rootstocks can be used for size control.
7. Rootstocks can be used to protect from nematodes and other soilborne organisms.
8. Rootstocks can be used to enable plants to tolerate poorly drained soils.
9. Advective freezes occur on cool, clear nights.
10. Radiation freezes are accompanied by wind and can be overcome by proper site selection.
11. Pruning increases fruit quality such as color.
12. Fruit thinning eliminates problems with biennial bearing.
13. Fruit thinning enhances fruit size, shape, color, and overall quality.
14. Fruits to be used for processing are often picked based on acid, sugar, and color levels.

Multiple Choice

1. Rootstocks are typically used for the following purposes in fruit trees:
   A. Size control
   B. To protect against soil insects or soil diseases
   C. To enable plants to tolerate poorly drained soil conditions
   D. All of the above

Fill in the Blanks

Pomology is the branch of horticulture dealing with ____________________ and ____________________.

Activities

Now that we have completed our discussion of growing fruit and nut crops, you will have the opportunity to explore this area in more detail. In this activity, you will visit a local fruit producer, farmer's market, or a neighbor's yard where fruits are being grown and prepare a summary of what you learn. Some hints on what to look for:

- Find out what fruits and nuts are grown.
- If a commercial operation, find out what type of inputs there are and if the fruits and nuts are to be used for fresh market or processing.
- If a farmer's market, inquire about how the fruits and nuts were grown and compare the prices with a local supermarket. If you feel adventurous, buy fruits and nuts from the farmer's market and the local supermarket and then run a taste test to see which taste better to you.
- Go to your local extension cooperative extension office and find out what fruit and nut crops are recommended for your area.

In addition to information you were asked to gather, include your thoughts on the production practices used. If you do not have access to any of these locations, surf the Internet for two sites that present information on growing vegetable crops and then summarize these sites. Provide the Web site addresses for the sites where you found your information.

References

Childers, N., Moore, F. J. R., & Sibbett, G. S. (1995). *Modern fruit science*. Gainesville, FL: Horticultural Publications.

Krawczyk, G. Pennsylvania State University, College of Agricultural Sciences (2005). *Pennsylvania fruit tree production guide*. University Park, PA: Publications Distribution Center. Retrieved from January 4, 2006. http://tfpg.cas.psu.edu/

Glossary

A

abiotic – nonliving factor.

abiotic disease – disease or disorder that is noninfectious, or disorders.

abscisic acid – a plant hormone that is produced in response to water stress and is directly involved in stomatal opening and closing.

abscission – the separation of a plant part from the parent plant.

accent planting – an area of particular beauty or interest established in a landscape.

active ingredient (a.i.) – the actual amount of pesticide in a formulation that is responsible for killing the pest or the actual amount of nutrient available.

advective freezes – freezes that are accompanied by wind and can be overcome by the proper site selection or through the use of windbreaks.

aggregate fruits – fruits that are comprised of a single receptacle with masses of similar fruitlets.

Agricultural Adjustment Act – the overproduction or surplus of goods became a major problem during the Depression, which occurred in the 1930s. To correct this problem the U.S. Congress passed this act, which was directed at the expansion of utilization research.

agricultural experiment station – established by Morrill Land Grant College Act, with the first state agricultural experiment stations located in California and Connecticut in 1875.

agricultural limestone – used to raise the pH of a soil and contains calcium.

Agricultural Marketing Act – this act was passed by the US Congress in order to change the imbalance between production and postproduction research.

agriculture – the production of plants and animals to meet basic human needs.

agronomy – the cultivation of grains and forage crops.

air layering – propagation method that involves removing a portion of the bark on the stem, placing moist material such as sphagnum moss around the wounded site, wrapping with clear plastic, and sealing both ends to hold in moisture.

air pollution – when harmful or degrading materials contaminate the air.

alternate leaf arrangement – the alternate pattern of leaf arrangement occurs when leaves are staggered along the length of the stem.

angiosperms – a class of flowering plants with seeds that develop in fruits.

annuals – plants that germinate from seed, grow to full maturity, flower, and produce seeds in one growing season; in other words, from seed to seed in one season.

anther culture – a technique in which immature pollen is induced to divide and generate tissue, either on solid media or in liquid culture.

antisense technology – involves putting a known gene sequence into the plant backward to block a process in plants.

apical meristem – the primary growing point of the stem.

apomixis – a form of asexual propagation in which seeds are produced without fertilization; for example, because there is no fusion of male and female gametes, these seeds are solely maternal in origin.

approach grafting – when two independent self-sustaining plants are grafted together.

arboretum – a collection of trees arranged in a naturalized fashion.

arboriculture – the study of trees, their growth, and culture.

arthropods – organisms that have exoskeletons and jointed legs.

asexual propagation – the reproduction of new plants from the stems, leaves, or roots taken from the parent plant.

asymmetrical balance – when different numbers of plants are on both sides of the landscape.

atmospheric environment – the aboveground portion of a terrestrial plant's environment.

attached greenhouse – a greenhouse that is connected to a building.

auxein – Latin, meaning to grow.

auxin – a class of plant hormones with activity similar to indole-3-acetic acid (IAA).

available water – the difference between soil moisture at field capacity and the wilting point.

axillary or lateral bud – located along the side of the stem at the base of the entire leaf below the terminal bud.

azalea pot – container slightly shorter than the standards or three-fourths as high as it is wide.

B

bag culture – when materials such as peat, sawdust, wood chips, or bark are put in bags and watered using a drip irrigation system.

Bakanae disease – foolish seedling.

balance – the physical and visual stability of a floral design or a landscape design principle, which uses equal weight of design elements to show uniformity.

balled and burlapped – a production system starting with field-grown trees that are dug keeping the ball of soil containing the root system that is covered with burlap material.

band treatments – fertilizer treatments made in a narrow band around the crop row.

bare root – a seedling plant without a ball of soil around the roots.

bare root method – a method that describes trees grown in the field and dug without taking soil.

bare root system – a hydroponic system where no physical support is given to the root system, which is suspended in the nutrient solution.

bark slips – when the bark separates easily from the wood (xylem) during the time of year when cambium cells are actively dividing.

base temperature – the lowest temperature at which growth can occur.

bedding plants – refers to a wide range of plants, mostly annuals, including herbs, vegetables, and flowering ornamentals; however, there are garden designs that use perennials, bulbs, rhizomes, corms, and tubers for continual bloom throughout the year.

biennials – plants that complete their life cycle in two growing seasons.

biological pest control – an IPM method that uses living organisms that are predators to control pests or uses naturally occurring chemicals extracted from plants.

bioremediation – when living organisms are used to remove pesticide spills, heavy metals, and other pollutants from the environment.

biotechnology – the manipulation of living organisms or substances obtained from living organisms for the benefit of humanity.

biotic – living factor.

biotic disease – disease that is caused by parasites or pathogens (organisms that cause disease) that is infectious and transmissible.

bolting – rapid stem elongation.

botanical garden – a plant collection that forms a habitat.

bracts – modified leaves located just below the flower.

brassinolide – the active component in brassins.

brassinosteroids – a class of plant hormones with activity similar to brassinolide.

brassins – a crude lipid extract from rape pollen.

breeder seed – a small amount of seed with desirable traits produced by the breeder and used for the production of foundation seed.

broadcast treatments – treatments that cover the entire area uniformly.

broadcasting – when fertilizer is evenly spread either manually or mechanically on the soil surface.

bud scale – a tiny leaf-like structure that covers the bud and protects it.

budding – similar to grafting except that the scion is reduced to a single bud with a small portion of bark or wood attached.

bulb pan – a container that is half as high as its width; also called half pots.

bush planting – when an untrained person tries to landscape without any knowledge of plant materials.

C

callus – an undifferentiated mass of cells.

callus culture – the growth of callus on a solid medium.

calyx – the term used to describe all the sepals on one flower. The function of the calyx is protection. For example, spines on the calyx deter animals from feeding on the plant.

cambium – the area in stem tissue where new plant cells are formed. It is very important to know where the cambium is located when grafting because if the cambium layers do not match, the graft union will not be successful.

Camerarius, Rudolph – demonstrated sexuality in plants, thereby providing the roots of genetics.

career goal – the level of accomplishment you want to reach in your work.

cation exchange capacity (CEC) – the measure of total exchangeable cations a soil can hold.

cell suspension culture – a method in which plant cells are suspended in a liquid media under continuous agitation to provide aeration.

central leader – the main upright shoot of the tree.

central leader method – pruning system where one strong upward-growing branch is identified and trained to grow as the central axis of the tree.

certified seed – the progeny of registered seed that is sold commercially.

chemical dormancy – caused by germination inhibitors, which accumulate in the fruit or seed coverings during development.

chemical pest control – an IPM method which uses pesticide chemicals to control pests.

chewing insect – an insect that uses mandibles for chewing plant parts such as leaves, stems, roots, fruits, flowers, and petals.

chilling injury – injury that occurs mainly in tropical and subtropical crops stored at temperatures above their freezing point.

chlorosis – when a plant or plant part begins to yellow due to poor chlorophyll development or destruction of chlorophyll caused by mineral deficiency or pathogen attack.

climacteric fruits – fruits that show large increases in CO_2 and ethylene (C_2H_4) production rates coincident with ripening and will ripen in response to exogenous applications of ethylene.

climate – a yearly pattern of weather factors.

clone – a plant that is grown from a piece of another plant and is genetically identical to the parent.

colere – Latin, meaning culture.

complete fertilizer – a fertilizer that contains all three primary fertilizer nutrients—nitrogen, phosphorous, and potassium—and may have select micronutrients.

complete flower – a flower that contains all four major flower parts, including sepals, petals, stamens, and pistils.

complete metamorphosis – a complete change in an insect's form from egg to adult.

compound leaf – a leaf that has two or more leaflets.

conduction – heat loss by transmission through the greenhouse.

conifer cutting – cuttings containing tissues that are hardwood obtained from conifer plants in early winter.

connected-greenhouse – greenhouses that are joined. Various styles of connected greenhouses include ridge-and-furrow, barrel-vault, and saw-tooth greenhouses.

conservation tillage – also called minimal tillage; involves tillage practices that leave 30 percent or more crop residues on top of the soil to prevent soil erosion.

container-grown plants – plants that are grown in different-sized containers.

container nursery – growing nursery crops to a marketable size in containers that differ in size and shape according to the species and the marketable size desired.

continual bloom – flowers in the landscape year round.

controlled atmosphere (CA) storage – regulation of CO_2 and O_2 levels during storage.

corner planting – arrangement of plants in a landscape to create the frame of the outdoor room.

cortex – primary tissue of the stem or root that is located between the epidermis and the vascular region.

cultivar – cultivated variety.

cultural pest control – an IPM method that uses management techniques to control pests.

cut flowers – flowers that are grown for the sole purpose of removing them from the parent plant and displaying them in containers or for other uses such as corsages.

cuticle – an impermeable, waxy material on the outside layer of leaves and stems that prevents water loss.

cuttings – detached vegetative portions of the plant that are used to produce a new plant.

cytokinins – a class of plant hormones with activity similar to kinetin.

D

damping-off – a fungal disease that causes stems to rot at the soil line.

Dark Ages – medieval time period between the fall of the Roman Empire and the Renaissance.

Darwin, Charles – first to describe plant movement in response to light and gravity.

day-neutral plants – plants that flower in response to the genotype with no specific light requirement.

dead heading – removing dead or dying flowers from annuals so the plant will continue to live and bloom for a longer period of time.

dead zone – in evergreens, it is the area 6 to 12 inches below the green needles on the branch.

deciduous hardwood cutting – a mature woody stem cutting.

deciduous perennials – the root systems and stems of this type of plant live throughout the winter months.

deciduous plants – plants that lose their leaves during a portion of the year, usually the winter months.

degree-days – calculated by adding the daily high and low temperature together, dividing by two, and then subtracting the base temperature for a specific crop.

denitrification – occurs when nitrate is released into the atmosphere as N_2, N_2O, NO, or NO_2.

detached (freestanding) greenhouse – a greenhouse that is separate from other buildings or greenhouses.

detached scion grafting – when a detached scion is inserted into the apex, side, bark, or root of the understock.

dicot – plant that is characterized by two cotyledons (seed leaves) and has reticulate leaf venation.

dicotyledonae – a subclass of angiosperm plants that has two cotyledons and reticulate leaf venation.

DIF – using differences in day and night temperatures to modify plant height.

Dioscorides – wrote the authoritative book *De Materia Medica*

direct seeding – the placement of seed directly into the field.

dirt – soil that is out of place.

division – a method of propagation in which parts of plants are cut into sections that will grow into new plants naturally.

dolomitic limestone – used to raise the pH of a soil and contains calcium and magnesium.

dominance – one design element or characteristic should be more noticeable than others, thereby creating a focal point.

dormancy – the temporary suspension of visible plant growth.

down lighting – when light shines down on a tree or interesting object, creating an interesting shadow pattern on the ground.

dry fruit – a fruit that has hard seeds enclosed in a fruit wall, such as a sunflower.

E

ECe Number – units of electrical conductivity designated by milliSiemens/cm at 25°C.

ecodormancy (quiescence) – the state in which the plant is not growing and will not grow until external conditions are satisfied, usually with warm temperatures.

edaphic environment – the soil and area where plant roots are located.

Egyptian – ancient civilization whose major accomplishments include developing drainage and irrigation systems, refining the hoe, and perfecting the plow. The Egyptians also developed technologies associated with baking, wine making, and food storage.

electroporation – a method of gene insertion where plant protoplasts are exposed to a sudden electrical discharge that opens up pores in the plant cell, enabling DNA to enter.

embryo culture – when the embryo is removed from the seed aseptically and grown on solid gel medium under optimal environmental, nutritional, and hormonal conditions to promote growth of the embryo, which would not germinate within the seed; also called embryo rescue.

endodormancy (rest) – dormancy of seeds and buds imposed by internal blocks, which are removed by winter chilling, and regulated by endogenous levels of plant hormones.

environment – involves all the factors that affect the life of living organisms.

epidermis – the outer cell layer of plant parts.

epigeous seed germination – when the hypocotyl elongates and brings the cotyledons above ground (for example, cherry).

epinasty – downward movement of the petioles.

ethylene – a plant hormone that is a simple unsaturated hydrocarbon generally accepted to be the fruit-ripening hormone.

etiolation – exaggerated growth of the stem caused by low light levels.

eutrophication – pollution that occurs in lakes or streams that have too many nutrients in their water due to excessive nutrient runoff.

evapotranspiration – the combination of water lost from the soil surface by evaporation and by transpiration from the leaf surface.

evergreen perennials – the root system, stems, and leaves of this type of plant live throughout the winter months.

evergreen plants – plants that retain their leaves all year.

explants – pieces of plant material used in tissue culture.

F

facultative fungi – fungi that can survive on both dead and living tissue.

facultative parasites – normally live as saprophytes but can live as parasites under the proper conditions.

facultative saprophytes – parasitics that attack living tissues but can also live on dead tissues provided the proper conditions exist.

fallow – leaving the land uncultivated for a period of several years to rejuvenate naturally.

fast-release fertilizer – nutrients are immediately available.

fertigation – the application of fertilizer through the irrigation system.

fertilization – occurs when the male sex cells fuse with the egg cell to form a new plant.

fertilizer – any organic or inorganic material used to provide the nutrients plants need for normal plant growth and development.

fertilizer analysis – the proportion of nutrients supplied by a fertilizer formulation.

fertilizer with herbicide – a fertilizer with herbicide added; the herbicide commonly used is 2,4-dichlorophenoxyacetic acid (2,4-D), which is an auxin compound that selectively kills broadleaf weeds.

fibrous root system – a root system that has as a large number of small primary and secondary roots.

field moisture capacity – when all gravitational water has drained out of the large pore spaces leaving only the small pore spaces containing water.

field nursery – grows nursery crops to a marketable size in the field.

fleshy fruit – a fruit that has soft fleshy material with or without seeds enclosed, such as a watermelon or orange.

floating system – a hydroponic system where the plant is put on a raft that floats on the nutrient solution and the roots are completely immersed below.

floral foam – a block of absorbent material used to provide support for floral arrangements.

floral garland – a wreath that has not been circularized; it can be made of flowers, foliage, or a combination.

floriculture – branch of ornamental horticulture which deals with the production, transportation, and use of flower and foliage plants.

florists – individuals who use cut flowers in their trade.

flower initiation – internal physiological changes in the apical meristem that lead to the visible initiation of floral parts.

fluid drilling – a method that uses seeds that have been treated, pregerminated, suspended in a gel, and placed in precision seeders that deposit a designated number of seeds per hole in the soil by an automated process.

focalization – a landscape design principle creating an accent in a particular arrangement.

foliar analysis – similar to soil analysis. The main difference between the two is that soil testing provides information on the availability of nutrients for uptake by the plant, whereas foliar analysis provides information on the amount of nutrients taken up by the plant.

foot candles – units for measuring illumination.

forestry – typically focuses on tree production for timber.

form – the three-dimensional shape of the outline of the floral design, or in landscape design, a three-dimensional element; for example, a tree shape can be round, weeping, horizontal, and a variety of other shapes.

foundation planting – plantings located along the walls or foundations of buildings.

foundation seed – seed produced under the supervision of agricultural research stations to assure genetic purity and identity.

freestanding greenhouse – a greenhouse that is separate from other buildings or greenhouses.

freezing injury – injury that occurs in some commodities stored at temperatures below freezing.

fresh market fruits – fruits that are packaged and sold as soon as possible after harvest.

fresh market vegetables – vegetables that are packaged and sold as soon as possible after harvest.

frost-free period – the average number of days from the last spring frost to the first frost in the fall.

fruit – defined botanically as a mature ovary of a flowering plant or tree that includes the seed, its covering, and any parts closely connected to it.

fungi – typically multicellular plants that lack chlorophyll.

G

genetic engineering – in its broadest sense, the isolation, introduction, and expression of foreign DNA into the plant.

genetic pest control – an IPM method that uses plant breeding and genetic engineering to manipulate plants to make them more resistant to specific pests.

genetically modified organism (GMO) – organism that carries a foreign gene or genes that were inserted by laboratory techniques into all of its cells (see transgenic plant).

germination – a series of events whereby the seed embryo goes from a dormant to actively growing state.

gibberellin – a class of plant hormones with activity similar to gibberellic acid (GA_3).

girdling – the process of blocking phloem transport by removing a ring of bark from the stem.

goal setting – describes what you want to achieve in life.

grafting – the process of connecting two plants or plant parts together in such a way that they will unite and continue to grow as one plant.

gravitational water – water that moves from the large pore spaces due to the pull of gravity.

gravitropism – the movement of a plant or plant organ in response to gravity.

Greek – Ancient civilization involved in agriculture in only a minor way; however, the Greeks did dedicate a lot of attention to the study of botany and were concerned about the nature of things.

green manure – the use of a leguminous cover crop for the main purpose of plowing it under to increase soil fertility.

greenhouse – structure that is covered with a transparent material that allows sufficient sunlight to enter for the purpose of growing and maintaining plants.

greenhouse range – when two or more greenhouses are located together.

Grew, Nehemiah – responsible for the initiation of basic studies in plant anatomy and morphology.

growing medium – the material in which the roots of plants grow.

gymnosperms – a class of plants that are primarily evergreen trees and usually have naked seeds borne in cones.

H

habitat – a place where wildlife lives in nature.

Hale, Stephen – published *Vegetable Staticks,* which was the first significant publication in plant physiology.

hand pruners (floral shears) – a cutting tool that is used to cut thick or woody stems that cannot be cut readily with a pocketknife.

hanging pots – containers made of wood, wire, ceramic, or plastic that are used for hanging plants from above.

hardened-off – gradually subjecting plants to cooler temperatures with less frequent watering.

hardpan – occurs when soil is compressed into a very dense mass.

hardscaping – the permanent landscaping structures, including fences, patios, walks, driveways, retaining walls, and other inanimate features in the landscape.

hardy plants – plants that are less sensitive to temperature extremes than tender plants.

Hatch Act – an act passed by the U.S. Congress that provided yearly support to agricultural experiment stations in each state.

headheight – the height of the scaffold branches aboveground at the adult stage.

headhouse – a central building that is used for offices, storage, and work space with attached greenhouses.

heading back – the removal of the tips of terminal branches to promote secondary branching.

heat unit requirement – the number of hours of warmth needed for a certain growth phase to occur.

herbaceous cutting – cuttings with tissues that are not lignified.

herbaceous perennials – plants whose tops die while the roots live throughout the winter months.

herbaceous plant – a plant that has soft stems; simply put, a nonwoody plant.

herbicides – chemicals that are used to control weeds; they are the most commonly used method of control on a large scale.

heteroauxin – Latin, meaning other auxin.

heterozygous – plants having different genes of a Mendelian pair present in the same organism, such as tall pea plants that contain the genes for both tallness and dwarfness; these plants are not true to type.

homozygous – the plant has similar genes of a Mendelian pair present, such as the dwarf pea plants that contain only the genes for dwarfness.

Hooke, Robert – found that living things were made of cells; this led to the future of cytology.

horizons – layers of soil.

horticultural characteristics – characteristics such as color, size, flavor, shape, and appearance.

horticultural garden – a garden that contains an assortment of plants represented by many horticultural varieties, arranged to achieve a desirable aesthetic effect.

horticulture – the culture of plants for food, comfort, and beauty.

horticulture career – the path a person takes through life relating to work in the field of horticulture; it includes a series of horticulture occupations and jobs.

horticulture job – a specific type of work that a person in a horticulture occupation performs.

horticulture occupation – a specific work that has a title and general duties that a person in an occupation performs.

horticulture therapist – an individual who helps people with health problems or disabilities by using plants.

hortus – Latin, meaning garden.

host plant – a plant that provides a pest with food.

humidity – the water content of the air.

hybrid – an improved plant developed by crossing parents of different genotypes for a beneficial trait.

hydrologic cycle – the cycle of water in the environment.

hydroponics – a method of growing plants that provides nutrients needed by the plant via nutrient solution in the absence of soil.

hygrometer – an instrument used to measure humidity.

hypogeous seed germination – when the epicotyl emerges and the cotyledons remain below the soil surface.

I

Ikenobo design – Japanese floral design when each flower has a specific meaning and an exact location in the arrangement.

imperfect flower – a flower that lacks either stamens or pistils; an example of a plant containing this type of flower is corn.

incomplete fertilizer – a fertilizer that is lacking one or more of the three primary elements found in a complete fertilizer.

incomplete flower – a flower that lacks one or more of the major flower parts.

incomplete metamorphosis – a gradual change in the insect's size.

indicator plants – plants that will show signs of wilting prior to crop plants thereby signaling when crop plants should be watered.

indirect seeding – planting seeds indoors and growing seedlings prior to being transplanted to a larger container or to a permanent location outdoors.

inert ingredient – the carrier or filler ingredient in a fertilizer.

infiltration – heat loss through cracks or holes that occur in a structure.

inorganic fertilizer – inorganic fertilizer contains synthetic nutrient compounds that are derived from mineral salts.

insects – organisms with three distinct body parts (head, thorax, and abdomen); three pairs of legs; and one, two, or no pairs of wings.

Integrated Pest Management (IPM) – an interdisciplinary approach that uses environmentally friendly methods to control pests.

interiorscaping – a branch of ornamental horticulture which uses ornamental plants for functional and aesthetic purposes to create pleasing and comfortable areas in buildings.

internode – region located between the nodes.

J

juvenile or vegetative phase – the initial period of growth when the apical meristem will not typically respond to internal or external conditions to initiate flowers.

L

landscape architect – a professional who establishes design plans for the installation of plants and hardscaping features outdoors to fulfill aesthetic and functional purposes.

landscape construction – the execution of the planting plan and hardscaping features.

landscape contractor – the person who uses architectural plans to install plants and hardscaping features into the landscape.

landscape design – the preparation of a landscape plan for the beautification of a site; this includes drawings and supporting information with known specifications.

landscape horticulture – a branch of ornamental horticulture which deals with producing and using plants to make an outdoor environment more appealing.

landscape planning – the preparation of details of how a site will be landscaped, including both the art and science involved.

landscaping – the use of plants to make outdoor areas more attractive.

layering – a simple method of asexual propagation in which roots are formed on the stem while it is still attached to the parent plant.

leaf apex – the tip of the leaf blade, which can be pointed, rounded, and a variety of other shapes.

leaf base – the bottom of the leaf blade, which can be rounded, pointed, and a variety of other shapes.

leaf blade – the flat, thin portion of the leaf.

leaf-bud cutting – consists of a leaf, petiole, and a short piece of stem with a lateral bud.

leaf covering – leaf surface or coverings, including hairy, not hairy, waxy, not waxy, and others.

leaf cutting – consists of a portion of a leaf blade, a leaf blade, or a leaf blade with the petiole attached.

leaf margin – the outer edge of the leaf blade, which can be lobed, smooth, toothed, or various combinations of these three.

leaf scar – a scar that is left when the leaf drops.

leaflets – leaves that do not have axillary buds.

lenticel – tiny pores located in the stem for gas exchange.

light duration – involves the length of exposure to light over a 24-hour period.

light intensity – the actual quantity of light.

light quality – the actual color or wavelength of light.

line – the movement between two points within a floral design; can be created with round or linear materials. In landscape design, a one-dimensional effect produced by arranging three-dimensional objects in a certain fashion.

line planting – arrangement of plant material in a landscape to create the walls of the outdoor room.

Linnaeus – developed a simple yet elegant system for the classification of plants called binomial nomenclature.

long-day plants – plants that will flower only when the dark period is shorter than a certain critical length.

lux – the metric unit for expressing the illumination falling on all points of a surface measuring one meter square, each point being a meter away from a standard light source of 1 foot candle (1 lux = 0.093 foot candles).

luxury consumption – the point where uptake of nutrients does not correspond to an increase in plant growth.

M

macronutrients – chemical elements that are required in large amounts for normal growth and development of a plant.

Malpighi, Marcello – responsible for the initiation of basic studies in plant anatomy and morphology.

manorial system – an important component of the social structure in the Middle Ages. This system divided plant cultivation practices into horticulture, agronomy, and forestry.

mechanical dormancy – caused by the seed-enclosing structure being too strong to permit expansion of the embryo even though water can penetrate it.

mechanical pest control – an IPM method that uses tools, equipment, or other physical means for the control of pests.

Mendel, Gregor – founder of modern genetics.

meristem – a region of the plant consisting of undifferentiated tissue whose cells can divide and differentiate to form specialized tissues.

meristem culture – a technique that uses the smallest part of the shoot tip as an explant, which includes the meristem dome and some of the leaf primordia.

metamorphosis – the gradual development of the insect.

micronutrients – chemical elements that are essential for normal plant growth and development but are needed in much smaller amounts than macronutrients.

midrib – the largest vein located in the middle of the leaf.

mineralization – a process that occurs when a plant decomposes and the organic form of nitrogen is converted into the inorganic form.

modified atmosphere (MA) storage – regulation of CO_2 and O_2 levels during storage.

modified central leader system – a pruning system that is the same as the central leader tree form system in the early stages to permit the formation of strong scaffold branches. After all the scaffold branches are established, the central leader is removed to create an open center.

modified roots – roots that serve as a food storage system.

monocots – plants that are characterized by one cotyledon (seed leaf) and have parallel leaf venation.

monocotyledonae – a subclass of plants that has only one cotyledon and parallel leaf venation.

morphological dormancy – occurs when seeds are shed from the parent plant when their embryos are not fully developed.

morphology – the plant's form and structure.

Morrill Land Grant College Act – passed in 1862 by Abraham Lincoln to offer each state free public land that could be used for institutions of higher learning.

mound layering – a form of propagation where the parent plant is first cut back to slightly above ground level in late winter and covered with soil.

multiple fruits – fruits that are comprised of ovaries of many separate but closely clustered flowers, for example, pineapple.

mutualism – a plant and fungi association in which both the host plant and fungi benefit.

mycoplasma-like organism – small parasitic organism lacking constant shape and intermediate in size between viruses and bacteria.

N

negative DIF – when the day temperature is lower than the night temperature.

negative gravitropism – upward bending of the stem against gravity.

nematode – appendage-less, nonsegmented worm-like invertebrate organism with a body cavity and a complete digestive track that is found in soil.

nitrogen cycle – the circulation of nitrogen in nature.

nitrogen fixation – when microbes fix atmospheric nitrogen. The process involves the symbiotic relationship between a bacteria known as *Rhizobia* and legume roots.

no metamorphosis – when the insect exhibits no change in shape after the egg hatches, it goes from egg to full size.

node – the point along the length of the stem where leaves or stems are attached.

nonarthropods – organisms that do not have exoskeletons or jointed legs.

nonclimacteric fruits – fruits that show no change in their generally low CO_2 and C_2H_4 production rates during ripening and will not ripen in response to ethylene.

nonpoint sources of pollution – pollution that comes from many sources which cannot be specifically identified.

nonselective herbicide – herbicides that destroy all vegetation.

nutrient film technique (NFT) – a hydroponic system that involves recirculating a shallow stream of nutrient solution with no solid rooting medium.

nutrient solution – a hydroponic solution that contains water with dissolved nutrient salts.

nutrients – the substances that roots absorb from the growing medium with water, are not carbohydrates, and supply the plant with either energy or essential mineral elements.

nuts – hard shelled dry fruits or seeds, such as a walnut.

O

obligate parasites – parasites that attack only living tissues.

obligatory fungi – fungi that survive on dead or living tissue.

olericulture – the branch of horticulture dealing with the production, storage, processing, and marketing of vegetables.

open center tree form – a pruning system which leaves the center of the tree open to permit good sunlight penetration.

opposite leaf arrangement – the opposite pattern of leaf arrangement occurs when two leaves are directly across from each other.

organic fertilizer – a fertilizer that contains naturally occurring materials that are derived from plants or animals, for example, bloodmeal, bonemeal, different types of manures, and others.

organic matter – the decayed remains of plants and animals in the soil.

organic vegetable production – method of growing vegetables without chemical pesticides or inorganic fertilizers.

ornamental horticulture – branch of horticulture which involves growing and using plants for their natural beauty.

P

parasitic fungi – fungi that can only survive on living tissue.

parent material – material that is used to form soil.

parthenocarpy – the development of fruits without pollination or fertilization, resulting in seedlessness.

pathogen – a disease-causing organism.

pathogenic – refers to an organism that causes diseases.

pattern – the physical characteristics of plant material, such as the arrangement of leaves and petals.

peat – partially decayed plant material and composed mainly of organic matter; provides some nutrients, including nitrogen.

peat moss – a specific form of peat that is obtained from moss plants growing in bogs.

peat pellets – compressed peat moss that expands when moistened to become small peat-filled pots.

peat plugs – peat moss compressed into plugs that expand when moistened to become small peat-filled plugs.

peat pots – compressed peat moss pots.

percent germination – the percentage of seeds that will sprout and grow.

perennials – plants that may be herbaceous or woody and live for more than two growing seasons.

perfect flower – a flower that contains both stamens and pistils.

perlite – an inorganic material that is made from light rock and is volcanic in origin.

pest – anything that causes injury or loss to a plant.

pesticides – chemicals used to control pests.

petals – the brightly colored portion of the flower used to attract pollinators.

petiole – the leaf stem or stalk that attaches the leaf to the stem.

pheromones – chemicals many female insects secrete to attract male partners.

phloem – composed of tiny tubes that transport manufactured food and carbohydrates from the leaves down to other parts of the plant such as the roots and shoots.

photoperiod – the length of the dark period that influences plant growth.

photoperiodism – the total light energy that may affect plant growth and development.

photosynthesis – a series of chemical reactions in which carbon dioxide and water are converted in the presence of light to carbohydrate (sugar) and oxygen.

phototropism – the movement of a plant or plant organ in response to directional fluxes or gradients in light.

physical dormancy – occurs when the seed covering is impervious to water.

physiological dormancy – refers to a general type of dormancy in freshly harvested seeds from herbaceous plants.

phytoremediation – when plants are used to remove pesticide spills, heavy metals, and other pollutants from the environment.

piercing and sucking insect – an insect that uses the labrum to pierce the leaf and then sucks out the juices.

pistils – the female reproductive part of the flower. It consists of the stigma, which is the sticky surface for collecting pollen; style, which is the tube that connects the stigma and ovary; and ovary, which contains ovules or eggs.

pith – the area located in the center portion of the stem where food and moisture are stored.

plant breeding – the science and art of controlled pollination for modifying plants in a way that is advantageous to humans.

plant diseases – abnormal conditions in plants that interfere with their normal appearance, growth, structure, or function.

plant environment – the above-ground and below-ground surroundings of the plant.

plant growth regulator – organic compounds other than nutrients (materials that supply either energy or essential mineral elements) that in small amounts promote, inhibit, or otherwise modify any physiological process in plants. A plant growth regulator can be synthetic or naturally occurring.

plant growth retardant – an organic compound that retards cell division and cell elongation in shoot tissues and thus regulates height physiologically without causing malformation of leaves and stems.

plant hormone – a compound that is chemically characterized, biosynthesized within the plant, broadly distributed in the plant kingdom, performs specific biological activity at extremely low concentrations, and plays a fundamental role in regulating physiological phenomena *in vivo* in a dose dependent manner and/or due to changes in sensitivity of the tissue during development. All plant hormones are naturally occurring.

plant propagation – the reproduction of new plants from seeds and vegetative parts of the plant such as leaves, stems, and roots.

plant selection – a good strategy to select species that are tolerant to specific environmental stresses, thereby enabling the plant to withstand an adverse environment.

planting plan – a drawing that uses symbols to specify the types and names of plants, their quantities, and locations in the landscape.

Pliny the Elder – wrote a book entitled *Historia Naturalis* that described Roman agriculture.

plug – a small block or square of turf containing plant material and soil.

point sources of pollution – pollution that has a definite and identifiable source.

pollination – the transfer of pollen grains from the anther to the stigma.

pollution – when harmful or degrading materials contaminate the environment.

Polo, Marco – one of the more well-known Polos; his great contribution was the descriptions of the East in his book *Il Milione*, known in English as the *Travels of Marco Polo*.

pomology – the branch of horticulture dealing with the production, storage, processing, and marketing of fruits and nuts.

positive DIF – when the day temperature is higher than the night temperature, promoting internode elongation.

positive gravitropism – downward bending of the root with the force of gravity.

postemergence treatment – herbicide treatments that are made after the emergence of crop plant seedlings, weed seedlings, or both.

pot-in-pot method – a method where larger pots (holder pots) are placed in holes in the ground and buried up to their lip. Containers with plants are then placed into the holder pot.

preemergence treatment – herbicide treatments that are applied to the soil surface after the crop is planted but before the emergence of the weed seedlings, crop seedlings, or both.

preplant treatment – herbicide treatments that are made to the soil prior to planting the crop.

Priestly, Joseph – showed that plants purify air.

primary root – the main root that first emerges from the seed. Starting from the tip of the primary root, there is the root cap, just behind the root cap is the area of cell division, followed by the areas of cell elongation and cell differentiation.

private area – the area of a landscape out of the view of the general public.

processed fruits – fruits that are canned, dried, or frozen.

processed vegetables – vegetables that are canned, dried, or frozen.

product mix – the mixture of crops that are grown at a nursery.

proportion – the floral design principle that is the relationship among size and shape, specific characteristics of the flowers, and the container used; the table or shelf and the room they are in all relate to proportion of a floral design. In the landscape design principle, it is the relationship between the sizes of the different types of plants used in a landscape design.

protoplasts – plant cells without cell walls.

public area – the area of a landscape seen from the street.

Q

quiescence – dormancy caused by external factors, also referred to as ecodormancy.

R

radiation – the radiation of heat from a warm surface, such as a plant leaf, to a cooler surface, such as a greenhouse covering.

radiation freezes – freezes that occur on still, clear nights.

registered seed – foundation seed that is expanded by a certified grower.

relative humidity – the ratio of the weight of water vapor in a given quantity of air to the total weight of water vapor that a quantity of air can hold at a given temperature; expressed as a percentage.

Renaissance – period in European civilization after the Middle Ages characterized by the rebirth of the arts and sciences.

renewal pruning – a method of pruning used to rejuvenate old plants by removing old unproductive branches, thereby promoting vigorous growth.

repair grafting – used as a specialized way of repairing a plant that has been damaged.

reproductive or mature phase – refers to qualitative changes that allow the plant or organ to express its full reproductive potential.

Research Marketing Act – act passed by the U.S. Congress to change the imbalance between production and postproduction research. In addition, this act put in place the mechanism for contracting with private research facilities.

respiration – the process by which stored organic materials (carbohydrate, protein, fat) are broken down into simple end products (such as H_2O and CO_2) with a release of energy for plant growth and development. Oxygen is used in this process and carbon dioxide is liberated, which is the opposite of photosynthesis.

rest – dormancy caused by internal factors, also referred to as endodormancy.

rhythm and line – a landscape design principle that deals with flow throughout the landscape.

Roman Empire – ancient civilization that followed the Greek civilization and then lasted for about 1,000 years. The Romans made great improvements in the areas of grafting and budding and used of many kinds of cultivated varieties of fruits and vegetables, legume rotation, fertility analysis, and methods for storing fruits.

root cap – the portion of the root located at the tip that consists of several layers of cells and protects the root as it grows through the soil.

root cutting – pieces of roots taken from young plants as a source of material for propagation.

root hairs – single cells that absorb the greatest amount of water and minerals. Improper handling during transplanting can cause the loss of many root hairs, which causes a decrease in water uptake by the plant resulting in transplanting shock.

rootstock – a root system and base of the tree on which the fruiting top or scion cultivar is budded or grafted.

S

saprophytic fungi – fungi that can live only on dead tissue.

saturated soil – when all the pore spaces in the soil are filled with water.

scaffold branches – the main branches growing from the trunk of the tree.

scale – a part of proportion dealing with the relative size among things only, not shapes.

scarification – the process of breaking the seedcoat.

scientific classification – a method that uses the morphology of plants as a means of classification.

scion – a short piece of stem with one or two buds.

scout – an employee of the person holding a patent who buys plants at various outlets and runs analysis of these plants to see if they are being used legally.

secondary root – the root that arises from the primary root.

seed – a ripened ovule, which consists of an embryo, stored food reserves, and a seedcoat or covering.

seed purity – the percentage of pure seed of an identified species or cultivar present in a particular lot of seed.

seed viability – the percentage of seed that is alive and will germinate under standard conditions.

selective herbicide – herbicides that are effective in controlling a limited number of plant species, for example, 2,4-dichlorophenoxyacetic acid (2,4-D) kills broadleaf weeds and does not affect grasses.

semihardwood cutting – cuttings that contain lignified tissues.

senescence – a general failure of many biosynthetic reactions that precede cell death; the phase of plant growth that extends from full maturity to death and is characterized by chlorophyll, protein, or nucleic acid degradation as well as many other factors.

sense technology – involves putting a known gene sequence into the plant in the correct orientation to stimulate a process in plants.

sepals – green leaf-like structures located beneath the petals.

separation – a propagation method in which natural structures are removed from the parent plant and planted to grow on their own.

serpentine layering – a propagation method where the stem of a plant is gently curved at several locations, nicked at each bend, placed in a shallow hole, anchored, and the terminal end of the shoot being buried is left exposed.

service area – the area of a landscape near the rear entrance that is relatively isolated from the public and private areas, where utility accessories and unattractive items such as garbage cans are stored.

sexual propagation – the reproduction of plants with the use of seeds.

shape – the two-dimensional term for form.

short-day plants – plants that will flower only when the dark period is greater than a certain critical length.

side dressing – when fertilizer is placed in bands along both sides of a row of plants.

simple fruit – fruit that is comprised of a single ovary with or without some other flower parts that have developed as part of the fruit.

simple layering – a propagation method where a stem of a plant is gently curved, nicked at the bend, placed in a shallow hole, and the terminal end of the shoot being buried is left exposed.

simple leaf – a leaf that consists of one blade per petiole.

simplicity – a landscape design principle that uses a number of strategies to reduce high levels of variation and general distractions in the landscape.

site analysis – the survey of a site to determine the presence, distribution, and characteristics both natural and synthetic along with environmental conditions at the site.

site plan – a drawing indicating the location of plants and hardscaping objects in the landscape.

sitescaping – landscaping a small part of a larger area, which is used to meet the particular needs of people; for example, a small Chinese garden in a portion of the backyard.

slow- or controlled-release fertilizers – fertilizers with nutrients that are available over a long period of time.

sod – the surface layer of turf, including plants and a thin layer of soil used in propagating new turf.

soil – the top few inches of earth's surface consisting of minerals, air, water, and organic matter that provides for plant growth.

soil aeration – the movement of atmospheric air into the soil.

soil compaction – occurs when soil is compressed into a relatively dense mass.

soil pH – the measurement of alkalinity or acidity in the soil based on the hydrogen ion concentration.

soil profile – the vertical section of soil at a particular location showing the various layers, also called horizons, that have developed over a period of time.

soil salinity – the amount of salt found in the soil.

soil structure – the physical arrangement of soil particles.

soil test – a test used to determine which nutrients are present in the soil.

soil triangle – used as a method of classifying soil on the basis of mineral content (texture) of sand, silt and clay.

soilless medium – a plant growth medium that contains no soil.

soluble fertilizers – contain powders or crystals of fertilizers that have been dissolved in water first and then applied to the soil as a liquid.

somaclonal variation – when somatic embryos derived from single cells are grown into mature plants, the plant's characteristics exhibit this type of variability.

somatic embryogenesis – a pathway of differentiation in plants, induced in undifferentiated cell, tissue, or organ cultures by appropriate control of nutritional and hormonal conditions, that results in the formation of organized embryo-like (embryoid) structures.

space – the distance between plant materials found in a floral arrangement.

spot treatments – when a herbicide is applied to a specific location, for example, to weeds growing through a crack in the cement.

sprig – part of the grass plant without soil, such as rhizomes or stolons.

stamens – the male reproductive part of the flower. It consists of a filament that supports the anther, which produces the male sex cells.

standard-size pot – containers equal in width and height.

stem cutting – a propagation method which uses portions of the stem containing terminal or lateral buds.

stomata – the tiny openings in the leaf blade in which gases enter and exit the leaf; they can be located on the top, bottom, or both on the top and bottom portion of the leaf.

stratification – when seeds are placed on a moist medium at temperatures between 32 and 50°F to promote germination.

strip tillage – the process of leaving strips of vegetation between tilled areas to prevent soil erosion.

substrate systems – a hydroponic system where plant roots are surrounded by either inert or organic materials that provide support for the plant.

surfactant – a material that helps in dispersing, spreading, wetting, or emulsifying a pesticide formulation.

suspension culture – the aseptic growth of plant cells in a submerged liquid culture.

sustainable horticulture – a production system that uses IPM and sound management practices to reduce inputs.

symbiosis – when a mutually beneficial plant and bacteria association occurs, as is the case in legumes.

symmetrical balance – where the same number and types of plants are on both sides of the landscape or in a floral design when the size and shape on both sides of a central axis are equal.

T

taproot system – a root system with a primary root that grows down from the stem with only a few secondary roots; both carrots and dandelions are examples.

taxon (Taxa) – a taxonomic group name applied to organisms making a hierarchy within a formal system of classification.

taxonomy – the study of scientific classification and nomenclature.

tender plant – a plant that cannot tolerate cool weather.

tendrils – modified leaves that are appendages produced by certain vines that wrap around a support and allow the vines to climb.

tensiometer – instrument used to measure soil moisture.

terminal bud – the bud positioned at the tip of the stem containing an undeveloped leaf, stem, flower, or mixture of them all.

terminal bud-scale scar – the scar left when the terminal bud begins growth in the spring, representing one year's growth.

terrarium – miniature landscape growing in a covered glass or plastic container that has a high capacity to retain moisture and is used for displaying and storing plants.

texture – the surface quality and placement of plant parts in the design, for example, smooth and rough.

thatch – the layer of organic residue above the soil surface and just below the green leaves of the host plant.

Theophrastus – referred to as the father of botany and wrote *History of Plants and Causes of Plants*, which included topics such as plant classification, propagation, forestry, horticulture, pharmacology, viticulture, plant pests, and flavors from plants.

thermoperiodism – the plant's response to changes in day and night temperature.

thermotropism – the plant's response to temperature.

thigmotropism – the movement of a plant in response to touch.

thinning out – the removal of excess vegetative growth to open the plant canopy and reduce the number of fruiting branches with the goal of promoting large and overall better-quality fruits.

tissue culture – a method for producing new plants from single cells, tissues, or pieces of plant materials called explants on artificial medium under sterile conditions.

top-dressing – the placement of fertilizer uniformly around the stem of the plant.

topiary – used to describe a plant that has been trained and pruned into formal geometric or abstract shapes.

topography – the natural and man-made surface features of an area and their relationships.

total active ingredient – the total percentage of pesticides or nutrients being applied.

totipotency – the capability of a single mature plant cell to produce an entire organism.

transformed – when genes are transferred from a single bacterium to a single plant cell and are integrated into the chromosome of a plant cell.

transgenic plant – a plant containing a foreign gene or genes.

transition zone – the area between definitive climatic zones; it favors some warm season and cool season turfgrasses but is not ideal for either.

transpiration – the loss of water from the plant through the leaves in the vapor form.

transplanting shock – a one- to two-week period following planting where no growth occurs due to shock.

trench layering – a propagation method where the middle portion of a flexible stem is buried in the soil after nicking it at several locations.

triple response – the first bioassay used to quantify ethylene consisted of the following: suppression of stem elongation, increase in radial expansion (lateral expansion), and promotion of bending or horizontal growth in response to gravity.

true to type – identical to the parent plant.

turf – a collection of plants in a ground cover and the soil in which roots grow.

turfgrass – a collection of grass plants that form a ground cover.

turfgrass blend – a combination of different cultivars from the same species.

turfgrass mixtures – a combination of two or more different turfgrass species.

turgid – when plant cells are full of water.

typiness – elongated fruit and prominent calyx lobes.

U

understock – the bottom part of the graft union that is in contact with the soil.

up lighting – when light shines up at the base of the tree or interesting object highlighting that particular tree or object in the landscape.

U.S. Department of Agriculture – created in 1862 by Abraham Lincoln to help America's farmers and ranchers.

V

vase life – the amount of time a cut flower retains its desirable qualities prior to deteriorating.

veins – tubes used to transport water and nutrients throughout the plant.

vermiculite – an inorganic material made from heat-treated mica that can be coarse or fine.

vernalization – the use of cold treatment to induce flowering.

viable – the embryo is alive and capable of germination.

vine-ripe – when fruits or vegetables are picked after they are ripe and ready for immediate use.

visual diagnosis – a system used to determine the nutrient status of the plant based solely on visual observation and requiring a trained eye.

W

water stress – when the plant is unable to absorb an adequate amount of water to replace that lost by transpiration.

wear – the physical deterioration of a plant community resulting from excessive stress.

weather – the combined effect of complex interactions between temperature, rainfall, wind, light, and relative humidity at a specific location.

weeds – any plant growing out of place or an unwanted plant.

wetland – swamps, ponds, and other places where water often stands.

whorled leaf arrangement – when three or more leaves are attached to the same portion of the stem.

wildlife – plant or animal organisms that live in the wild and rely on natural food and habitat to survive.

wilting point – the point at which water can no longer be absorbed by the plant, resulting in moisture stress, which causes wilting.

winter chilling requirement – calculated in most areas of the United States by how many hours at and below 45°F (7.2°C) are required during the winter for the plant to break its winter resting period and develop normally when temperatures rise in the spring.

winter overseeding – planting one grass in another established grass without destroying the established grass.

woody plant – a plant that produces wood and has buds that survive above ground level during the winter months.

X

xeriscaping – a form of landscaping that uses plants based on their water consumption.

xylem – composed of tiny tubes that transport water and nutrients up from the roots to other parts of the plants. The annual rings in a tree are made of xylem.

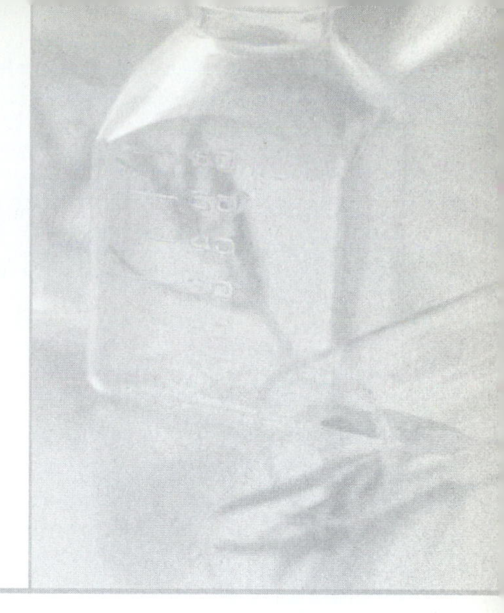

Index

Note: Figures and tables are denoted by an "f" or "t" following the page number.

D

G

H

I

J

Q

R

S